AF616421

UNIVERSITY OF BRADFORD
Library

Accession No. 0340190614

This book should be returned not later than the last date stamped below. The loan period may be extended on request provided there is no waiting list.
FINES ARE CHARGED ON OVERDUE BOOKS

NOT FOR LOAN UNTIL 1 3 JUL 1992
UNIVERSITY OF BRADFORD LIBRARY

10 DEC 1993
27 APR 1995
24 SEP 1998
- 8 JUN 2001
- 8 JUN 2001

L45 350

BD 034019061 4

# Plant Root Growth
## AN ECOLOGICAL PERSPECTIVE

# Plant Root Growth

## AN ECOLOGICAL PERSPECTIVE

SPECIAL PUBLICATION NUMBER 10 OF THE
BRITISH ECOLOGICAL SOCIETY

EDITED BY
D. ATKINSON
University of Aberdeen

OXFORD
BLACKWELL SCIENTIFIC PUBLICATIONS
LONDON EDINBURGH BOSTON
MELBOURNE PARIS BERLIN VIENNA
1991

© 1991 by the British Ecological Society
and published for them by
Blackwell Scientific Publications
Editorial offices:
Osney Mead, Oxford OX2 0EL
8 John Street, London WC1N 2BL
23 Ainslie Place, Edinburgh EH3 6AJ
3 Cambridge Center, Cambridge,
Massachusetts 02142, USA
54 University Street, Carlton
Victoria 3053, Australia

Other Editorial Offices:
Arnette SA
2, rue Casimir-Delavigne
75006 Paris
France

Blackwell Wissenschaft
Meinekestrasse 4
D-1000 Berlin 15
Germany

Blackwell MZV
Feldgasse 13
A-1238 Wien
Austria

All rights reserved. No part of this publication may be reproduced, stored in a retrieval system, or transmitted, in any form or by any means, electronic, mechanical, photocopying, recording or otherwise without the prior permission of the copyright owner.

Authorization to photocopy items for internal or personal use, or the internal or personal use of specific clients, is granted by the British Ecological Society for libraries and other users registered with the Copyright Clearance Center (CCC) Transactional Reporting Service, provided that the base fee of $03·00 per copy is paid directly to CCC, 21 Congress Street, Salem, MA 01970, USA. 0262–7027/91/$03·00.

This consent does not extend to other kinds of copying, such as for general distribution, for advertising or promotional purposes, for creating new collective works, or for resale. Special requests should be addressed to the British Ecological Society, Burlington House, Piccadilly, London.

First published 1991

Set by Excel Typesetters, Hong Kong
Printed and bound in Great Britain
by the University Press, Cambridge

DISTRIBUTORS

Marston Book Services Ltd
PO Box 87
Oxford OX2 0DT
(*Orders*: Tel. 0865 791155
Fax. 0865 791927
Telex. 837515)

USA
Blackwell Scientific Publications, Inc.
3 Cambridge Center
Cambridge, MA 02142
(*Orders*: Tel: (800) 759 6102)

Canada
Oxford University Press
70 Wynford Drive
Don Mills
Ontario M3C 1J9
(*Orders*: Tel. (416) 441 2941)

Australia
Blackwell Scientific Publications
(Australia) Pty Ltd
54 University Street
Carlton, Victoria 3053
(*Orders*: Tel. (03) 347 0300)

British Library
Cataloguing in Publication Data

Plant root growth.
1. Plants. Roots. Growth. Regulation
I. Atkinson, D. (David) 1944– II. British Ecological Society III. Series
581.10428
ISBN 0-632-02757-6

Library of Congress
Cataloging-in-Publication Data

Plant root growth: an ecological perspective/edited by D. Atkinson.
p. cm. (Special publication number 10 of the British Ecological Society)
Includes bibliographical references.
ISBN 0-632-02757-6
1. Roots (Botany)—Ecology—Congresses. 2. Roots (Botany)— Growth—Congresses. I. Atkinson, D. II. Series: Special publication of the British Ecological Society; no. 10.
QK 644.P53 1991
581.1′0428—dc20

270163
UNIVERSITY OF BRADFORD
LIBRARY
15 JUN 1992
ACCESSION No.
CLASS No.
0340190614
R581.43 ATK
WITHDRAWN
DISCARDED
LIBRARY

# Contents

## ROOT RESPONSES TO SOIL CONDITIONS



## THE ROOT SYSTEM OF DIFFERENT VEGETATION TYPES

## ROOT SYSTEMS AND PLANT INTERSPECIFIC EFFECTS

# Preface
# Plant Roots: An Ecological Perception

Reference to the entries in standard floras indicates that soil characteristics are often associated with specific species, e.g. *Fragaria vesca* L. woods and scrub on base-rich soil and in basic grassland (Clapham, Tutin & Warburg 1968).

The frequency of the association suggests that the root–soil associations of given species are both common, predictable and deterministic in relation to vegetation composition on particular sites. Aspects of this have been discussed by Grime (1979) who has been able to define the restricted range of soil conditions in which many species grow. In addition, observation of species and plant communities such as chalk grassland or even arable fields not being treated with pesticides indicates that within a short distance and under conditions where the above-ground microclimate seems unlikely to vary substantially that a number of species appear to co-exist. While history, the result of random factors which results in species A rather than species B, C or D inhabiting a particular site, may account for some of these differences many more seem likely to be due to major or subtle differences in soil factors or to subterranean species interactions. The effects of soil factors such as levels of mineral nutrients, water-holding properties and the presence of allelomorphic substances, symbionts or pathogens have been assessed for a limited number of major arable crops, although these factors have received rather less attention in respect of their roles in determining the composition of natural vegetation, especially at a mechanistic level. At this level these soil factors must operate either by influencing the growth of the plant root systems or by influencing its functioning and the perception of the various stimuli present in the soil environment.

This subject was addressed by a symposium of the Society held in 1968 (*Ecological Aspects of the Mineral Nutrition of Plants*, ed. I.H. Rorison) and followed by a further symposium in 1984 (*Ecological Interactions in Soil*, ed. A.H. Fitter *et al.*).

Neither of these meetings and the resulting symposium volumes addressed the question of the role played by the root system in determining the composition of vegetation and the dynamics of changes with time. The meeting of which this is the written proceedings aimed to assess methods currently being used to assess the growth and performance of roots, the functioning of the roots of species of ecological importance and some of the factors which lead to root properties influencing the dynamics, composition and structure of vegetation. These were clearly ambitious aims which, in the light of current knowledge, could not be wholly realized. The papers presented here therefore represent the current state of the art in respect of the above question rather than a definitive exposition.

The volume is subdivided into sections, each of which covers an area of study of particular importance to ecologists interested in the factors which control the dynamics of vegetation composition and structure. The evolution of roots and of root structure is a subject not often considered by ecologists but clearly one with important implications for those currently aiming to understand the relationships between root structure and physiology and functioning.

Harper *et al.* (pp. 3–22) comment that in recent years as much effort has been devoted to developing methods to study roots as to the actual study of roots themselves. While this may be undesirable there can be no doubt that the study of plant root systems under realistic conditions is limited by the absence of methods fully applicable to natural situations rather than to the situations found in lowland arable agriculture. The papers here concentrate on such methods and on those which contain an element of novelty and those, such as observation methods and the use of ingrowth bags, which have particular relevance to ecological studies. The recent advent of two rhizotrons/biotrons built specifically for the study of natural plant communities and the dynamics of fine roots are documented here.

Although the literature on root functioning is substantial few published reports relate to the functioning of the roots of natural species. The papers in this volume thus cover the allocation of resources to the root system, its allocation within the root system and the ecological significance of these various strategies. Soil condition not only affects the growth of roots but is in turn affected by the activities of roots and their associated micro-organisms.

An understanding of the relationships between structure and function can be studied both in detailed laboratory investigations and by assessing the features of root systems produced in extreme conditions such as tundra, hot deserts and salt marshes and by relating these to one of the best understood communities — temperate grassland. Papers in this volume allow this type of analysis to be made.

These sections provide the building blocks for the study of communities which involve the interaction of species, the sharing of resources and the effects of other organisms such as pathogenic or symbiotic fungi or soil herbivores. The papers presented here are far from presenting a definitive picture of this important area but they identify a number of subjects which require future research activity. It is hoped that the volume will help to stimulate interest in this subject area and to draw attention to the potential importance of an understanding of roots and root–root interactions in understanding vegetation dynamics and composition.

D. ATKINSON

## REFERENCES

**Clapham, A.R., Tutin, T.G. and Warburg, E.F. (1968).** *Excursion Flora of the British Isles*. Cambridge University Press, Cambridge.

**Grime, J.P. (1979).** *Plant Strategies and Vegetation Processes*. Wiley, Chichester.

# Roots in an Ecological Context

# The evolution of roots and the problems of analysing their behaviour

J.L. HARPER*, M. JONES† AND
N.R. SACKVILLE HAMILTON†
*Unit of Plant Population Biology, School of Biological Sciences, University College of North Wales, Bangor, Gwynedd LL57 2UW, UK*

## SUMMARY

1 The development of roots was crucial to the evolution of a land flora, yet the evolutionary origin of roots has rarely been considered and remains shrouded in mystery. By the time roots evolved, soils already had a well-developed fauna and flora.

2 Roots show no homologies with shoots and it is difficult to conceive of any half-way stage between roots and any other plant organs. The fossil record contains no intermediates between rootless soils and soils with fully developed root systems and we conclude that roots must have evolved *de novo* from a single major change. We propose that genes for root development originated by plasmid transfer after wounding by microbial infection, much as occurs today with the Ri plasmid in *Agrobacterium rhizogenes*.

3 Anchorage of a terrestrial plant and the uptake of water and nutrients all depend on the architecture of the root system. Optimal architecture for exploring and exploiting the soil depends on which resources are in short supply and the patchiness of their distribution. Root architecture is therefore a compromise of conflicts. Plasticity and the ability to make extremely localized branching responses to soil heterogeneity are vital attributes of an efficient root system.

4 The main problem in the way of studying the behaviour of roots in nature is that it is impossible to observe a root growing in soil. Most students grow roots in artificial media, ranging from mini-rhizotrons to water culture, where their behaviour, although easily observed, may have little bearing on their behaviour in real soil. Studies on roots in soil involve tedious and destructive sampling, making it impossible to study root dynamics in real soil. In this paper an attempt to construct the three-dimensional architecture of roots in soil, and one attempt to study the demography of whole roots of *Trifolium repens* in a permanent pasture are described.

5 Because of the invisibility of roots in soil, it is probable that more energy has been spent on developing technologies than on studying roots.

---

* Present address: Cae Groes, Glan-y-Coed Park, Dwygyfylchi, Penmaenmawr, Gwynedd LL34 6TL
† Institute of Grassland and Environmental Research, Plas Gogerddan, Aberystwyth, Dyfed SY23 3EB

## INTRODUCTION

The development of roots was probably the most dramatic event in the evolution of the plant kingdom. It made possible the extensive colonization of terrestrial environments. Traditional wisdom has it that multicellular plants evolved in aquatic habitats and subsequently invaded the land. A strong case has however been made (Stebbins & Hill 1980) that the first land plants were unicellular forms that inhabited moist interstices in soil and that multicellular bodies evolved in association with a terrestrial life habit. No matter what their origin, only something like a root system could have allowed terrestrial plants access to the water and nutrients necessary for large plant bodies to develop. It is surprising that this point has scarcely been considered in the various theories for the evolution of a land flora which have concentrated instead on the difficulties in the movement of motile gametes on land (e.g. Bower 1908). Yet it is clearly the case that large plants established successfully on land even though they continued to depend on motile gametes; the tree ferns, cycads and Gingko all have well-developed root systems but still depend on motile gametes.

Roots must have evolved in organisms that already possessed organized meristems and vascular systems. One expects, in phylogenetic argument, to account for the diversity of organs and structures as the consequence of progressive modification and specialization of pre-existing structures. It is necessary, therefore, to search for homologies between organs. It is relatively easy to develop a theory of homologies for the parts of a shoot system; the leaf (with its axillary bud and associated axis) can readily be interpreted as homologous with the carpel, bract, petal, sepal and stamen. It has been a morphologists' delight to point, for example, to the 'hochblättern' of *Paeonia* to illustrate the ontogenetic (and presumed phylogenetic) continuum between leaves and floral organs. Roots do not fit into any such pattern of homologies and it has been convenient for classical morphologists frequently to ignore them.

Three features set roots quite apart from 'normal' plant morphology.

1 They arise endogenously and (with the exception of the primary root of some dicots) emerge as a wounding eruption from within the cortex — most commonly from within the pericycle.

2 Their meristems are organized quite differently from those of shoots. They are subterminal and bidirectional — contributing cells both upwards to the root proper and downwards to a unique protective structure, the root cap.

3 Their vascular systems are organized in a manner quite different from that found in stems and leaves — notably with exarch xylem and a central stele.

In addition, roots *normally* differ from shoots in other qualities, but there are sufficient exceptions for the distinctions to be slightly blurred. For example, roots rarely form chloroplasts, but occasionally do so when exposed to light. When they do form chloroplasts these are often restricted to particular cell files (e.g. only in the pericycle and pith of *Aesculus hippocastanum*) and they are slow to form (Whatley 1983). Roots normally lack cuticle and their photo- and geotropisms are normally in the opposite direction to those of shoots.

Two forms of root are commonly recognized — primary and adventitious. Groff & Kaplan (1988) point out that this terminology is seriously misleading. Adventitious means 'an addition from without; supervenient, accidental, casual' (*Oxford English Dictionary*). In fact they are an addition *from within* and, far from being accidental or casual, are the *only* type of root found in pteridophytes, monocots and a great many dicots. They suggest that we should use the term 'shoot-born roots' to distinguish them from the rarer primary roots. Textbooks have spread the view that the primary (or tap) root is in some way the norm and roots that arise from stems are the exception. Rather it appears that the primary or tap-root system is a later evolved and derived morphology and far from being the commonest type of root system is restricted to gymnosperms and some dicots (Groff & Kaplan 1988). The earliest rooted flora was most probably composed of clonal rhizomatous plants (Tiffney & Niklas 1985) with a life cycle of the 'strawberry-coral' type (Williams 1975). The evolution of the tap root probably involved a sacrifice of clonal growth but allowed the evolution of large bodied trees and the 'elm tree-oyster' life cycle (Stebbins & Hill 1980).

## THE EVOLUTION OF ROOTS

The evolution of the root is shrouded in mystery. This is partly because palaeobotanists have been preoccupied by the problems posed by sexuality and the alternation of generations and partly because most of the fossil record was formed from detritus — bits of plants that fell to the ground and were preserved where they accumulated. It would have been rare for roots to contribute to this detritus. It does seem quite clear however that the earliest land plants had no roots, e.g. *Rhynia* (Rhyniophyta) and *Asteroxylon* (Lycophyta). Rootless plants continue to form a specialized part of the land flora, mosses and liverworts (restricted to a diminutive habit) and the epiphytes *Psilotum* and *Tmesipteris* in which mycorrhizal fungi apparently play the role of root substitutes.

The anchorage of rootless terrestrial plants is normally effected by rhizoids — hairs that differ profoundly from roots in being exogenous, lacking a meristem and vascular system and commonly being only single cells. They differ from root hairs in being cellular, whereas root hairs are outgrowths from single cells. Water and mineral nutrients may be absorbed through rhizoids but, especially in the bryophytes, may more often be absorbed directly through the general surface of the whole plant body. It is never seriously suggested that roots evolved from rhizoids.

Two special cases are commonly referred to in any of the rare discussions of the evolution of roots. Among extant species, plants of *Selaginella* bear rhizophores, organs of curious morphology which are in some respects intermediate (a missing link?) between shoots and roots. The structure develops exogenously at points of branching near the stem apex and sometimes produces leaves but more usually grows down into the soil where it develops normal endogenous roots. Rhizophores bear no root hairs and have no root cap or bidirectional meristem — however, as if to make the situation confusing, a cap does apparently form at a very early stage of rhizophore development in three species (*S. densa, S. kraussiana* and *S. wallacei*) (Wochock & Sussex 1974; Grenville & Peterson 1981).

A second special case is that of fossil trees such as the Devonian *Protolepidodendron* (probably more than 6 m high) of which the trunk ended in a bulbous base with many 'rootlets' radiating into the soil and *Lepidodendron* and *Sigillaria*, which flourished in Carboniferous forests, and bore forked, spreading, root-like structures (*Stigmaria*) which are found penetrating the clay layers under seams of coal. These were apparently not endogenous and had a 'phyllotaxy' in their exogenous laterals. They seem more likely to have been modified absorbing and supportive leaves, rather than roots or their precursors. It is hard to see how terrestrial plants of the stature of trees could have absorbed water and nutrients, or indeed stayed upright, without major structures serving as root substitutes.

An Australian moss, *Dawsonia superba* Grev., grows to over 1 m tall (Scott & Stone 1976), but rootless plants are generally restricted in stature. Rigid stems and trunks can be achieved by accumulated leaf bases (as happens today in many palms and tree ferns). However further limitations are imposed on stature by the need for anchorage and especially by the need for lateral meristems (cambium and cork cambium) to produce a vascular system that can grow and accommodate the increasing transport of water and nutrients in a growing tree. Although the evolution of a herbaceous land flora may have been made possible by the development of roots, the evolution of a tree flora depended also on the evolution of cambial secondary thickening. In our present floras this limitation persists and plants that depend wholly on a primary or tap-root system are strictly limited in size unless they possess a cambial system that allows the connection between the root and shoot to expand and keep pace with their growth (Groff & Kaplan 1988).

## ROOTS IN FOSSIL SOILS

Whatever the process involved in the evolution of roots, once they evolved the consequences must have been profound. Only after roots evolved could large terrestrial biomasses develop which would reduce soil erosion, promote the development of clays and change the patterns of alluvial deposition. Only in the past 15 years has it been recognized that there are abundant fossil soils (Retallack 1985) which contribute evidence about the early evolution of a land flora. Such studies of fossil soils are in their infancy, but they greatly change the impression that we have of early terrestrial vegetation. Well-developed soils apparently existed long before roots were present to exploit them; quite large animals were inhabiting the soil before roots were present; large plants did not invade sterile bare earth, but added to pre-existing soil biota; when roots first evolved they entered an already rich biotic environment.

Very surprisingly there is evidence of extensive colonization of well-drained soils by burrowing animals before any mega-fossil plants were present. In a soil from the late Ordovician there were unbranched blunt-ended burrows thought to have been excavated by elongate invertebrates of 3–16 mm body width. However there were no discernible traces of roots or rhizomes and any plants present were unlikely to have been vascular or to have extended very far above the ground. Presumably there was some plant fodder for the animals, probably formed on the soil surface by non-vascular plants. A late Silurian clay also contained no megafossil plants though it contained tubules that might have been traces of animals or plants. Burrowing animals rather than plant roots may have been the first explorers into and disturbers of primitive palaeozoic soils and Retallack (1985) suggests that their burrows may have been a 'stepping stone for the greening of palaeozoic landscapes'.

A late Devonian soil (Peas Eddy Clay) presented a quite different picture. Large root traces and impressions of *Archaeopteris* leaves were present in the superficial shales and a B horizon was clearly developed. By the later Devonian, landscapes near streams appear to have become heavily forested although it is not clear how far the drier areas had been colonized before the early Carboniferous. Fungi may have been important in such primitive soils and acted as the equivalent of mycorrhizae in some transport of minerals and nutrients. (Strictly, mycorrhizae could not have existed before roots evolved as a mycorrhiza is a fungus–root association! We appear to have no word for fungal associations with rhizomes (e.g. in *Psilotum*) or with thalloid and leafy bryophytes).

## MICROBIAL STIMULATION OF ROOT FORMATION AND TRANSFORMATION

Neither the methods of classical morphology nor the fossil records of plants or soils help to answer the question of how roots evolved. We have to ask how a quite new type of structure might have arisen — not how existing structures were progressively transformed. The clue may come from recent developments in molecular biology. It has long been known that the bacterium *Agrobacterium rhizogenes* can cause the formation of roots on stems (Hildebrand 1934). This ability has been shown to be correlated with the presence of a large plasmid (Moore, Warren & Strobel 1979; White & Nester 1980). The plasmid is called Ri (root-inducing), distinguishing it from but indicating the parallelism with the Ti (tumour-inducing) plasmid of *Agrobacterium tumefaciens*. Whereas *A. tumefaciens* induces the formation of disorganized callus, *A. rhizogenes* induces characteristic endogenous roots. Chilton *et al.* (1982) and White & Nester (1980) confirmed, with carrot and *Nicotiana glauca* that *A. rhizogenes*, like *A. tumefaciens*, inserted a fragment of plasmid DNA into the plant genome. The transformed root tissue synthesized novel metabolites, opines, even in the absence of the bacteria.

Even more interesting is that transformed roots very readily regenerate shoot systems in culture and the resulting plants can be brought to flower and seed set. Tepfer (1983) inoculated wounded stems of *Nicotiana tabacum, Convolvulus sepium*, *C. arvensis* and discs of carrot with *A. rhizogenes* and these all formed roots, whereas the uninoculated controls failed to do so. The roots differed from normal roots of the species; they grew faster, were more highly branched and plagiotropic. The transformed roots of *N. tabacum* and of *C. arvensis* regenerated whole plants spontaneously and these could be induced in carrot. The whole plants, grown in soil, exhibited the same characteristic highly branched plagiotropic root systems as transformed roots in culture. The leaves of the transformed plants were wrinkled ('waffled') in a way that distinguished them from non-transformed plants. Tepfer crossed transformed with non-transformed plants of *Nicotiana*, crossed transformed carrot plants with each other, and selfed transformed plants of *C. arvensis*. The progeny of all three species included the transformed and non-transformed phenotypes bearing all the markers. Transformed carrots behaved as annuals; the progeny contained both annual and biennial phenotypes.

It does not greatly stretch the imagination to hypothesize that the evolution of the root involved a process not dissimilar to that now known to result from infection by *A. rhizogenes*. The induction of extra roots as a result of infection by *A. rhizogenes* has already been shown to confer drought tolerance on apple seedlings (Moore, Warren & Strobel 1979). If some early rootless invaders of the land were stimulated to pericyclic activity by pathogenic infection the increased contact with water and nutrients in the soil would have brought immediate potential for more rapid growth. The hypothesis that the first roots evolved as a response to pathogens was weak, until the new evidence (Tepfer 1983), that, through adventitious shoots, the transformed genotypes could enter the gene pool of the populations. It would be interesting to test this hypothesis for the evolution of roots by infecting wounded shoots of the rootless *Psilotum* or *Tmesipteris*!

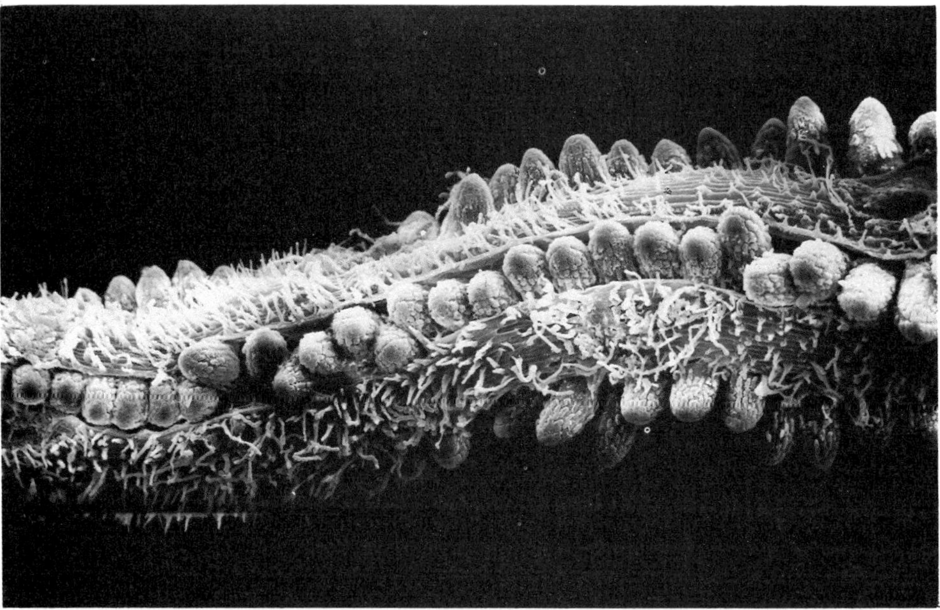

Fig. 1. Proteoid rootlets of *Leucadendron salicifolium* (Cape, S. Africa) breaking the surface of the parent root in longitudinal rows. Photograph kindly provided by B. Lamont.

The formation of nodules on the Leguminosae by *Rhizobium* has close analogies with the formation of tumours by *A. tumefaciens* and of roots by *A. rhizogenes*, but neither elicit the production of such highly organized tissues. Microbial initiation of true roots appears still to be important in some primitive angiosperms, e.g. Proteaceae (*Banksia* spp. and at least 67 species of *Hakea*). The 'proteoid' roots of these species are dense clusters of rootlets formed opposite all the protoxylem poles (Fig. 1) contiguous with the axis of the root. They were first recognized by Purnell (1960) and are thought to bring special advantages in conditions of extreme nutrient deficiency (Jeffrey 1967; Lamont 1972a, b; Lamont & McComb 1974). Plants of *Banksia grandis* grown in sterile soil failed to produce 'proteoid' roots but these were formed in abundance when the soil was inoculated with rhizosphere organisms (Malajczuk & Bowen 1974).

In nature proteoid roots are most strongly developed near the soil surface where the organic matter content is highest but on soil types where the organic matter content is generally low. Their development is promoted in soils deficient in nutrients — though they are not, as originally suggested by Jeffrey (1967), peculiarly responsive to phosphate deficiency. There are intriguing parallels between the localized development of these specialized, finely-branched roots of the Proteaceae and the localized fine-scale branching reported for many other species when a normal root encounters a local zone of organic matter or a layer of clay in a sandy soil (Weaver 1926).

## THE ARCHITECTURE OF ROOTS

The three roles of root systems, water absorption, nutrient uptake and anchorage all depend on their architecture. As in the architecture or a shoot system, that of a root system depends on four variables: (i) the extent of elongation; (ii) branching frequency; (iii) branch angle; and (iv) the mortality of axes and apices. These together determine how the absorptive membranes of the roots are distributed through the soil and the nature of the support provided against the shearing forces that act on the above-ground parts.

Water and some nutrients (especially nitrates) can move relatively freely in the wetter soils. Very close proximity of the root surface to the location of the resource is then relatively unimportant and branching at a fine scale will bring little reward. Under these conditions extension growth will tend to be favoured over fine-scale branching. However, for nutrients with low diffusion rates in soil (e.g. phosphates) and for water and water-borne nutrients in drier soils, close physical contact between the root surface and the source of the resource becomes increasingly important. If we were to design an ideal root system for growth in a phosphate-limited environment, or for one in which the water status was regularly close to the wilting point, it would have to have a fine-scale and intimate branching pattern and probably have to sacrifice extensive elongation (or develop mycorrhizal association).

Roots do not grow in isolation from each other. For the roots of a single plant the architecture of the root system will be inefficient if it results in a root entering a resource zone already depleted or being depleted by another root. An ideal architecture ensures a spacing of resource-depletion zones that maximizes access to the most limiting resources and minimizes the overlap of resource-depletion zones (RDZs). No single programme of branching can be repeated indefinitely without overlap occurring (Harper 1985). This means that a root system cannot be fully efficient unless the branching programme changes during growth *or* there is sufficient time delay before a zone is reoccupied for its resource level to be recharged. One change in programme occurs during plant development when the strong positive geotropism of those roots first formed by a seedling becomes weaker in later formed roots. However we seem to have no idea how most changes in branching pattern are brought about — whether through integrated control from the whole plant, controlling where its roots will grow, or as a purely localized response of individual roots to neighbouring roots (of their own plant or of another).

Some aspects of root architecture are under genotypic control. The ability of an embryo to generate a tap root or not is clearly deeply seated in the phylogeny of species and species differ widely in their ability to produce roots from the stem nodes (e.g. some prostrate spreading annuals such as *Polygonum aviculare* appear never to produce nodal roots and others such as *Veronica persica* do so freely). Seedlings of *Sorghum* spp. usually produce only one seminal root, whereas maize usually produces three, and wheat produces five. Roots commonly appear from the stem above ground in maize (forming buttresses) but rarely in wheat and barley. There are many other examples of species specific differences between the architecture of root systems but, in general the architecture of roots is much less tightly canalized than that of shoots: root systems are much more plastic than shoots.

Perhaps the most striking aspect of the plasticity of root development is their extremely localized branching response to local environments within the soil — in particular their ability to branch profusely within very local pockets of organic matter and regions of high-nutrient status (even in buried human corpses). Proteoid roots are a special case of this phenomenon and we do not know whether local microbial stimulation of such branching occurs widely outside the Proteaceae. However, local profuse branching is a very odd phenomenon. It implies that cells in the pericycle react, by very local cell division, to develop lateral branches in response to a signal that is received many cell layers away in the root epidermis or in the root hairs. The initiation of root nodules by *Rhizobium* in the pericycle of legume roots also occurs in advance of rhizobial invasion of the dividing cells. The formation of proteoid roots also occurs without apparent penetration of micro-organisms to the pericycle. The ability of roots to respond to very local signals about soil heterogeneity must have profound consequences on the efficiency with which they exploit soil volumes. It means that some of the fine-scale branching of roots is a response to the conditions in which very local parts of the root find themselves rather than a response to some form of control by the parent plant. It might be interesting to see how far such local root responsiveness might be selected for in crop plants.

Extension growth and branching probably represent alternative strategies in the exploration of soil volumes — they certainly have different consequences. 'Thus, a linear system cannot make optimal use of resources encountered and, conversely, a clumped system cannot discover new resources. Roots . . . are weakly canalized in their development and may be linear or clumped, all in one plant. Root-bearing plants thereby possess both the ability to explore for resources (linear growth) and the ability to exploit resources efficiently when encountered (clumped growth) (Tiffney & Niklas 1985). The effectiveness of all root systems must depend on the balance between canalized genetic control and mechanisms that control the plastic response to local environmental signals. Even the most perfectly organized absorbing membranes are useless unless root architecture places them in the right places or mycorrhizal associations take on the role of fine-scale exploration. Unfortunately for the advance of a science of root biology we still lack simple effective technologies for studying this architecture. This subject is discussed further by Fitter (pp. 237–239).

## STUDYING ROOT SYSTEMS IN SOIL

For students of plant roots it is more than just inconvenient that they usually grow in soil — more energy may have been spent on developing technologies than on studying roots. Each method so far developed has serious shortcomings. The use of water culture as an alternative to soil as a growth medium is the easiest, and probably the least meaningful: a fundamental character of soil is that it is heterogeneous at the scale of a root and efficient water culture gives a homogeneous medium. Devices like glass fronted boxes, rhizotrons with glass viewing panels, plastic or glass viewing tubes sunk into the soil, all concentrate the observed root system into two dimensions or allow only very small regions to be observed. They may be very useful for observing what roots do when they meet obstructions (such as shiny stones or the edges of a flower pot) but must misrepresent what happens in the essential three dimensions of a real soil (Levan, Ycas & Hummel 1987; Vos & Groenwold 1987; McMichael & Taylor 1987). A root that grows until it meets a glass plate must then either change direction or stop growth!

Tables 1 and 2 list a variety of the procedures devised to gain access to the inaccessible including some new developments, such as the use of NMR, a promising technology in its infancy. No one technique, of the many so far developed, informs us about all aspects of root growth and structure. Most of the current methods can be put into two groups. The first group (Table 1) includes those most suited to obtaining data on the distribution and, more rarely, the structure, of root systems in the soil at a single point in time. The second group (Table 2) includes methods designed to study root growth over time (e.g. rhizotrons).

Our own attempts to study root systems have concentrated on two problems: (i)

TABLE 1. Structure and distribution of root systems

| Technique | Type of data obtained | Disadvantages | References |
|---|---|---|---|
| Excavation of whole plant root systems | Information on whole root system structure of individual plants | Limited data about precise distribution of roots. No data concerning the interactions between roots of neighbouring plants | Weaver (1926) |
| Profile wall<br>Root distribution mapped or recorded on surface of trench wall | Information on vertical and horizontal distribution of roots | Only part of root system studied. No data on structure — cannot tell which roots are connected to which plants | Schuurman & Goedewaagen (1971)<br>Perry, Lyda & Bowen (1983) |
| Pinboard<br>Wooden board with grid made up of metal pins driven into surface of trench wall. Soil around board cut away to give block of soil and roots held on board in natural position | As profile wall, with additional data about root length, etc. in different parts of the profile | Data limited to a 'slice' of the root system. Roots have to be separated from the soil | Schuurman & Goedewaagen (1971)<br>Kirby & Rackham (1971) |
| Coring<br>Soil samples taken using auger | Information about root length/weight in soil samples taken from various areas | No insight into structure or neighbour effects. Roots must be separated from the soil | Barber (1971) |
| Isotopes<br>Solution containing radioactive element (usually P or S) injected into plant or into soil around plant | Relative amounts of radioactive roots in soil sampled in the vicinity of plants fed with isotope, or amount of radioactivity in tops of plants when soil is labelled. With dual isotope techniques, some information can be obtained about the interpenetration of the root systems of neighbours | No information on structural parameters | Baldwin & Tinker (1972)<br>Fusseder (1983, 1985) Caldwell *et al.* (1985) |
| Resin embedding<br>Technique developed by soil micromorphologists in which soil samples are impregnated with liquid resin which hardens to give a solid | Information on the precise spatial distribution of roots in the soil. Serial sections give data on the 3-d structure of root systems. Intermingling of the root systems of neighbours | Time consuming. Requires special equipment | Gadgil (1963)<br>Melhuish (1968)<br>Tippkötter *et al.* (1986) |

| | | | |
|---|---|---|---|
| block, preserving soil structure. Block is cut and surface polished to show details of pore sizes, distances between roots, etc. | can be studied in great detail | | |
| Nuclear magnetic resonance<br>Use of static and radio-frequency magnetic fields to detect the relatively mobile protons of H in water molecules. Roots contain a high proportion of water so stand out from the background soil | Images of roots in soil can be analysed to give data on root system parameters. A non-destructive technique, so growth of individual plants can be followed over time | Cannot detect roots of less than 1 mm in diameter. Image can be affected by ferro-magnetic particles in the soil and by soil water. Specialized and expensive equipment required. Works only when soil is dry | Bottomley *et al.* (1986)<br>Rogers & Bottomley (1987) |

TABLE 2. Growth and dynamics of root systems

| Technique | Type of data obtained | Disadvantages | References |
|---|---|---|---|
| Rhizotrons<br>Underground chambers with windows against the soil | Changes in lengths and numbers of roots over time. Information on root grazing by soil animals | Expensive to build. Artifacts — root growth against a glass wall may not be very representative of that in the bulk soil | Rogers (1969)<br>Atkinson (1983) |
| Mini-rhizotrons<br>Glass or plastic tubes inserted into soil. Roots observed using variations on periscope method or with cameras | As for rhizotrons | Only net changes in growth of roots can be followed. Possible artifacts | Sanders & Brown (1978)<br>Upchurch & Ritchie (1983)<br>Meyer & Barrs (1985) |
| Dyeing<br>Different coloured dyes applied sequentially to soil. Parts of root system which are of different colours indicate amount of growth made | Changes in root length over time | Dyes have to be applied as a soil drench and leached after a short time hence some disturbance of rooting medium likely. Method works well only with porous media. Need to separate roots from the soil to obtain the data | Carman (1982) |
| Mesh bags<br>Soil cores taken from field and bags filled with soil free from roots put in place | Growth of root system into 'rootless' soil | Data limited to net changes in growth over time. Roots must be separated from the soil | St John, Coleman & Reid (1983)<br>Persson (1983) |

how may the three-dimensional architecture of a root system be described and analysed so that we may begin to discover whether and how it changes in response to environmental factors such as the presence of neighbouring root systems? and (ii) how can the demographic parameters of roots (rates of birth and death) be measured and how do these relate to the colonization of roots by other organisms?

## THE SPATIAL DISTRIBUTION OF ROOTS IN SOIL AND THREE-DIMENSIONAL RECONSTRUCTION FROM SERIAL SECTIONS

Students of soil micromorphology have developed techniques for embedding soils in resin with virtually no structural distortion. The resin blocks are then cut with diamond saws and polished with abrasives. Pretreatment with glutaraldehyde prevents shrinkage of the roots (Tippkötter, Ritz & Darbyshire 1986). Undistorted roots and even root hairs can be seen on the polished surface. By repeatedly grinding away 100 μm of the embedded soil surface, polishing it, staining the surface and photographing it we can build up a complete set of serial images. To reconstruct the root system in three dimensions requires matching the corresponding sections of roots and other objects in the soil in successive images. The technique is destructive and laborious but can be speeded by using image analysis (e.g. Magiscan, made by Joyce–Loebl Ltd.) to automate the identification and location of roots. The iterative process of grinding, polishing and photography might possibly be effected by robotics as a further sophistication.

We have grown plants of *Lolium perenne* and *Trifolium repens* both singly and in pairs in pots containing compost or field soil. In resin-embedded sections the roots of the two species can be readily distinguished. In the case of conspecific pairs of plants the identity of each root may be established, after reconstruction of the three-dimensional images, when each root can be traced back to its parent shoot. Development of this technology is at an early stage but should make it possible to analyse the three components that determine the architecture of a root system: (i) the number of branches; (ii) the distance between branches; and (iii) the angles of branching. None of these features can be adequately approached by techniques, such as rhizotrons, which force the studied part of a root system to grow in two dimensions. However each of these features of root architecture is crucial if we are to understand how the root systems of species differ, how they respond to soil types and if the plant breeder is to manipulate genotypic variation in the root systems of crops.

## THE DEMOGRAPHY OF ROOTS AND THEIR COLONIZATION

All organs of a plant have a life history in which they pass from birth to death. The rate at which a plant or its canopy or its root system grows is determined by the birth rate of its parts minus their death rate. Other papers in this volume (e.g. Mackie-Dawson & Atkinson, pp. 36–37) show clearly that the life span of individual roots may be very brief, both birth and death rates may be very rapid. The demographic properties of roots determine the ways in which a root system forages within the soil and also the speed with which it provides resources and substrates for the complex biota that live on or in it. However the application of demographic methods to root systems provides problems very much more difficult than the application of the same methods to canopies. A whole leaf during its life passes through an ontogenetic drift from birth to death — because it is a system of limited growth. The demography of leaves can therefore be treated like that of whole unitary organisms (such as *Drosophila* or man) by analysing birth and death rates (Harper & Sellek 1987; Harper 1988). In contrast, roots are systems of (at least theoretically) unlimited growth from apical meristems. A single root may contain all the phases from newly-born meristematic cells (at the apex and in the pericycle) through phases bearing active root hairs, phases in which the cortex has sloughed off or decomposed and phases in which little remains except a conducting strand. Moreover, when the oldest part of a root dies (from old age or from being damaged) the whole of the rest must also die.

A demography of roots, to be a complete descriptor, would need to analyse: (i) the rate at which whole roots are born and die; and (ii) the rate at which different parts of a single root make the transition from one phase and type of activity to another (i.e. a single root has its own demography and its own age structure). A root is a microcosm for a variety of root inhabitants, surface dwellers, parasites, saprophytes and mutualists. Their activity must clearly depend on the age structure of the root system and without an adequate understanding of root demography it is hard to see how the successional processes in the root community can be understood. This is a daunting task. We have made a very preliminary approach to this problem in a study of the demography of whole roots of white clover (*Trifolium repens*) (Sackville Hamilton & Harper 1989) and their in- and exhabitants (Freeman 1988) in a permanent pasture. We have not attempted the more difficult task of analysing the demographic processes within single roots.

The special advantage of white clover for a study of root demography is that, after the seedling stage, all new roots are developed from nodes on the plagiotropic shoots. Roots do not form at every node, but if a root develops this usually happens while the leaf is present. The leaf on a node lives for only 1–6 weeks but the root may continue to live and grow for as long as its parent node survives (up to 2 years). The birth date of each leaf therefore gives an estimate of the birth date of each root. A shoot of clover may bear roots ranging from 0–2 years old. One consequence of this growth habit is that the leaves of clover may be foraging for light in the presence of quite different neighbours from those met by the roots foraging for water and nutrients! New roots are formed in all seasons of the year (though at varying rates, Freeman 1988) and so it is possible to compare the effects of changing season with the effects of age.

Clover plants were repeatedly mapped in the field and the birth of each node was recorded. Plants were dug up at intervals through a year and the age of each of its rooted nodes was determined. In this way it was possible to obtain roots of all possible ages at all times of the year. These roots were washed free of soil and examined to determine the age of root sequence and the seasonal sequence of: (i) nodule formation; (ii) infection by vesicular arbuscular mycorrhizae (VAM) fungi; and (iii) colonization by nematodes and the abundance of their eggs. Detailed results are given by Freeman (1988). Figures 2–8 show examples of the sort of findings that emerged from the study. The longest roots were present in autumn but during the rest of the year there was little variation (Fig. 2). If we ignore the age of roots, there was little variation in the intensity of mycorrhizal infection, nematode infestation, or in the number of nodules per root (Figs 3, 4 and 5). However, the mean age of rooted nodes (equivalent to age of their roots) showed marked seasonal variation (Fig. 6), reaching a maximum of 250 days old in May. From October to April only a few new roots were produced and only a few old ones died and with each passing day all the existing roots became one day older. In spring, roots start to elongate and new roots were initiated. The mean age of rooted nodes then decreased. However the mean length of roots did not change because the increase in length of existing roots was counterbalanced by the initiation of new (shorter) roots. This seasonal variation in the age distribution of the root population masked underlying seasonal variation in root growth. Using an analysis of

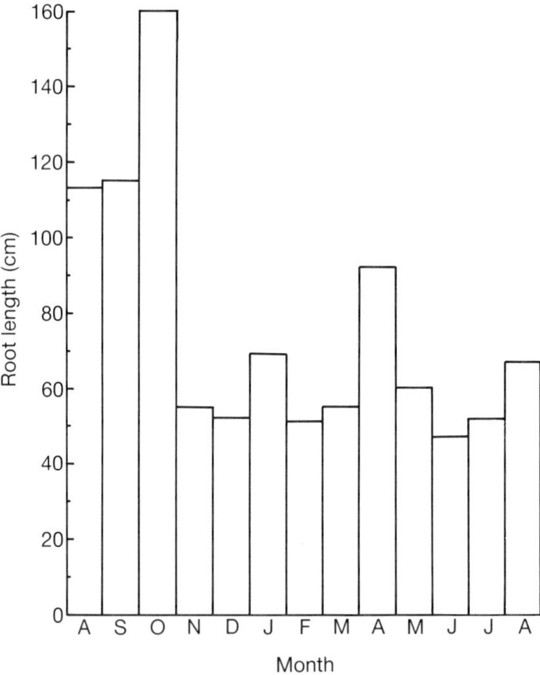


FIG. 2. Seasonal variation in the length of roots of *Trifolium repens* in a permanent grassland at Henfaes, near Bangor, North Wales.

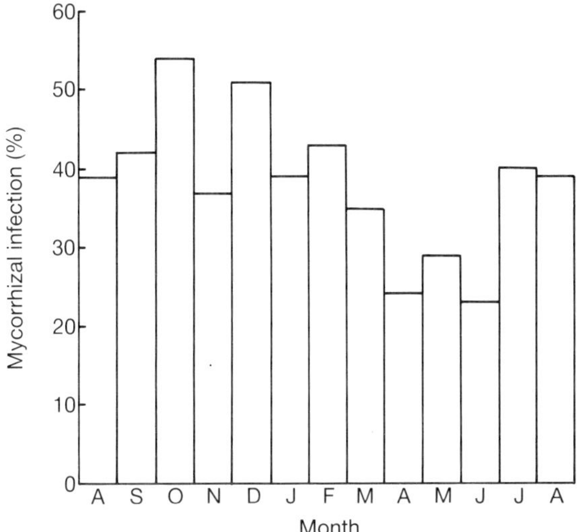


FIG. 3. Seasonal variation in the percentage infection of roots of *Trifolium repens* by vesicular–arbuscular mycorrhiza in a permanent grassland at Henfaes. The level of infection is expressed as the percentage of microscopic fields of view ($1\cdot72\ mm^2$) in which VAM were present at intervals along the roots.

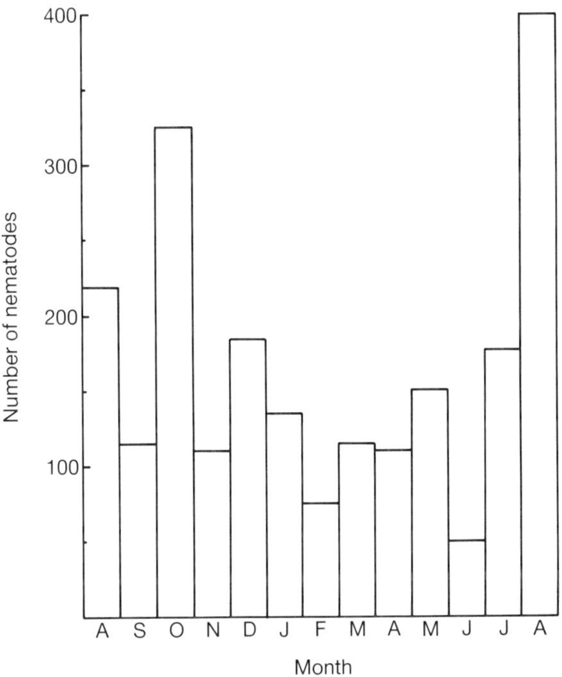


FIG. 4. Seasonal variation in the numbers of nematodes on the roots of *Trifolium repens* in a permanent grassland at Henfaes.

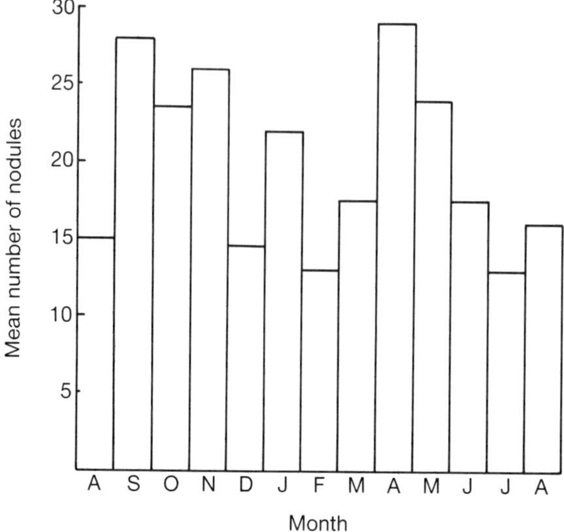


FIG. 5. Seasonal variation in the mean number of nodules on the roots of *Trifolium repens* in a permanent grassland at Henfaes.

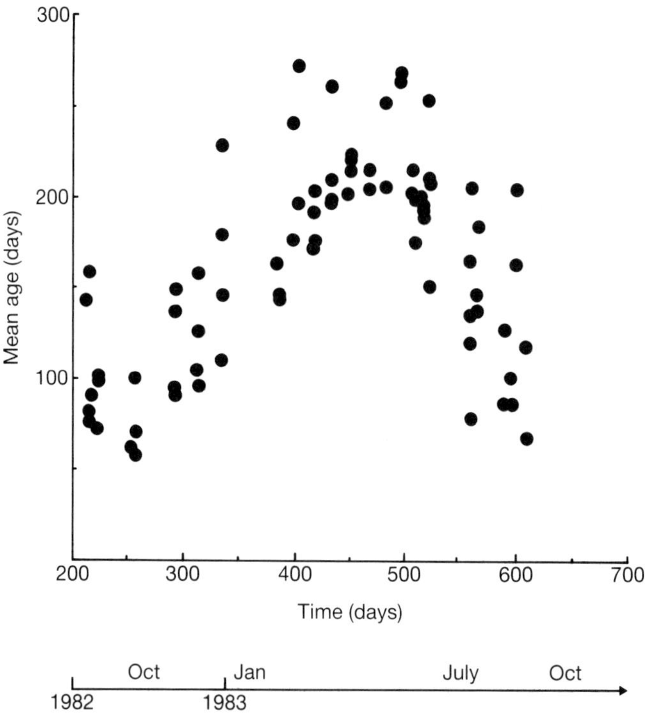


FIG. 6. Seasonal variation in the mean age of the roots of *Trifolium repens* in a permanent grassland at Henfaes.

TABLE 3. Analysis of variation in root lengths of *Trifolium repens*. Plants were harvested at intervals from a permanent pasture during one year, and each plant bore a number of roots of different ages. The effects of age were approximated by a quadratic regression

| Source of variation | DF | MS | F | P |
|---|---|---|---|---|
| Plants | 79 | 3·891 | 3·599 | <0·001 |
| Age of root | 2 | 21·261 | 19·667 | <0·001 |
| Plant age | 158 | 1·553 | 1·436 | <0·01 |
| Residual | 758 | 1·081 | | |

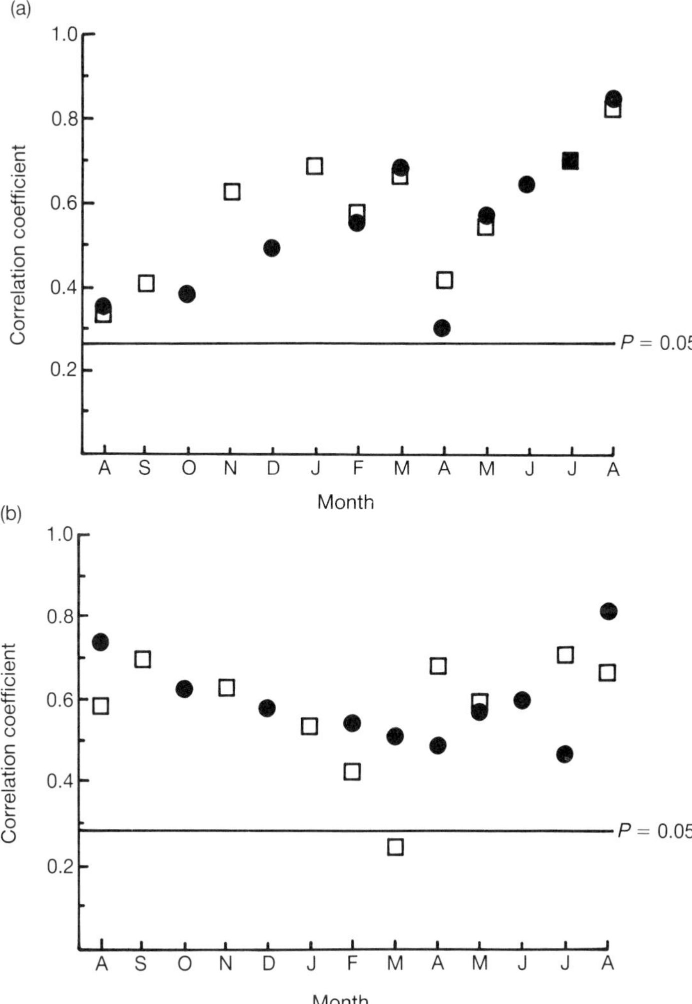


FIG. 7. (a) The correlation coefficients of root length of *Trifolium repens* and the number of nematodes per root in samples taken at different times of year in a permanent grassland at Henfaes. (b) The correlation coefficients of the number of nematodes and the intensity of mycorrhizal infection on roots of *Trifolium repens* sampled at different times of year in a permanent grassland at Henfaes. ● and □ refer to values from replicate plots within the field.

variance to separate the effects of season and age, we found highly significant effects of age on, for example, root length (Table 3).

Correlation coefficients between root length, and the number of nematodes, nodules, and mycorrhizal infection, calculated separately for each harvest, also showed seasonal fluctuations. The correlation between root length and nodule numbers was significant at all times of the year. The correlation between root length and number of nematodes per root was always significant, but fluctuated, reaching a maximum of almost 0·9 in late summer of the second year, when the total number of nematodes also reached a maximum (Fig. 7a). Such a high correlation at the time of maximum number of nematodes indicates that nematodes were evenly distributed throughout the root system at that time, and suggests that the number of nematodes was becoming limited by the length of root available.

The number of nematodes was also significantly but variably correlated with the level of infection by mycorrhizae (Fig. 7b). The correlation coefficient increased from a minimum of 0·4 in early spring to a maximum of 0·8 in late summer. This increase occurred before the strong correlation developed between root length and nematodes, and it is tempting to speculate that wounds in the roots caused by infection with mycorrhizae facilitate infestation by nematodes, and that, once nematodes had penetrated the root, their numbers increase until limited by the length of root available.

## CONCLUSION

Plant root systems present the research worker with many of the greatest unsolved problems in the plant sciences. Their evolution, development, form and function are all sources of open questions that demand new technologies and imaginative insights. Later papers in this volume contribute to unearthing some of the information that we need to answer questions about a system that is so inconveniently buried.

## REFERENCES

**Atkinson, D. (1983).** The growth, activity and distribution of the fruit tree root system. *Plant and Soil*, **71**, 23–35.

**Baldwin, J.P. & Tinker, P.B. (1972).** A method for estimating the lengths and spatial patterns of two interpenetrating root systems. *Plant and Soil*, **37**, 209–213.

**Barber, S.A. (1971).** Effect of tillage practice on corn (*Zea mays* L.) root distribution and morphology. *Agronomy Journal*, **63**, 724–726.

**Bottomley, P.A., Rogers, H.H. & Foster, T.H. (1986).** NMR imaging shows water distribution and transport in plant root systems *in situ*. *Proceedings of the National Academy of Sciences*, **83**, 87–89.

**Bower, F.O. (1908).** *The Origin of a Land Flora*. Macmillan Publishers Ltd, London.

**Caldwell, M.M., Eissenstat, D.M., Richards, J.H. & Allen, M.F. (1985).** Competition for phosphorus — differential uptake from dual isotope labelled soil interfaces between shrub and grass. *Science*, **229**, 384–386.

**Carman, J.G. (1982).** A non-destructive stain technique for investigating root growth dynamics. *Journal of Applied Ecology*, **19**, 873–879.

**Chilton, M-D., Tepfer, D., Petit, A., David, C., Casse-Delbart, F. & Tempé, J. (1982).** *Agrobacterium rhizogenes* inserts T-DNA into the genomes of the host plant root cells. *Nature*, **295**, 432–434.

**Freeman, A.C. (1988).** *Coevolution of white clover to associated mycorrhizae*. Ph.D. thesis, University of Wales.

**Fusseder, A. (1983).** A method for measuring length, spatial distribution and distances of living roots *in situ*. *Plant and Soil*, **73**, 441–445.

**Fusseder, A. (1985).** Verteilung des Wurzelsystems von Mais im Hinblick auf die Konkurrenz um Makronährstoffe. *Zeitschrift für Pflanzenernährung und Bodenkunde*, **148**, 321–334.

**Gadgil, P.D. (1963).** Soil sections of grassland. *Soil Organisms. Proceedings of the Colloqium on Soil Fauna, Microflora and their Relationships, Oosterbeek Sept 10–16 1962* (Ed. by J. Doeksen & J. van der Drift), pp. 327–332. North Holland Publishing Co, Amsterdam.

**Grenville, D.J. & Peterson, R.L. (1981).** Structure of aerial and subterranean roots of *Selaginella kraussiana* A Br. *Botanical Gazette*, **142**, 73–81.

**Groff, P.A. & Kaplan, D.R. (1988).** The relation of root systems to shoot systems in vascular plants. *Botanical Review*, **54**, 387–422.

**Harper, J.L. (1985).** Modules, branches and the capture of resources. *Population Biology and Evolution of Modular Organisms* (Ed. by J.B.C. Jackson, L.W. Buss & R.E. Cook), pp. 1–34. Yale University Press, New Haven.

**Harper, J.L. (1988).** Canopies as populations. *Plant Canopies: their Growth, Form and Function. Society for Experimental Biology Seminar Series 31* (Ed. by G. Russell, B. Marshall & P.G. Jarvis), pp. 105–127. Cambridge University Press, Cambridge.

**Harper, J.L. & Sellek, C. (1987).** The effect of severe mineral nutrient deficiencies on the demography of leaves. *Proceedings of the Royal Society (London), Series B*, **232**, 137–157.

**Hildebrand, E.M. (1934).** Life history of the hairy-root organism in relation to its pathogenesis on nursery apple trees. *Journal of Agricultural Research*, **48**, 857–885.

**Jeffrey, D.W. (1967).** Phosphate nutrition of Australian heath plants. I The importance of proteoid roots in *Banksia* (Proteaceae). *Australian Journal of Botany*, **15**, 403–411.

**Kirby, E.J.M. & Rackham, O. (1971).** A note on the root growth of barley. *Journal of Applied Ecology*, **8**, 919–924.

**Lamont, B.B. (1972a).** The effect of soil nutrients on the production of proteoid roots by *Hakea* spp. *Australian Journal of Botany*, **20**, 27–40.

**Lamont, B.B. (1972b).** The morphology and anatomy of proteoid roots in the genus *Hakea*. *Australian Journal of Botany*, **20**, 155–174.

**Lamont, B.B. & McComb, A.J. (1974).** Soil microorganisms and the formation of proteoid roots. *Australian Journal of Botany*, **22**, 681–688.

**Levan, M.A., Ycas, J.W. & Hummel, J.W. (1987).** Light leak effects on near-surface soybean rooting observed with minirhizotrons. *Minirhizotron Observation Tubes: Methods and Applications for Measuring Rhizosphere Dynamics* (Ed. by H.M. Taylor), pp. 89–98. Special Publication No. 50. ASA, Madison.

**Malajczuk, N. & Bowen, G.D. (1974).** Proteoid roots are microbially induced. *Nature* **251**, 316–317.

**McMichael, B.L. & Taylor, H.M. (1987).** Applications and limitations of rhizotrons and minirhizotrons. *Minirhizotron Observation Tubes: Methods and Applications for Measuring Rhizosphere Dynamics* (Ed. by H.M. Taylor), pp. 1–13. Special Publication No. 50. ASA, Madison.

**Melhuish, F.M. (1968).** A precise technique for measurement of roots and root distribution in soils. *Annals of Botany*, **32**, 15–22.

**Meyer, W.S. & Barrs, H.D. (1985).** Non-destructive measurement of wheat roots in large undisturbed and repacked clay soil cores. *Plant and Soil*, **85**, 237–247.

**Moore, L., Warren, G. & Strobel, G. (1979).** Involvement of a plasmid in the hairy root disease of plants caused by *Agrobacterium rhizogenes*. *Plasmid*, **2**, 617–626.

**Perry, R.L., Lyda, S.D. & Bowen, H.H. (1983).** Root distribution of four *Vitis* cultivars. *Plant and Soil*, **71**, 63–74.

**Persson, H.A (1983).** The distribution and productivity of fine roots in boreal forests. *Plant and Soil*, **71**, 87–101.

**Purnell, H.M. (1960).** Studies of the family Proteaceae. I Anatomy and morphology of the roots of some Victorian species. *Australian Journal of Botany*, **8**, 38–50.

**Retallack, G.J. (1985).** Fossil soils as grounds for interpreting the advent of large plants and animals on land. *Philosophical Transactions of the Royal Society (London), Series B*, **309**, 105–142.

**Rogers, H.H. & Bottomley, P.A. (1987).** *In situ* nuclear magnetic resonance imaging of roots: influence of soil type, ferromagnetic particle content, and soil water. *Agronomy Journal*, **79**, 957–965.

**Rogers, W.S. (1969).** The East Malling root observation laboratories. *Root Growth* (Ed. by W.J. Whittington), pp. 361–376. Butterworth, London.

**Sackville Hamilton, N.R. & Harper, J.L. (1989).** The dynamics of *Trifolium repens* in a permanent pasture. I. The population dynamics of leaves and nodes per shoot axis. *Proceedings of the Royal Society (London), Series B*, **237**, 133–173.

**Sanders, J.L. & Brown, D.A. (1978).** A new fiber optic technique for measuring root growth of soybeans under field conditions. *Agronomy Journal*, **70**, 1073–1076.

**Schuurman, J.J. & Goedewaagen, M.A.J. (1971).** *Methods for the Examination of Root Systems and Roots*. 2nd edn. Pudoc, Wageningen.

**Scott, G.A.M. & Stone, I.G. (1976).** *The Mosses of South Australia*. Academic Press, London.

**Stebbins, G.L. & Hill, G.J.C. (1980).** Did multicellular plants invade the land? *American Naturalist*, **115**, 342–353.

**St John, T.V., Coleman, D.C. & Reid, C.P.P. (1983).** Growth and spatial distribution of nutrient-absorbing organs: selective exploitation of soil heterogeneity. *Plant and Soil*, **71**, 487–493.

**Tepfer, D. (1983).** The potential uses of *Agrobacterium rhizogenes* in the genetic engineering of higher plants: Nature got there first. *Genetic Engineering in Eukaryotes* (Ed. by P.F. Lurquin & A. Kleinhofs), pp. 153–164. Plenum Press, New York.

**Tiffney, B.H. & Niklas, K.J. (1985).** Clonal growth in land plants: a palaeobotanical perspective. *Population Biology and Evolution of Clonal Organisms* (Ed. by J.B.C. Jackson, L.W. Buss & R.E. Cook), pp. 35–66. Yale University Press, New Haven.

**Tippkötter, R., Ritz, K. & Darbyshire, J.F. (1986).** The preparation of soil thin sections for biological studies. *Journal of Soil Science*, **37**, 681–690.

**Upchurch, D.R. & Ritchie, J.T. (1983).** Root observations using a video recording system in mini-rhizotrons. *Agronomy Journal*, **75**, 1009–1015.

**Vos, J. & Groenwold, J. (1987).** The relation between root growth along observation tubes and in bulk soil. *Minirhizotron Observation Tubes: Methods and Applications for Measuring Rhizosphere Dynamics* (Ed. by H.M. Taylor), pp. 39–49. Special Publication No. 50. ASA, Madison.

**Weaver, J.E. (1926).** *Root Development of Field Crops*. McGraw-Hill, New York.

**Whatley, J.M. (1983).** The ultrastructure of plastids in roots. *International Review of Cytology*, **85**, 175–219.

**White, F.F. & Nester, E.W. (1980).** Hairy root: plasmid encodes virulence traits in *Agrobacterium rhizogenes*. *Journal of Bacteriology*, **141**, 1134–1141.

**Williams, G.C. (1975).** *Sex and Evolution*. Princeton University Press, Princeton.

**Wochok, Z.S. & Sussex, I.M. (1974).** Morphogenesis in *Selaginella*. II Auxin transport in the root (rhizophore). *Plant Physiology*, **53**, 738–741.

# Methodology

# Methodology for the study of roots in field experiments and the interpretation of results

L.A. MACKIE-DAWSON AND D. ATKINSON*
*Macaulay Land Use Research Institute, Craigiebuckler, Aberdeen AB9 2QJ, UK*

## SUMMARY

1 The range of methods which is currently available for the study of plant root systems in the field is reviewed especially in relation to those methods which are appropriate to studies of natural plant communities or where major advances in technology have recently occurred and since the subject was last comprehensively reviewed.

2 Available methods are divided into three major groups: (i) soil sampling methods (total and potential root system removal, profile wall methods, soil coring and other monolith techniques); (ii) observation methods (root observation laboratories/rhizotrons/soil biotrons, mini- and micro-rhizotrons); and (iii) indirect methods (depletion of soil moisture, uptake of stable or radio-isotopes).

3 Recent advances seem to be greatest in respect of techniques for separating and measuring the lengths of root in soil cores and visualizing the growth of roots adjacent to mini- or micro-rhizotron tubes and of digitizing the resultant video-recorded images. This combination of methods should allow soil coring to generate spatial and quantitative data on root density and observation methods to give temporal and developmental information. The advantages and disadvantages of the methods are discussed in respect of the biological and ecological significance of the data generated.

## INTRODUCTION

Any papers on methodology run the risk of reading like a recipe book and with the same limitations but without many of the important decisions inherent in cookery, being obvious. Before consulting a recipe book the decisions of what to make and why that rather than something else have usually been made. This is not always the case with studies of plant roots where making any measurement is reasonably difficult. This may be especially true for studies of the root systems of species in non-agricultural situations where remoteness and difficult soils compound prob-

* Present address: Scottish Agricultural College, 581 King Street, Aberdeen AB9 1UD

lems. However, despite these difficulties it is critical to decide why the measurement is being made and the use to which the data will be put. This decision is complicated by a lack of clear basic understanding of the relationships between root morphology, anatomy and physiology and root and root system functioning. Similar decisions with respect to leaves are much easier as they serve fewer functions, and basic relationships, even under field conditions, are better understood. The amount of root present, expressed as length, weight, volume or surface area, can be related to absorption; as leaf area can be to light or $CO_2$ interception, but the distributions and mobilities of the elements to be absorbed from the soil are neither uniform, identical to each other and are not needed in the same absolute or relative amounts at different times in a season. Physiological properties in roots, as in leaves, vary with age but as roots absorb a much greater range of elements than leaves and as the physiology of the root interacts with the different elements, links between amount and activity in roots are difficult to interpret from limited samples of root mass or simple soil measurements. Precise understanding of root functioning must be essentially dynamic so as to allow for fluctuating demands by the plant for resources and the fluctuating ability of the soil to supply resources in relation to demand.

This paper reviews and discusses the range of methods which are currently available to study the root systems of plants growing in soil under field conditions. Many methods can give information on a range of parameters although some provide more appropriate results than others, e.g. when a root is observed directly, through a mini-rhizotron tube, measurements can be made of diameter, longevity and length but not of dry weight or nutrient concentration. On the other hand, when roots are collected from soil cores, information may be obtained on mass, diameter, length and nutrient concentration but not on longevity or periodicity. The chosen method has therefore to reflect the object(s) of the experiment and the level of explanation which it is hoped to obtain from the data. Root measurements must be interpreted in conjunction with other, e.g. leaf area, measurements.

The supply of soil reserves to the above-ground part of the plant is dependent on the length of root available for uptake at particular times in relation to demand, although this is modified by carbon allocation from the leaves to the roots. In a natural ecosystem, as opposed to a high fertilizer agricultural system, there are often less resources to manipulate and so it may be more important to understand where, how and how much the roots are exploiting the soil.

## *The range of techniques available*

Despite developments in the use of imaging nuclear magnetic resonance spectrometry (Bottomley, Rogers & Foster 1986) there are no *field* methods which allow roots to be directly observed without their removal from the soil or the establishment of an *in situ* observation surface in the soil. Three main groups of methods can be used to assess root growth in the field.

1 Root system extraction.
2 Observation methods.
3 Indirect methods.

An indication of the main root parameters which can be measured by these field techniques is given in Table 1. These are described and examples of the types of results obtained discussed.

## *Root system extraction*

Here either the complete root system (excavation) or a part of the root system (soil monolith, soil cores or needleboards) is either removed from the soil and examined subsequently or is assessed *in situ* (profile wall). The range of methods are discussed here although emphasis is placed on areas where there have been recent developments in either methodology or interpretation.

### *Total excavation*

Total excavation is useful for determining the standing root biomass and may be essential to estimating the amount and distribution of the few very large roots which in perennial species account for much of total biomass (Fig. 1). However, the

FIG. 1. The excavated root system of a mature apple tree (26 year Fortune/M9) as seen from above.

TABLE 1. The main root parameters which can be measured by groups of field techniques. (* Requires assumptions to be made)

| | Method | | | | |
|---|---|---|---|---|---|
| | Root system removal | | | Observation | Indirect |
| Parameter | Extraction | Profile wall | Coring | Rhizotron/ mini-rhizotron | Activity |
| Mass | + | | + | | |
| Length | | +* | + | +* | |
| Number | | + | + | +* | |
| Density | + | +* | + | +* | + |
| Distribution | + | + | + | + | + |
| Diameter | + | + | + | + | |
| Turnover | | | +* | + | |
| Longevity | | | | + | |
| Periodicity | | | | + | + |
| Growth rate | | | | + | + |
| Nutrient concentration | + | | + | | |

loss of fine roots during excavation (Atkinson 1985), means that excavation cannot be used to estimate root length. Excavations, other than on sandy soils where the soil can be washed away with a power hose, are time consuming. They have been used however to provide allometric relationships between root biomass and that of other plant components (see pp. 89–90). Although visual comparisons of root systems are non-quantitative they can be valuable in developing conceptual models and as a basis for planning future studies. For herbaceous species complete excavations intergrade to monolith sampling methods. Despite the time involved large numbers of excavations, especially for tree species, have been carried out. A recently published volume (Tamasi 1986) details the excavation of several hundred trees and represents a valuable resource in terms of gross morphology. Methods of excavation remain essentially conservative as are detailed by Rogers and Vyvyan (1934). Advances in computer technology (Diggle 1988) may make it easier to use 'pictures' from root excavations as the basis for whole system growth models.

When dealing with large plants, such as trees, partial excavations can be used. This may involve the excavation of a section of the root system (Coker 1958) or a combination of stump pulling and root excavation (Dudney 1972). However, unless root distribution is uniform then excavation of a quadrant will not quantitatively reflect total system weight. Variance can be very high. Partial excavation will only give information on dry weight and distribution for large roots, and only then if the section sampled is representative.

### *Profile wall method*

In this technique, a trench is cut and a layer of soil is removed from the surface of the trench to expose a soil profile from which records of the partially exposed roots can be made. Root records can be either of 'cut ends' or, for perennial species, of the regrowth of new roots from the cut ends (Abercrombie 1990). The soil trench, dug either by hand or with a mechanical digger, can be sequentially cut back towards plants if information on horizontal distribution is required. This method has the advantages of being suitable for stony soils and 'immediately visible' but is, however, labour intensive and time consuming and, as with total excavation, causes extensive soil disturbance. It can also be difficult to obtain a statistically meaningful number of replicates although data from this type of study can be combined with that from observation studies to give total root system values (Atkinson 1983). The method may have a favourable ratio of information gained to labour expended compared to other methods (Kopke 1981).

An example of the method is described by Mackie-Dawson *et al.* (1988). A mechanical digger was used to dig the trench and faces were prepared using a profile knife. A grid (50 × 50 mm) in a metal frame (1 × 1 m) was nailed against the profile face and soil horizon boundaries, depths and roots recorded. The number of roots inside a grid square was recorded immediately after exposure. Root count data was expressed as the number of roots per 100 cm$^2$ for 5 cm depth intervals down the profile. This is shown against depth in Fig. 2 which provides examples of three contrasting crops. Information is based on twenty-one pits for grass, forty-eight pits for winter barley and twenty-eight pits for spring barley. All three were significantly different ($P < 0{\cdot}05$) at a depth of 0–20 cm, reflecting the more prolific root system of the grass and the winter barley. The maximum depth of rooting was also significantly different, with spring barley having a much shallower root system (82 cm) than either winter barley (100 cm) or grass (120 cm). Using the profile wall method effects of soil type on root distribution can also be identified (Anon 1986).

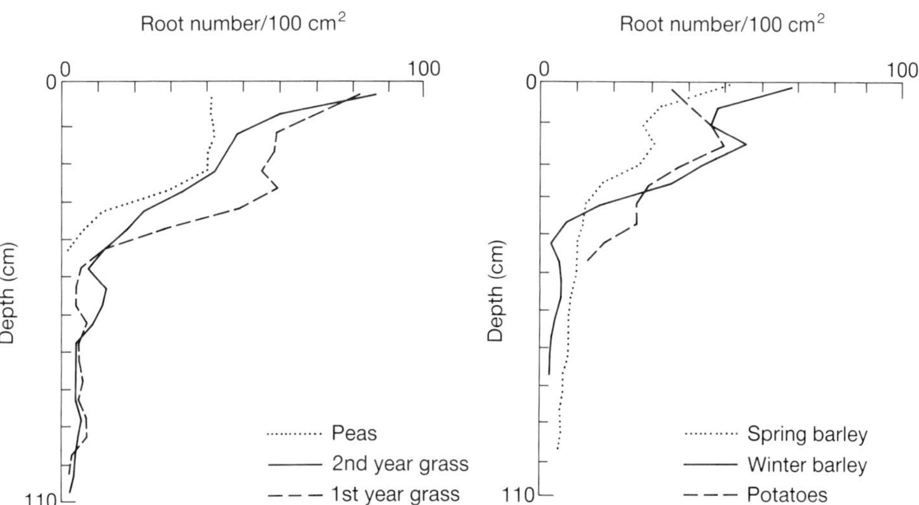


FIG. 2. The distribution of the roots of a range of crops as assessed by the profile wall method.

Under most of the crops studied, roots grew significantly deeper on silty loam soils then they did on sand or sandy gravel soils.

Root distribution can be obtained with this method indirectly, as root counts v. depth (Fig. 2), or directly, by mapping roots onto paper or transparent plastic sheet placed over the profile (Abercrombie 1990). In this way root distribution can be related to local soil features, such as channels or cracks, soil horizon, depth or stones and to soil physical properties, e.g. compaction (Abercrombie 1990). Root diameter can be measured directly on the exposed roots using a lens, a micrometer gauge or callipers. Estimates of total root length or root-length density can be made using the assumption that roots present in the profile wall go back into the soil for at least the depth exposed. For example, Bohm (1976) set one root unit to 5 mm for a profile which had been cut back 5 mm, roots 10 mm in length were counted as 2 root-length units. Estimates of root length using this method are however lower than those usually obtained by washing roots from undisturbed blocks of soil (Bohm 1976). This may be related to the difficulty of seeing very fine roots in natural light and the breaking of fine roots during the exposure of the profile. The method appears to work best where roots are horizontally distributed and of a large average diameter (e.g. trees), and less well where most roots are vertically distributed (e.g. grasses). For trees the change in root length with distance from the trunk complicates estimation. A normal straight trench will provide a relatively large sample of roots adjacent to the trunk but a much smaller sample at the periphery of the system. This problem can be overcome by the use of a spiral trench (Huguet 1973).

### *Monolith methods*

Monolith methods involve the removal of a sample of soil to represent the whole or part of the rooting volume of a plant. After soil removal, samples are either washed to remove roots from the soil or the roots can be held in something resembling their original position in the soil by a series of pins or needles; this provides information on distribution. The basic pin-board method has been described in detail in Schuurman & Goedewaagen (1971) and a modification of the method, combining cylindrical coring with the addition of a network of nylon line through the holes of an acrylic core container by Gooderham (1969). This method grades into soil coring and only differs from it in terms of the size of samples extracted.

### *Soil coring*

Soil coring is probably the most frequently used method of root sampling and allows repeated sampling of restricted experimental plot areas and the sampling of plots with limited soil depth and difficult geography. Cores can be taken with a range of techniques ranging from simple hand-held augers to power-driven systems (e.g. Welbank *et al.* 1974). Figure 3 shows a petrol-powered motor hammer coring system in operation. The coring tubes are fitted with split liners to help in the

FIG. 3. The use of a power-driven motor hammer to obtain a soil core.

removal of the soil-root samples. The internal diameter of the cutting edge is slightly less than that of the liners to avoid compaction of the soil core. Tubes are removed using a tripod and hoist. This method can be used on stony or indurated soils. The sample core can be cut into appropriate depth sections, measurements based on core samples are commonly used to validate other methods.

The soil extracted from cores is usually separated from roots by washing on sieves. A modification of this method which uses pressurized water spray jets and air flotation has been described (Smucker, McBurney & Srivastara 1982); the hydropneumatic elutriation system (Fig. 4). A comparison of careful hand washing with the above system for a range of crop species (potato, barley, wheat, grass, tree) and the same mesh size (930 and 500 μm) for primary and secondary sieves showed 15, 20, 25, 25 and 20% more respectively, root dry weight recovered using the automated system than by hand washing. A system using water jets and root washing cans was described by Welbank *et al.* (1974). These types of systems work best for mineral soils and are often ineffective at separating roots from the organic layers of forest soils.

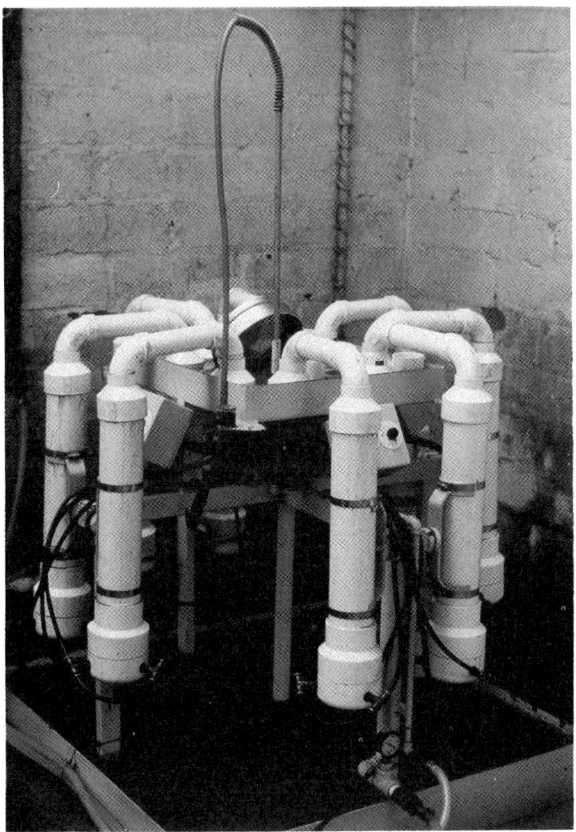

FIG. 4. The hydropneumatic elutriation soil core washing system.

Root length is most frequently measured on variable samples obtained as described above, as this is a better indicator of nutrient absorbing capacity, and is more sensitive to soil factors than mass (Atkinson & Chauhan 1987). It can be measured directly for fresh samples by placing roots on graph paper although this is only practical for small samples. For samples with a larger root length, measurements are more conveniently made by counting the number of intersections between roots and a random or regular pattern of lines (Newman 1966; Marsh 1971; Tennant 1975). With these types of estimates organic debris does not have to be separated from samples. Computer programs exist which either directly or with the aid of a mouse, e.g. C.MAP.ROOT from Michigan State University (Pregitzer, personal communication), allow roots to be digitized on the basis of the theory developed by Newman (1966). In one commercially available system, roots are spread out on a transparent rotating turntable which is traversed by a light beam and detector. Interruptions of the beam by roots are converted to length as a direct readout. Estimates using this method are sensitive to sample size with the optimum being 20–40 m. Greater root lengths tend to be underestimated as a result of overlapping and lengths <20 m overestimated. Systems using a computer mouse can overcome this as a result of operator involvement. Length measurement can also be automated using high resolution scanning cameras and image analysis by computer. Here limitations in the size of the field of view mean that only small clean samples can be scanned while there can be limitations as to the minimum diameter of root detected. Roots of varying diameters may require different camera settings for different classes of roots.

In an experiment to assess nutrient cycling in grass and clover plots, root samples were collected with a powered coring system, washed with the hydro-pneumatic elutriation method and root length determined using a 'comair' root-length scanner. Six replicate cores were required to obtain statistically significant differences in root length for a range of fertilizer treatments. In contrast only two replicate cores were necessary to detect differences in available soil phosphorus. Comparisons of grass and clover root distribution at two times in the season, July and October, both show a sharp decline in root-length density with depth (Fig. 5). In October root decay was greater than the rate of root growth. Available phosphorus content showed a decline similar to root length with soil depth.

In species with relatively low root-length densities and substantial spatial variation, e.g. trees (Atkinson 1985), large numbers of cores may be needed to

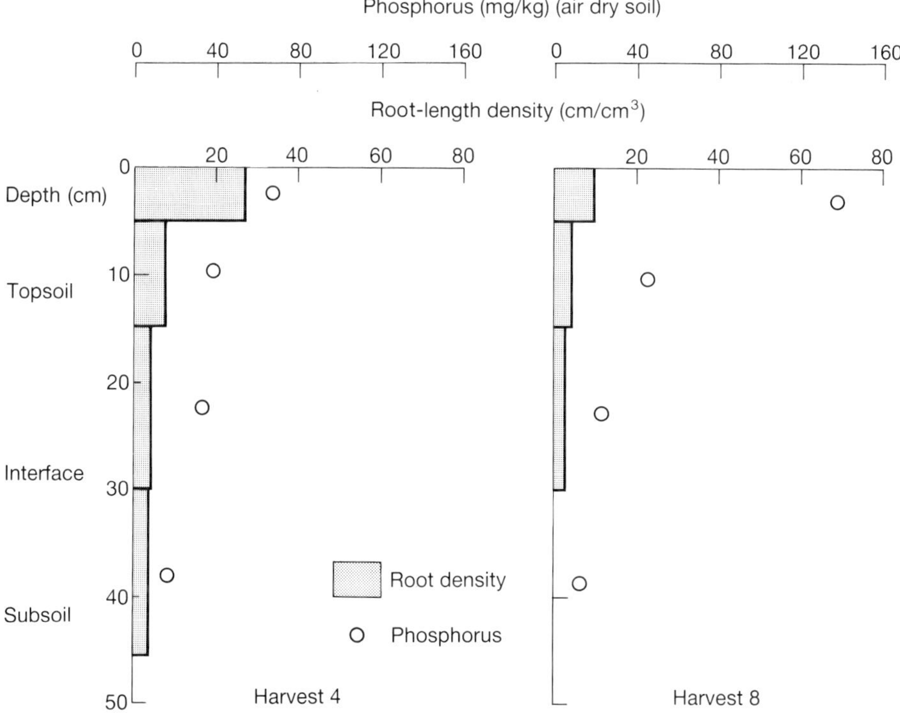


FIG. 5. Changes in soil phosphorus and root-length density, as determined from soil cores, with depth.

compare root density at different depths. In a study where 3·4 cm core samples were removed from around mature apple trees, 50 or 150 cm from the trunk, both absolute root densities and variance were high at the greater distance, variance increased with depth. Data from studies of this type can be difficult to analyse as a result of the large number of samples containing no roots. Atkinson & Wilson (1980) found that at a distance of 150 cm and a depth of 60 cm as many as 80% of core samples were root free.

Other parameters that can be measured or estimated from washed out root material are listed in Table 1, e.g. the number of root tips per unit volume of soil has been used by some workers as a criterion for root distribution and activity (Weller 1971). Nutrient analyses can be performed on dried root material and if this is combined with analysis of samples of other plant materials, e.g. leaves and stems, and growth analysis then values for nutrient uptake with time by the roots and, where length has also been determined, inflow rates ($I$) can be calculated as:

$$I = \frac{T}{L_a} \times \frac{(\mathrm{d}X)}{(\mathrm{d}E)},$$

where $L_a$ is root length per unit ground area averaged over time $T$, $X$ is the content of nutrient in plant material per unit ground area, and $E$ is the time after emergence of the crop.

Where a series of cores have been removed from an experiment over a period of time then total root system productivity can be estimated if information is available on the rate of root turnover and the amounts of live and dead roots in a sample. This approach has been discussed by Perrson (1983) and Fogel (1983, 1985). The time interval between the taking of samples is critical and must relate to the root-turnover period. For example in apple most roots exist in a primary (white) state for around 4 weeks and so if samples are taken at intervals much exceeding this, estimates of total productivity based on core sampling may be unreliable as some roots will have been produced, died and decomposed within the sampling interval.

Another related method to coring is the 'ingrowth bag'. Here a mesh bag filled with soil is placed in the hole left by the removal of a soil core. The bag is left *in situ* for a period and then recovered and the amount of root within the bag assumed to represent productivity over the time period. This method is discussed in detail by Steen (see pp. 75–78).

Diameter measurements can also be made directly on the washed-out roots, although many replicates may be required, and have been used to estimate root surface area and root length from volume (Bhat 1983).

## *Observation methods*

With all observation methods, a viewing surface is inserted into the soil. This may either be a small observation window (Asamoah 1984), an observation tube (Waddington 1971) or a large walk-in facility (Rogers 1939). The development of the root system *in situ* can then be seen through this window onto the soil. It allows the same area of volume of soil and population of roots to be observed continuously although there are major questions about the representative nature of the sample. Despite these problems this type of method has shown the greatest advances over recent years. Root distribution, diameter, longevity, periodicity, and turnover can all be potentially measured directly through the 'viewing window' and can be related to other aspects of rhizosphere ecology. No other single method permits this to happen.

To allow data recorded from an observation surface to be used, certain conditions must be satisfied. The major conditions are as follows.

1 The presence of the observation surface must not result in a concentration of roots either at the observation surface or away from the surface, i.e. the surface must represent a transect through the soil rather than a root collection or inhibition zone. This is the most stringent condition and must be satisfied if the two-dimensional data from an observational surface is to be converted into three-dimensional data, e.g. a root-length density. Where this condition is not satisfied, observation data can be used to assess the periodicity of new growth, root longevity, root survival times and amounts and used to compare treatments or species on a relative basis. Meeting this condition will be affected by the geometry of the arrangement of the observation surface relative to the form of the root system. This is discussed in more detail in relation to the different types of observation method.

2 Information must be available on the volume of soil being 'sampled' by the observation surface if length at that surface is to be converted into root-length density data. This distance, which can be calculated (Atkinson 1985), will vary for species with different root diameters and probably for different soil types. This is discussed in more detail later.

3 The sample size represented by replicate observation windows must be large enough to cope with the variation which is inherent in all estimations of root length. If this is not the case then although estimates may represent the amount at that site they will not provide a basis for calculating amounts elsewhere. Sample size is less important where observations are to be used only to assess parameters based on the fate of individual roots or the periodicity of growth. The size of samples provided by observation windows of various types is detailed in Table 2. A 1 $m^2$ observation window represents a similar volume to a 5 cm diameter, 1 m long soil core (1963 ml)

TABLE 2. The volume of soil 'sampled' by different observation types assuming 2 mm of soil depth is visible

| Method | Volume (ml) |
|---|---|
| 1 m × 1 m observation window | 2000 |
| 20 cm × 20 cm field observation unit | 80 |
| 56 mm OD mini-rhizotron tube, 1 m length | 352 |
| 15 mm OD micro-rhizotron tube, 30 cm long | 28 |

although a single mini-rhizotron tube represents around one fifth of this. A small observation window or a micro-rhizotron tube represents a small sample compared to a 30 cm length of a 5 cm diameter core (589 ml). Although recording observation units can be time consuming it is important to have sufficient recorded 'samples'.

**4** The position of the sampling unit gives a reasonable representation of growth in the whole soil volume or alternatively is in a position which allows a whole soil volume estimate to be calculated on the basis of a known spatial distribution. With tree species (Fig. 1) this is a major problem as although density decreases with distance from the trunk the volume of soil available for exploitation by the tree increases (and so as a result do factors used to convert per unit volume to total length). With other species this is less of a problem.

The extent to which the observed roots represent a valid sample of the whole population will vary for different plant species and ages of plant material.

Regardless of the accuracy with which root density can be estimated by observation methods this remains the method of choice for the estimation of root system dynamics. Although root turnover can be estimated from frequent soil coring this requires assumptions about the peaks and troughs in root length and their significance (Perrson 1983). In contrast to sampling methods, observation methods allow sequential measurements to be made of the same root or roots so that determinations of the length of time taken for a root to turn brown and of root survival from season to season will always be absolute. Determinations will only be influenced by root to root variability (Atkinson 1985). Determinations of the longevity of white roots, around 4 weeks in apple, and of the survival of new roots as components of the permanent woody root system, 20–25% in apple, allow total root length to be estimated (Fig. 6) for a single season and across a series of years. Measurements of survival seem to be relatively constant (Atkinson 1985) so that relatively small populations can be used to determine root longevity. With large or small observation windows roots can be marked directly on the window or on a plastic overlay. With mini- or micro-rhizotron tubes this must be done by photographing selected areas of the tube at frequent intervals or by filming and videotaping the same areas. For this to be valid the tube must be firmly anchored so that any reference lines or squares are not moved relative to the root(s) under study. With large windows it is possible to make assessments of the fate of whole roots while with the 'micro' methods this can usually only be done in relation to sections of root.

Quantification of root length by observation methods is more difficult than measurements of turnover. The number of individual roots that arrive at a window surface can be counted but may be independent of any property expressed by the root after intersection with the 'window'. Counting only 'first intersections' may minimize the effect of the interface on results, although growth adjacent to the observation surface uses resources and so will have an effect on growth elsewhere, including new intersections, especially in species with a high degree of apical dominance. The density and types of roots in the surrounding soil will influence and may determine the number of roots arriving at the glass–soil interface, but will

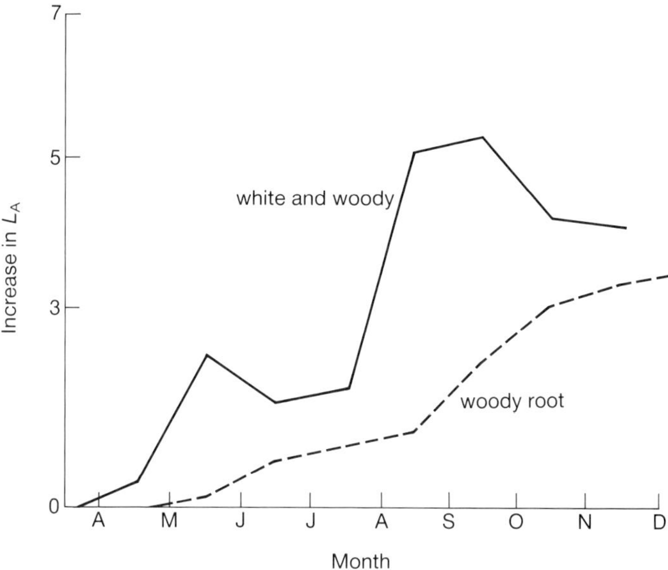


FIG. 6. The changes in root-length density ($L_A$) during a single season as a consequence of the production of new white root and its survival as brown/woody root for apple trees Worcester/MM104. Survival of 25% of root length and a time to brown of 4 weeks are assessed.

have a smaller effect on the length of root at the interface. The length at the interface is a function both of root arrival and growth after the initial intersection. However, the length of root at the interface is often recorded as a means of quantifying both the intersection of roots and as a means of measuring root-length density. Generally an intersect method derived by Head (1966) and numerically identical to that devised by Newman (1966) is used. A recorded image may be projected onto a grid or a grid may be inscribed on a tube or a window. Three approaches have been used to convert window or tube observations to root-length density: (i) empirical conversion coefficients; (ii) length/volume factors; and (iii) theoretical conversion coefficients. Root intersections at the window and root-length density in the surrounding soil are determined simultaneously and from this, a regression equation obtained, which relates window observations to root-length density (Table 3). The higher correlation with root fresh weight than length suggests that length on the glass is related more to the amount of woody roots than to the total length. Generally, for mini-rhizotrons the conversion equation includes an intercept, which implies that zero intersections do not indicate a total lack of rooting (Upchurch 1987). At low rooting intensities it is probable that no roots will intersect a particular tube, which leads to a positive intercept when a small number of observations are made.

A new conversion equation may be required for each new crop, soil combination and time of the season. In an experiment using mini-rhizotron tubes in a well-

TABLE 3. The relationship between root length assessed at the surface of a glass tube and the length and weight of barley roots grown within the tubes

| Correlates | Correlations | *P* |
|---|---|---|
| Fresh weight | 0·619 | <0·01 |
| Dry weight | 0·524 | <0·05 |
| Root length | 0·595 | <0·01 |

structured clay soil, the calibration equation for winter wheat changed considerably from December/January to May/June.

A number of studies have related root-length measurement at the surface of observation tubes to root-length density. Usually comparisons suggest that mini-rhizotrons underestimate length near the surface and overestimate it at depth. As the relationship is empirical this situation could be due to compaction at the soil surface during installation together with inadequate resolution of the fine roots present at the surface. This latter problem has been greatly reduced as a consequence of the introduction of high-resolution miniature colour TV cameras rather than the original black and white types. These cameras are able to resolve objects of around 25 μm diameter and to distinguish fine roots from the soil background more accurately. Similarly, for many species higher counts are obtained from observation windows when they are examined with a magnifying lens compared to the unaided eye. Temperature differences at the glass–soil interface may also reduce root intersections near the surface of the mini-rhizotron (McMichael & Taylor 1987).

Calculation of root density from observation measurements requires the assumption that all roots within a specific distance of the wall of the tube or of the rhizotron wall are seen and that no roots can be seen beyond that distance. The assumed distance seen into the soil for mini-rhizotrons has ranged from 1 to 3 mm (Upchurch 1987). A volume of soil is calculated from this distance and the observed area of the face. The volume is then divided into the root length observed at the interface.

Calculation of root length density can also be performed using a conversion factor, usually based on root diameter, to give a generalized length from the window length (Atkinson 1985). In studies with apple trees the mean lengths of root adjacent to an observation window and at a distance from the window were not significantly different (Atkinson 1985).

With newly planted trees this may not be the case but in any case the relationship between the tube surface and bulk soil is likely to vary with plant age.

An alternative approach is to use a theoretically derived coefficient involving geometrical probability (e.g. Meyer & Barrs 1985). The above considerations apply to all observation methods. Individual considerations apply to some of the types of methods. The advantages and limitations of the various observation methods are summarized in Table 4.

TABLE 4. The characteristics of different types of observation methods. (* Limit to the number which can be recorded before operator stress/boredom becomes a problem)

| Method | Approximate cost (£) Field unit | Equipment | Replication | Portability | Use for ecosystem studies | Ease of use |
|---|---|---|---|---|---|---|
| Permanent observation laboratories | 1 000–100 000 | 1 000–20 000 | Difficult—limited by window number | No | Yes | Easy |
| Miniature windows* | 5 | — | Yes | Yes | Limited | Can be uncomfortable |
| Mini-rhizotrons | 10–20 | 15 000 | Yes | Yes | Limited | Large amounts of equipment to move |
| Micro-rhizotrons* | 0.5–1 | 2 000 | Yes | Yes | Limited | Eye tiredness limits |

*Mini-rhizotrons*

The use of glass tubes inserted into the soil to study root growth was first described by Bates (1937). Sections of lamp glass were viewed with a mirror monitored on a rod and illuminated by a small electric bulb. Fibre optic probes have been used (e.g. Waddington 1971), as have rigid boroscopes (Atkinson, Crisp & Gurung 1982). Upchurch & Ritchie (1983) used a boroscope equipped with a miniaturized video camera for recording root images adjacent to 50 mm diameter glass tubes. The use of the video camera allows an increased number of tubes to be examined. A number of materials have been used for mini-rhizotron tubes. Glass gives good visibility but tends to fracture under cold conditions whereas acrylic tubes are easier to insert and last longer; scratching can be a problem and charges on the acrylic surface may have some effects on the interface.

The angle of insertion of tube has been suggested as having a significant effect on the results. Bragg, Govi & Cannell (1983) found for cereals that root growth down the side of a tube occurred in vertical tubes, but not in tubes angled at 45°. The effect of angle of insertion on root counts may depend on the counting method adopted and the orientation of root growth in the study crop. In a laboratory study using an *in situ* glass tube method, tracking was found to occur to the same extent both on vertical and 45° angled tubes (Mackie-Dawson *et al.* 1989). However, there was some evidence that the variance associated with the measurement of root length was greater for vertical tubes. Irrespective of angle of tube, care has to be taken in insertion. A special jig used for inserting angled tubes is shown in Fig. 7. In addition, for the method to be successful, a 'settling-in' period following installation is needed.

Coefficients of variation are high when studying root systems by any method.

FIG. 7. An auger jig system used to install angled mini-rhizotron tubes.

The appropriate number of tubes required depends on the needs of each experiment and resources available. In a study performed in winter wheat, in the month of May, variability of results from rhizotron tubes showed that sixty-nine tubes would be needed per plot for a 10% difference between treatment means to be significant ($P < 0.05$) (Mackie-Dawson & Goss, personal communication). Atkinson, Crisp & Gurung (1982) found that with only five replicate tubes per treatment, coefficients of variation could be as high as 189% of the mean so that only differences of around 95% or higher values were detected. When small differences are important, or when rooting intensity is low, a large number of samples is needed (Upchurch 1987).

### *Micro-rhizotron*

The term mini-rhizotron has come to be synonymous with the use of large diameter tubes viewed with a television system. This type of system is relatively expensive and cannot easily be used for studies of plants growing in pots nor for field studies of specific herbs growing in mixed communities. These require the use of smaller diameter tubes (micro-rhizotrons) where growth can be visualized by the use of a small diameter (8–10 mm) boroscope (Atkinson 1989; Mackie-Dawson *et al.* 1989). The restrictions which limit all observation methods clearly apply to these and the volume of soil sampled (Table 2) is very small. Measurements can be made directly by eye or a photographic record can be taken with the use of a suitable adaptor and a still or video camera. Positioning of records is normally by means of a grid inscribed on the tube which can be used either to record the site of first contact or in the determination, via the intersection method (Head 1966), of length adjacent to the tube. Information produced in this way is similar to that obtained with a permanent root laboratory. Atkinson (1989) found the seasonal pattern of root growth of trees of *Prunus avium* grown in pots and measured with the micro-rhizotron method to be similar to that described for other *Prunus* species by Atkinson & Wilson (1980) on the basis of conventional root laboratory measurements.

*Miniature windows*

The basic root laboratory method has been modified by the use of small observation windows set into the soil adjacent to a growing crop. A plate of glass or perspex is held against the soil with a wooden frame (Asamoah 1984) (Fig. 8). Root growth can be estimated directly without the need for a viewing device which reduces the time needed for measurements. However it is often difficult to photograph these windows and their positioning in the ground can make them uncomfortable to record. They are subject to most of the limitations of permanent root laboratories except that they can be installed on field plots and at any spatial position in relation to a fixed object like a tree. Usually miniature windows are inserted to a depth of around 20–50 cm although in natural ecosystems and with many crop species this is less of a restriction than it might seem as most growth is found in this depth range. Most of the measurements which can be made in permanent facilities can be made with installations of this type although the microscopic examinations which are needed to assess inter-relationships between roots and the soil fauna and flora are difficult. Their ease of installation and relative cheapness allows reasonable replication using units of this type. Asamoah (1984) was able to follow the periodicity of root growth of trees receiving a range of water supply treatments although the variation he found from tree to tree was high.

*Root observation laboratories (rhizotrons)*

Despite the expense of building permanent observation laboratories, facilities of this type continue to be built and two of the most recent are described in this volume (Fogel & Lussenhop, pp. 61–68 and Sackville Hamilton *et al.*, pp. 49–59). The limitations inherent in the use of this type of facility and the assumptions which need to be made to facilitate estimations of root-length density and turnover have been discussed by Atkinson (1985). The advantages of facilities of this type are permanence, ease of working, the ability to study interactions between roots and other elements of the soil flora and fauna at a range of scales (including the microscopic), the ability to access soil adjacent to a specific event so as to make

FIG. 8. A miniature observation window installed in the soil adjacent to young cherry trees (after Asamoah, 1984).

associated measurements and the ability to influence the situation by the addition of inputs, e.g. nutrients or fungal inoculum. Many of these features are important to studies of long term ecosystem productivity or nutrient cycling. The installation of a root laboratory will cause major soil disturbance to the surrounding area (Fogel & Lussenhop, pp. 61–68), although with care this can be minimized and in a long-term study presents an opportunity to study site recolonization. Replacement of soil in a satisfactory physical, chemical and biological condition will require care, although results from the Michigan Soil Biotron (Fogel & Lussenhop) suggest that even where soil is removed from a 1 m band from the windows and replaced, that root growth of existing forest trees will occur within months and biological activity due to earthworms resume within days. The use of this type of facility for the study of multi-species ecosystems emphasizes the need for a key to separate the roots of different plant species. Measurements of root system growth and dynamics must be capable of being assigned to identified species rather than to a vegetation type. Recent studies (Atkinson & Fogel, unpublished) have suggested that the range of potential variation is such that a classic dichotomous key will not be effective, but that separation via a computer database holding information on the possible ranges of variation under specified conditions for properties, such as colour, diameter, texture, branching pattern, etc., may be possible. Information from root laboratories can be combined with that from mini- or micro-rhizotron measurements made at specific points in relation to particular species or site features or from access points of this type inserted into the soil through the windows of the root laboratory.

Permanent root laboratories represent the easiest method of taking advantage of developments in video-recording techniques which allow time-lapse video images to be made of specific small areas of the window.

### *Indirect methods*

The presence and the activity of the root system has often been inferred from indirect measurements related to their functioning, usually the supply of water and mineral nutrients to other parts of the plant. Under good conditions the relationship between root length and the uptake of water or the absorption of a specific nutrient, frequently $^{32}P$, can be good (Atkinson 1989). For example in a study of root activity, both via the windows of a root observation laboratory and the injection of $^{32}P$ to 15 cm depth, Atkinson, (1974) found a close relationship over a period of 2 months. These data are reproduced in Fig. 9. Similarly, Evans (1978) found a relationship between root-length density and maximum soil-moisture depletion. Variation in soil-moisture potential with depth has often been used as a means of estimating root activity and on a gross scale (Atkinson 1978) it can be effective. Some of the problems relating to its use have been discussed by Atkinson (1980). Similarly the injection of radio-isotopes in soil and the estimation of relative activity by counting samples of foliage at intervals has been used by several workers on a range of plant species, most frequently tree crops, e.g. Huxley *et al.* (1974). Under conditions where root density is appreciable and relatively uniform this method can indicate changes in activity related to changing soil condition or shoot activity. More exacting standards in respect of the use of unsealed sources of radio-isotopes and their release into the environment together with the limited mobility of phosphorus which leads to high variance, if root density is low or variable, have restricted the use of this method. The use of $^{15}N$ as a tracer overcomes some of these disadvantages (Atkinson 1977) but $^{15}N$ is expensive and the dynamics of its movement in soil and reactions with other soil nitrogen sources can make $^{15}N$ concentration data difficult to interpret. Where major differences exist between treatments (Atkinson & White 1980) the pattern of root activity given by the uptake of $^{15}N$, by root observation methods and the profile wall method can be very similar. As with observation methods changes in isotope concentration in the plant with time give an indication of changes in the amounts and distribution of root activity.

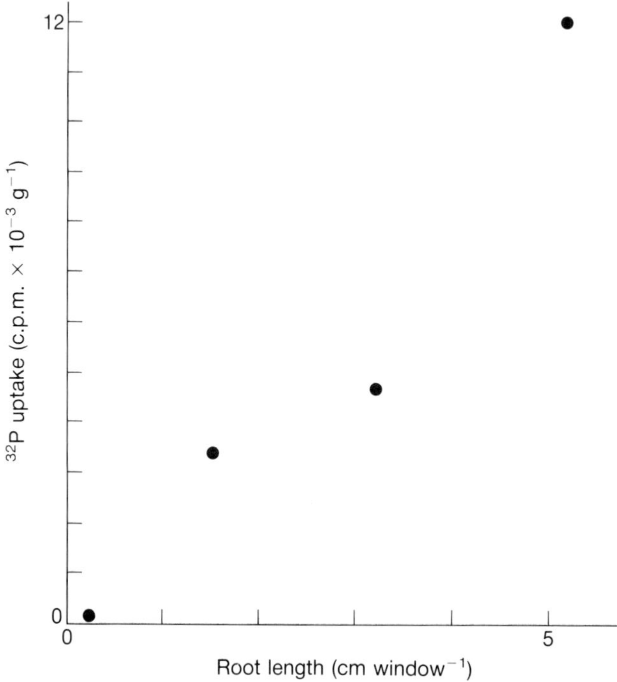


FIG. 9. The relationship between uptake of $^{32}$P from 15 cm depth and white root length, as seen in the window of a rhizotron, for 2 year Cox/M9 apple trees. Redrawn from Atkinson (1974).

## DISCUSSION

Although not comprehensive, this paper has outlined most of the main methods currently available for studying root distribution in the field. No one single method is applicable to all situations. In field plots or sites where only limited space is available or when there is a need to sample over a long period and it is important to identify the component of the root system which needs to be measured, the use of general methods for studies will generate general information. The main criterion influencing the selection of a method is probably whether information is needed on changes with time or whether spatial data on distribution at one moment in time will be adequate. Where time is involved an observation method will be indicated (although frequent soil coring will also give this). Where spatial distribution is the key issue a direct soil sampling method, e.g. profile wall and coring, will be the best choice. Most methods provide information on root distribution down the soil profile, although the profile wall method provides it in a most cost-effective manner, provided that measurement at one time is satisfactory. Total or partial excavation also provides good information on root distribution, but is limited to the larger diameter roots, and is very time consuming.

Of the many factors to consider when choosing a method of study, the main ones will probably be the availability of equipment, facilities and resources, the soil and plant types to be investigated and the type, distribution and heterogeneity of the root systems. As in studies of above-ground plant growth, where a range of measurements are often made, a range of methods are required to characterize factors such as activity, growth, nutrient uptake and soil exploitation.

## REFERENCES


**Abercrombie, R.A. (1990).** A root distribution study of Avocado trees on a sandy loam soil. *Acta Horticulturae*, **275**, 505–512.

**Anon (1986).** Investigation into the proposed water abstraction from the River Spey. *Agricultural Studies*, Grampian Regional Council.

**Asamoah, T.E.O. (1984).** *Fruit tree root system. Effect of nursery and orchard management and some consequences for growth, nutrient and water uptake.* Ph.D. thesis, University of London.

**Atkinson, D. (1974).** Some observations on the distribution of root activity in apple trees. *Plant and Soil*, **40**, 333–342.

**Atkinson, D. (1977).** Some observations on the root growth of young apple trees and their uptake of nutrients when grown in herbicidal strips in grassed orchards. *Plant and Soil*, **49**, 459–471.

**Atkinson, D. (1978).** The use of soil resources in high density planting systems. *Acta Horticulturae*, **65**, 79–89.

**Atkinson, D. (1980).** The distribution and effectiveness of the roots of tree crops. *Horticultural Review*, **2**, 424–490.

**Atkinson, D. (1983).** The growth, activity and distribution of the fruit tree root system. *Plant and Soil*, **71**, 23–26.

**Atkinson, D. (1985).** Spatial and temporal aspects of root distribution as indicated by the use of a root observation laboratory. In *Biological Interactions in Soil* (Ed. by A.H. Fitter, D. Atkinson, D. Read & M.B. Usher), pp. 43–65. Blackwell Scientific Publications, Oxford.

**Atkinson, D. (1989).** Root growth and activity: current performance and future potential. *Aspects of Applied Biology*, **22**, 1–13.

**Atkinson, D. & White, G.C. (1980).** Some effects of orchard soil management on the mineral nutrition of apple trees. In *Mineral Nutrition of Fruit Trees* (Ed. by D. Atkinson, J.E. Jackson, R.O. Sharples & W.M. Waller), pp. 241–254. Butterworth, London.

**Atkinson, D. & Wilson, S.A. (1980).** The growth and distribution of fruit tree roots: some consequences of nutrient uptake. In *The Mineral Nutrition of Fruit Trees* (Ed. by D. Atkinson, J.E. Jackson, R.O. Sharples & W.M. Waller), pp. 137–150. Butterworth, London.

**Atkinson, D. & Chauhan, J.S. (1987).** The effect of paclobutrazol on the water use of fruit plants at two temperatures. *Journal of Horticultural Science*, **62**, 421–426.

**Atkinson, D., Crisp, C.M. & Gurung, H.P. (1982).** The effect of mecoprop on shoot and root growth and mineral nutrition of young apple trees. *Proceedings British Crop Protection Conference — Weeds*, 281–284.

**Bates, G.H. (1937).** A device for the observation of root growth in the soil. *Nature*, **139**, 966–1067.

**Bohm, W. (1976).** *In situ* estimation of root length at natural soil profiles. *Journal of Agricultural Science, Cambridge*, **87**, 365–368.

**Bhat, K.K.S. (1983).** Nutrient inflows into apple roots. *Plant and Soil*, **71**, 371–380.

**Bottomley, P.A., Rogers, H.H. & Foster, T.H. (1986).** NMR imaging shows water distribution and transport in plant root systems *in-situ*. *Proceedings National Academy of Science, USA* **83**, 87–89.

**Bragg, P.L., Govi, G. & Cannell, R.Q. (1983).** A comparison of methods including angled and vertical mini-rhizotrons for studying root growth and distribution in a spring oat crop. *Plant and Soil*, **73**, 435–440.

**Coker, E.G. (1958).** Root studies, XII. Root systems of apple on Malling rootstocks on five soil series. *Journal of Horticultural Science*, **33**, 71–79.

**Diggle, A.J. (1988).** Root morphology. A model in three dimensional co-ordinates of the growth and structure of fibrous root system. *Plant and Soil*, **105**, 169–178.

**Dudney, P.J. (1972).** On the estimation of root biomass in a growth pattern experiment on apples. *Report of East Malling Research Station for 1971*, 66–67.

**Evans, P.S. (1978).** Plant root distribution and water use patterns of some pasture and crop species. *New Zealand Journal of Agricultural Research*, **21**, 261–265.

**Fogel, R. (1983).** Root turnover and productivity of Coniferous forests. *Plant and Soil*, **71**, 75–86.

**Fogel, R. (1985).** Roots as primary producers in below-ground ecosystems. In *Ecological Interactions in Soil* (Ed. by A.H. Fitter, D. Atkinson, D.J. Read & M.B. Usher), pp. 23–36. Blackwell Scientific Publications, Oxford.

**Gooderham, P.T. (1969).** A simple method for the extraction and preservation of an undisturbed root system from a soil. *Plant and Soil*, **31**, 201–204.

**Head, G.C. (1966).** Estimating seasonal changes in the quality of white untuberised root on fruit trees. *Journal of Horticultural Science*, **41**, 197–206.

**Huguet, J.G. (1973).** A new method of studying the rooting perennial plants by means of a spiral trench. *Annals of Agronomy*, **24**, 707–731.

**Huxley, P.A., Patel, R.Z., Kabaara, A.M. & Mitchell, H.W. (1974).** Tracer studies with $^{32}P$ on the distribution of functional roots of Arabica coffee in Kenya. *Annals of Applied Biology*, **77**, 159–180.

**Kopke, U. (1981).** A comparison of methods for measuring root growth of field crops. *Zeitschrift für Acker und Pflanzenbau*, **150**, 39–49.

**Kutschera, L. (1960).** *Wurzelatlas mitteleuropaischer Ackerunkrauter und Kulturpflanzen*. DLG-Verlag, Frankfurt/Main, 574.

**Mackie-Dawson, L.A., Walker, A.D., Atkinson, D. & Bibby, J.S. (1988).** Water abstraction from the River Spey area for domestic and agricultural purposes and its effects on agriculture. *Scottish Geographical Magazine*, **104**, 91–96.

**Mackie-Dawson, L.A., Buckland, S.T., Duff, E.T., Pratt, S.M. & Reid, E.J. (1989).** The use of *in situ* techniques for the investigation of root growth. *Aspects of Applied Biology*, **22**, 349–356.

**Marsh, B. (1971).** Measurements of length in random arrangements of lines. *Journal of Applied Ecology*, **8**, 265–267.

**McMichael, B.L. & Taylor, H.M. (1987).** Applications and limitations of rhizotrons and mini-rhizotrons. *Mini-rhizotron Observation Tubes: Methods and Applications for Measuring Rhizosphere Dynamics* (Ed. by H.M. Taylor), pp. 1–13. Special Publication No. 50. ASA, Madison.

**Meyer, W.S. & Barrs, H.D. (1985).** Non-destructive measurements of wheat roots in large undisturbed and repacked clay soil cores. *Plant and Soil*, **106**, 16–22.

**Newman, E.I. (1966).** A method of estimating the total length of root in a sample. *Journal of Applied Ecology*, **3**, 139–145.

**Perrson, H.A. (1983).** The distribution and productivity of fine roots in boreal forests. *Plant and Soil*, **71**, 87–102.

**Rogers, W.S. (1939).** Root studies, VIII. Apple root growth in relation to rootstock, soil, seasonal and climatic factors. *Journal of Pomology and Horticultural Science*, **17**, 99–130.

**Rogers, W.S. & Vyvyan, M.C. (1934).** Root studies v. rootstock and soil effect on apple root systems. *Journal of Pomology and Horticultural Science*, **12**, 110–150.

**Schuurman, J.J. & Goedewaagen (1971).** *Methods for the Examination of Root Systems and Roots*. Pudoc, Wageningen.

**Smucker, A.J.M., McBurney, S.L. & Srivastara, A.K. (1982).** Quantitative separation of roots from compacted soil profiles by the hydropneumatic elutriation system. *Agronomy Journal*, **74**, 500–503.

**Tamasi, J. (1986).** *Root Location of Fruit Trees and its Agrotechnical Consequences*. Akademiai Kiado, Budapest.

**Tennant, D. (1975).** A test of a modified line intersect method of estimating root length. *Journal of Ecology*, **63**, 995–1002.

**Upchurch, D.R. (1987).** Conversion of mini-rhizotron root intersections to root length density. In *Mini-rhizotron Observation Tubes: Methods and Applications for Measuring Rhizosphere Dynamics* (Ed. by H.M. Taylor), pp. 61–65. Special Publication No. 50. ASA, Madison.

**Upchurch, D.R. & Ritchie, J.T. (1983).** Battery operated color video camera for root observations in mini-rhizotrons. *Agronomy Journal*, **76**, 1015–1017.

**Waddington, J. (1971).** Observations of plant roots *in situ*. *Canadian Journal of Botany*, **49**, 1850–1852.

**Welbank, P.J., Gibb, M.J., Taylor, P.J. & Williams, E.D. (1974).** Root growth of cereal crops. *Report of Rothamsted Experimental Station for 1973*, **2**, 26–66.

**Weller, F. (1971).** A method for studying the distribution of absorbing roots of fruit trees. *Experimental Agriculture*, **7**, 351–361.

# A modular rhizotron for studying soil organisms: construction and establishment

C.A.G. SACKVILLE HAMILTON, J.M. CHERRETT, J.B. FORD, G.R. SAGAR AND R. WHITBREAD

*School of Biological Sciences, University College of North Wales, Bangor, Gwynedd LL57 2UW, UK*

## SUMMARY

1 A modular prefabricated rhizotron was built in a pasture. It had enough windows to allow replicated treatments. Three of the windows were easily removable to allow access to the soil profile. On the north side mesh cages were constructed to exclude different components of the soil fauna.

2 The soil profile was removed and maintained in discrete layers and reinstated with the old turf inverted under the topsoil, to simulate ploughing. Soil for the animal exclusion cages was sterilized.

3 On the south side two levels of fertilizer and two frequencies of cutting were imposed.

4 The effect of soil sterilization was short lived. Thirteen weeks after reinstating the soil around the rhizotron, there were similar numbers of soil fauna in sterilized and unsterilized soils. The initial colonization of both soils was by the same taxa. Some taxa did not invade until spring. Late invasion was usually a phenological effect of the particular taxa.

5 The soil profile compacted on average 15 cm in 16 months. On the north side the compaction was greatest in the control plots, from which only moles were excluded.

## INTRODUCTION

Rhizotrons [from Greek *rhiza* a root (Liddell & Scott 1871) and *-tron* a Greek suffix indicating instrument (Hanks 1982)] are subterranean glass walled chambers from which soil and soil organisms can be studied *in situ*. They have mainly been used for the study of roots (Bohm 1979) including roots of grasses (Garwood 1967; Atkinson 1977; DiPaola, Beard & Brawand 1982), but much less commonly to investigate soil fauna (Harding 1968; Carpenter 1985). The most extensive study to date is that of Carpenter (1985), who used a small rhizotron with nine windows on four sides, to make preliminary observations on the fauna of an amenity grassland in North Wales. This facility however did not allow adequate replication for experimental treatments. There was thus a requirement for a cheap rhizotron extensive enough to allow for the replication of experiments on the manipulation of the soil fauna, by physical exclusion using mesh of different sizes. It was also desirable to be able to access the soil behind the glass to add baits or fertilizer to specific horizons in the soil, or to remove animals or fungi for identification. Windows had to extend to about 10 cm above the soil surface to allow the litter layer to be studied. For the experiments currently planned the windows needed to be deep enough to observe the whole rooting depth of *Trifolium repens* L. and *Lolium perenne* L.

The requirement for cheapness, combined with the strength necessary for safety at depth, suggested the adaptation of an existing building module. The need for minimum disturbance of the habitat suggested prefabricated units. The requirement for soil sterilization and exclusion cages necessitated soil removal and replacement immediately around the rhizotron.

## METHODS

### *Site and Orientation*

The rhizotron is situated in a small enclosure within an area of amenity grassland unploughed for at least 40 years (Carpenter *et al.* 1985). The long axis of the building is orientated east–west.

### *Design*

#### *The building*

The rhizotron is 16·49 m long × 2·42 m wide × 2·24 m deep, of which 0·57 m protrudes above ground level (Fig. 1). It is constructed of sixteen prefabricated L-shaped concrete sections, normally used in the construction of silage pits. These face one another to form the sides and the floor of a trench, which is completed by a concrete roof. PVC window frames were cast into the sections, and subsequently glazed with 10 mm glass. Three of the windows were each glazed with five vertical slats to allow easy access to the soil profile. In the other windows the glass is entire but removable. On each long side of the building are sixteen windows each 90 cm deep × 50 cm wide and on the west wall are two slatted windows each 90 cm deep × 55 cm wide. At the east end is a door reached from ground level by a staircase. The rhizotron has mains water and electricity and is drained by a gutter running the entire length of the building. The cost of construction was £36 000 in 1987.

#### *Cages*

Attached to the north side of the building adjacent to each window are fifteen 1 $m^3$ cages which protrude 15 cm above ground level. Each cage is formed by two 1 m × 1 m aluminium sheets at right angles to the building, with mesh on the bottom and distal side. The side proximal to the building is formed by the window and the top of the cage is open. Three treatments, each with five replicates (see Fig. 2) were imposed on one side of the rhizotron by using different gauge mesh to physically exclude different size classes of the soil fauna. Control cages have a Netlon mesh of

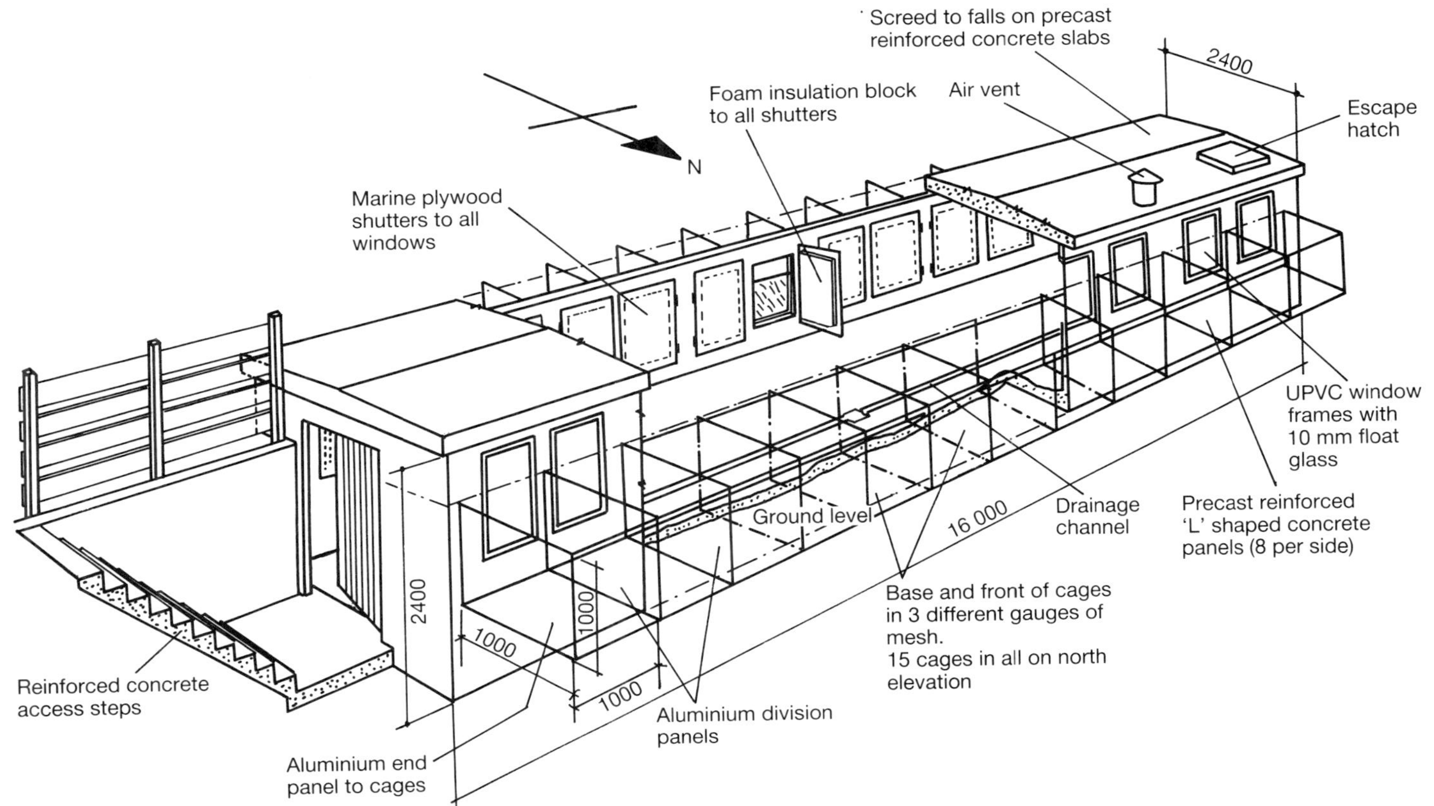


FIG. 1. An exploded view of the rhizotron. All measurements are in millimetres.

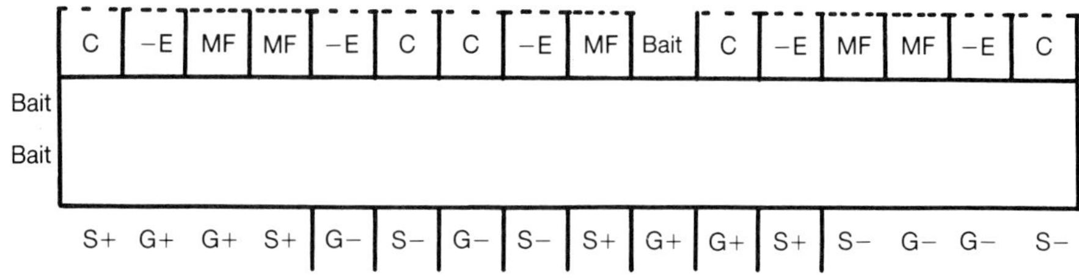


FIG. 2. Plan view of the rhizotron showing the design of experimental treatments. C = control (only moles excluded); −E = earthworms excluded; MF = minimal fauna; S = silage, plots cut three times year$^{-1}$; G = grazing, plots cut fortnightly March–November; + = added fertilizer; − = no added fertilizer; Bait = slatted windows suitable for studies on bait consumption.

10 mm allowing all the soil fauna except moles to enter from the surrounding pasture; earthworm exclusion cages have a 425 μm mesh, and minimal fauna cages a 35 μm mesh on the side and a 200 μm mesh on the bottom. The larger mesh was used on the bottom of the latter treatment to prevent drainage from being impeded. In the plots with fine mesh support is given by Netlon. The physical exclusion of different sizes of soil fauna using mesh cages is a modification of the mesh bag technique used by Edwards and Heath (1963).

### *Management treatments*

On the south side of the rhizotron two levels of fertilizer and two frequencies of cutting were imposed in a split–split plot design (Fig. 2). There were four replicates of each treatment and plots were given either no added fertilizer or NPK (20 : 10 : 10) at normal agricultural rates of 18·83 g m$^{-2}$ at planting and 37·66 g m$^{-2}$ the following spring. Plots were either cut frequently or three times a year, to simulate cutting for silage. Silage plots were top dressed with fertilizer after each cut.

### *Soil conservation*

On 9 March 1987 the soil, a heavy clay loam, pH 6·3, was removed from the site in four discrete layers: turf, A horizon (0–15 cm), B horizon (15–30 cm), and C horizon (below 30 cm). Within each layer the soil was mixed and then half was sterilized with methyl bromide and stored separately. All the soil was protected from the rain with tarpaulins to minimize damage to its structure. Six weeks after sterilization (8 May 1987) samples of the sterilized soil were hand sorted and examined with a binocular microscope. No animals were found in any of the samples.

### *Reinstatement of soil*

On 22 May 1987 soil was replaced by hand in front of each of the windows in four horizons. Sterile soil was used only on the north side. To allow for compaction, the horizons were deeper than ultimately required and initially filled the whole window. The soil was replaced layer by layer and after each inch it was tamped down gently by foot to give the following depths: C = 40 cm; B = 15 cm; buried, inverted turf = 5 cm; A = 30 cm. This process was intended to simulate ploughing and reseeding of an old grassland, with a deep plough layer. The sterilized soil on the north side was reinoculated with soil micro-organisms by eluting each layer, in proportion to the volume of soil in that layer, with an inoculum made by adding unsterilized soil (horizon A), from the same site, to water (1 : 10 v/v) stirring the mixture and sieving the supernatant with a fine mesh sieve. This allowed micro-organisms to be replaced without replacing the meso- and macro-fauna.

The turf had to be resterilized using steam sterilization (68 °C for 4 hours) as the methyl bromide failed to kill all the animals in it.

### *Planting*

*Trifolium repens* L. 'Grasslands Huia' and *Lolium perenne* L. 'S24' were broadcast by hand at normal agricultural rates (0·4 and 2·0 g $m^{-2}$ respectively) on 15 July 1987 in each of the plots. The 26 m × 16 m enclosure surrounding the plots was sown similarly. Adjacent to each window eleven seeds of *T. repens* and ten of *L. perenne* were planted alternately at 2·5 cm spacing. All plots except those in the 'no fertilizer' treatment, were fertilized at sowing. The perimeter of the enclosure was kept bare by bimonthly applications of paraquat to prevent the invasion of stoloniferous plants into the enclosure.

### *Insulation*

The soil against the windows is kept dark and buffered from fluctuations of temperature within the rhizotron by internal marine plywood shutters backed by 10 mm thick polyurethane foam which abuts the glass. Transmission of sunlight down the thickness of the glass is minimized by aluminium tape stuck to the outside of the glass above soil level. As the soil consolidates, more tape is added.

### *Observations within the rhizotron*

Roots and soil fauna are observed through the glass using either a hand lens or a binocular microscope. The latter is suspended from a rope and pulley system. A modified binocular microscope can be hooked onto the rope in any position and pulled down to observe a window. A counter weight in a corner of the building, attached to one end of the rope ensures that the microscope does not move from wherever it is placed. The pulley and weight system is finely adjusted so that it takes little effort to move the microscope to another position. The microscope has rubber suckers attached to its feet so that the window is not scratched.

## RESULTS

### *Appearance of taxa*

There were no long-term effects of soil sterilization with methyl bromide (Fig. 3). Initially the appearance of taxa was slower in sterilized than in non-sterilized soil, but after 13 weeks when 50% of the total taxa were present, the rate of arrival of taxa in the two soils was similar, and overall followed a sigmoidal curve. The curves for cumulative maxima of individuals in each taxa in both soils were also sigmoid with an initial lag phase in the sterilized soil, after which the two curves were similar. The speed of invasion of taxa and the time at which each reached its maximum can be seen in Fig. 4a for unsterilized soil and Fig. 4b for sterilized soil. Most taxa reached their maxima in spring. In both soils the same taxa invaded early, and those not present until spring 1988 were again generally similar. Late invasion was usually a phenological effect, for example young slugs are not present in soil until early spring. Carabidae, Geophilidae, Collembola, Diplura and Chilopoda (*Lithobius* spp.) appeared at the same time in both soils. These taxa move rapidly in the soil and are ubiquitous at the site.

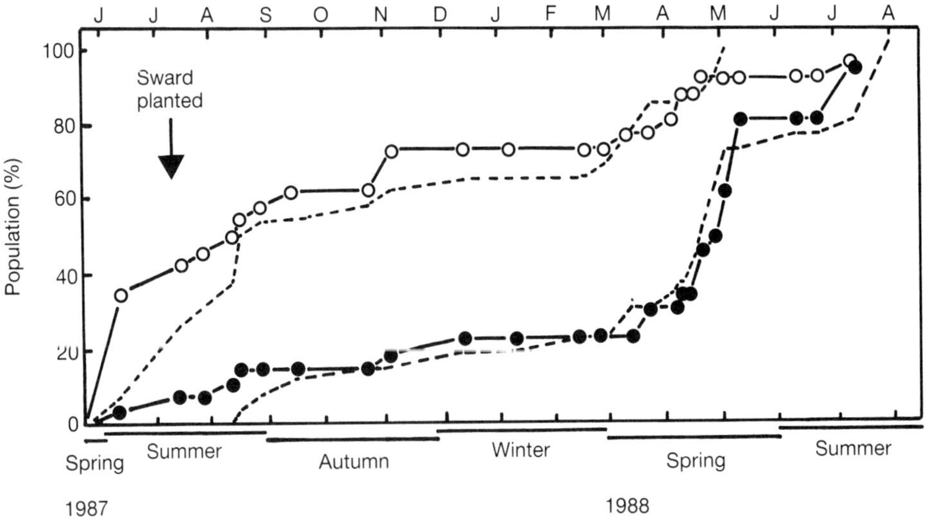


FIG. 3. Rate of invasion of animal taxa into sterilized (broken line) and unsterilized (solid line) soils. Cumulative percentage of all the taxa to have arrived by July 1988 which had arrived by the date shown (○). Percentage of all the taxa which had achieved the maximum number of individuals ever observed at one time for that taxa up to July 1988 (●).

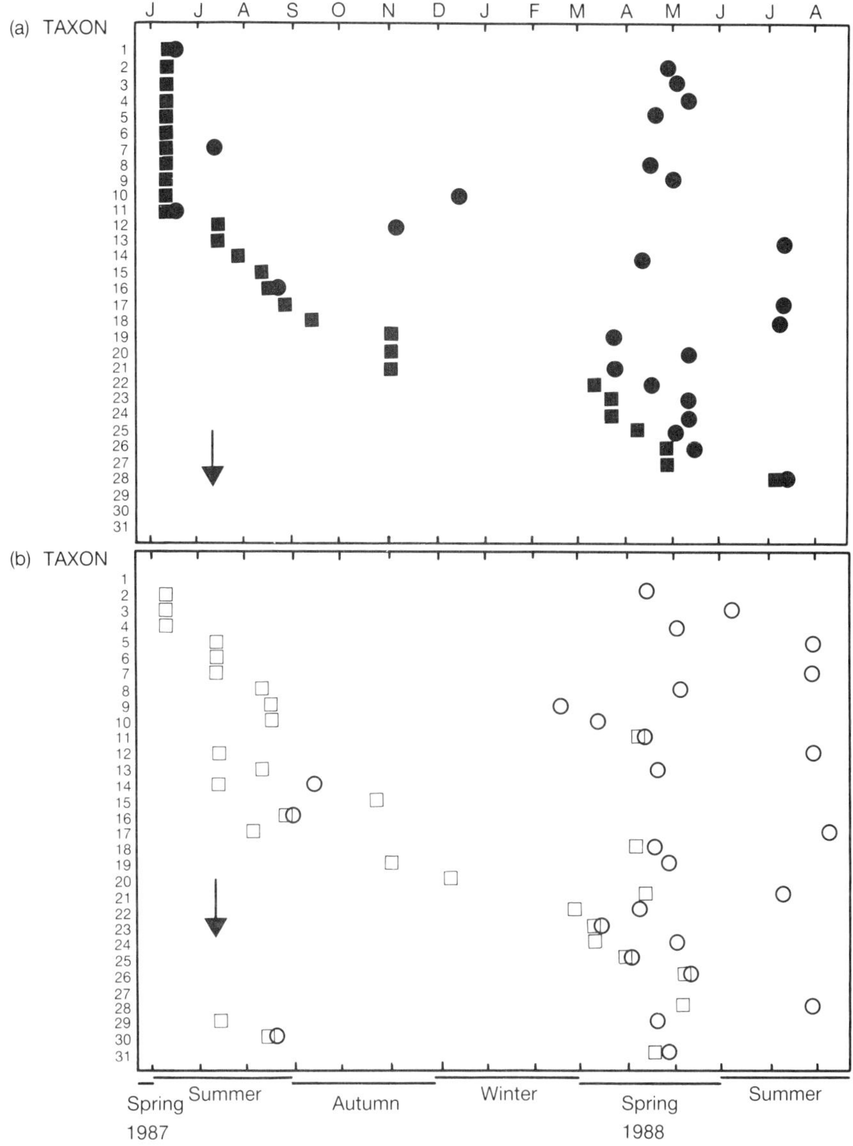


1 Nematoda 2 Collembola 3 Coleoptera: Carabidae 4 Chilopoda: Geophilidae
5 Mollusca: slugs 6 Earthworm burrows 7 Acarina 8 Annelida: Lumbricidae
9 Annelida: Enchytraeidae 10 Diptera: other larvae 11 Seedlings
12 Diplura: *Campodea staphylinus* Westwood 13 Coleoptera: other larvae
14 Coleoptera: Staphylinidae 15 Mollusca: slug's eggs 16 Diptera: other nematoceran larvae
17 Crustacea: woodlice 18 Coleoptera: Elateridae larvae 19 Chilopoda: *Lithobius* spp.
20 Platyhelminthes 21 Mollusca: *Boettgerella pallens* Simroth 22 Symphyla
23 Diplopoda: *Blaniulus guttulatus* Simroth 24 Mollusca: young slugs
25 Diplopoda: *Cylindroiulus londinensis* Simroth 26 Diptera: *Lonchoptera* spp.
27 *Rhizobium* nodules on *Trifolium repens* L. 28 Diptera: Tipulidae larvae
29 Diplopoda: *Polydesmus* spp. 30 Diptera: web-spinning larvae 31 Diptera: pupae

FIG. 4. Details of the time of arrival (■, □) and of achieving maximum population size (●, ○) of various taxa in (a) unsterilized and (b) sterilized soil over a period of 62 weeks from May 1987.

All animal taxa, except nematodes, that were found in the unsterilized soil were also found in the sterilized soil. Three taxa, *Polydesmus* spp., a web-spinning dipteran larva and dipteran pupae were found only in the sterilized soil. Seedlings were only present in the unsterilized soil, the weed seed bank being destroyed by fumigation. A few seedlings appeared in spring 1988 from seeds which had invaded during 1987. No nodules were seen on *T. repens* in the treated soil and they were not apparent until late spring in the unsterilized soil. The differences observed between the fauna in the two soils could in part be attributed to their different aspects. The sterilized soil was on the north side of the rhizotron whilst the unsterilized soil was on the south side. There were differences in illumination, growth of above-ground material and soil temperature on the two sides. On 27 April 1988 the soil temperature at 5 cm depth was the same (13°C) on both sides 1 m from the rhizotron but immediately adjacent to the windows the temperature was 10°C on the north and 17°C on the south side, as a result of the shading and reflecting effects of the concrete walls.

### *Soil horizons*

The four horizons replaced in May 1987, could still be distinguished in all the windows except those on the west face in August 1988. After 16 months *in situ* there were no differences overall between the profiles on the north and south sides. On the north side the soil consolidated so that the whole profile shrank down the window by about 15 cm. This was caused by the A and C horizons compacting and the control compacting more than the treated plots, $P < 0{\cdot}05$ (Fig. 5). There was an observable difference between cages containing earthworms and those without: this remains to be quantified. The soils containing earthworms had been worked extensively so that after 1 year most of the topsoil had passed through an earthworm's gut, and had a finer tilth.

## DISCUSSION

The applications and limitations of rhizotrons and mini-rhizotrons have recently been reviewed (McMichael & Taylor 1987). The main advantage of a glass wall method is that roots can be observed continuously *in situ*. The recent trend in rhizotron design is to minimize the disturbance to the soil profile and to build the roof as close to the ground as possible to minimize the effect on the micro-environment. The rhizotron in Ohio, USA, has a roof at soil level (Karnok & Kucharski 1982), but the design constraints of the rhizotron described here did not allow for this type of construction. Soil had to be removed to sterilize it adequately with methyl bromide, so that all animals could be removed from the north side. This was useful at the construction stage as manoeuvring the concrete modules into position required additional space. Storing the soil required for the 1 m proximal to the rhizotron in piles away from the site protected the soil structure from trampling and damage by heavy machinery. The roof could not be level with the ground as we

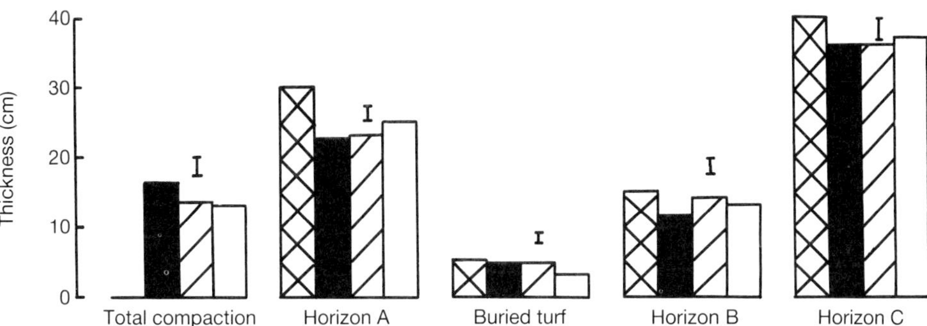


FIG. 5. The extent of soil compaction on the north side of the rhizotron 1 year after reinstating the soil profiles. Soil profile immediately after reinstatement (☒); full fauna present (■); earthworms excluded (▨); minimal fauna (□). Bars denote least significant difference ($P = 0{\cdot}05$).

particularly wished to study the litter and uppermost soil layers. However, as Carpenter *et al.* (1985) pointed out the need to prevent light from coming in through the windows prevents the observation of the top of the litter layer so that the part of the window that protrudes above soil level is wasted. It would be adequate to have only 1–2 cm above ground, if the extent of soil settlement could be more accurately forecast. It must be remembered that this rhizotron was designed specifically to observe the interactions of the soil fauna and the roots of pasture plants and the early stages in soil development are pertinent.

There are a number of disadvantages in the use of rhizotrons. There may be higher (Taylor & Bohm 1976) or lower concentrations of roots at the glass wall compared with the rest of the soil. Artifacts such as these cause problems in extrapolating data obtained from a rhizotron to the soil as a whole although large objects projecting above the soil surface, such as rocks do occur naturally in soil. It is common to find roots concentrated around stones in the soil — this may be a purely physical effect or may be because the stone is a source of nutrients and or water. As the intention was to observe any relative differences in root growth between cages in which different size classes of animals had been removed, the presence of an atypical number of roots next to the window would not necessarily invalidate this investigation. We were pleased at the rapidity with which the soil fauna reinvaded from the surrounding untouched pasture over 1 m away. Earthworms reintroduced into grassland on reclaimed polder soils in the Netherlands were found to disperse at a species specific rate of 9 m year$^{-1}$ (*Allolobophora caliginosa* Savigny) or 4.5 m year$^{-1}$ (*Lumbricus terrestris* L.) (Hoogerkamp, Rogaar & Eijsackers 1983). There was a higher proportion of mature worms at the periphery. Pioneers were predominantly adult, producing cocoons from which the immature worms developed. Both population development and dispersal showed a lag phase after which there was a linear increase in the dispersal distance and a log-linear increase in earthworm numbers. This suggests that from a faunal point of view, rhizotrons can give valuable data within a year of soil reinstatement.

In the cages with a full complement of soil fauna the soil consolidated more than in the plots with a reduced soil fauna. A possible explanation is that there was a more rapid breakdown of organic matter in the control plots because the soil fauna could reach their food source to break it down. The main source of organic matter in these plots was the buried turf which was 30 cm below the soil surface. Without earthworms there are few routes through the soil for non-burrowing animals; this suggestion has been discussed more fully elsewhere (Sackville Hamilton & Cherrett, pp. 295–299). However, Malone & Reichle (1973) have shown that in the year following the application of pesticides to reduce soil fauna there was an increase in the number of micro-organisms breaking down organic matter, so that its rate of disappearance was not reduced in the absence of meso-fauna.

## ACKNOWLEDGEMENTS

Invaluable help and expertise in the building of the rhizotron were given by Edward Evans and the staff of the estates department. Figure 1 was drawn by G. Wyn Jones. Nigel Brown and the staff of the Botanic Gardens were most helpful with the preparation and maintenance of the site. The programme is funded by AFRC.

## REFERENCES

**Atkinson, D. (1977).** Some observations on the root growth of young apple trees and their uptake of nutrients when grown in herbicided strips in grassed orchards. *Plant and Soil*, **49**, 459–471.

**Bohm, W. (1979).** *Methods of Studying Root Systems.* Springer-Verlag, Berlin.

**Carpenter, A. (1985).** *Studies on the invertebrates in a grassland soil.* Ph.D. thesis, University of Wales.

**Carpenter, A., Cherrett, J.M., Ford, J.B., Thomas, M. & Evans, E. (1985).** An inexpensive rhizotron for research on soil and litter-dwelling organisms. In *Ecological Interactions in the Soil* (Ed. by A.H. Fitter, D. Atkinson, D.J. Read & M.B. Usher), pp. 67–71. Blackwell Scientific Publications, Oxford.

**DiPaola, J.M., Beard, J.B. & Brawand, A. (1982).** Key events in the seasonal root growth of Bermuda grass and St. Augustine grass. *Horticultural Science*, **17**, 829–831.

**Edwards, C.A. & Heath, G.W. (1963).** The role of animals in the breakdown of leaf material. In *Soil Organisms* (Ed. by J. Doeksen & J. van der Drift). North-Holland Publishing Company, Amsterdam.

**Garwood, E.A. (1967).** Seasonal variation in appearance and growth of grass roots. *Journal of the British Grassland Society*, **22**, 121–130.

**Hanks, P.** (Ed.) **(1982).** *Collins Dictionary of the English Language.* Collins, London & Glasgow.

**Harding, D.J.L. (1968).** A preliminary survey of the relationships between the soil fauna and roots of fruit trees. *Annual Report of East Malling Research Station for 1967*, pp. 169–172.

**Hoogerkamp, M., Rogaar, H. & Eijsackers, H.J.P. (1983).** Effect of earthworms on grassland on recently reclaimed polder soils in the Netherlands. In *Earthworm Ecology* (Ed. by J.E. Satchell). Chapman & Hall, London.

**Karnok, K.J. & Kucharski, R.T. (1982).** Design and construction of a rhizotron–lysimeter facility at the Ohio State University. *Agronomy Journal*, **74**, 152–156.

**Liddell & Scott. (1871).** *Greek–English Lexicon.* Abridged version (1980). Clarendon Press, Oxford.

**Malone, C.R. & Reichle, D.E. (1973).** Chemical manipulation of soil biota in a fescue meadow. *Soil Biology and Biochemistry*, **5**, 629–639.

**McMichael, B.L. & Taylor, H.M. (1987).** Applications and limitations of rhizotrons and mini-rhizotrons. In *Mini-rhizotron Observation Tubes: Methods and Applications for Measuring Rhizosphere Dynamics* (Ed. by H.M. Taylor), pp. 1–13. Special Publication No. 50. ASA, Madison.

**Taylor, H.M. & Bohm, W. (1976).** Use of acrylic plastic as rhizotron windows. *Agronomy Journal*. **68**, 693–694.

# The University of Michigan Soil Biotron: a platform for soil biology research in a natural forest

R. FOGEL* AND J. LUSSENHOP†

*University of Michigan Herbarium, North University Building, Ann Arbor, Michigan 48109, USA; and
†Department of Biological Sciences, University of Illinois at Chicago, Chicago, Illinois 60680, USA

## SUMMARY

1 Construction of a root observation laboratory, Soil Biotron, has recently been completed at the University of Michigan Biological Station in northern lower Michigan. The facility is located in a mixed northern hardwood forest about 70 years old, dominated by *Populus grandidentata* Michx., *Acer rubrum* L. and *Quercus rubra* L.

2 The laboratory houses thirty-four window bays each containing sixteen removable glass panes permitting observation and direct access to roots and soil biota from the ground surface to 1.2 m depth.

3 Equipment available for documenting activity of roots and soil invertebrates includes a trolley-mounted stereo-microscope, 35 mm still camera with macrolens, and a time-lapse video recording system.

4 Ancillary data collected include precipitation, wind speed and direction, relative humidity, wet and dry fall, soil temperature (four depths) and soil water content. In addition, the forest abutting the Biotron has been permanently gridded and the species of trees and their diameters recorded for allometric reconstruction of the forest.

## INTRODUCTION

The importance of soil organisms and processes in regulating ecosystem processes is gaining increasing recognition and study in both forest and agricultural systems. Progress beyond simple estimates of the regulatory role has been hampered by an old, perennial problem in soil research — technique. The soil coring method substantially improved quantification of the importance of root and mycorrhizal processes, but these processes are so dynamic other techniques need to be explored if factors limiting forest production are to be elucidated in sufficient detail to predict the consequences of perturbations, etc. The need to sample new areas in time-course studies employing coring or other destructive techniques contributes significantly to our inability to reduce variance to an acceptable level due to the large spatial variability in the distribution of roots and soil biota. Another serious limitation inherent in destructive sampling techniques is the loss of information on spatial relationships of roots and associated rhizosphere biota. One is left with the impression that the Heisenberg Uncertainty Principle should have been proposed first in biology rather than quantum physics.

Direct observation techniques, especially rhizotron and mini-rhizotron techniques, originally developed for observing root growth, appear the most attractive for studying the interactions of roots and soil biota in an ecosystem context. Admittedly there are concerns about the artificial state of these approaches (e.g. concentration of roots at windows), but they offer the great advantage of observing the same population over time, eliminating the variance due to spatial hetero-

Fig. 1. University of Michigan Soil Biotron. (a) View of tunnel and above-ground laboratory from the south. (b) View from the west. Note white pine stump left after logging and burning in about 1917. (c) Interior of tunnel showing window bays covered with insulated shutters. (d) Close-up of glass, wire re-reinforced, 6 mm thick window pane. Note fungal rhizomorphs. The wire grid is about 2 cm × 2 cm.

geneity inescapable in destructive techniques (Atkinson 1985; Huck & Taylor 1982; McMichael & Taylor 1987). The desire to study interactions among soil organisms, roots and mycorrhizas *in situ* provided the impetus to construct a root observation laboratory (Figs 1–4) modelled on the East Malling rhizotrons (Rogers 1969). A major goal in designing the facility was that it be large enough for replicated studies and that direct access to the soil be possible for experimental manipulation at the root level. Another goal was that the facility provide a focus and forum for soil scientists from different disciplines and institutions to interact and hopefully synthesize new approaches to soils research.

## DESIGN AND CONSTRUCTION

### *Site description*

The Biotron is located at the University of Michigan Biological Station (UMBS), an International Biosphere Reserve site near Pellston, Michigan (45°34′ N latitude, 84°40′ W longitude). UMBS occupies a 4000 ha tract in northern lower Michigan and a 1250 ha tract on Sugar Island in the Upper Peninsula of Michigan. The main tract of 4000 contiguous hectares has been almost completely undisturbed since lumbering ceased 90 years ago and fire suppression was imposed 60 years ago. The forest surrounding the Biotron is characteristic of one type developing on hundreds of thousands of hectares in the region following logging, repeated fires, and reforestation by natural means. *Populus grandidentata*, *Quercus rubra*, and *Acer rubrum* dominate the forest with *Fagus grandifolia* Ehrh. (American beech) and *Amelanchier* spp. (serviceberry) being subdominant. A number of small *Pinus strobus* L. (white pine), <3 m in height, are present and will eventually replace the hardwoods. *Pteridium aquilinum* L. 'Kuhn' (bracken fern) dominates the herbaceous layer.

The soil surrounding the Biotron developed from an excessively drained glacial outwash plain of Rubicon sand — a poor quality soil due to its extreme porosity, rapid leaching, and the absence of clay lenses or a clay layer in the subsurface horizons. The coarse texture (98% sand) facilitates observation of roots and soil biota.

### *Design*

The Biotron (Figs 1–3) consists of a chamber 2·5 m × 2·2 m × 31 m, oriented from north to south, and contains thirty-four window bays. Each window bay contains sixteen small, 28 cm × 31 cm, removable windows; 544 windows total or 136 vertical profiles from ground level to 1·2 m deep. The window frames are similar to those used in the Auburn rhizotron (Taylor 1969). The windows provide access to 49 $m^2$ of vertical soil profile. The roof extends 46 cm above ground level, to minimize interference with the microclimate of adjacent plants. Rainwater from a flat roof is channeled off the south end in soakaways. Entrance to the chamber is via an above-ground, 24 $m^2$, instrument room/office at the north end.

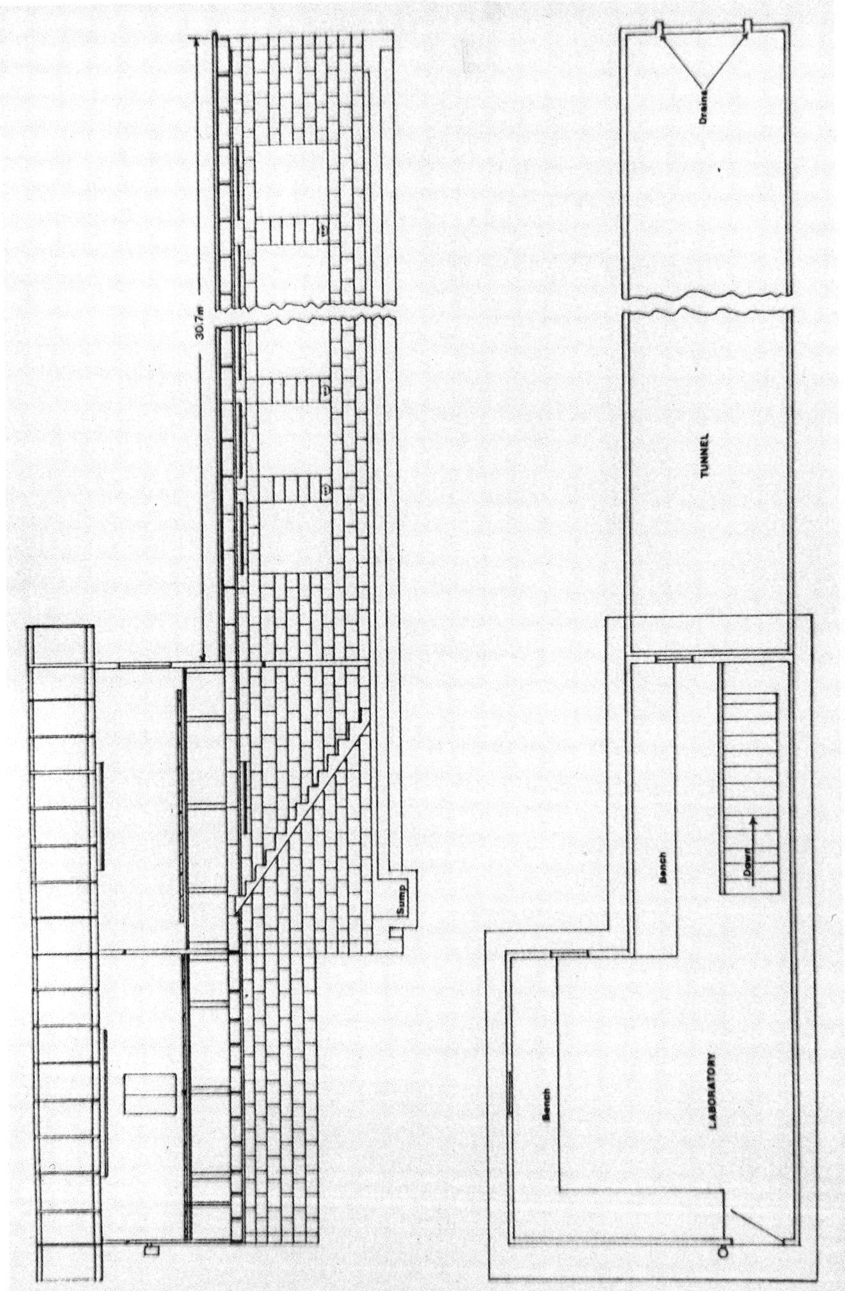


FIG. 2. Side and top views of Soil Biotron. Laboratory is heated, tunnel unheated. Tunnel is equipped with fluorescent lights, 110 V outlets between every other window bay. Two 220 V electrical outlets are also provided. Data can be recorded with a Wild M5A stereo-microscope suspended from an overhead track or by time-lapse video recording.

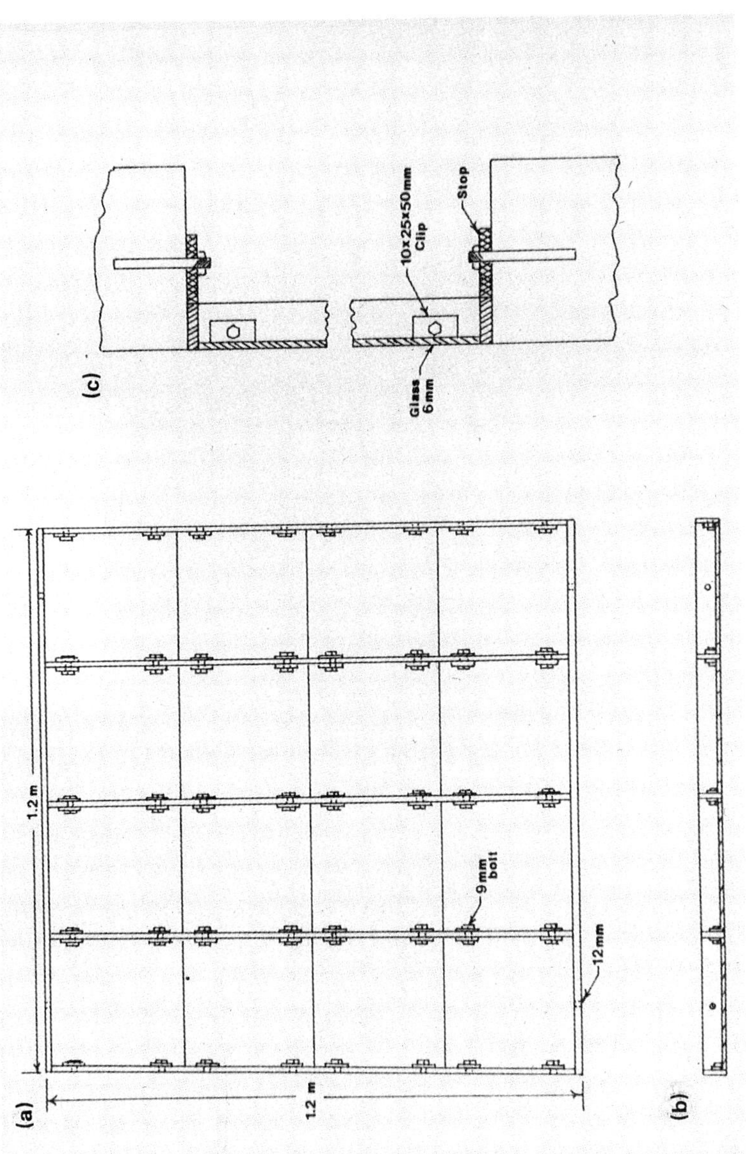


FIG. 3. (a) Front view of frame holding sixteen window panes. Frame constructed of 6 mm × 100 mm mild steel and coated with epoxy powder paint. (b) Top view of window frame. (c) Detail of removable clips used to hold glass panes in position. Frames are held in position by anchor bolts and 6 mm × 100 mm steel stops at the top and bottom of each frame. Gaps between glass panes are sealed with silicone caulk.

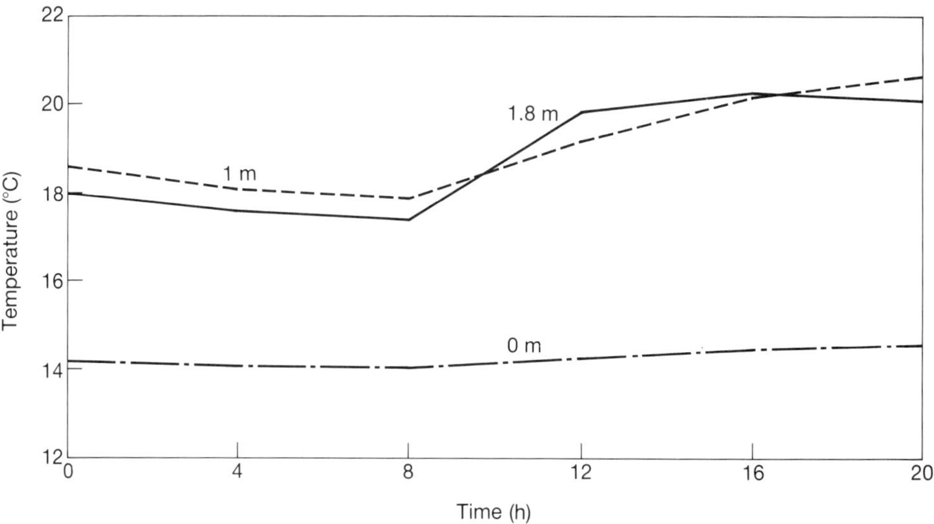


FIG. 4. Soil temperature adjacent to Biotron during first growing season following construction. Root recolonization at windows had just started. Temperatures recorded from 20 cm depth on 7 July 1988. Higher soil moisture adjacent to the Biotron has depressed temperatures compared to reconstructed profile (1 m) and undisturbed forest soil (1·8 m).

## *Construction*

Construction started in April 1986 with removal of trees from the excavation area and clearing of access roads. All materials entering and leaving the site were channelled through the north and south ends of the excavation to minimize damage. Excavated soil was loaded directly into trucks from a power shovel starting at the far end of the site, and hauled off-site. A wood retaining wall was built to allow construction to proceed as close to the native soil profile as possible. A 15 cm thick concrete pad with a central floor drain and a sump at one end was then poured, 20 cm thick concrete block walls laid, block cavities filled with concrete and steel rebar, and an insulated, flat roof rimmed by a 15 cm high parapet constructed. Block construction was used due to the costs of forms for poured concrete construction, plus the logistic difficulties in working from one end of the 35 m long excavation. Thirty-four steel frames coated with epoxy powder paint were glazed using silicone caulk and then installed. The glass is 6 mm thick and contains embedded wire mesh to facilitate mapping of roots and to increase strength. The embedded wire mesh also helps resolve parallax problems caused by the thick glass in measurements made through windows. Individual window panes can be removed as required by cutting the bead of silicone caulking, attaching a rubber sucker unit (type used by glaziers) to the pane, and easing gently backwards. The windows are covered with light-proof, insulated (5 cm rigid foam) shutters to exclude light and moderate the temperature gradient from the tunnel to the soil.

Sieved soil from each soil horizon was repacked by horizon into the 1 m space between the windows and the native soil horizon after the shoring was removed. Soil used in repacking the windows was excavated by horizon 50 m from the Biotron site. Soil removed during tunnel excavation could not be used due to mixing by the power shovel. A few centimetres of soil was added at one time, and then compacted with a vibrator until the bulk density of the native soil (about $1{\cdot}3\,g\,cm^{-3}$) was attained as measured by a pentometer. The surface organic layer was restored to re-establish a realistic water infiltration pattern.

Illumination in the tunnel is provided by a strip of fluorescent lights on the ceiling. The fluorescent lights are only used when the windows not being observed are shuttered to avoid possible alterations in root growth and development, inhibition of soil animal activity, and algal growth on the windows. Electrical outlets, 110 V, are positioned between every other window bay; two 220 V outlets are also present in the tunnel as well as one in the above-ground laboratory.

Two dehumidifiers are used in the tunnel to reduce condensation. A sump pump stands ready to remove any water that manages to infiltrate the chamber.

## ASSOCIATED DATA

UMBS has a long history of research that provides useful background information for research in the Biotron. Nearly a thousand scientists have maintained research projects at UMBS during its 75 year history; over 1500 publications have resulted from this work (D.M. Gates, personal communication). The data resulting from studies in the Biotron will supplement the UMBS forestry database that contains long-term information on nutrient losses associated with clear cutting, carbon dioxide exchange in aspen forests, dimensional analyses of aspens on sites of differing quality, and contributions of structural, fine roots, and mycorrhizae to net primary production of aspen and red pine ecosystems.

The proximity of the Biotron to Stockard Lakeside Laboratory, 200 m, permits ready access to a direct computer link to the mainframe computer in Ann Arbor, microcomputers, photographic darkrooms, flame photometer, Technicon autoanalyser and other equipment and utilities expected in a modern research facility. A recent addition has been a culture facility containing laminar flow hood, autoclave, etc., as well as a freeze-dryer and ultracold freezer (−80 °C) for DNA extraction and storage.

Meteorological data is available from an installation adjacent to the Biotron. Data are recorded all year for precipitation, air temperature, wind speed and direction, solar radiation and relative humidity.

The meteorological data collected by UMBS are augmented with soil moisture and temperature data collected with thermocouple psychrometers at 2, 5, 20 and 100 cm depth at 0, 1·0 and 1·8 m from the windows (Fig. 4). The moisture data is also being supplemented with tensiometer data at 15, 60 and 120 cm depth at 1, 5 and 20 m from the windows. Collection of neutron probe data will begin in 1989 for three depths at 1 m and 20 m from the windows, a total of forty neutron probe tubes.

Permanent 10 m × 10 m quadrats have been established in the forest abutting the Biotron. Nearly 400 trees have been permanently tagged and their diameters recorded for allometric reconstruction of the forest. Tagged species include the dominants: *Populus grandidentata*, *Acer rubrum*, *Quercus rubra*, and subdominants: *Betula papyrifera* Marsh. (paper birch), *Fagus grandifolia*, *Amelanchier* spp., and *Pinus strobus*.

Several systems are available for recording data on root growth and invertebrate activity and several preliminary data sets are currently being compiled. Colour photographs (35 mm) of all 544 window panes were taken before significant root recolonization to provide background data on spatial heterogeneity in soil texture. Root recolonization on selected windows was recorded during the 1988 growing season on acetate sheets as were the dimensions of earthworm burrows. Data, at 0–50 × magnification, on the distribution of invertebrates was recorded using a Wild M5A stereo-microscope and fibre optic light source. Additional data on invertebrates and root growth has been obtained using a video camera. The video system made photographs at 0–35 × and either continuous or time-lapse recording possible. A TARGA board and associated software facilitate dimensional analysis of digitized video images.

## ACKNOWLEDGEMENTS

Construction of the Soil Biotron was supported by a grant from the National Science Foundation (BSR–8615565). The University of Michigan Biological Station kindly provided space and technical support.

## REFERENCES

**Atkinson, D. (1985).** Spatial and temporal aspects of root distribution as indicated by the use of a root observation laboratory. In: *Ecological Interactions in Soil* (Ed. by A.H. Fitter, D. Atkinson, D.J. Read & M.B. Usher), pp. 43–66. *Special Publication Series British Ecological Society No.* 4. Blackwell Scientific Publications, Oxford.

**Huck, M.G. & Taylor, H.M. (1982).** The rhizotron as a tool for root research. *Advances in Agronomy*, **35**, 1–35.

**McMichael, B.L. & Taylor, H.M. (1987).** Applications and limitations of rhizotrons and minirhizotrons. In *Mini-rhizotron Observation Tubes: Methods and Applications for Measuring Rhizosphere Dynamics* (Ed. by H.M. Taylor), pp. 1–13. Special Publication No.50. ASA, Madison.

**Rogers, W.S. (1969).** The East Malling root-observation laboratories. In: *Root Growth*. (Ed. by W.J. Whittington), pp. 361–376. Plenum Press, New York.

**Taylor, H.M. (1969).** The rhizotron at Auburn, Alabama — a plant observation laboratory. *Auburn University Agricultural Experiment Station Circular No. 171*, 1–9.

# Automated analysis of roots in bulk soil

N.R. SACKVILLE HAMILTON*, M. JONES* AND J.L. HARPER†

*Unit of Plant Population Biology, School of Biological Sciences, University College of North Wales, Bangor, Gwynedd LL57 2UW, UK*

## SUMMARY

1 A technique is illustrated for image analysis of roots in natural soil, using photographs of soil impregnated in resin. The position and size of roots can be determined automatically with an accuracy of 30–50 μm.

2 High-contrast photographs have been obtained of the surface of impregnated blocks of soil, obviating the need for preparing thin sections and allowing the preparation of serial surfaces through the soil as close as required.

3 The photographs are analysed using a Magiscan (Joyce-Loebl Ltd) Image Analyser. It has been trained to recognize roots, determine their position, orientation, size and shape, and to distinguish between grass and clover roots by their anatomy. Once trained, minimal intervention by the operator is required.

4 The technique is intended for three-dimensional reconstruction of roots from serial images through an entire block of soil, and hence for an analysis of the spatial relationships of competing root systems.

## INTRODUCTION

The opacity of soil makes roots invisible and therefore difficult to study in their normal environment. All attempts to make the invisible visible involve either destroying the roots to determine their state, or placing them in an unnatural environment to determine their dynamics (Harper, Jones & Sackville Hamilton, pp. 11–14). By analogy with the Principle of Uncertainty (according to which it is impossible to understand the Universe because it is impossible to measure precisely and simultaneously both the position and velocity of a particle), it is impossible, with the technology currently available, to fully understand the behaviour of roots in soil. The best we can do is to follow two parallel, complementary approaches, to study either the dynamics or the architecture of roots.

This chapter is concerned with the second of these approaches: the measure-

* Present address: Institute of Grassland and Environmental Research, Plas Gogerddan, Aberystwyth, Dyfed SY23 3EB

† Cae Groes, Glan-y-Coed Park, Dwygyfylchi, Penmaenmawr, Gwynedd LL34 6TL

ment of roots in bulk soil. A method is described which, at the expense of destroying roots, can be used to quantify the size, orientation and position of all roots in a block of soil to an accuracy of 30–50 μm, thereby including all fine roots. It can be used with any rooting medium, whether from a glasshouse or the field, and whether the soil is stony, sandy, loamy or organic. The method is based on a combination of basic techniques already well developed by others (Fitzpatrick 1984; Murphy 1986).

## BASIC TECHNIQUES

Soil micromorphologists have developed a technique of impregnating desiccated blocks of soil with resin (e.g. Ringrose-Voase & Bullock 1984), that fixes the soil in place with virtually no effect on its fine-scale morphology. A diamond saw, followed by a series of aluminium oxide and diamond pastes, are used to cut, grind and polish the block to produce a smooth surface regardless of the presence of stones or pores. To obtain adequate contrast between the soil components, it is usually necessary to prepare thin sections for viewing under the microscope.

The process of desiccation does, however, shrivel roots and other organisms. Soil biologists have overcome this problem by prior impregnation with glutaraldehyde and by desiccating by acetone replacement rather than by air- or freeze-drying (Tippkötter, Ritz & Darbyshire 1986). Cellular structure is then preserved intact even for such fine structures as root hairs, root apices and hyphae. Stains can be applied after preparation of the surface. The resulting surface contains a complex of sections through roots, soil particles, soil pores, organic matter, etc.

Manual analysis of the objects on the surface would be very laborious. Scientists in other fields have already developed techniques of computerized image analysis to identify, count and measure objects in images (Joyce-Loebl, 1985). Such techniques can, in principle, be applied to images of roots in soil.

This paper describes some results from the application of image analysis to photographs of serial surfaces through a block of resin-impregnated soil. It forms the first initial stages of an attempt to reconstruct and analyse root systems in soil in three dimensions.

## METHOD

Using a diamond saw, a 5 cm × 5 cm × 5 cm cube is cut from a block of soil previously treated with glutaraldehyde, desiccated in acetone and impregnated with resin. Grooves left by the saw are ground away from the surface of the block by holding it on a revolving diamond studded metal disc. The block is then fixed on to a glass slide (76 mm × 51 mm) using epoxy resin. Surface scratches left by the diamond disc are removed by grinding the block with an aluminium oxide slurry (Abralap C15) on a lapping machine (Logitech PM2). During operation of the machine sample slides are held by vacuum on the face of a rotating lapping jig which is placed sample side down on to a revolving cast iron plate. Grinding slurry is fed on to the plate. The face of the jig can be set at any position, thus allowing any chosen thickness to be ground off the sample. Once the required thickness has been attained, further grinding of the sample is prevented by the presence of a stop ring made of diamond encrusted steel at the bottom of the jig — when the sample is flush with the stop ring the jig begins to grind the plate. Standard thin section jigs (e.g. Logitech PLJ2) can deal with samples having a maximum thickness of 5 mm. Our investigations require much thicker samples, so a jig which can hold samples of up to 70 mm was built (Beaumaris Instrument Co, Ltd). The smooth surface left after lapping produces images of sufficient quality for analysis, so the final stages of thin section preparation, polishing with diamond pastes, is omitted. The surface is stained with methylene blue — which we have found to give the best contrast between roots and soil — and photographed. A 100 μm thick layer of soil is then removed by grinding on the lapping machine, and the new surface is prepared and photographed. The process of grinding, staining and photographing is repeated until a complete set of photographs of serial surfaces through the block of soil is obtained.

The photographs are taken with a Sony 3-chip colour CCD camera and Elicar macro-lens attached to a Magiscan (Joyce-Loebl Ltd) image analyser, a 'true colour' image analyser with three primary colours (red, green and blue) and sixty-four levels of brightness for each primary colour. Each image is displayed and stored as a square array of 512 × 512 points, or 'pixels'. Each complete colour photograph occupies three-quarters of a megabyte of storage space on disc. We use a 5 mm field of view, which is adequate for the reliable measurement of the shape of roots down to 100 μm diameter, and the reliable detection of roots down to 40 μm diameter.

Image analysis is done on Magiscan's Image Processing Unit, controlled by an Opus PC V Turbo AT microcomputer running Microsoft Pascal. The first stage in image analysis is to enhance the contrast between the different types of object in the image. The better the initial quality of the image, the less enhancement is necessary. Here the only enhancement necessary to highlight roots is 'colour saturation', i.e. white light, of equal brightness in all three primary colour channels, is subtracted from each pixel until the darkest channel is black. Dull colours, including black, white and intermediate shades of grey and brown, all become dark, while bright colours remain bright. Photographs of roots, stained with methylene blue, remain a bright cyan (bright in the blue and green channels, black in the red channel) after colour saturation, while everything else becomes dark.

In principle, further improvements could be made by transforming the three colour channels to the principal components of their values in roots or, better still, to the canonical axes discriminating between their values in roots and elsewhere. In practice, this has been found to be unnecessary; colour saturation is highly effective, and further improvements have not been obtained with additional transformations.

The second stage in image analysis is to train the computer to recognize the objects of interest. With Magiscan, this is done by pointing a light-pen at a selection of pixels which represent the objects of interest, i.e. roots. Based on the range of brightnesses of the selected pixels, the computer then slices the colour image into a binary image, a perfect black-and-white image in which the white pixels are the ones that the computer has decided are points in roots.

With the binary and colour images superimposed on the same screen, the operator can verify that roots have been correctly identified by the computer. Provided sufficient care is taken in preparing, staining and photographing the soil surface, excellent agreement has been obtained between what the computer declares as roots and what the operator knows to be roots.

The third stage in image analysis is the analysis of the objects in the binary image. The computer recognizes each connected set of white pixels as one object, and can measure its position, size, shape and orientation.

Using these measurements, the computer is 'trained' to distinguish grass from clover roots. These differ anatomically in that there is a layer of lysigenous lacunae surrounding the endodermis only in grass roots (Soper 1959). The lacunae are not stained by methylene blue, so grass roots appear as a central blue cylinder, surrounded by unstained lacunae, surrounded in turn by a thin blue-stained epidermis. A grass root therefore appears to the computer as two objects, one nesting inside the other. In contrast, a clover root appears as a single solid object. Since the computer recognizes the extent to which an object is nesting within other objects, it can be given a simple rule to distinguish between grass and clover roots.

In addition, the distribution of root diameters is measured. This is easiest for roots that are straight and unbranched at the point where they pass through the surface of the block of soil. The section through these roots will be elliptical; for an ellipse, the relationship between the area, length and breadth is:

$$\text{area} = \pi/4 \times \text{length} \times \text{breadth}. \tag{1}$$

Curved or branched segments of roots will generally have more invaginated boundaries, i.e. their cross-sectional area will be less. It can, however, be greater if the root is curved normal to the plane of the photograph.

Area, length and breadth are three simple measurements that the computer can give for each object. As a first approximation, we can say that if eqn (1) is true for a root section, then that section may be regarded as an ellipse. The minor diameter, or breadth, of the ellipse is the actual diameter of the root. Thus, the size distribution of roots is obtained from the breadths of elliptical objects. If a root is curved or branched at the surface of the block, more complex formulae are required to measure its diameter, and these have not yet been implemented.

Attempts are also being made to reconstruct the root systems in three dimensions from serial binary images. Superimposition of binary images in pseudocolour gives a visual three-dimensional reconstruction. Quantitative reconstruction, in which corresponding roots in adjacent images are matched up by the computer, is more complex. It is necessary to be able to tell the computer, given a section through a root in one image, approximately where the same root will appear in the next image.

As with root diameter, the procedure is simplest with straight, unbranched segments of roots, of which the sections are ellipses. For these segments, the angle at which the root passes through the section is given (Lang & Melhuish 1970) by

$$\alpha = \sin^{-1}(\text{breadth/length}). \tag{2}$$

Given the distance $d$ between adjacent images, then the position of the centre of the root in the second image is offset from its position in the first image by

$$\text{offset} = d/\tan(\alpha). \tag{3}$$

The direction of this offset is the orientation of the long axis of the (elliptical) section through the root. The implementation of these equations in a computer program for searching adjacent images for corresponding sections of roots is currently in progress.

## REFERENCES

**Fitzpatrick, E.A. (1984).** *Micromorphology of Soils.* Chapman & Hall, London.

**Joyce-Loebl (1985).** *Image Analysis: Principles and Practice.* Joyce-Loebl Ltd, Gateshead.

**Lang, A.R.G. & Melhuish, F.M. (1970).** Lengths and diameters of plant roots in non-random populations by analysis of plane surfaces. *Biometrics,* **62**, 421–431.

**Murphy, C.P. (1986).** *Thin Section Preparation of Soils and Sediments.* AB Academic, Berkhamsted.

**Ringrose-Voase, A.J. & Bullock, P. (1984).** The automatic recognition and measurement of soil pore types by image analysis and computer programs. *Journal of Soil Science*, **35**, 673–684.

**Soper, K. (1959).** Root anatomy of grasses and clovers. *New Zealand Journal of Agricultural Research,* **2**, 329–341.

**Tippkötter, R., Ritz, K. & Darbyshire, J.F. (1986).** The preparation of soil thin sections for biological studies. *Journal of Soil Science*, **37**, 681–690.

# Usefulness of the mesh bag method in quantitative root studies

E. STEEN
*Department of Ecology and Environmental Research, Swedish University of Agricultural Sciences, Box 7072, S-750 07 Uppsala, Sweden*

## SUMMARY

1 Cylindrical nylon mesh bags filled with sieved soil incubated in field plots were used in annual and perennial plant stands as well as in shrub and tree stands to estimate the effect of experimental treatments on root biomass and its seasonal dynamics, root mortality and the chemical composition of roots.
2 The method revealed significant differences in root biomass between plots treated with different doses of nitrogen fertilizer and between plots with different degrees of soil compaction. Herbicide treatment caused a significant decrease in root biomass. Differences in the annual pattern of variation in carbohydrates and lignin between roots and rhizomes were also demonstrated. Seasonal changes in root mortality in a Portuguese stand of *Eucalyptus globulus* were clearly illustrated.
3 The mesh bag method is a good alternative to soil coring for studying seasonal dynamics of root biomass, and chemical composition of roots, especially in the topsoil in annual and perennial plant stands if the bags are incubated just after seed germination. If inserted in established stands complications arise, which are discussed.
4 The method is also discussed in comparison with soil coring when root production including root mortality and root turnover is studied.

## INTRODUCTION

The rate at which roots grow into a mesh bag which can be cylindrical, a rectangular container, or a basket filled with a certain volume of soil incubated in the field, was first documented by Hendrickson & Veihmeyer (1931) in a study of sunflower roots. Later, Lund, Pearson & Buchanan (1970) investigated the influence of soil-acting herbicides on the ingrowth of roots in mesh bags. In recent years the method has been used in Sweden by Persson (1979, 1983) and Ahlström, Persson & Börjesson (1988) in Scots pine stands and by the agro-ecology group at the Swedish University of Agricultural Sciences in a number of studies of annual and perennial agricultural crops (Steen 1984, 1985, 1987; Steen, Persson & Al-Windi 1984; Steen & Larsson 1986; Steen & Håkansson 1987; Larsson & Steen 1984, 1988; Hansson & Andrén 1986).

The mesh bags are filled with sieved soil. Thus, all roots, dead and living, found in a bag after a period of incubation are assumed to have grown into the soil in the bag after its placement in the field. Occasionally a few coarser fragments of organic material are introduced by animals dwelling on the soil surface or in the topsoil. However, these fragments are easily sorted out during the sieving procedure. Thus, the method measures the root mass (living and dead roots) existing in the bags at the time of sampling. The length of the period during which a given root mass is produced is equal to the incubation time. The upper limit of root age is set by the length of the incubation period.

The method has a potential value in field trials where experimental treatments are being compared. It yields a relative estimate of living and dead root mass for each of the treatments. Calculations are more complicated, however, when the objective is to get absolute estimates of root production, since root decomposition during the incubation period is not considered. Artefactual effects — e.g. pruning of roots, drainage of water along the walls of the drilled hole, differences in the degree of soil compaction between the bag soil and the surrounding soil, may also have an effect, at least initially.

As the mesh bag method has several apparent advantages in field trials on agricultural land it was used in a number of studies designed to determine how soil compaction as well as treatment with nitrogen fertilizer, herbicides, and other agrochemicals affect root growth dynamics. It was also used in studies of seasonal variation in root carbohydrates and in root decomposition (for references see above) and in studies in Portugal and Tunisia of Mediterranean trees and shrubs.

## MATERIALS AND METHODS

### *Principle*

The mesh bag method (Steen 1984), also called the ingrowth core method (Persson 1983), refers to a procedure in which a given volume of sieved, root-free soil contained in a mesh bag is incubated in a hole drilled in the soil. Plant roots can easily pass through the mesh and grow into the new soil volume. The mesh bag can be removed from the ground after a predetermined period of time in order to measure the amount of living and non-decomposed, dead root mass.

### *Field procedure*

Mesh bags for use in the field are constructed of nylon net with a mesh size of 5–7 mm. They are cylindrical, 50 cm long, and have a diameter of 7 cm. The bags are pulled over an open-ended plastic pipe of the same diameter. The procedure is no different than putting on a silk stocking. A hole of the same diameter and 30–40 cm deep is drilled in the soil with a motor-driven soil corer or a hand-held auger. The plastic tube with the mesh bag is inserted into the hole by careful screwing and pressing until the end of the pipe reaches the bottom of the hole. Soil from the field is sifted over a sieve with a mesh size of 3–4 mm. This yields a non-aggregated soil or a soil consisting of small aggregates but free of roots and organic debris. In principle, larger aggregates can also be used so long as the aggregates do not contain dead roots.

The plastic pipe is filled stepwise with sieved soil, with 5 cm in each step. The soil is compacted with a wooden dowel (diameter 6·5 cm) to approximately the same degree of compaction as that of the surrounding soil. This can be checked in the laboratory by determining bulk density (Steen & Håkansson 1987). After each 5 cm filling the plastic pipe is pulled up an additional 5 cm. This procedure is repeated until the hole is filled to the top. The pipe is then removed, and the mesh bag with the soil remains in the hole.

Since the mesh bags are gradually grown over by vegetation they can be difficult to locate. Long, thin sticks or other types of non-disruptive markers are therefore needed. When a bag is removed the roots must be cut around the bag to a depth of at least 15 cm. Normally the bag can thereafter be pulled up from the hole. Roots on the outside of the bag which are dragged out from the surrounding wall of the hole are cut with a pair of scissors.

### *Processing of samples*

The sampled mesh bags are put in labelled plastic bags and placed in a freezer as soon as possible. The samples are allowed to defrost for one day before any further handling. Initially, the non-aggregated soil can be sieved dry to separate most of the mineral particles from roots and organic debris. Some roots pass through the sieve, however. Therefore both fractions are then washed over another sieve (1 mm, mesh). Both dead and living roots are sorted out from the rest of the materials. In samples with aggregated soil the process of separating out roots is limited to washing and wet sieving. In most types of studies it is necessary to either separate the root mass into living and dead roots or to make a visual estimate of the living/dead ratio.

### *Methods for estimating the loss of dead-root mass*

Estimates of root production must take both living and dead roots into account as well as the rate of decomposition of dead roots. The latter can be estimated in a mesh bag by placing it into another bag made of fine mesh ($<0 \cdot 1$ mm) nylon cloth which is then inserted back into the hole. The rate of dead-root decomposition can then be estimated by periodically sampling the cloth bags, washing the samples and weighing the root residues. Similar techniques have been used by Titlyanova (1987), Fiala (1979) and Dickinson (1982). Traditional litter bags, with and without soil, can also be used (Larsson & Steen 1988). This procedure provides data on all three of the necessary components: living-root mass, dead roots and rate of root decomposition. Consequently, root production over time can be estimated (Hansson & Steen 1984; Santantonio & Grace 1987).

### *Layout of field experiments*

The studies were carried out in field trials on agricultural soils in central Sweden. The four to six treatments were replicated four times in blocks with a plot size of 20–25 $m^2$. The following trials were carried out.

1 Annual plant species, mesh bags inserted shortly after sowing. Assessments of:
   (a) effects of herbicides on sugar beet roots;
   (b) effects of herbicides on potato roots;
   (c) effects of herbicides on barley roots;
   (d) effects of CCC on roots of barley and summer wheat; and
   (e) effects of soil compaction on roots of rapeseed and summer wheat.

2 Perennial plant species (grass and clover leys). Mesh bags inserted shortly after sowing and in the established ley as long-, medium- and short-term bags. Assessments made of the effects of nitrogen-fertilizer on roots and rhizomes.

3 Trees and shrubs. Assessments of:
   (a) dynamics of short-term root mortality of *Eucalyptus globulus* in Portugal; and
   (b) relationships between soil hydraulic conductivity and root mass in new plantations and old stands in Tunisia (see Petersson, Messing & Steen 1987).

The results from these trials illustrate the use of the method.

## RESULTS

### *Sugar beet*

In sugar beet the method revealed a decrease in fine (lateral) root growth over the growing season associated with the use of herbicides (lenacil, metamitron). This reduction was more accentuated in sandy loam soil than in clay soil.

### *Potatoes*

A reduction in root growth was observed in potatoes on a sandy soil following use of the herbicide metribuzin. Root growth remained suppressed throughout the growing season in rows and furrows and at different soil depths. In recent field trials (Steen & Andrèn 1990) similar effects were also observed in sandy soil of low (1.2% C) but not high (5% C) organic matter content.

### *Cereals*

MCPA caused a reduction in the root mass of barley in a trial on a sandy loam soil. There were significant differences on two out of three occasions. No significant effects on root biomass were observed in barley after using seed dressings (mercury as well as non-mercury compounds) in a loamy sand (Steen 1987). In further trials there were no consistent effects of CCC on root growth of barley and summer wheat on loamy clay soil, but there was a significant reduction in the length of wheat straw.

The method can also demonstrate the relationships between soil compaction and root growth (root biomass). For example there was a significant reduction in root growth in rapeseed and summer wheat after compaction of a sandy soil.

*Grass*

A comprehensive program of studies carried out on grass leys is briefly summarized here. The application of nitrogen-fertilizer caused a reduction in the ingrowth of roots (Fig. 1) but an increase in rhizome growth. This was reflected in the measurements made on long-, medium- and short-term bags inserted in third-year ley (Fig. 2). The comparison between long-term bags inserted shortly after sowing and soil core samples revealed that these two methods yield comparable results in grass but not in legume leys, owing to the legume tap roots appearing in the soil cores (Fig. 3 and Table 1). Long-term bags demonstrated the rapid ingrowth of roots but slow ingrowth of rhizomes in an established grass ley with rhizomatous grasses (Fig. 4).

Qualitative dynamics can be studied in long-term as well as short-term bags. This is illustrated for the effect of nitrogen concentration. The great differences in carbohydrate and lignin between roots and rhizomes and their seasonal variation are illustrated in Fig. 5. More details about non-structural and structural

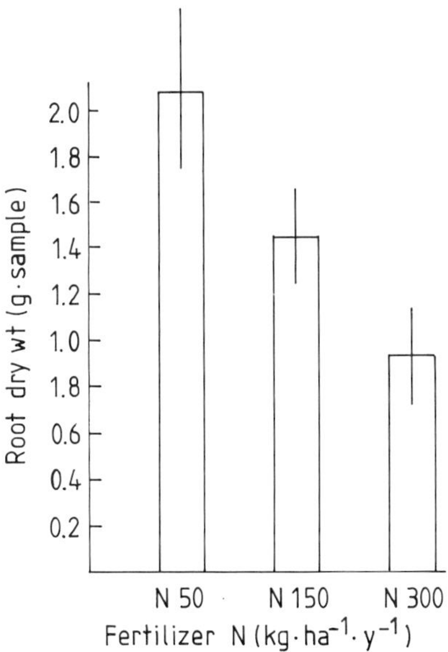


FIG. 1. Grass ley, 2nd year, 1982. Ingrowth of roots in relation to level of nitrogen fertilization. Steen (1984).

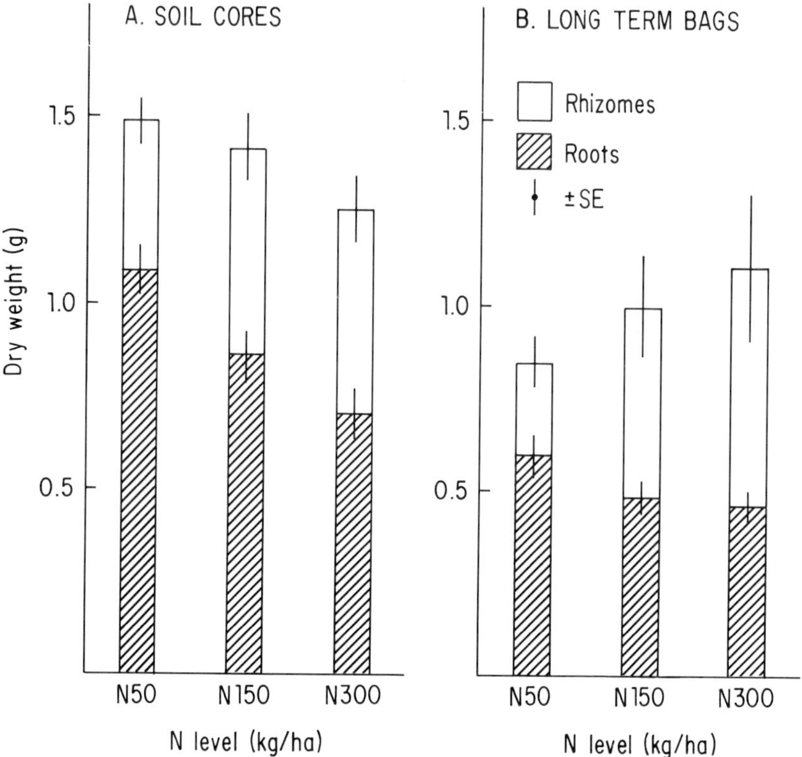


FIG. 2. Grass ley, 3rd year, 1983. Ingrowth of roots and rhizomes in relation to level of nitrogen fertilization. Steen (1985).

TABLE 1. Lateral roots and tap roots of red clover. (a) Under plants and between plants, October 1986. Soil cores, dry organic matter, g/sample, mean and standard error, $n = 5$. (b) Random soil cores, dry organic matter, g/sample, mean and standard error, $n = 10$. Steen (1989)

(a)

| Root type | Under plant | Between plants |
|---|---|---|
| Lateral roots | 0·824 ± 0·131 | 0·574 ± 0·147 |
| Taproots | 1·998 ± 0·352 | 0·158 ± 0·158 |
| Total | 2·822 | 0·732 |

(b)

| Time | Lateral roots | Tap roots | Sum |
|---|---|---|---|
| 1985 | | | |
| May | 0·405 ± 0·072 | 0·040 ± 0·021 | 0·445 |
| Oct | 0·229 ± 0·047 | 0·017 ± 0·012 | 0·246 |
| 1986 | | | |
| Oct | 0·535 ± 0·049 | 0·235 ± 0·109 | 0·760 |

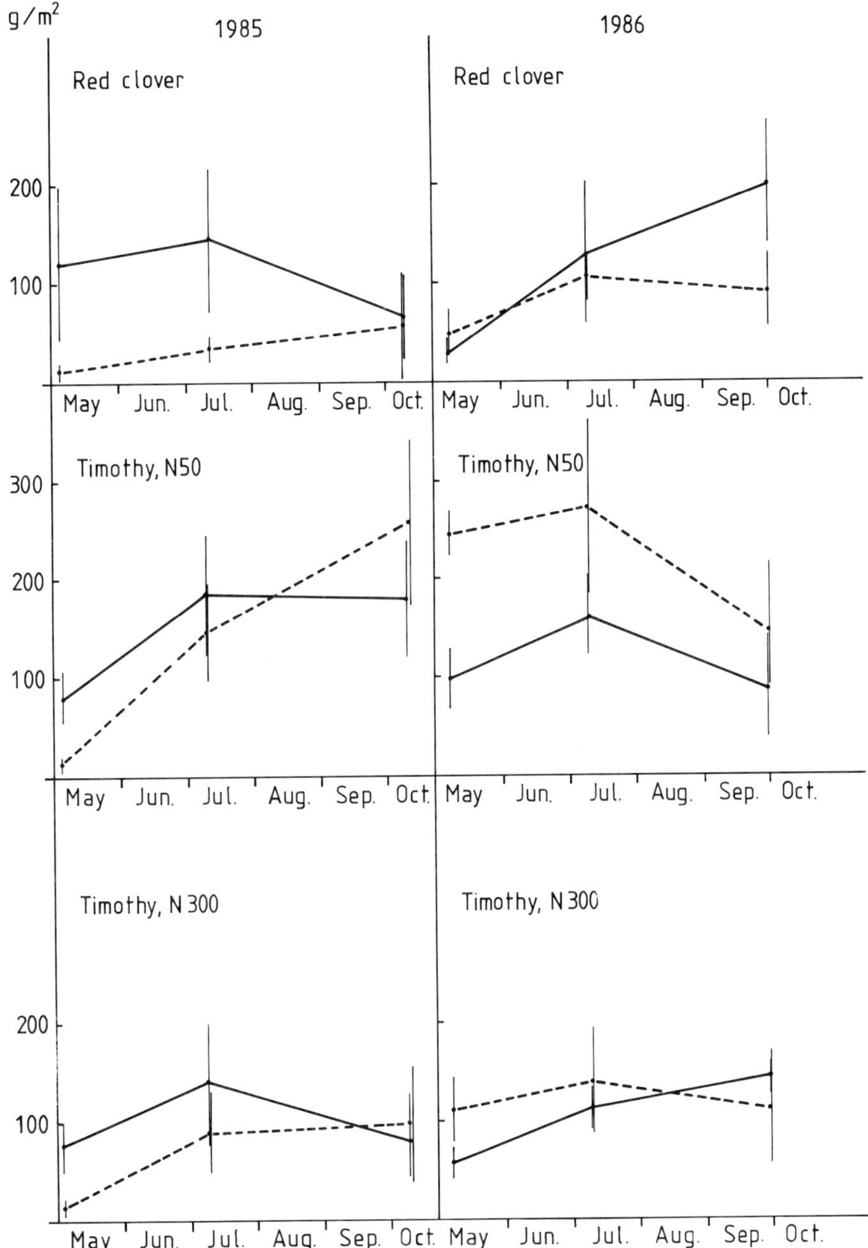


FIG. 3. Red clover and grass ley, 1st and 2nd year, 1985 and 1986. Root mass in soil cores and long term mesh bags (incubated at ley establishment) over the growing season. Soil cores (——); long term bags (----).

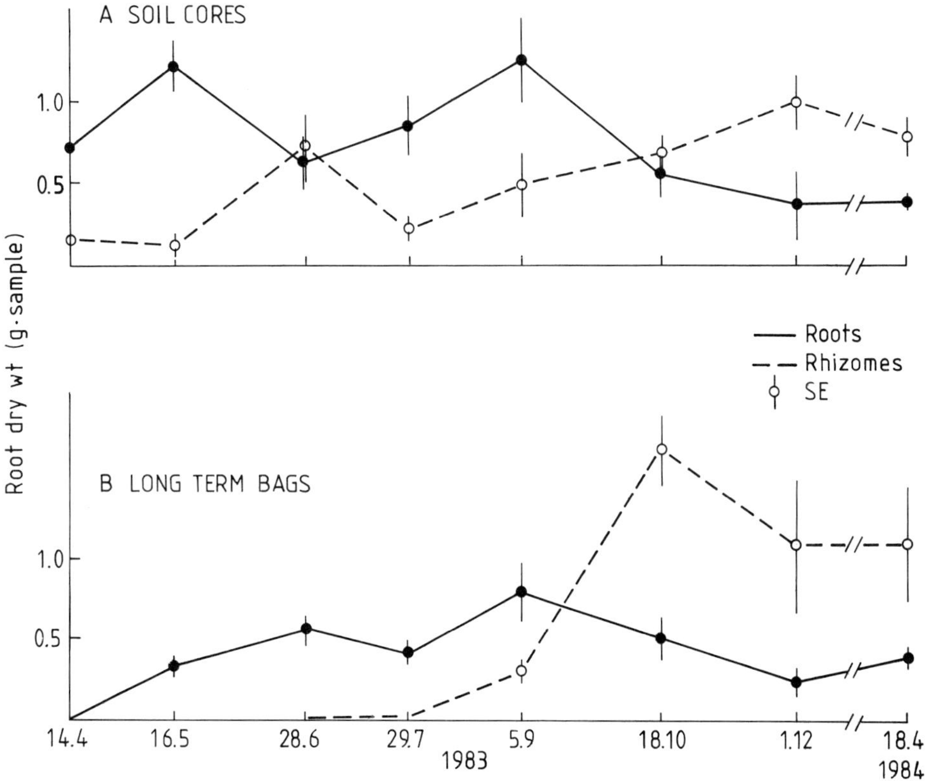


FIG. 4. Grass ley, 3rd year. Changes in root and rhizome mass over a year (1983–84) in soil cores and long term bags. Steen (1985).

carbohydrates are reported by Larsson & Steen (1984) and Steen & Larsson (1986), while dead, decomposing roots are treated by Larsson & Steen (1988).

### *Tree crops*

Mesh bags are also of value in studies of root mortality, as demonstrated by data from stands of *Eucalyptus globulus* in Portugal. There was also considerable root mortality in the short-term bags where no roots were older than 2 months, and there was great variation in mortality over the year (Fabiao, Persson & Steen 1985).

## DISCUSSION

The data presented demonstrate the applicability of the mesh bag technique in annual and perennial stands of plants in agricultural land and in stands of shrubs and trees. The method has advantages and disadvantages, being useful for one purpose but impractical for another. It seems most valuable when used to study

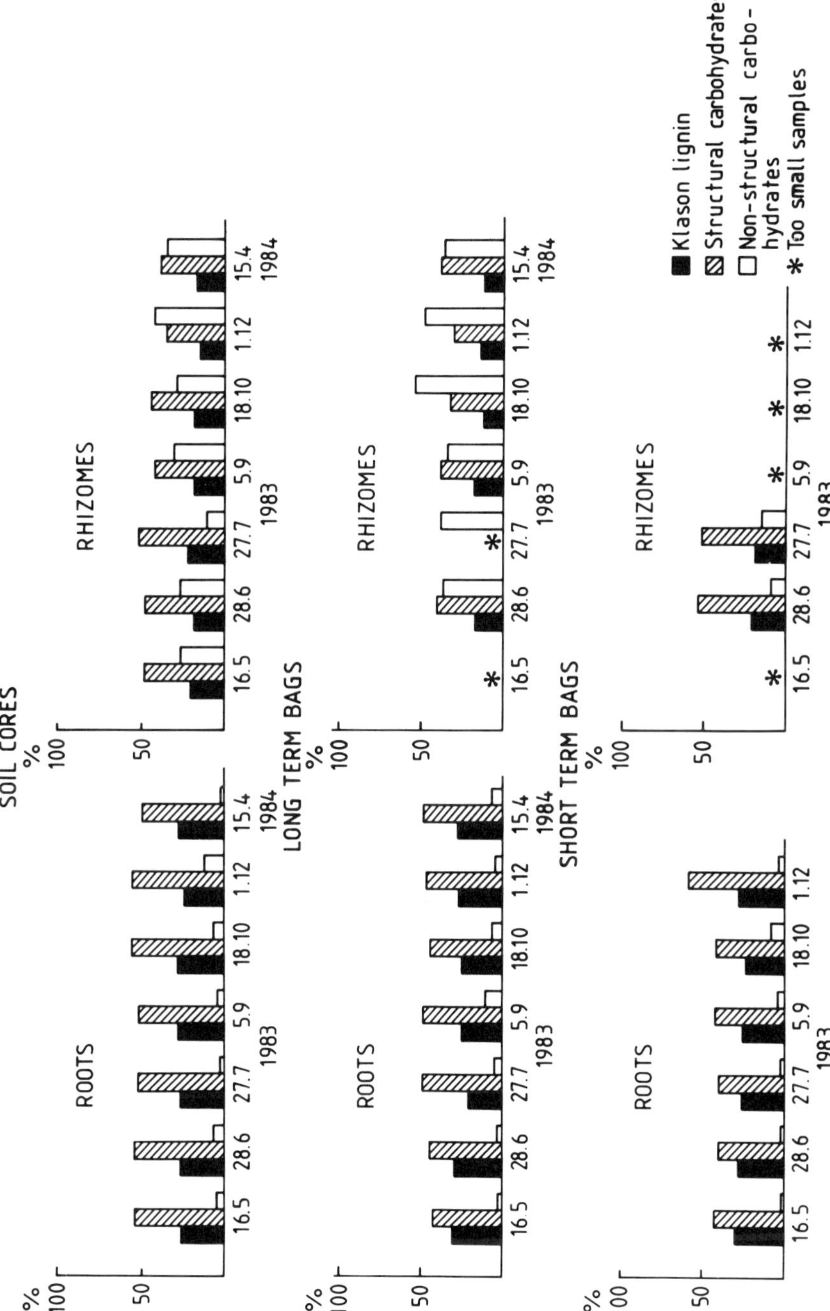


FIG. 5. Variation of carbohydrates and Klason lignin in grass roots and grass rhizomes over a year (1983–84) in soil cores, long term bags and short term bags. Steen & Larsson (1986).

stands of annual plants where the mesh bags are installed at sowing or germination. In such a situation the bags produce results comparable to those of soil cores although the bags do not contain plant residues from the previous year, thus eliminating the need for separation. The lack of crop residues does not seem to affect root growth although this conclusion needs to be confirmed by additional testing.

For perennial species the situation is more complicated, but if the bags are inserted after the emergence of newly established plants the bags yield results comparable to those from soil core samples, as with the annual species. More difficult problems arise when the bags are inserted into an older stand. Nevertheless, the method can still be used for comparisons between treatments in field trials when relative instead of absolute estimates are sufficient. For absolute estimates, i.e. for root production studies, soil core sampling would still be the preferred method although mesh bags can be used if they are incubated for a long time before the production estimates start. Other comparative aspects of methods are discussed by Mackie-Dawson & Atkinson (pp. 25–47). The period of initial root ingrowth into the soil core has then passed and artificial effects of pruning, voids along the wall of the drilled hole, etc., are avoided; the bags can be inserted during one year and production estimates started in the next.

All quantitative methods for obtaining absolute estimates of root production must address the problem of short-term root turnover (Martin 1987; Helal & Sauerbeck 1987; Hansson 1987; Santantonio & Grace 1987; Santantonio & Santantonio 1987). In this respect the mesh bag technique is no worse than soil coring if used in the proper manner. However, neither the mesh bag nor the soil coring method can quantify rapid turnover, e.g. for deciduous trees and perennial grasses root longevity may be only 10–20 days and so in a 3-week period the whole population of fine roots may turn over (Atkinson 1985; Larsson & Steen 1988). In stands of annual plants a large decline in root biomass occurs during the second half of the growing period after a maximum in the middle of this period, i.e. 60–80 days after germination although almost no dead roots are observed in mesh bag samples during this second period. Dead roots of annuals evidently are either decomposed very rapidly or are so fragile and fragmented that they pass the sieve during the wet sieving procedure.

Moreover, the mesh bag method has a great advantage; field tests can be designed so as to yield results that can be readily analysed with statistical methods. The bags can be randomly placed out in standardized field trials with five to six treatments, which are sampled five to ten times. This is easier and less laborious than soil coring, if only roots in the topsoil are being considered.

## REFERENCES

**Ahlström, K., Persson, H. & Börjesson, I. (1988).** Fertilization in a mature Scots pine (*Pinus sylvestris* L.) stand — effects on fine roots. *Plant and Soil*, **106**, 179–190.

**Atkinson, D. (1985).** Spatial and temporal aspects of root distribution as indicated by the use of a root observation laboratory. In *Ecological Interactions in Soil. Plants, Microbes and Animals* (Ed. by A.H. Fitter, D. Atkinson, D.J. Read & M.B. Usher), pp. 43–65. Special Publication No. 4 of the British Ecological Society. Blackwell Scientific Publications, Oxford.

**Dickinson, N.M. (1982).** Investigations and measurement of root turnover in semi-permanent grassland. *Revue d'Ecologie et de Biologie du Sol*, **19**, 307–314.

**Fabiao, A., Persson, H.Å. & Steen, E. (1985).** Growth dynamics of superficial roots in Portuguese plantations of *Eucalyptus globulus* Labill studied with a mesh bag technique. *Plant and Soil*, **83**, 233–242.

**Fiala, K. (1979).** Estimation of annual increment of underground plant biomass in a grassland community *(Polygalo-Nardetum). Folia Geobotanica Phytotaxonomia*, Praha **14**, 1–10.

**Hansson, A.-C. (1987).** *Roots of arable crops: Production, growth dynamics and nitrogen content.* Ph.D. thesis. Swedish University of Agricultural Sciences, Department of Ecology and Environmental Research, Report No. 28.

**Hansson, A.-C. & Steen, E. (1984).** Methods of calculating root production and nitrogen uptake in an annual crop. *Swedish Journal of Agricultural Research*, **14**, 191–200.

**Hansson, A.-C. & Andrén, O. (1986).** Below-ground plant production in a perennial grass ley *(Festuca pratensis* Huds.) assessed with different methods. *Journal of Applied Ecology* **23**, 657–666.

**Helal, H.M. & Sauerbeck, D. (1987).** Direct and indirect influences of plant roots on organic matter and phosphorus turnover in soil. *Intecol Bulletin*, **15**, 49–58.

**Hendrickson, A.H. & Veihmeyer, F.J. (1931).** Influence of dry soil on root extension. *Plant Physiology*, **6**, 567–576.

**Larsson, K. & Steen, E. (1984).** Nitrogen and carbohydrates in grass roots sampled with a mesh bag technique. *Swedish Journal of Agricultural Research*, **14**, 159–164.

**Larsson, K. & Steen, E. (1988).** Changes in mass and chemical composition of grass roots during decomposition. *Grass and Forage Sciences*, **43**, 173–177.

**Lund, Z.F., Pearson, R.W. & Buchanan, G.A. (1970).** An implanted soil mass technique to study herbicide effects on root growth. *Weed Science,* **18**, 279–281.

**Martin, J.K. (1987).** Carbon flow through the rhizosphere of cereal crops — a review. *Intecol Bulletin,* **15**, 17–23.

**Persson, H. (1979).** Fine-root production, decomposition and mortality in forest ecosystems. *Vegetatio,* **41**, 101–109.

**Persson, H. (1983).** The distribution and productivity of fine roots in boreal forests. *Plant and Soil,* **71**, 87–101.

**Petersson, H., Messing, I. & Steen, E. (1987).** Influence of root mass on saturated hydraulic conductivity in arid soils of Central Tunisia. *Arid Soil Research and Rehabilitation*, **1**, 149–160.

**Santantonio, D. & Grace, J.C. (1987).** Estimating fine-root production and turnover from biomass and decomposition data: a compartment-flow model. *Canadian Journal of Forest Research*, **17**, 900–908.

**Santantonio, D. & Santantonio, E. (1987).** Effect of thinning on production and mortality of fine roots in a *Pinus radiata* plantation on a fertile site in New Zealand. *Canadian Journal of Forest Research*, **17**, 919–928.

**Steen, E. (1984).** Variation of root growth in a grass ley studied with a mesh bag technique. *Swedish Journal of Agricultural Research*, **14**, 93–97.

**Steen, E. (1985).** Root and rhizome dynamics in a perennial grass crop during an annual growth cycle. *Swedish Journal of Agricultural Research*, **15**, 25–30.

**Steen, E. (1987).** Effects of agrochemicals on fine root growth and turnover. *Intecol Bulletin*, **15**, 91–99.

**Steen, E. (1989).** Root biomass in timothy and red clover leys estimated by soil coring and mesh bags. *Journal of Agricultural Science*, **19**, 241–247.

**Steen, E. & Larsson, K. (1986).** Carbohydrates in roots and rhizomes of perennial grasses. *New Phytologist*, **104**, 339–346.

**Steen, E. & Håkansson, I. (1987).** Use of in-growth soil cores in mesh bags for studies of relations between soil compaction and root growth. *Soil and Tillage Research*, **10**, 363–371.

**Steen, E. & Andrén, O. (1990).** Effects of metribuzin on potato root growth. *Swedish Journal of Agricultural Research*, **20**, 127–133.

**Steen, E., Persson, H. & Al-Windi, I. (1984).** Effects of herbicides on the ingrowth of lateral roots of sugar beet (*Beta vulgaris* L.) in mesh bags. *Swedish Journal of Agricultural Research*, **14**, 103–105.

**Steen, E., Andrén, O. & Al-Windi, I. (1987).** Reduced growth of potato roots caused by metribuzin. *Swedish Journal of Agricultural Research*, **17**, 41–46.

**Titlyanova, A.A. (1987).** Ecosystem succession and biological turnover. *Vegetatio*, **50**, 43–51.

# Root Demography and Functioning

# Root system demography and production in forest ecosystems

R. FOGEL
*University of Michigan Herbarium, Ann Arbor, MI 48109–1057, USA*

## SUMMARY

1 Root/mycorrhiza studies have reached a critical juncture.
2 The soil coring method for quantifying root demography has substantially improved our knowledge of the importance of root processes, but these processes are so dynamic other approaches have to be explored if factors limiting production are to be studied in detail.
3 In parallel with the new approaches, models integrating data on the myriad interactions among roots and soil organisms have to be developed.

## INTRODUCTION

The application of demographic methods to roots consisting of portions of unrelated species is accompanied by conceptual as well as practical problems. The interest in root demography arises from efforts to improve estimates of root production, root turnover, and the proportion, or cost, of photosynthate needed to grow and replace roots. As earlier reviews have indicated, an evolution in the sophistication of the methods used in root research has occurred during the last 20 years (Fogel 1983, 1985).

Early studies often ignored root production entirely or estimated it indirectly by assuming that the ratio of production to mass must be similar for root and shoot systems. The ratio of root production to mass was estimated to be from 20–100% of the above-ground ratio (Harris, Santantonio & McGinty 1980; Pastor & Bockheim 1981; Whittaker & Marks 1975).

The harvest approach, estimating production by measuring differences in biomass on successive sampling dates and correcting for losses, provided the first direct estimates of root production (e.g. Fogel & Hunt 1979; Ford & Deans 1977; Harris, Kinerson & Edwards 1977; Persson 1978; Santantonio, Hermann & Overton 1977; Vogt *et al.* 1980). The model for this approach takes several forms, but one equation adopted from population dynamics can be expressed as:

$$NPP = Bt + 1 - Bt + L,$$

where $NPP$ is net primary production, $Bt$ is the standing crop at time $t$, and $L$ is losses, such as tissue lost in herbivory, rhizodeposition, translocation, etc. The adoption of this approach changed our concept of the importance of roots and mycorrhizae in ecosystem processes. Production and turnover of roots and mycorrhizae was much more dynamic than originally thought, with estimates of 40–73% of net primary production being used to replace and increase root biomass (Fogel 1985).

It was soon recognized that production estimates derived by the harvest method can be improved substantially if roots can be separated into 'live' and 'dead' categories rather than using the difference in minimum and maximum standing crops of all roots (Santantonio 1979; Fairley & Alexander 1985). Such an approach is often termed 'balancing transfers' and permits the calculation of mortality and decomposition rates. In practice, distinguishing between live and dead roots except in the most obvious cases, unsuberized root tips, is difficult. Fogel (1985) discusses a number of methods that have been used and the problems associated with each.

Improved estimates resulting from the balancing transfers approach will be invaluable in comparing root processes in different ecosystems, but soil coring because of labour-intensive effort involved in sampling frequently and the disruption of spatial relationships during root separation cannot provide the fine detail needed to advance beyond a general comparative picture of root dynamics. Due to the effort involved most coring studies have concentrated on a single forest ecosystem (e.g. Fogel & Hunt 1979; Ford & Deans 1977; Persson 1978; Santantonio & Hermann 1985). Core samples have generally been taken at monthly intervals in these studies. Studies comparing more than one stand (e.g. Harris, Kinerson & Edwards 1977; Vogt *et al.* 1980; Vogt, Edmonds & Grier 1981) generally employ seasonal soil core sampling, or even a single sample (Vogt *et al.* 1983). Direct observations of root growth indicate that seasonal sampling is too infrequent for precise estimates of root standing crops and that even monthly sampling may be too infrequent. A large percentage of apple roots (75%) may become moribund within 4 weeks of initial root growth (Atkinson 1985). Similarly, Roberts (1976) reports that the mean life of Scots pine fine roots (≤1 mm diameter) is 4 weeks, but varies between 2 to 9 weeks.

Optimization of the number of soil cores and timing of sampling is further complicated by the effects of temperature and moisture on root growth, plus the effects of nutrient availability, carbohydrate supply, root hormone levels, and genetic factors (Persson 1983). Root growth in a Scots pine plantation in East Anglia peaked in April–May and then declined (Roberts 1976). In contrast, no distinct seasonal pattern was observed in a young Scots pine stand in central Sweden (Persson 1978). Santantonio & Hermann (1985) report a general pattern for peaks in numbers of new root tips in the spring and autumn of years with dry summers in western Oregon Douglas fir stands, but this pattern was disrupted when rain occurred during the normal summer drought season.

A more serious limitation of the soil coring method for ecosystem studies is the loss of information during the separation of roots for biomass determination. Nearly all of the research employing the harvest approach ignores losses to herbivory, exudates, etc., because of the labour-intensive effort and time needed to separate roots from soil cores. Given the magnitude of the net primary produc-

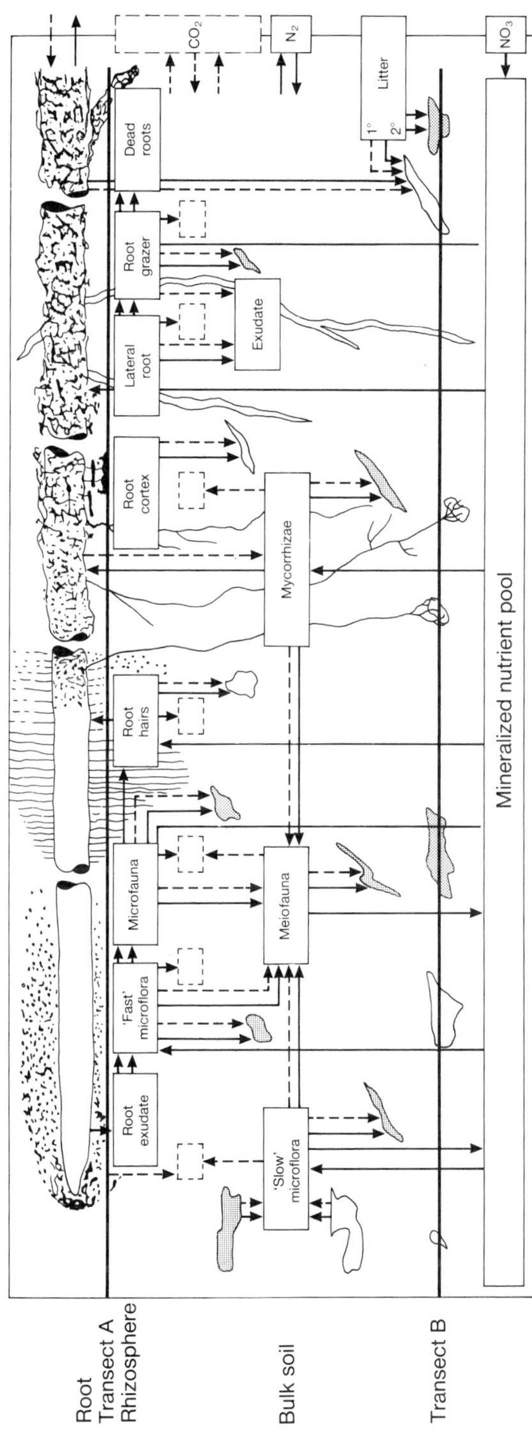


FIG. 1. Conceptual diagram of root (transect A) and bulk soil (transect B), showing organisms and principal processes (from Coleman *et al.* 1983). Regions of the root from left to right: meristem/elongation zone, absorption or root hair zone, maturation zone, branching zone, and moribund zone.

tion utilized in root processes and the brief life span of fine roots and mycorrhizae, a more detailed understanding of the factors controlling root longevity and transfers from roots to soil biota is imperative if the factors limiting productivity of trees are to be understood.

Of the available techniques, direct observation in rhizotrons and mini-rhizotrons coupled with experimental manipulation appears the most attractive for studying roots in an ecosystems context. Models incorporating current knowledge of root dynamics and interactions among soil biota are needed to guide research. Coleman *et al.* (1983) present a conceptual model (Fig. 1) relating different developmental segments to nitrogen mineralization in the rhizosphere as compared to soil that may represent a starting point for model development. Their model is based on the 'classic' growing root tip, but includes vesicular arbuscular mycorrhizae (VAM). Below, I attempt to characterize their model in more detail than originally presented and to examine modifications required by ectomycorrhizal associations (ECM).

## MODELS OF ROOT GROWTH

### *Coleman* et al. *model of growing root apex*

The classical concept of the root apex divides the root into several zones: meristematic, elongation, maturation (Russell 1977). Coleman *et al.* (1983) use this basic approach in developing their conceptual model, but expand it by adding a branching zone and a zone of moribund roots. They also show a connection to VAM mycorrhizae, although they do not present the mycorrhizal association as a distinct root zone. These zones are assumed to have different metabolic rates, rhizodeposition products and consequently different microbial and faunal communities. These differences result in different rates of carbon and nitrogen mineralization in the rhizosphere of each root zone. These zones may be characterized as follows:

1 Meristem/elongation zone: size of meristem generally correlated with growth rate. Growth rates of 0·5–2 cm day$^{-1}$ are common for cereals grown under favourable laboratory conditions (Russell 1977). Rhizodeposition includes mucigel and root cap lysates deposited at a rate of 2–5 g per 100 g of root dry weight increase. Soluble exudate production ranges from 1 to 10 g per 100 g of dry weight increase (Newman 1985). Bacterial numbers increase 100-fold in this zone during the first 5 days and only ten-fold during the second 5 days (Newman 1985). The plateau in growth is attributed to carbon shortage by Newman (1985).

2 Absorption or root hair zone: root hairs only persist for 2–3 days before they are destroyed by microbes (Russell 1977). Root hairs may produce some exudates and be important in water and mineral absorption and/or anchorage during drought. This zone may also be the zone of greatest VAM mycorrhizal infection.

3 Maturation zone: as growth of the root slows 'browning' or the deposition of polyphenols occurs. Usually occurs 4 weeks (3–5 weeks) after initial root growth in

apple (Atkinson 1985). After 4 weeks, brown roots 'lose' their cortex (40–50% of root volume). Nearly 75% of the brown roots die at this point and the remaining 25% undergo secondary thickening to become part of the permanent root system (Atkinson 1985). All apple roots less than 1 mm in diameter die.

**4** Branching zone: very little information is apparently available on this zone other than details of secondary root initiation.

**5** Moribund zone: may occur as early as 4 weeks after initial root growth in apple, especially in roots less than 1 mm in diameter. Newman (1985) indicates that intact moribund roots are the main source of plant carbon entering the soil rather than exudates. His calculations indicate that moribund root loss may reach 0·10–0·25 g per 100 g of root.

Several hypotheses result from the Coleman *et al.* model: (i) rhizodeposition is primarily plant in origin; (ii) rhizodeposition and biotic diversity are greatest at the root tip (zone 1); (iii) fast growing microflora will immobilize nitrogen and phosphorus in zone 1; and (iv) root grazer populations develop on mature (suberized) roots (zone 4) remineralizing nutrients. Direct observation of microbes on root surfaces of crop plants and microcosm studies of simplified communities in artificial soils offer conflicting support for these hypotheses.

Direct observation of roots with light and electron microscopy has shown that the root tip is almost always devoid of bacteria, mucigel is sparsely colonized, bacteria first appear as isolated cells or small colonies in the zone of elongation, and that older portions of the root are densely colonized (Rovira & Davey 1974). Colonization is not uniform, bacteria and often fungi are found at the junction of cell walls — sites of greater exudation. Colonization of sloughed cells is similar. Sloughed root cap cells, each surrounded by mucilage, of French bean are virtually axenic for the first 24 h. By the third day, the cells are coated with a layer of bacteria three to four cells deep (Rovira & Davey 1974). Similarly, fungi are seldom observed in the root tip region, but become more abundant on older portions of the root.

Microcosm studies, in contrast, provide support for the hypotheses. Trofymow, Coleman & Cambardella (1987) used a multiple split root chamber and artificial soil to demonstrate that the highest rates of rhizodeposition of carbon, and nitrogen depletion, occurred along the terminal 10 mm of the root tip of 26-day-old axenic oat roots. Rhizodeposition of soluble and insoluble carbon compounds comprised up to 50% of the standing root biomass carbon. Rhizodeposition consisted of 44% mucigels and exudates, and 56% sloughed root cap cells, loose cortical cells, and mucigels adhering to soil particles.

A long series of microcosm experiments also provide indirect support for a scenario in which a large carbon input just behind the root tip stimulates bacterial growth (Clarholm 1985). The bacteria mineralize enough nitrogen for their growth, protozoa are attracted, protozoan grazing of the bacteria releases bacterial nitrogen as ammonium, and then some of the ammonium is taken up by the root. In Clarholm's experiments, protozoa were needed to make bacterial nitrogen available to roots.

This very attractive scenario could be quite different in woody root systems.

Nitrogen mineralization from sloughed cortical cells may be equal to or greater than that from the root tip. Time-lapse photography of apple roots in the East Malling rhizotron shows large numbers of nematodes appearing in the rhizosphere when the cortex is being sloughed, feeding, and then moving away from the root.

### *VAM modifications of the growing root apex*

The classical concept of the root tip consisting of a root cap, meristematic zone, elongation zone, and a zone of maturation supporting root hairs is modified slightly in VAM associations. The fungal symbiont consists of two phases — an internal mycelium within the cortex of the root and an external mycelium spreading through the soil. The basis for the mutualism is the provision of carbohydrate to the fungus in exchange for a more cost-effective system than root hairs for obtaining nutrients. While root hairs may be up to 2000 μm long and 10–15 μm in diameter, they may only persist for 2–3 days before being destroyed by microbes (Baylis 1975; Russell 1977). Estimates of the quantity of external mycelium of VAM fungi vary from 1·3–1·4 m $cm^{-1}$ of infected root length in clover and ryegrass compared to 0·8 m $cm^{-1}$ in onion (Sanders & Tinker 1973; Tisdall & Oades 1979). Hyphal lengths of this magnitude would double the potential absorbing surface of roots 0·5–1 mm in diameter (Gianinazzi-Pearson & Gianinazzi 1983). Rhodes & Gerdemann (1975) have experimentally shown that the zone of phosphate absorption in VAM plants is greatly expanded beyond the 1–2 mm zone explored by root hairs to at least 8 cm in mycorrhizal onions. The cost to the plant of maintaining the VAM association is difficult to estimate, but is assumed to be about 10% of the fungal biomass, plus carbon respired by the increased metabolic activity of root cells following infection (Gianinazzi-Pearson & Gianinazzi 1983). Growth depression in some VAM plants has been attributed to host–fungus competition for carbon. The external mycelium of the VAM fungus also provides interplant, even interspecies connections, for direct transfer of carbon, phosphorus and perhaps other nutrients among neighbouring plants (Chiariello, Hickman & Mooney 1982; Read, Francis & Findlay 1985; Whittingham & Read 1982).

In addition to the production lost in supporting the VAM association, VAM species have relatively short-lived fine roots as the apple example cited above illustrates. After 4 weeks apple roots lose their whole cortex, nearly 75% of the roots die at this time, but 25% survive to become part of the permanent root system (Atkinson 1985). All apple roots less than 1 mm in diameter die after 4 weeks. This scenario suggests that root turnover and the related investment in photosynthate to grow and replace these roots is high. The proportionally lower investment in roots on nutrient rich sites than on poor sites may offset this cost to some extent in horticultural cropping systems (Marks, Ditchburne & Foster 1968).

'Root exudates' at the rhizoplane are probably primarily host in origin since the development of fungal hyphae in this type of association is not extensive. Sloughed cortical cells may represent the single largest input of rhizodeposition to the rhizosphere.

*ECM modifications of the growing root apex*

Ectomycorrhizal associations alter the classical concept of the root tip more profoundly than the VAM symbiosis. In the ECM symbiosis, typified by pine ECM, the classical root is modified by the absence of an epidermis and associated root hairs, the root cap is reduced, and the apical meristem is very small (Harley & Smith 1983). The result is limited growth producing a short root for the association. In addition to the anatomical changes in short roots, the classical root is further modified by the presence of a three-phase mycelial system. An internal mycelium (Hartig net) grows intercellularly around cells of the primary cortex, a sheath or mantle of hyphae covers the entire root surface, and external hyphae extend into the soil. The internal mycelium functions in exchange of materials between fungus and host while hyphae of the mantle, lacking in VAM, act as storage cells, particularly for phosphorus and glycogen (Harley & Smith 1983). The external mycelium functions the same as the external mycelium of VAM extending the zone of nutrient absorption and providing interplant connections. An individual hypha may extend more than 2 m from an ECM and form more than 120 lateral branches; 200 to over 2000 individual hyphae have been counted emerging from a single ECM (Trappe & Fogel 1977).

In contrast to VAM, ECM tend to be long-lived, from several months to a reported 13 years (Fogel 1983). In addition to serving as a storage organ, the ECM mantle may provide some protection against drought (Harley & Smith 1983). The relatively long life of ECM coupled with their drought resistance may provide an advantage for the host over VAM for nutrient absorption on dry, nutrient-poor sites. The cost in maintenance respiration will be higher however, because more hyphal biomass has to be supported for a longer period. The energetic cost of this association may be quite high. ECM respiration comprises 40–50% of the root respiration of beech (Harley & Smith 1983).

One implication of the ECM mantle is that 'root' exudates are primarily fungal in origin and the differences in exudate chemistry will affect the species composition of the biota near the ECM creating a mycorrhizosphere. Species differences in the mycorrhizosphere bacteria and/or fungi in turn will affect the species of soil animals present. A number of predictions of the factors determining species diversity in the mycorrhizosphere, nearly all untested, arise from this scenario. These predictions include: (i) mycorrhizosphere bacteria will utilize fungal produced mannitol or trehalose rather than sucrose, glucose or fructose as simple carbon sources; (ii) fungal exudates will be produced over a period of months compared to a few days in actively growing non-mycorrhizal roots; (iii) senescent ectomycorrhizae are the main substrate available for microbes; (iv) ectomycorrhizal fungi as k-selected species will produce a broad range of secondary compounds to inhibit mycovores; (v) nematodes parasitizing root hairs will be few in number; (vi) species diversity of mycorrhizal fungi will be higher than in VAM systems; (vii) the diversity of ectomycorrhizal fungi present will produce a great variety of mycorrhizospheres due to strain and species differences in production of secondary compounds, strand or rhizomorph formation, wall thickness of hyphae, and cell wall chemistry, e.g. fungal melanin; and (viii) the mycorrhizospheres produced by the same mycorrhizal fungus on quite different hosts, e.g. pines and oaks, will be more similar than mycorrhizospheres produced by different fungi on the same host.

Very little evidence exists to show a quantitative difference between classical rhizospheres and mycorrhizospheres. Most of the research on metabolites of ectomycorrhizal fungi has concentrated on compounds affecting the host, e.g. growth hormones, or inhibiting microbes (e.g. antibiotics, rather than simple sugars, amino acids, or organic acids). Most of the ectomycorrhizal fungi tested produce auxins in culture and some but not all have been shown to produce substances capable of promoting the growth of cytokinin-dependent soyabean callus (Harley & Smith 1983). Graham & Linderman (1979) report the production of ethylene by ectomycorrhizal fungi in liquid culture. *Suillus variegatus* produces ethanol, isobutanol, isoamyl alcohol, acetoin, and isobutyric acid in culture (Krupa & Fries 1971). Production of bacteriostatic and fungistatic substances is strain specific in ectomycorrhizal fungi and consequently highly variable (Marx 1969, 1973). Of twenty-six species examined by Krywolap (1971), six produce substances active against both Gram-positive and Gram-negative bacteria, eleven are active against Gram-negative bacteria only, no inhibitory activity is found in the culture media of four species. The active compounds *Leucopaxillus cerealis* var. *piceina* produces have been identified as diatretyne nitrile and diatretyne-3. The nitrile is very inhibitory against both bacteria and fungi, whereas the diatretyne-3 has only a bacteriostatic action (Marx 1969). Ectomycorrhizal fungi have also been shown to produce hydroxamate siderophores that enhance iron absorption at neutral and alkaline pH (Szaniszlo *et al.* 1981), acid phosphatases (Dighton 1983), phytase (Harley & Smith 1983), and polyphenol oxidases (Giltrap 1982). Cellulases and pectinases may be produced by some ectomycorrhizal fungi (Harley & Smith 1983). No references apparently exist describing sugars exuded by ectomycorrhizal fungi even though it is known that these fungi convert host sugars, i.e. glucose and fructose, to fungal sugars and alcohols, i.e. trehalose and mannitol, that are unavailable to the host and are not generally found in non-mycorrhizal roots (Harley & Smith 1983). The extrapolation of *in vitro* studies to field conditions is difficult because production of secondary compounds is often very sensitive to culture conditions.

One type of fungal exudation that has been studied in an ecosystems context is the production of calcium oxalate. Crystals of calcium oxalate are found covering older hyphae of some ectomycorrhizal fungi and may be present in the ECM mantle (Malajczuk & Cromack 1982). Fungal mats of the ectomycorrhizal basidiomycete, *Hysterangium crassum*, accumulate twenty times more calcium oxalate in the soil A-horizon than in adjacent uncolonized soil. *H. crassum* mats are one of the few locations in which earthworms are found in the acid forest soils of western Oregon, presumably because of their requirement for calcium (Fogel, unpublished). Although calcium oxalate cannot be utilized as a sole carbon source by most organisms, it is interesting to note that calcium oxalate decomposing bacteria

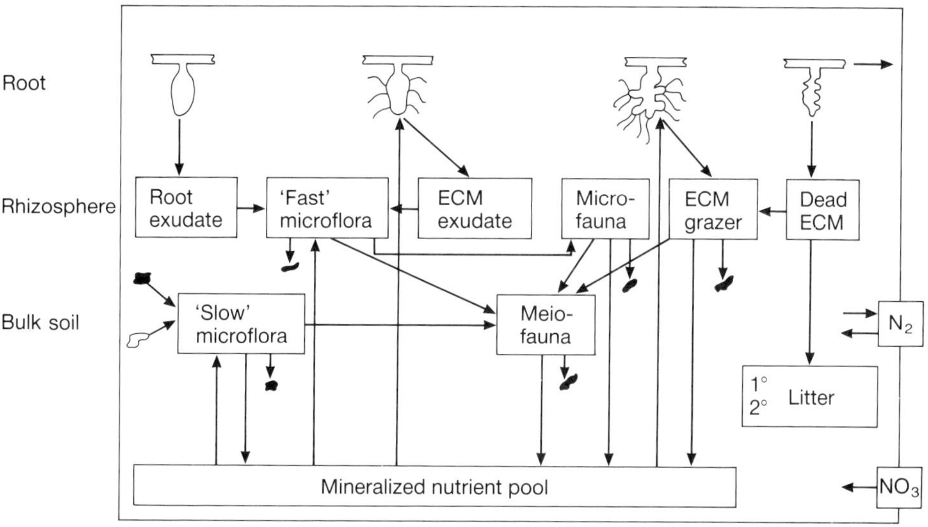


Fig. 2. Conceptual diagram of ectomycorrhizal root, showing organisms and principal processes of nitrogen mineralization. Regions of the mycorrhizosphere from left to right: uninfected root, simple ECM, ECM cluster, moribund ECM.

have been isolated from the gut and casts of earthworms. Oxalate production by ectomycorrhizal fungi may explain increased rates of weathering and nutrient release by ECM as compared to non-mycorrhizal roots (Cromack *et al.* 1977). Oxalic acid is the only fungal exudate that has also been reported in tree root exudates (Bowen & Theodorou 1973; Smith 1977).

Clearly, ECM associations are different to either the classical or VAM root association. Coleman, Reid & Cole (1983) used as a basis for their model, although a small number of lateral root tips will fit the VAM model. The effect of these differences on the Coleman *et al.* model is shown in Fig. 2. The result is a simplified model in which fungal hyphae and intact moribund ECM are the major energy sources for soil biota. The implication is that 'fast' microflora (bacteria) and microfauna (protozoa, nematodes) will be much less important than in VAM dominated systems due to fungal exudates replacing root exudates and a large reduction in rhizodeposition of lysates and cells compared to the amounts produced during secondary thickening of VAM roots. Nitrogen mineralization will be slower than in VAM systems and will be localized near the apex of lateral roots.

## CONCLUSIONS

Root/mycorrhiza studies have reached a critical juncture. The practical limit of the soil coring approach for improving our basic knowledge of root production and turnover has been reached, although it will continue to be useful in comparative studies of different ecosystems. The method was very valuable in demonstrating that root processes are very dynamic and more important energetically than previously thought. Root processes are so dynamic however, that other approaches will have to be used to unravel the factors or combination of factors controlling root processes, and in turn, limiting forest productivity.

Of the available methods, direct observation using mini-rhizotrons and rhizotrons, in conjunction with radio-labelling experiments, appear to have the greatest potential for improving our understanding of interactions among roots and soil biota. Both of these methods represent an improvement over soil coring in reducing sample variance, increased frequency of measurement, ease of computerized image analysis, and perhaps most important, minimal disruption of the spatial and temporal relationships among roots and associated biota (McMichael & Taylor 1987).

Development of models encompassing the myriad interactions among soil organisms, as well as internal processes regulating root growth, must parallel the use of new approaches for studying roots. The Coleman *et al.* model represents a starting point for conceptual modelling of non-mycorrhizal and VAM root systems, but will have to be modified for ECM systems. The brief review presented above indicates that even for systems very similar to the 'classic' root apex, present knowledge of the spatial and temporal distribution of soil biota and patterns in root associated processes is very fragmentary.

The rudimentary status of our knowledge is apparent in the discrepancy between the nitrogen mineralization scenario suggested by the microcosm studies and the microbe distribution data obtained by direct microscopy. The apparent lack of data on the distribution of micro- and meio-fauna along roots is another indication. Perusal of the literature quickly reveals that sampling in most soil invertebrate studies is restricted to the top 10–12 cm of the soil, although soil animals are known to occur at greater depths. The agroecosystem literature indicates, for instance, that collembolan species may be found at considerable depths in sandy soils (Curl & Truelove 1986) and that collembola are more abundant in the rhizosphere of cotton than 20 cm distant (Wiggins *et al.* 1979). Studies in a fumigated tomato field showed that a phytoparasitic nematode can migrate 120 cm vertically to infect tomato roots (Johnson & McKeen 1973). Despite these and other glimmerings on the distribution of soil fauna, I was unable to locate any quantitative data relating the distribution of soil invertebrates and the phenological stage of roots in forest ecosystems.

Direct observation alone may prove inadequate to unravel dietary preferences of soil biota without experimental manipulation or use of labelled substrates. Many of the food items are very small, e.g. bacteria, fungal hyphae, sloughed cells, and may not be discernible after ingestion unless marked food items are employed, or animals are harvested for examination of gut contents. Mouth parts may not be a reliable indicator as Walter *et al.* (1986) found *in vitro* that even when 'fungivorous' mites were offered yeast and algal cells, 11–56% of the mites fed on nematodes. Another challenge will be to develop methods, e.g. fluorescent antibodies, for visualizing or otherwise distinguishing saprophytic, parasitic, and mycorrhizal fungi. Despite these difficulties, some would say unparalleled opportunities, quantitative data garnered by direct observation on root growth, senescence, abundance of soil biota, and dietary preferences of soil animals will provide initial answers to the fundamental questions of which processes, e.g. mycophagy of mycorrhizal hyphae, grazing of roots, or grazing of sloughed cortical cells, impose the most severe limits on forest productivity.

## REFERENCES

**Atkinson, D. (1985).** Spatial and temporal aspects of root distribution as indicated by the use of a root observation laboratory. In *Ecological Interactions in Soil* (Ed. by A.H. Fitter, D. Atkinson, D.J. Read & M.B. Usher), pp. 43–65. *Special Publication Series of the British Ecological Society No. 4.* Blackwell Scientific Publications, Oxford.

**Baylis, G.T.S. (1975).** The magnolioid mycorrhiza and mycotrophy in root systems derived from it. In *Endomycorrhizas* (Ed. by F.E. Sanders, B. Mosse & P.B. Tinker), pp. 373–389. Academic Press, New York.

**Bowen, G.D. & Theodorou, C. (1973).** Growth of ectomycorrhizal fungi around seeds and roots. In *Ectomycorrhizae: Their Ecology and Physiology* (Ed. by G.C. Marks & T.T. Kozlowski), pp. 107–150. Academic Press, New York.

**Chiariello, N., Hickman, J.C. & Mooney, H.A. (1982).** Endomycorrhizal role for interspecific transfer of phosphorus in a community of annual plants. *Science*, **217**, 941–943.

**Clarholm, M. (1985).** Possible roles for roots, bacteria, protozoa and fungi in supplying nitrogen to plants. In *Ecological Interactions in Soil* (Ed. by A.H. Fitter, D. Atkinson, D.J. Read & M.B. Usher), pp. 355–365. *Special Publication Series of the British Ecological Society No. 4.* Blackwell Scientific Publications, Oxford.

**Coleman, D.C., Reid, C.P.P. & C.V. Cole (1983).** Biological strategies of nutrient cycling in soil systems. *Advances in Ecological Research*, **13**, 1–55.

**Cromack, K., Jr., Sollins, P., Todd, R.L., Fogel, R., Todd, A.W., Fender, W.M., Crossley, M.E. & Crossley, D.A., Jr. (1977).** The role of oxalic acid and bicarbonate in calcium cycling by fungi and bacteria: some possible implications for soil animals. In *Soil Organisms as Components of Ecosystems* (Ed. by U. Lohm & T. Persson) Vol. 25, pp. 246–252. Ecological Bulletin, Stockholm.

**Curl, E.A. & Truelove, B. (1986).** *The Rhizosphere.* Springer Verlag, Berlin, 288 pp.

**Dighton, J. (1983).** Phosphatase production by mycorrhizal fungi. *Plant and Soil*, **71**, 455–462.

**Fairley, R.I. & Alexander, I.J. (1985).** Methods of calculating fine root production in forests. In *Ecological Interactions in Soil* (Ed. by A.H. Fitter, D. Atkinson, D.J. Read & M.B. Usher), pp. 37–42. *Special Publication Series of the British Ecological Society No. 4.* Blackwell Scientific Publications, Oxford.

**Fogel, R. (1983).** Root turnover and productivity of coniferous forests. *Plant Soil*, **71**, 75–85.

**Fogel, R. (1985).** Roots as primary producers in below-ground ecosystems. In *Ecological Interactions in Soil* (Ed. by A.H. Fitter, D. Atkinson, D.J. Read & M. Usher), pp. 23–36. *Special Publication Series of the British Ecological Society No. 4.* Blackwell Scientific Publications, Oxford.

**Fogel, R. & Hunt G. (1979).** Fungal and arboreal biomass in a western Oregon Douglas-fir ecosystem: distribution patterns and turnover. *Canadian Journal of Forest Research*, **9**, 245–256.

**Ford, E.D. & Deans, J.D. (1977).** Growth of Sitka spruce plantation: spatial distribution and seasonal fluctuations of lengths, weights and carbohydrate concentrations of fine roots. *Plant and Soil*, **47**, 463–485.

**Gianinazzi-Pearson, V. and Gianinazzi, S. (1983).** The physiology of vesicular–arbuscular mycorrhizal roots. *Plant and Soil*, **71**, 197–209.

**Giltrap, N.J. (1982).** Production of polyphenol oxidases by ectomycorrhizal fungi with special reference to *Lactarius* spp. *Transactions of the British Mycology Society*, **78**, 75–81.

**Graham, J.H. & Linderman, R.G. (1979).** Ethylene production by ectomycorrhizal fungi and *Fusarium oxysporum* spp. *pini in vitro* and by aseptically synthesized mycorrhizal and *Fusarium* infected roots of Douglas fir. *Proceedings of the 4th North American Conference on Mycorrhizae*, Fort Collins.

**Harley, J.L. & Smith, S.E. (1983).** *Mycorrhizal Symbiosis*. Academic Press, London.

**Harris, W.F., Kinerson, R.S., Jr. & Edwards, N.T. (1977).** Comparison of belowground biomass of natural deciduous forest and loblolly pine plantations. *Pedobiologia*, **17**, 369–381.

**Harris, W.F., Santantonio, D. & McGinty, D. (1980).** The dynamic below-ground ecosystem. *Forests: Fresh Perspectives from Ecosystem Analysis* (Ed. by R.H. Waring), pp. 119–129. Oregon State University Press, Corvallis.

**Johnson, P.W. & McKeen, C.D. (1973).** Vertical movement and distribution of *Meloidogyne incognita* (Nematodea) under tomato in a sandy loam greenhouse soil. *Canadian Journal Plant Science*, **53**, 837–841.

**Krupa, S. & Fries, N. (1971).** Studies on ectomycorrhizae of pine. I. Production of volatile organic compounds. *Canadian Journal of Botany*, **49**, 1425–1431.

**Krywolap, G. (1971).** Production of antibiotics by certain mycorrhizal fungi. In *Mycorrhizae* (Ed. by E. Hacskaylo), pp. 210–222. USDA, Forest Service Miscellaneous Publication 1189.

**Malajczuk, N. and Cromack, K., Jr. (1982).** Accumulation of calcium oxalate in the mantle of ecto-mycorrhizal roots of *Pinus radiata* and *Eucalyptus marginata*. *New Phytologist*, **92**, 527–531.

**Marks, G.C., Ditchburne, N. & Foster, R.C. (1968).** Quantitative estimate of mycorrhiza populations in radiata pine forests. *Australian Forestry*, **32**, 26–38.

**Marx, D.H. (1969).** The influence of ectotrophic mycorrhizal fungi on the resistance of pine roots to pathogenic infections II. Production, identification, and biological activity of antibiotics produced by *Leucopaxillus cerealis* var. *piceina*. *Phytopathology*, **59**, 411–417.

**Marx, D.H. (1973).** Mycorrhizae and feeder root diseases. In *Ectomycorrhizae: Their Ecology and Physiology* (Ed. by G.C. Marks and T.T. Kozlowski), pp. 351–382. Academic Press, New York.

**McMichael, B.L. & Taylor, H.M. (1987).** Applications and limitations of rhizotrons and minirhizotrons. In *Mini-rhizotron Observation Tubes: Methods and Applications for Measuring Rhizosphere Dynamics* (Ed. by H.M. Taylor), pp. 1–13. Special Publication No. 50. ASA, Madison.

**Newman, E.I. (1985).** The rhizosphere: carbon sources and microbial populations. In *Ecological Interactions in Soil* (Ed. by A.H. Fitter, D. Atkinson, D.J. Read & M.B. Usher), *Special Publication Series of the British Ecological Society No. 4*. Blackwell Scientific Publications, Oxford.

**Pastor, J. & Bockheim, J.G. (1981).** Biomass and production of an aspen-mixed hardwood-spodosol ecosystem in northern Wisconsin. *Canadian Journal of Forest Research*, **11**, 132–138.

**Persson, H. (1978).** Root dynamics in a young Scots pine stand in Central Sweden. *Oikos*, **30**, 508–519.

**Persson, H. (1983).** The distribution and productivity of fine roots in boreal forests. *Plant and Soil*, **71**, 87–101.

**Read, D.J., Francis, R. & Findlay, R.D. (1985).** Mycorrhizal mycelia and nutrient cycling in plant communities. In *Ecological Interactions in Soil* (Ed. by A.H. Fitter, D. Atkinson, D.J. Read and M.B. Usher), pp. 193–217. *Special Publication Series of the British Ecological Society No. 4*. Blackwell Scientific Publications, Oxford.

**Rhodes, L.H. & Gerdemann, J.W. (1975).** Phosphate uptake zones of mycorrhizal and non-mycorrhizal onions. *New Phytologist*, **75**, 555–561.

**Roberts, J. (1976).** A study of root distribution and growth in a *Pinus sylvestris* L. (Scots pine) plantation in East Anglia. *Plant and Soil*, **44**, 607–621.

**Rovira, A.D. & Davey, C.B. (1974).** Biology of the rhizosphere. In *The Plant Root and Its Environment* (Ed. by E.W. Carson), pp. 155–204. University Press of Virginia, Charlottesville.

**Russell, R.S. (1977).** *Plant Root Systems: Their Function and Interaction with the Soil*. McGraw Hill, London, 298 pp.

**Sanders, F.E. & Tinker, P.B. (1973).** Phosphate flow into mycorrhizal roots. *Pesticide Science*, **4**, 384–395.

**Santantonio, D. (1979).** Seasonal dynamics of fine roots in mature stands of Douglas-fir of different water regimes: A preliminary report. In *Root Physiology and Symbiosis* (Ed. by A. Riedacker & J. Gagnaire-Michard), Vol. 6. *Proceedings of IUFRO Symposium on Root Physiology and Symbiosis, Nancy, France, Sept. 1978*. CNRF, Champenoux.

**Santantonio, D. and Hermann, R.K. (1985).** Standing crop production, and turnover of fine roots on dry, moderate, and wet sites of mature Douglas-fir in western Oregon. *Annales des Sciences Forestieres*, **42**, 113–142.

**Santantonio, D., Hermann, R.K. & Overton, W.S. (1977).** Root biomass studies in forest ecosystems. *Pedobiologia*, **17**, 1–31.

**Smith, W.H. (1977).** Tree root exudates and the forest soil ecosystem: exudate chemistry, biological significance and alteration by stress. In *The Belowground Ecosystem: A Study in Plant-Associated Processes* (Ed. by J.K. Marshall), pp. 289–302. *Range Science Department Science Series No. 26.* Colorado State University, Fort Collins.

**Szaniszlo, P.J., Powell, P.E., Reid, C.P.P. and Cline, G.R. (1981).** Production of hydroxamate siderophore iron chelators by ectomycorrhizal fungi. *Mycologia*, **73**, 1158–1174.

**Tisdall, J.M. and Oades, J.M. (1979).** Stabilization of soil aggregates by the root systems of ryegrass. *Australian Journal of Soil Research*, **17**, 429–441.

**Trappe, J.M. and Fogel, R. (1977).** Ecosystematic functions of mycorrhizae. In *The Belowground Ecosystem: A Synthesis of Plant-associated Processes* (Ed. by J.K. Marshall), pp. 205–214. *Range Science Department Science Series No. 26.* Colorado State University, Fort Collins.

**Trofymow, J.A., Coleman, D.C. & Cambardella, C. (1987).** Rates of rhizodeposition and ammonium depletion in the rhizosphere of axenic oat roots. *Plant and Soil*, **97**, 333–344.

**Vogt, K.A., Edmonds, R.L., Grier, C.C. & Piper, S.R. (1980).** Seasonal changes in mycorrhizal and fibrous-textured root biomass in 23- and 180-year-old Pacific silver fir stands in western Washington. *Canadian Journal of Forest Research*, **10**, 523–529.

**Vogt, K.A., Edmonds, R.L. and Grier, C.C. (1981).** Seasonal changes in biomass and vertical distribution of mycorrhizal and fibrous-textured conifer fine roots in 23- and 180-year-old subalpine *Abies amabilis* stands. *Canadian Journal of Forest Research*, **5**, 681–690.

**Vogt, K.A., Moore, E.E., Vogt, D.J., Redlin, M.J. & Edmonds, R.L. (1983).** Conifer fine root and mycorrhizal root biomass within the forest floors of Douglas-fir stands of different ages and site productivities. *Canadian Journal of Forest Research*, **13**, 429–437.

**Walter, D.E., Hudgens, R.A. & Freckman, D.W. (1986).** Consumption of nematodes by fungivorous mites, *Tyrophagus* spp. (Acarina: Astigmata: Acaridae). *Oecologia*, **70**, 357–361.

**Whittaker, R.H. & Marks, P.L. (1975).** Methods of assessing terrestrial productivity. *Primary Productivity of the Biosphere* (Ed. by H. Leth & R.H. Whittaker), pp. 55–118. *Ecological Studies, No. 14.*

**Whittingham, J. & Read, D.J. (1982).** Vesicular–arbuscular mycorrhizas in natural vegetation systems, III. Nutrient transfer between plants with mycorrhizal interconnections. *New Phytologist*, **90**, 277–284.

**Wiggins, E.A., Curl, E.A. and Harper, J.D. (1979).** Effects of soil fertility and cotton rhizosphere on populations of collembola. *Pedobiologia*, **19**, 75–82.

# Roots and resource fluxes in plants and communities

D. ROBINSON
*Soil-Plant Dynamics Group,*
*Cellular and Environmental Physiology Department,*
*Scottish Crop Research Institute, Invergowrie, Dundee DD2 5DA, UK*

## SUMMARY

1 The effectiveness with which resources (nutrients and water) are captured by the roots of plants can have major effects on the plants' capability for spreading their genes in a population.
2 Progress in understanding the ecological importance of resource acquisition by roots is unlikely to come from technological advances alone. New ideas and imaginative approaches are needed as well.
3 Some of these approaches might include: cost-benefit analyses of alternative phenotypic strategies; techniques for exploring the consequences of non-uniformities in the operation of root systems; and estimations of the effects on gene reproduction of the transport of resources between the root systems of closely related and distantly related plants within communities.

## INTRODUCTION

*It is, indeed, not a grand ecological revelation that the scientist should seek from an awareness of the evolutionary process, but rather an enlargement of the understanding made possible by a new or wider angle of vision, a clue here and apt analogy there, and a general sense of evolutionary depth in contexts in which it might otherwise be lacking.* (Paraphrased from Medawar 1982).

This chapter is a personal and selective view of how roots and root systems acquire resources for plants growing in a community. For a comprehensive review of what the roots of different plants do in contrasting environments, see the other contributions to this volume. The main purpose of this chapter is to emphasize that new ideas are needed as urgently as novel technology in the study of root systems. Ultimately, any new ideas or approaches that are applied in this field must, as Hardwick (1986) said, 'make evolutionary sense' if they are to be of any lasting value to plant ecology. This means genes.

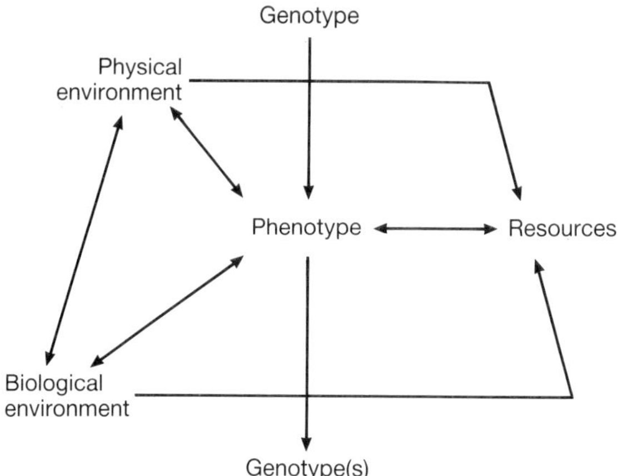


FIG. 1. A representation of the relations between the environment, plant phenotypes and their reproduction. Arrows indicate flows of information and material. The vertical arrows represent the 'evolutionary plane' which passes through the 'ecological plane' of the diagram.

## GENES, PHENOTYPES AND RESOURCES

### *Designs for survival*

'The phenotype is essentially a gene's way of making more genes' (Calow & Townsend 1981). This aphorism is illustrated in Fig. 1. Events in the 'ecological plane' of Fig. 1 cannot be understood fully without reference to its 'evolutionary plane': genetic reproduction by phenotypes and the subsequent spread (or not, as the case may be) of those genes through a population. Phenotypes operate by using resources to make genes. The effectiveness with which an individual acquires and uses resources can have a big impact on the extent to which its genes become represented in the population of which it is a member. In other words, alternative phenotypic patterns of the acquisition and use of resources can confer different selective advantages on an individual's genes (see Robinson 1991, for example). If we look at the diversity of phenotypic strategies by which genes are reproduced from their raw materials, we get to what Williams (1966) called 'the central biological problem . . . design for survival'. This problem can be summarized by the following questions. Why does one set of phenotypic mechanisms allow greater reproduction in a certain environment compared with alternative sets? Why is there a certain diversity of mechanisms in one environment and a different one elsewhere? And, in the present context, what do roots contribute to reproduction?

In most free-living vascular plants, the problem of acquiring resources has been solved by the evolution of leaves and roots (Fig. 2). There are some oddities that acquire carbon through their roots: the amphibian plant *Stylites andicola* is one (Keeley, Osmond & Raven 1984). Leaves often absorb significant amounts of nitrogen as ammonia or $NO_x$ and sulphur as $SO_2$. Nevertheless, it is generally

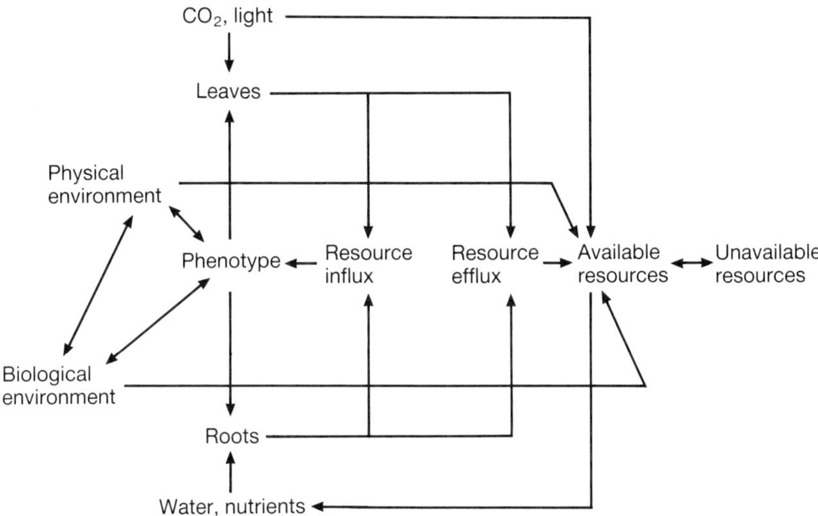


Fig. 2. An expansion of the 'ecological plane' of Fig. 1, illustrating the fluxes of resources between plant phenotypes and their environment.

true that leaves evolved to absorb $CO_2$ and to intercept solar radiation, and roots evolved to absorb nutrients and water *and* to anchor the plant to its substrate. Roots are natural selection's 'design solution' (not by any means a *perfect* solution) to the problem of obtaining these resources from a heterogeneous, porous, semi-compressible medium containing solid, liquid and gaseous phases. The possession of roots (or root-like organs) is one of the characters on which a fundamental classification of phototrophs can be based. In this classification, to the chagrin of terrestrial biologists, all non-aquatic vascular plants are merely a subdivision of the 'rhizophytes' (Raven 1986).

### *Resources: money in the lives of all roots*

According to Tilman (1982) a resource is 'any substance or factor which can lead to increased growth rates as its availability in the environment is increased, and which is consumed by an organism'. This definition includes, for autotrophic plants, the familiar 'substances' water, carbon, nutrients, oxygen and photons, and 'factors' like space and temperature. I would also include the living components of a plant's environment such as potential mates, symbionts, hosts (in the case of parasitic plants), and agents of pollination or dispersal. While these are hardly 'consumed', they certainly have the potential to influence the growth rate of a plant or the likelihood of it reproducing its genes effectively.

Roots obtain nutrients and water from the soil, but they consume nutrients, water, carbon and oxygen in the process. Significant amounts of resources, particularly nutrients and carbon can also be lost from the plant, deliberately or accidentally, via its root system (Fig. 2). Because resources are ultimately involved in genetic reproduction, the contribution made towards this by a particular resource and by the phenotypic mechanism(s) responsible for its acquisition can be quantified, at least notionally (Maynard Smith 1982). In this sense, resources such as nutrients and water play the same role in plant communities as money does in ours. And, as with money, it is necessary to know how resources are supplied to, acquired by, and transferred between the members of a community if its behaviour is to be fully understood (Harper 1977; Bloom, Chapin & Mooney 1985).

## RESOURCES, ROOTS AND DEPLETION ZONES

### *Fluxes from picosystem to ecosystem*

Fluxes of solutes into (and out of) roots can be studied from the molecular level (Fig. 3a) to that of the plant community (Fig. 3d). All genetic differences between plants' capacities to transport resources must have molecular bases. But, as Clarkson (1985) has said, 'more is involved in getting nutrients into plants than ion transport across cell membranes'. Genetic superiority in nutrient acquisition may have more to do with the developmental control of root growth and of demographic processes within root systems than with the presence of greater numbers of ion 'carrier' molecules in plasma membranes. Most of this chapter will concentrate on the fluxes of nutrients into single roots, into root systems and between individual plants (Figs 3c and d).

First, however, it is worth pointing out that there are many ecological mysteries that may be explicable in molecular terms, should plant molecular biologists ever tire of *Arabidopsis thaliana*. For example, there is good evidence (Gigon & Rorison 1972) that some plants (e.g. *Deschampsia flexuosa*) indigenous to acidic, ammonium-rich soils can absorb ammonium effectively over a wide range of pH. Nitrate, however, can be taken up only at an acidic pH. Conversely, other plants (e.g. *Scabiosa columbaria*) are restricted in the British Isles to neutral or calcareous soils in which nitrate is usually the predominant form of available nitrogen. *S. columbaria* can take up nitrate over a wide range of pH, but ammonium only near neutrality. No one knows the basis of these differences. One possible, speculative explanation might be the evolution of genotypes having different capacities to transport protons across their cell membranes. Proton fluxes drive the uptake of anions and cations from the soil solution (Fig. 3a). The potential connection between events at the molecular level — the operation of proton pumps against pH gradients obtaining in acidic or neutral soils that allow nitrate or ammonium ions to be absorbed by roots — and the clear differences in response to pH and nitrogen source by whole plants is obvious, but it remains unexplored.

### *Real and imaginary roots*

The production of leaves and roots is the botanical analogue of the running around that animals do when they forage for food. The appeal of this analogy is growing in

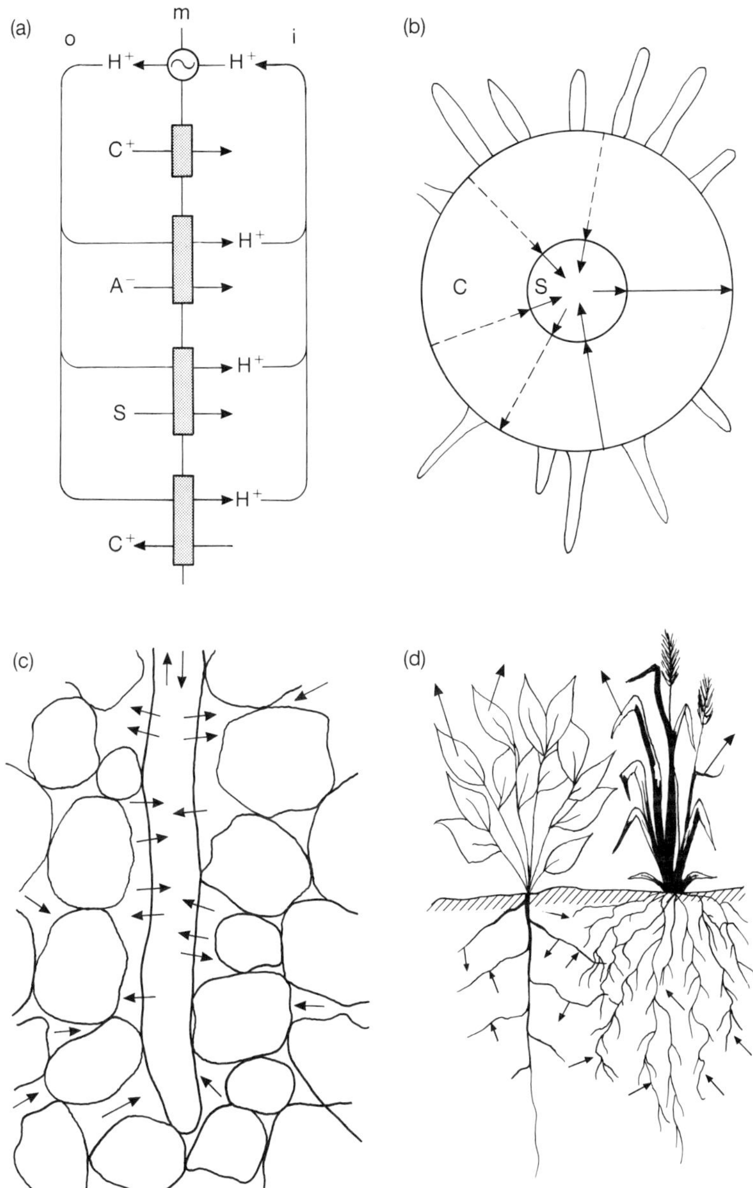


Fig. 3. Hierarchies of nutrient and water flux in roots and root systems. (a) Transmembrane fluxes of cations ($C^+$), anions ($A^-$) and uncharged solutes (S) and their relation to active proton efflux (modified from Smith & Smith 1986); m, membrane; o, outside; i, inside. (b) Symplastic (solid arrows) and apoplastic (broken arrows) fluxes within and between cells in a transverse section of a root cortex (C) and stele (S). (c) Fluxes in the soil solution to and from a single root. (d) Fluxes to and from root systems, and within plants growing in a community.

plant ecology, but so far only at an experimental level. This contrasts with the deluge of theoretical work that burdens the zoological literature (see Stephens & Krebs 1986). A big problem with the development of theories of foraging by plants is how to make mathematical sense of what is happening in an entire leaf canopy or a root system during the course of their active lives. Root systems, in particular, have got to be simplified conceptually. Little can be gained scientifically by reconstructing in exact detail, at enormous cost of time and effort, the three-dimensional dynamic mess that is a real root system. The same point has been made by Stephens & Krebs (1986) in defence of simple modelling approaches to the study of the complex foraging behaviour of animals.

Fortunately, there is a large body of work devoted to the simplification of plant roots and how they take up resources from the soil. This is based on theoretical thermodynamic models of solute and water transport in a homogeneous, geometrically simple, thermodynamically ideal system. Even though the soil is the last medium one would associate with these properties, assuming it does behave in this unrealistic way has advanced enormously our understanding of how roots work (see Nye & Tinker 1977).

## *Depletion in the root zone*

When a root absorbs water or nutrients from the soil, ions and molecules move from the soil towards the root. This movement occurs down gradients of free energy potential created by differentials in ionic concentration, hydrostatic pressure or matrix potential between the surface of the root and the bulk soil. If water is absorbed relatively faster than ions dissolved in it, the latter will accumulate at the root surface and tend to diffuse away from the root. The more usual case, though, is for ion uptake to proceed relatively faster than that of water. The ionic concentration at the root surface then falls progressively. Ions will diffuse from the soil towards the root. Thus a zone is created around an absorbing root in which ions are depleted in concentration relative to their concentration in the bulk soil. These depletion zones are the equivalent of shadows — zones of photon depletion — cast by leaves.

Depletion zones around roots can be visualized by autoradiography following the uptake by a root of a radio-isotopically labelled nutrient (e.g. Nye & Tinker 1977). Unfortunately, the autoradiographs that have been published give a poor impression of the *dynamic* nature of nutrient depletion. For this, we must turn to theoretical relations between concentration *versus* radial distance from a cylindrical root (Nye & Tinker 1977). Confusion can arise from these models however. The effect of root growth on the development of depletion zones is usually ignored. Lacking from the literature — Passioura (1963) being an exception — are simple pictures encapsulating the interactions between nutrient supply, uptake, and root growth that lead to the formation of depletion zones around roots. I have attempted to produce such pictures in Figs 4, 5 and 6.

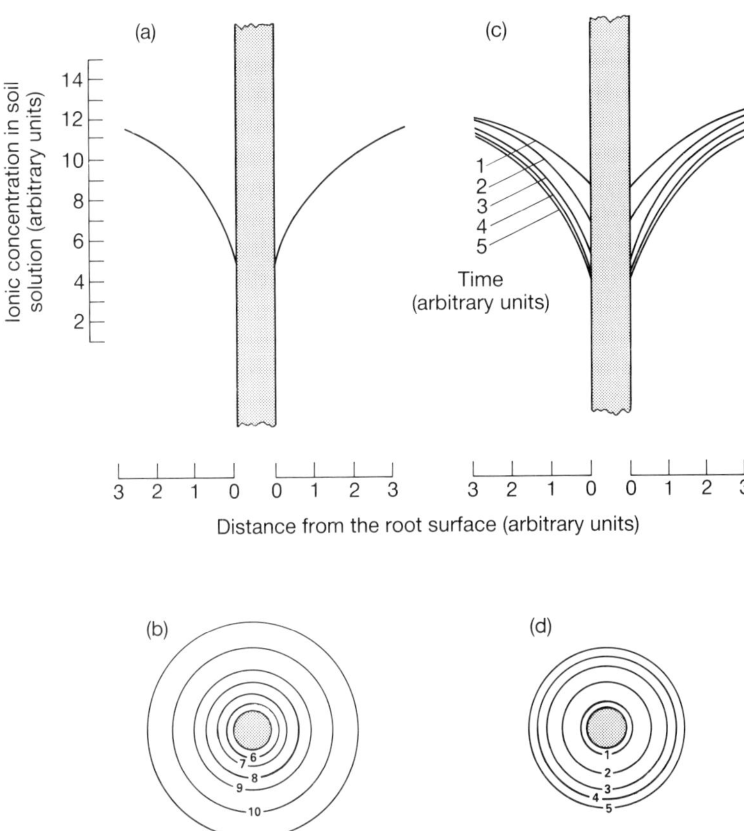


FIG. 4. Idealized ionic concentration profiles around a single cylindrical root in relation to distance from the root surface. The hatched areas indicate the position of the root. (a) At one position along the root, for all times. (b) A 'transverse section' across (a), showing concentration contours around the root. (c) The development of the concentration profile at five successive times. (d) A 'transverse section' across (c) showing the position reached of a single concentration (equal to 9 on the vertical scale) at five successive times.

## *A brief natural history of depletion zones*

Figure 4a shows an idealized concentration profile around a cylindrical root. Initially, concentration rises sharply with distance from the root surface, but then increases more slowly. If the root is assumed to grow in an infinite volume of soil so that its activity is not affected by neighbouring roots, the depletion zone eventually disappears at large distances from the root. Concentration then remains independent of distance. This non-linearity reflects the cylindrical geometry of the system, and the differential rates of ion movement into the depletion zone, and out of the depletion zone into the root.

Plotting the same relation between concentration and distance as a 'transverse section' across the root and soil emphasizes that the concentration gradient becomes increasingly steeper close to the root (Fig. 4b). For many purposes it is convenient to consider a 'steady state' in which the outward spread of the notional outer boundary of the depletion zone has effectively stopped. At this boundary, the rate of ion movement towards the root is balanced exactly by desorption or mineralization of ions, thereby buffering the concentration in the soil solution.

Concentration gradients around a root develop as shown in Fig. 4c (see also Nye & Tinker 1977). The distance from the root at which a certain concentration is reached varies with the square root of time (see Vogel 1988), giving the 'transverse section' shown in Fig. 4d. The concentration at the root surface falls rapidly at first, then more slowly, and eventually hardly at all as the rate of *net* ion uptake (gross uptake minus efflux) is balanced by the rate of ion movement to the root surface.

A point at a certain distance behind the tip of a growing root can be equated with the time that has elapsed since the cells at that point were formed, that is, with the *age* of those cells. Suppose that the concentration profile shown in Figs 4a and b is applicable to that point along a growing root that is 3 days old. If the root extends at a constant rate of 20 mm per day, say, the profile in Figs 4a and b is formed continually at 60 mm behind the root's tip. So, this profile 'travels' with that point on the root through the soil. Figures 4a and b apply to a *single* age of root tissue, but for *all times* after the start of root growth and nutrient uptake.

An observer located always 60 mm behind the root's tip would see only that depletion zone described by the profile in Figs 4a and b no matter how much time had passed. An observer at a fixed point in the soil, however, would see Figs 4a and b for only a moment in time, when the point 60 mm behind the root's tip passed by. Over a period of time, the soil-bound observer would see a progressive reduction in concentration as in Figs 4c and d. Relativity is, it seems, as relevant at the root level as at the cosmological level: from worm holes to black holes, perhaps!

Combining the various facets of Fig. 4 shows that the development of a depletion zone depends on the position along (or age of) a root and the position in the soil (Fig. 5). In Fig. 5, the single concentration that is plotted at four different times is that which occurs at a distance from the root surface of $x = 2(Dt)^{1/2}$ (see Nobel 1974; Nye & Tinker 1977). $D$ is the diffusion coefficient ($m^2\ s^{-1}$) of the ion in question (see Vogel 1988) and $t$, time (s). This concentration is usually taken as the notional outer boundary of the depletion zone of a root growing in isolation.

By performing a mathematical rotation around the root axis of the depletion zones shown in Fig. 5, the three-dimensional 'envelope' of nutrient depletion around the root can be shown (Fig. 6a). A clear implication of the progressive expansion of the depletion zone is that concentration gradients between soil and root will probably be steeper close to the root's tip than they are further back. This means that *potential* rates of nutrient uptake per unit length of root ('inflow rates') will be faster towards the root's tip. Variations in uptake rates along a root have previously been interpreted purely in physiological terms (see Nye & Tinker 1977), but they may also be a consequence of simple physical processes.

Real roots, unlike the imaginary ones beloved of theorists, usually branch. An emerging lateral root will initially be within the depletion zone of its parent (Fig.

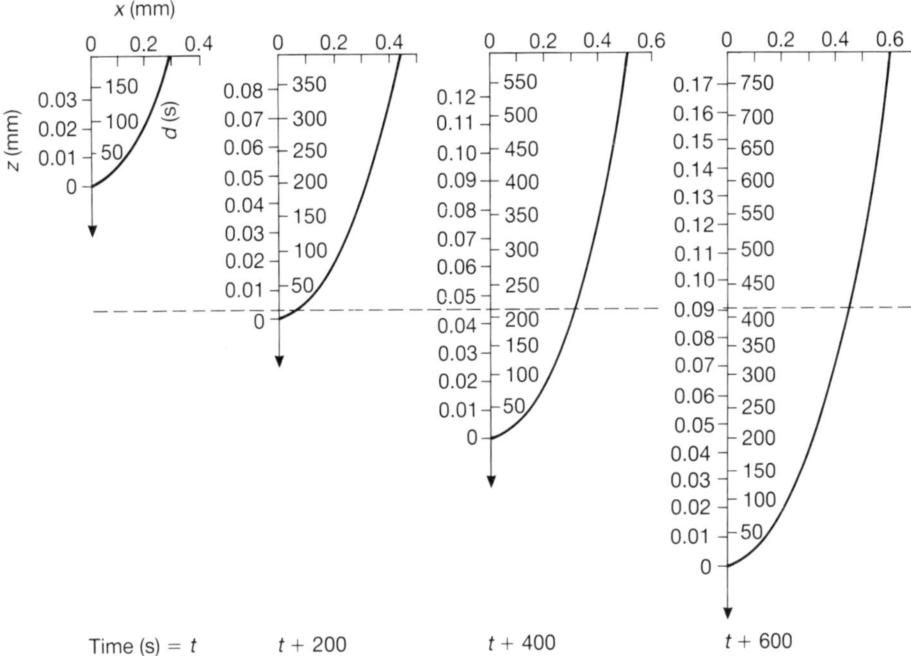


FIG. 5. The development of a depletion zone around a growing root. The curve is the position of that concentration given by $x = 2(Dt)^{1/2}$, where $x$ is the radial distance from the root surface. $D$ is the ionic diffusion coefficient in the soil, and $t$ is the time after the start of uptake. The scales on the vertical axes are distance behind the root's tip ($z$) and age of the root ($d$), equated by assuming a rate of root extension of 20 mm day$^{-1}$. The depletion zone is shown at four times, each 200 s apart, as the root grows vertically downwards (arrows). The broken line is an arbitrarily chosen fixed point in the soil.

6b). Note that a depletion zone is not necessarily a *depleted* zone. Mineralization, desorption and movement of ions in the soil solution ensure that a root does not deplete the available resources in the adjacent soil totally. So, an emerging lateral will have access to resources that it can take up from within its parent's 'shadow'. But the concentration gradients around the lateral will be much shallower than those around its parent. The potential rates of nutrient inflow to a lateral within a pre-existing depletion zone are likely, therefore, to be slower than those of an otherwise similar root growing into virgin soil. For such a lateral to make a significant contribution to the plant's acquisition of resources it must at some point 'burst through' the envelope of the depletion zone around its parent as illustrated in Fig. 6b. For this to happen, the rate of lateral elongation must be greater than the rate of radial spread of the depletion zone's outer boundary. Happily, as the latter varies as the square root of time (see above) but the former is (more or less) constant, lateral roots are able eventually to out-run the receding boundary of the depletion zones of even mobile resources such as water and nitrate, as a few back-of-the-envelope calculations show, (unlike, as Vogel (1988) mentions, the bacterium that cannot swim faster than the rate at which the depletion zone around itself

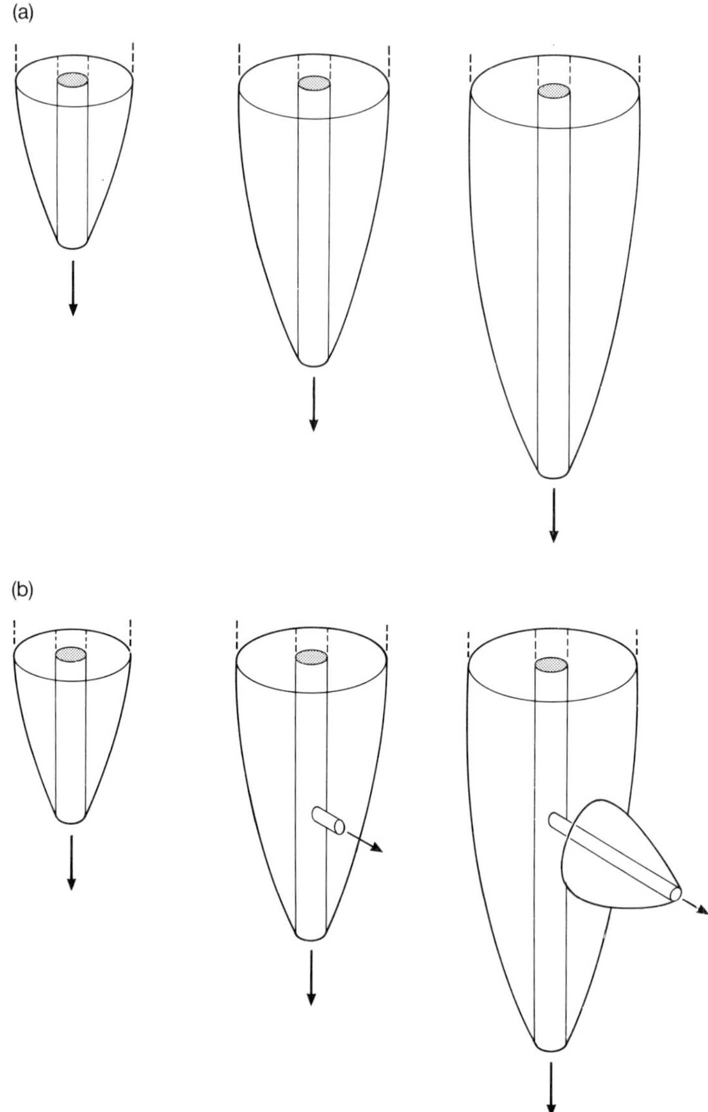


FIG. 6. Three-dimensional representations of the development of a depletion zone around (a) a single unbranched root and (b) a branching root, at three successive times. Roots grow in the directions indicated by the arrows.

spreads, so that food diffuses towards the bacterium faster than it can swim towards the food!)

Naturally, a root will age to a point at which it begins to senesce. Its capacity to take up nutrients and water declines through physiological decrepitude or because of physical disruption of the transport pathways between soil and root. Both of these can be a consequence of cortical senescence, particularly common in cereals and other grasses (see Robinson 1990). Aged roots are physiologically less capable of transporting ions and water across their cortices and into their xylem than their younger counterparts. This has been examined extensively in solution culture experiments. In soil, senescent cortices may collapse, reducing the hydraulic continuity between soil and root, and impairing uptake. The effects of these processes on the depletion of resources in soil have not been examined in any detail. One can imagine, however, that around an old root a depletion zone would progressively disappear, as supply exceeded demand at the root surface.

The picture which emerges from this and the previous discussion is one of a root system acquiring resources from 'hot spots' determined by the age of the root at any point in the system, by the local conditions in the soil governing the availability of resources, and by the rates at which these change in relation to one another. This has been visualized for water uptake by Caldwell & Richards (1986). It is a picture very different from that which assumes uniformity of supply and demand throughout a root system and in the soil in which it grows.

### *Competition in the depletion zone*

It is in depletion zones that competition happens in its most literal sense. Competition between roots for nutrients and water happens when the respective depletion zones around adjacent roots overlap (Nye & Tinker 1977). The above discussion can be extended to show that along two parallel, adjacent roots the intensity of competition will not be constant. The depletion zones will overlap to differing extents as they are not cylindrical, but are as shown in Fig. 6 and in Caldwell & Richards (1986).

If the selective influence exerted by competition for resources leads to the evolution in plants of certain phenotypic strategies, the depletion zones are the battlegrounds on which some of those strategies are put to the test. If competition for nutrients and water can shape the structure and dynamics of plant communities, it is clearly important to look at mechanisms by which the depletion zones created by roots come to interact with each other. The obvious practical difficulties of doing this mean that most experimental studies of competition examine it indirectly. They involve measurements of the short-term effects of competition on the growth of the *plants* that are thought to be in competition with one another, and on their acquisition of the resource for which they are supposed to be competing (Caldwell *et al.* 1987; Eissenstat & Caldwell 1988, 1989; Eissenstat & Newman 1990, for example). Otherwise, long-term effects are measured: the progressive suppression and displacement of one plant *species* by another (Tilman 1982). Few experiments have assessed explicitly whether roots were packed closely enough for their depletion zones to overlap and for competition — rather than for the *effects* of competition — to occur (see Newman & Andrews 1973). Theoretical aspects of competition within root systems have been covered superbly by Baldwin (1975) and Nye & Tinker (1977). The theory of competition between the root systems of neighbouring plants is less well developed: Baldwin (1976) and Nye & Tinker (1977) provide the best coverage to date.

As soon as competition begins, either within or between root systems, the scientific emphasis shifts from the study of individuals (roots) to that of populations (root systems).

## *Crowds of individuals*

Information about isolated roots can be scaled up to account for their behaviour as part of a root system. The ways in which this is done are related to the independence with which each element of a root system interacts with the immediate environment. The problem is analogous to that of relating the control that stomata exert over gas exchange from a leaf when that leaf is isolated from its neighbours, and when it forms part of a large, dense canopy: 'the individual *versus* the crowd' (Jarvis & McNaughton 1986).

Potentially, an isolated, intact root can acquire resources faster than an identical root that is competing with its neighbours (Nye & Tinker 1977; Robinson 1986; see above). The distinction between the operation of a single root, or of a defined part of a root, and that of a whole root system, or of a collection of roots, is an important one. It determines the uses to which information relevant to one scale of operation can legitimately apply to another.

## *Nutrient-inflow rates in roots and root systems*

This heading was (almost) the title of a previous paper (Robinson 1986) attempting to distinguish between the measurements made of nutrient-inflow rates on isolated, defined parts of root systems and those obtained as 'averages' for nutrient uptake by whole root systems. The latter are the easier measurements to make: estimate total root length and total quantity of a nutrient in the whole plant, at several harvests over a sensible time course; derive the *apparent* inflow rate of the nutrient as (rate of change of nutrient)/(root length content), using linear interpolation between harvests, or by curve-fitting routines. To obtain the former type of inflow rate, more subtle techniques are needed. First, isolate your root, but in such a way that it remains attached to the plant, and so that the amount of nutrient taken up *from soil* by a defined part of that root can be measured over a period of time (see, for example Clarke & Barley 1968; Drew & Nye 1969). The inflow rates derived from such measurements are *actual* inflow rates. These reflect what actually happened in terms of fluxes through the soil and of the physiological activity of a known part of the root system: 12 femtomoles of potassium really *did* pass into each millimetre of root every second!

Ideally, an actual inflow rate should equal the corresponding apparent one. If this was so, then apparent inflow rates, which are far less trouble to measure, would be true reflections of real events. But of course, this is not what happens. Apparent inflow rates no more reflect real events than do measurements of the mean number of scientific papers published annually per head of population in any research institute.

There is no problem in using either apparent or actual rates in the correct way. Apparent rates are useful comparators that relate net performance to a unit of resource invested in the production of that performance. This is a common approach in comparative plant ecology (e.g. Robinson & Rorison 1983; Poorter 1989). In the case of scientific papers, an *apparent* rate of production can be, and is, used to compare the effectiveness of different research institutes without, of course, pretending to identify whether one scientist writes twenty papers annually or whether twenty scientists write one apiece.

The analogy between roots and scientists breaks down because it is possible to identify directly which scientist amongst many writes how many papers. It is not possible to identify directly which part of a root system acquires how much of a resource unless, as described above, special techniques are used to isolate a small part of a root system from the rest.

Problems arise when apparent inflow rates are used as if they were actual ones. A case in point is their use to predict the minimum concentration differential between the root surface and the bulk soil required to drive the apparent inflow rate (Barraclough 1986, 1989). The problem with this approach is that the predicted concentration differentials are, for nitrate at least, far lower than those which have been found to limit *actual* inflow rates and rates of nitrogen uptake by whole plants. Basing apparent inflow rates on the total length of root is the probable reason for this discrepancy. As described above, actual inflow rates are very much dependent on the position in a root system, and not all of the root system will be equally active (or active at all) in nutrient uptake. If the actual inflow rate was known at every point in the root system, there would be no problem, but these rates cannot be measured directly (see above).

Robinson, Linehan & Caul (1991) have recently suggested a way around this problem. Actual inflow rates are used, rather than apparent ones which are best ignored entirely in such applications. If an actual inflow rate can be calculated (according to equations given in Robinson 1986, for example) from a knowledge of the ionic concentration in the bulk soil solution, it can be multiplied by the total root length to obtain the total net uptake of nutrient by the plant. This uptake is then compared with that actually measured experimentally, one of the few measurements involving nutrient uptake that can be made with much confidence and little tedium.

In plants other than very young ones, the predicted uptake is greater than that which was measured. The latter is the correct value, of course! The former is adjusted by, in our example, reducing the length of the root system by a small fraction. The calculations are repeated, with successive reductions in root length until equality between measured and predicted uptake is reached. At that point, the length of root used in the calculation may be equivalent to an 'effective' root length. 'Effective' is used in the sense that that amount of root can be 'held responsible' for the observed quantity of nutrient measured in the plant. We found that no more than 10% of the root system of spring wheat could be held responsible for the plants' acquisition of nitrate from the soil. We were unable to identify *which* 10%. Passioura (1980) used a similar scheme to estimate the fraction of a wheat root system effective in water uptake: 30%.

Increasingly it is becoming recognized that temporal and spatial nonuniformities are as fundamental to the operation of root systems as they are in real populations. Indeed, a 'root system is also in a real sense a population, a constantly changing number of foraging units' (Harper 1977), and it is on this basis that the functioning and ecological significance of root systems can be best understood.

## DEPLETION ZONES AND LOST RESOURCES

### *Strategic fluxes*

In this section the relation between the process of nutrient acquisition by roots to the loss of other resources from roots is considered. If the rate of ion transport from the soil to the root surface imposes a significant limitation on the acquisition of a nutrient, what strategies can a plant adopt to overcome it?

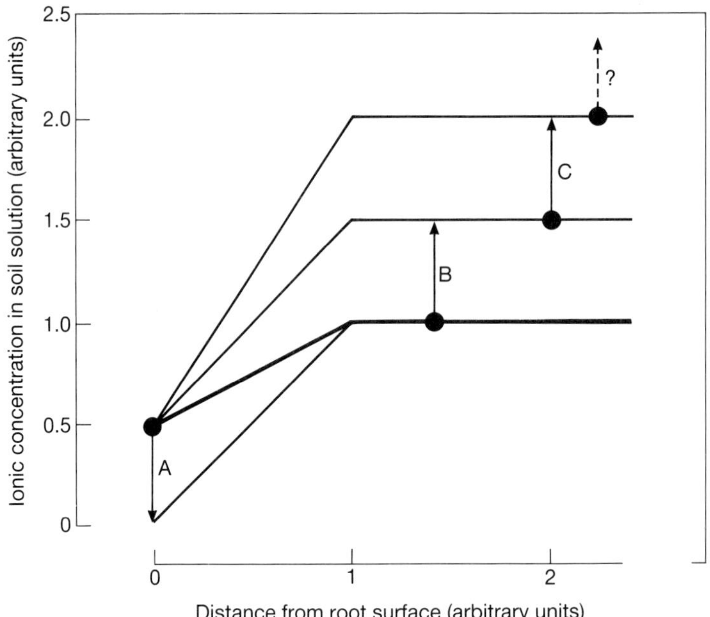


FIG. 7. Idealized linear concentration profiles around a root, at a steady state. The concentration in the soil solution at the root surface can be depleted to 50% of that in the bulk soil (bold line), and the depletion zone extends to unit distance from the root's surface. Possible strategies to improve the resulting diffusive flux into the root are indicated by arrows. A can deplete the concentration at the root surface by a further 0·5 units. B and C can increase the concentration in the bulk soil by 0·5 units and 1·0 units, respectively.

Suppose a population of plants is growing in an environment in which this nutrient was maintained at unit concentration in the soil solution (Fig. 7). Each root of the plants in this population can deplete the ionic concentration to the same degree, to a maximum of 50% of its concentration in the bulk soil, for example. Suppose also that a steady state prevailed, the width of the resulting depletion zone being maintained at unit distance from the root's surface. For convenience, assume that concentration changes linearly with distance from the root (as opposed to the non-linear reality of Fig. 4). If the ion moved towards the root by diffusion according to Fick's First Law (Nye & Tinker 1977; Vogel 1988), the net flux of an ion would be the concentration gradient $(1 - 0{\cdot}5)/1$ multiplied by its diffusion coefficient. To give the fluxes some semblance of realism, assume that $D = 10^{-10}$ arbitrary units. The maximum net flux into each root is therefore $5 \times 10^{-11}$ (in units of amount of substance per unit area of root per unit time).

Now, suppose that in this environment there was some selective advantage to be gained from the attainment of greater fluxes than this (cf. Jackson, Manwaring & Caldwell 1990). If the diffusion coefficient is not amenable to manipulation by the plant, the potential flux could be increased only by an increase in the concentration gradient between soil and root.

Mutant A arises in the population. The roots of plants of this genotype happen to have a capacity to reduce the ionic concentration at the root surface close to zero. They are, in effect 'infinite sinks' for the ion (Robinson 1986). No other mutant in the population could evolve greater powers of nutrient depletion. If the ionic concentration in the bulk soil solution is maintained at unity, mutant A would be capable of attaining a maximum flux of $10^{-10} \times (1 - 0)/1 = 10^{-10}$, a doubling of that attainable by its immediate ancestors.

Exactly the same improvement could be made by another mutant genotype, B. Such plants have the same powers of depletion enjoyed by their predecessors, but they can increase the ionic concentration gradient by increasing the concentration of diffusible ions in the bulk soil by 0·5 units (Fig. 7).

In evolutionary terms, it is obvious that the strategy of A has reached the end of the line. None of the descendants of A could do any better than the parental genotype. B, on the other hand, has not reached such an impasse: it is, potentially an open-ended strategy. If there is an evolutionary advantage to be gained by further increases in potential flux, B could give rise, in theory, to another genotype, e.g. C, capable of increasing the ionic concentration in the bulk soil to 2·0 (Fig. 7). The maximum potential flux in the roots of C would be an unprecedented $1{\cdot}5 \times 10^{-10}$.

The trend for the evolution of genotypes with even greater capacities for increasing local ionic concentrations presumably could continue until the costs of doing so become prohibitive, or until some other constraint was imposed. This type of strategy could potentially continue *ad infinitum*. Care must be taken here not to endow natural selection with the power of foresight (Williams 1966). Selection operates in response to immediate imperatives, not to long term ones. There may certainly be short-term advantages in the evolution of an A-type strategy.

Most plants do seem to have some physiological flexibility in the extent to which they can deplete ions at the surfaces of their roots. This can be seen as an *ad hoc* phenotypic response to compensate temporarily for transient deficiencies in nutrient availability (Robinson 1989; Jackson, Manwaring & Caldwell 1990).

### *Making available the unavailable*

To increase the ionic concentration locally in the soil, a root must change its environment by physical or chemical means. Roots release a rich variety of organic and inorganic material into the soil. Intact cells synthesize and exude specific molecules. $CO_2$ is released in respiration. Bits of root are rubbed off by the abrasive action of soil particles. Material is released into the soil when a root, or when part of it, dies. This is a non-specific type of loss, in contrast to the exudation of particular molecules.

The ultimate fate of all plant material, roots included, is to be eaten (Crawley 1983). If root-derived material is consumed by soil microbes their activities could potentially enhance the local availability of nutrients that have large organic reserves in the soil. This idea has been developed for nitrogen by Clarholm (1985; see also Coûteaux *et al.* 1988), who suggested that bacteria close to the root consume organic matter released from the plant, as well as any substrate present already in the soil. These bacteria are then, in turn, eaten by predatory microbes such as protozoa or nematodes. Any ammonium excreted by the predators is potentially available to be taken up by other nitrogen-limited microbes or by the roots of the plant that released the organic matter in the first place. Calculations (Robinson *et al.* 1989) indicate that for this process to be a realistic way of improving a plant's supply of inorganic nitrogen, the consumption of bacteria by the predator is an essential step; nitrogen mineralization by bacteria alone is not enough to meet a plant's demand for the element. This confirms what Clarholm and others have found in carefully controlled experiments.

However, attractive though this idea is, it conflicts with other evidence which raises serious doubts about its importance under natural conditions. For plants to evolve strategies to exploit this potential source of nitrogen in nitrogen-deficient soils, they must have high turnover rates of carbon in their root systems. This requirement disagrees completely with the conclusions reached by Sibly & Grime (1986). They maintained that carbon turnover in the roots of plants growing on nutrient-deficient soils should be *less* than that under fertile conditions. This view is supported by some very good data from field experiments (e.g. Nadelhoffer, Aber & Melillo 1985). Additionally our calculations suggest that not enough nitrogen could be mineralized by the availability of root-derived carbon to have a large impact on the nitrogen supply of a fast growing plant. Whether it would make a difference to the nitrogen requirements of slow growing species remains to be seen.

Some plants do successfully render unavailable nutrients more available by releasing substances from their roots. The mobilization of phosphorus by certain non-mycorrhizal herbs, e.g. rape (*Brassica napus*) and lupin (*Lupinus albus*), are the best documented species (Grinsted *et al.* 1982; Dinkelater, Römheld & Marschner 1989). These species are able to acidify the soil around their roots. They excrete organic acids (e.g. citrate) thereby shifting the sorption equilibria of phosphate so as to increase the concentration of diffusible ions in the soil solution. The concentration gradient of phosphate between the soil and roots is increased, allowing correspondingly greater amounts of phosphorus to be supplied. In lupins part of this strategy is the production of 'proteoid' roots in phosphorus-deficient soils. These are finely branched, densely clustered lateral roots from which large amounts of exudate can be released into a small volume of soil, and into which mobilized phosphate can be readily absorbed by virtue of their large surface area (Marschner 1986). Marschner (1986) gives other examples of which the chemical complexation of polyvalent metal cations (e.g. iron) as a necessary prelude to their absorption by roots is particularly important. Other, often bizarre, ways in which plants gain access to apparently unavailable nutrients have been described by Lamont (1982).

Missing from all studies of the likely importance of these strategies under natural conditions is a risk assessment. There is always a chance that even if the substances released from the roots of one plant did increase the availability of a nutrient by a significant extent the nutrient would not be acquired by that plant. Instead, its competitors, including soil microbes as well as other plants, could be the beneficiaries of the plant's activity. A high risk of a plant losing out in this way might reduce the probability that a strategy to make unavailable resources more available would become evolutionarily successful compared with one that allowed that plant merely to subsist on meagre resources but which gave no advantage directly to competitors. Studies of this facet of a root system's behaviour in a natural community might profit by approaching it in ways similar to those that zoologists have used to investigate successfully many apparently inscrutable aspects of animal behaviour.

## LOST RESOURCES, RELATIVES AND GENES

### *The selfish bean?*

An evolutionary theorist would view the gain of resources by one individual that resulted from the loss of resources by another as 'altruism' on the part of the latter. At one time zoologists were perplexed by certain patterns of animal behaviour that were apparently altruistic because they flouted a central tenet of Darwinian theory. Individuals should be evolutionarily successful if they act to promote the spread of their own genes rather than those of others. By this token, altruism should not be an evolutionarily stable characteristic. Theory was eventually reconciled with the evidence when the patterns of behaviour were, it was realized, explicable in terms of kin selection, reciprocal altruism and inclusive fitness (see Dawkins 1976, particularly Chapter 10).

Basically, the representation in a population of a gene (or of a 'strategy' in the sense used by Dawkins 1982) for any kind of altruistic behaviour is proportional to the probability that the survival and replication of that gene is improved by its phenotypic expression. This means that the individual towards which the altruism is directed should be a close relative of the one being altruistic. The closer the relative, the greater the probability that one of the altruist's genes is present also in the other individual. Dawkins (1976, 1982) describes with relish the many entertaining exceptions to, and variations on, this theme. But the basic idea that genetic 'selfishness' can paradoxically be expressed as altruism is now familiar almost to the point of cliché when applied to the sociobiology of animal communities.

There is no reason to suppose that a plant's genes are any less selfish than those of animals. But obviously the 'behaviour' of plants must be defined in a different way from that which is applicable to animals. The most general 'currencies' in which the latter is measured are the expenditure of time or energy (Stephens & Krebs 1986). The closest equivalent for plants would be the dynamics of resource acquisition and loss, an idea that can be traced to Mirmirani & Oster (1978). A sociobiology of plant communities could, in principle, be based on the exchange of resources between individuals.

### *Communication and exploitation*

For altruism, expressed in terms of the exchange of resources, to be understood as a manifestation of kin selection, two conditions must be met. First, the individuals concerned should be close relatives. Second, there must be a pathway between the individuals through which the resources can be transferred with a low risk of their being intercepted by an unrelated individual. This could, perhaps, be the botanical equivalent of the requirement that animals can reliably distinguish their relatives from other members of the community, known colourfully as the 'green beard' or 'green armpit' effect (Dawkins 1982).

Clearly the distinction between closely and distantly related individuals is important and one that is not easy to make in the case of plants (see Dawkins 1982; Sackville Hamilton, Schmid & Harper 1986; Tuomi & Vuorisalo 1989). In some species, what may appear superficially to be a number of different plants is in fact a population of shoot modules of the same genetic individual connected by underground rhizomes, stolons or similar structures: e.g. aspen (*Populus tremuloides*), bracken (*Pteridium aquilinum*) and clover (*Trifolium* spp.). Resources can be transferred between modules (see Watson & Casper 1984; Pitelka & Ashmun 1985). This is hardly suprising since (assuming insignificant somatic mutation), the modules are genetically identical; their relatedness is unity (Dawkins 1976; Haig 1987), as it is between monozygotic twins. In evolutionary terms, for one module to supply resources to its twin is no different from supplying them to itself (Williams 1966).

In some species, different parts of the same plant have apparently no direct means of communicating via the mutual exchange of resources (Watson & Casper

1984). Indeed, to avoid instability in the organized growth of a developmentally complex modular organism, the number of connections (e.g. phloem vessels) allowing communication between modules should not exceed some critical fraction of the total number of possible connections (Hardwick 1986). The degree of connection between modules is not always constant, but can be modified by environmental conditions and ontogentically. A part of the plant growing in a resource-deficient part of the environment could benefit by receiving supplies from another part connected to it and which happened to be in a resource-rich patch. Watson & Casper (1984) have reviewed the evidence for carbon transport in such circumstances. Baker & Van Bavel (1986) showed that the roots of Bermuda grass (*Cynodon dactylon*) growing in dry soil could receive water transported from other parts of the same root system growing in moist soil. Caldwell, Richards & Beyschlag (see p. 423) describe a similar phenomenon, 'hydraulic lift' in deep-rooting desert shrubs. Grass roots growing in nitrogen-deficient conditions can receive nitrogen from roots supplied externally with a high concentration of nitrate or ammonium (Robinson & Rorison 1983).

The transfer of resources between genetically identical tissues should be an evolutionarily stable feature of a plant's phenotype. It compensates for patchiness in the availability of resources or for the poor quality of the local environment. Such tissues do not have to be connected physically for this transfer to occur, but the reliability of the transport pathway is greatly reduced if they are separate. In a species that grows as a densely packed group of clones, resources released from roots (and leaves) into the surrounding soil have a relatively low risk of being acquired by a competitor. The best example of this is the tight seasonal cycling of phosphorus and other nutrients in tussocks of *Eriophorum vaginatum* in the Alaskan tundra (Chapin, Van Cleve & Chapin 1979; Fig. 8).

Not all transfer of resources occurs between genetically identical tissues. The most familiar example in sexually reproducing species is the provision of resources from the leaves and roots of the parent plant to its gametes and seeds. The mean relatedness of the plant's vegetative structures to an embryo in a seed is only 0·5 (Haig 1987). In theory there is no lower limit of relatedness below which some altruism should not, in certain circumstances, be evolutionarily stable. Even between distantly related individuals, some altruism can be found. For this to occur, however, the genome of the 'altruist' must be manipulated by the beneficiary so as to induce the expression of behaviour that could never in itself be evolutionarily stable. This is what happens when viral nucleic acid parasitizes its host cell (Dawkins 1982; Buss 1987). Most vascular plant species are not parasitized directly by others, which may suggest that the majority have successfully evolved defences against this form of exploitation. But have they?

### *Hyphal power?*

The transfer of resources from one genetic individual to another makes evolutionary sense if the two are closely related or if one can genetically manipulate

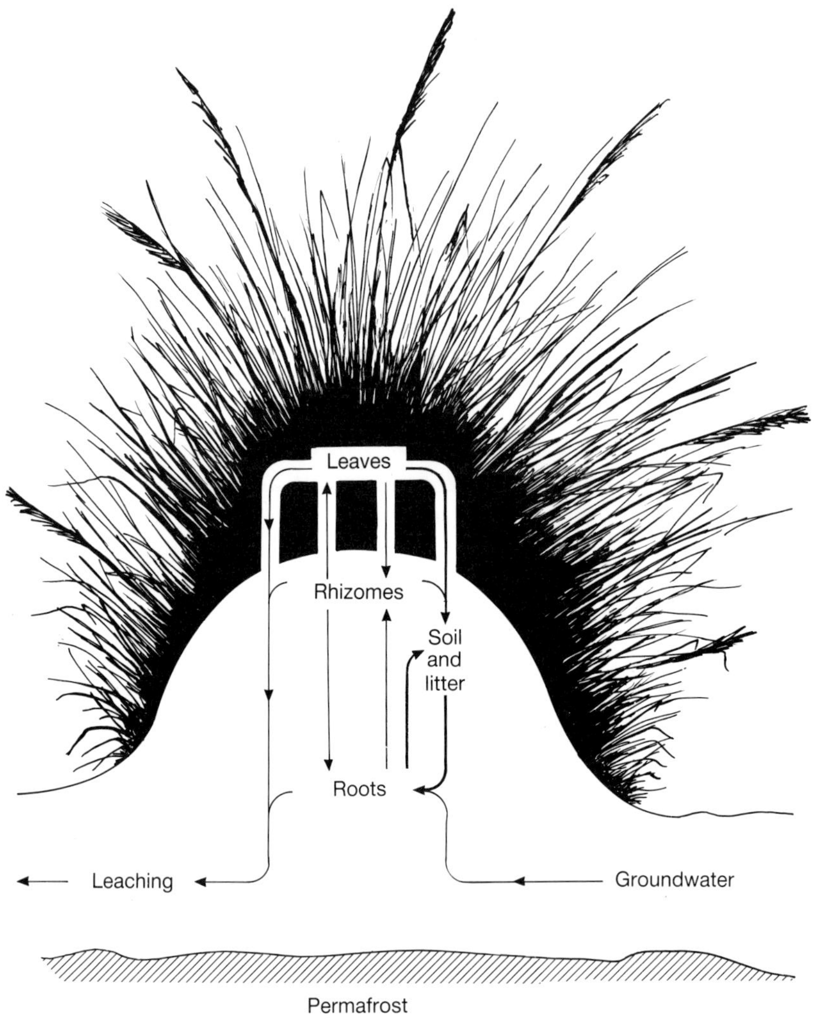


FIG. 8. Nutrient fluxes in tussocks of *Eriophorum vaginatum* (after Chapin, Van Cleve & Chapin 1979). Bold arrows represent the tight seasonal cycling between plant and soil.

the other. There are many examples of these in both botanical and zoological literatures. There is, however, no obvious zoological equivalent of the subject of this section: mycorrhizal networks.

Mycorrhizal fungi obtain carbon from their hosts at a cost to the latter of some reduction in growth compared with uninfected plants (see Koide & Elliott 1989). Sometimes the mycorrhizal associate enhances the plant's supply of nutrients, particularly phosphorus and micro-nutrients. The old idea of mycorrhizal plants having direct access to organic nitrogen in soils capable of only low rates of mineralization (e.g. in acidic woodlands) has received new support (Abuzinadah, Finlay

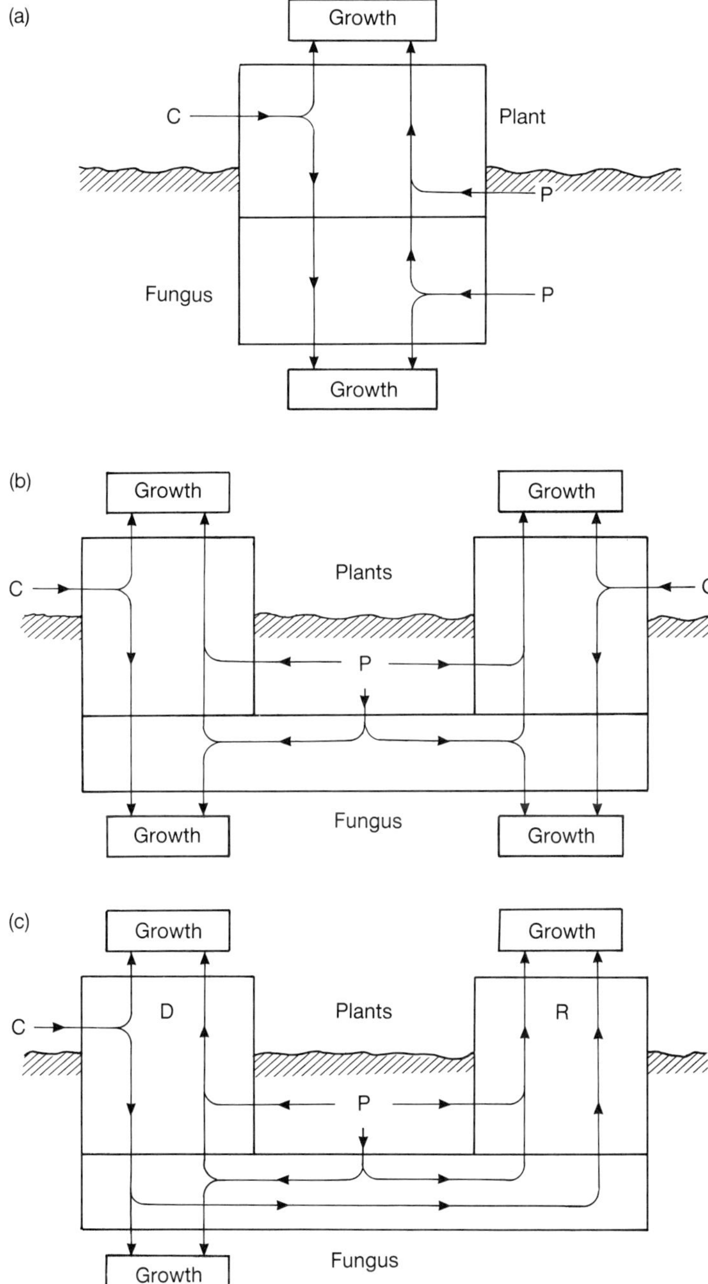


FIG. 9. (a) Fluxes of carbon (C) derived from photosynthesis and nutrient (phosphorus, P) absorbed from soil, within and between a single plant and a mycorrhizal fungus with which it is associated. (b) As for (a), but with the same fungus associated with more than one plant. (c) As for (b), but with a net flux of carbon from one plant (donor, D) to another (receiver, R) via the common mycorrhizal pathway. R is shown as being incapable of fixing any carbon photosynthetically, as in deep shade.

& Read 1986). A plant's tolerance of high concentrations of toxic metals can apparently be improved by its association with a mycorrhizal endophyte. The textbook picture of mutualism of such an association (Fig. 9a) cannot always be demonstrated, particularly in the field (McGonigle 1988). It is quite common for the fungus to parasitize its host (Fitter 1985; Grime *et al.* 1987). Even in phosphorus-deficient soils the active operation of a mutualistic association is not always needed to account for the continued growth, survival and coexistence of the indigenous plants but, of course, this possibility can rarely be ruled out.

There is little doubt that the roots of most plants can be interconnected by a common network of hyphae of mycorrhizal fungi. There is no doubt that if plants are connected in this way, carbon, nutrients and water can pass from one plant to another through the mycelium (although it is possible that no *net* movement of resources occurs between the plants) (Fig. 9b, which is simply an extension of that shown in Fig. 9a). Recent evidence that such networks might be ecologically important has been gathered most assiduously by Read *et al.* (e.g. Read, Francis & Finlay 1985; Finlay & Read 1986a and b; Grime *et al.* 1987; see also Jennings 1987; Newman 1988).

Read, Francis & Finlay (1985) suggested that mycorrhizal networks might have the most impact on the supply of nutrients and photosynthate to shaded seedlings in the understorey of an established community (Fig. 9c). This is an attractive idea. Tentative experimental evidence for this has been obtained by Grime *et al.* (1987). They found that infection by vesicular arbuscular mycorrhizal fungi of plants growing in artificial laboratory communities resulted in the apparent transfer of carbon ($^{14}C$) from a canopy dominant (*Festuca ovina*) to subordinate species such as *Scabiosa columbaria, Centaurea nigra* and *Plantago lanceolata*. The growth of these subordinates was increased considerably in the presence of the endophyte. This was not the case for another subordinate, *Rumex acetosa*, a species that does not form mycorrhizal associations. It apparently received comparatively small amounts of $^{14}C$ from *F. ovina*. This work is not without its critics, however, and as Newman (1988) rightly concluded, 'Evidence on almost every aspect of mycorrhizal links and their possible roles is inadequate'. Different interpretations of such experiments are always possible and more critical, well-crafted tests of the type carried out by Newman & Ritz (1986) and Eissenstat & Newman (1990) are needed.

It must be emphasized that all the available data on the transfer of resources from one plant to another via a common mycorrhizal network are qualitative. No one has any idea how much carbon, nutrients or water is, or could be, transported by this route. There are considerable technical difficulties that must be overcome if such data are to be obtained, and formidable criteria must be met if the data are to be unequivocal (Newman 1988). 'The basic problem . . . is the very common one of the easily measured variables not being the theoretically important ones' (Williams 1966) and 'what few data there are . . . only serve to show that whatever does happen is not straightforward!' (Crawley 1983).

*Numerical guesswork*

Meanwhile, it *is* possible to make some very simple calculations to get some idea of the scale of transport that *could* occur via a common mycorrhizal network. Are significant or trivial quantities of resource involved?

Suppose that we have two plants whose root systems are connected by a common mycelium. There are $10^6$ common hyphal entry points in their root systems: I have absolutely no idea whether this is a 'reasonable' or 'typical' value or not as no one has yet been able to make such measurements. Suppose that carbon is translocated through the mycelium as trehalose. Brownlee & Jennings (1981) reported that the concentration of trehalose in the mycelial strands of the dry rot fungus, *Serpula lachrymans*, was *c.* 100 $\mu g\,g^{-1}$ dry weight. Assuming a fresh : dry weight ratio of 10 : 1 gives 0·3 mol carbon $m^{-3}$. A net rate of water transport through a single hypha of a vesicular–arbuscular mycorrhiza of 28 $ng\,s^{-1}$ has been measured by M.F. Allen (see Jennings 1987). From these figures, it is calculated that the instantaneous, one-way, water-driven flux of carbon into one root system through each hyphal entry point is *c.* 0·01 $pmol\,s^{-1}$. This flux would probably be smaller if transport was assumed to occur solely by cytoplasmic streaming rather than by mass flow (Finlay & Read 1986b; Jennings 1987). For $10^6$ entry points, the flux is 10 $nmol\,s^{-1}$. Over a 16 h photoperiod, some 0·6 mmol of carbon could be transported from one plant to another via this route.

How does this compare with the photosynthetic acquisition of carbon? A high rate of net carbon fixation in a leaf of a $C_3$ plant might be *c.* 30 $\mu mol\,m^{-2}\,s^{-1}$. If this rate was sustained over 16 h by a canopy of 1 $m^2$ area, the net fixation of carbon would be 1·7 mol, some 2800 times greater than that calculated for the mycorrhizal flux. Of course, both sets of calculations are linearly sensitive to the chosen values. If the number of common hyphal entry points in each root system happened to be $5 \times 10^9$, say, instead of $10^6$, the mycorrhizal flux would rival that of photosynthesis. No allowances have been made for any carbon abstracted en route for the growth and maintenance of the fungus, nor for any respiratory losses or leakages into the soil, nor for spatial or temporal variations in the fluxes in the mycorrhiza or leaf canopy. Most importantly, no consideration has been given to the simultaneous transport of carbon in the *reverse* direction (Jennings 1987; Newman 1988). The mycorrhizal flux calculated above is a 'best case', and should be seen as a probably upper limit of a spectrum of possible fluxes based very much on numerical guesswork.

A net flux of 0·6 mmol carbon daily is the equivalent of up to *c.* 18 mg of dry matter if the total carbon concentration in plant dry matter is 40% and if all of the carbon was converted into dry matter. In a plant of 1 g dry weight, this daily increment could generate a relative growth rate (RGR) of up to 0·018 $day^{-1}$. This compares with 0·03 $day^{-1}$ in *Picea sitchensis* which was the slowest RGR measured in Grime & Hunt's (1975) survey of RGRs under 'standardized productive conditions'. If the above calculation of carbon flux via mycorrhizal networks is realistic, mycorrhizal fluxes of carbon from one plant to another could barely support the modest growth rates even of young tree seedlings, not to mention those of faster growing herbs. Again, it is important to emphasize the tentative, speculative nature of these calculations and that they are no substitute for good experimental data.

### *Extending the extended phenotype*

So, a mycorrhizal flux of carbon from one plant to another might be no more than a small fraction of the photosynthetic fixation, if the plants were growing in full sunlight. But as Read, Francis & Finlay (1985) noted, mycorrhizal fluxes will be important, if they are important at all, when one of the connected plants is shaded. The loss of carbon through a mycorrhizal network from an illuminated plant to one in the shade might be small compared with the former's photosynthetic fixation. However, the same mycorrhizal flux might represent a substantial input, indeed the only input, of carbon to a plant in deep shade.

This asymmetry may be equivalent to that known by evolutionary theorists as the 'life/dinner' principle (Dawkins 1982, see Chapter 4), from the idea that a fox runs only for its dinner whereas the rabbit that it chases runs for its life. Selection pressures for running speed are of different intensity in fox populations and in rabbit populations because the consequences of insufficient speed are rather more drastic for one than the other. Similarly, a mycorrhizal flux of carbon from an illuminated plant could diminish its eventual reproductive capacity by an insignificant amount, but it could sustain that of the shaded plant to which it was connected. Little selection pressure would be exerted on hosts to evolve ways of preventing what is (to them) only a trivial exploitation of their carbon resources by their shaded neighbours. But the selection pressures on those neighbours to evolve strategies to promote and maintain mycorrhizal networks would be more intense, provided that the quantities of photosynthate available through the network would make a significant difference to the receiver's carbon balance. Missing from this story is an explanation of the benefit(s) obtained by the fungus. Why should *it* maintain an association with a shaded plant from which, presumably, it obtains no photosynthate?

In evolutionary terms, the important thing about mycorrhizal networks is that they can link, and transfer resources between, distantly related plants. The likely widespread occurrence of such networks in undisturbed vegetation, as opposed to the rarity of interspecific root grafts in tree stands (Buss 1987; Loehle & Jones 1990), makes altruism between distantly related individuals potentially obligatory in such communities. This has no known zoological equivalent. 'Extended phenotype' hardly seems an adequate description of individuals linked by a common route through which they could access each other's resources.

Opportunities for parasitism would be rife in such a community (Loehle & Jones 1990). However, a would-be parasite could hardly depend on the fungal endophyte to maintain continuously an adequate degree of infection and supply of resources between it and its putative host(s). A parasitic member of such a community of plants could find itself being host instead, a source of resources rather than a sink, if the direction of the net flux through the mycelium was reversed. This could be initiated by the fungus, or by a change of circumstances of the ex-host. Some spectacular, short-term gains could be made by opportunistic, facultative parasitism via a mycorrhizal network, but the evolutionary stability of a strategy to specifically exploit this possibility seems unlikely. The application of game theory might be profitable here (Maynard Smith 1982). It may be that a plant connected to others by a mycorrhizal network is, at different times during its life, host *and* parasite. While altruism between distant relatives is not usually evolutionary stable, *reciprocal* altruism can be (Dawkins 1976, Chapter 10). But again, without knowing for certain the *amounts* of resources involved, it is difficult to come to firm conclusions.

The only other ways of explaining the apparent evolutionary stability of such resource transfer are to invoke a mechanism that involves some hitherto unsuspected benefit to the fungus. This might involve both the life/dinner principle (see above) and natural genetic manipulation. Even if carbon, nutrients and water cannot be transferred between plants in quantities large enough to directly influence their growth, could fragments of nucleic acid take this route . . . ?

## CONCLUDING REMARKS

I hope to have shown how new ideas and imaginative ways of looking at difficult problems can be useful in the ecological study of plant roots and root systems. Regrettably, I feel that an impression is often given in this field of biology that the 'sense of evolutionary depth' referred to in Medawar's observation at the start of this chapter is lacking in abundance, notable cases excepted (e.g. Caldwell 1979). It is important that the study of roots is not seen as a dull, peripheral area with no clear relevance to the mainstream of ecological theory.

Another unfortunate image is that 'documentation is the ultimate aim rather than problem solving and synthesis' (Hutchings 1986). There is a danger of concentrating obsessively on counting and measuring roots, to the exclusion of studies of what they actually do, and of their interactions and conflicts with other organs and organisms. There is more to the study of roots than digging holes in the ground, in short.

Finally, the interactions and conflicts between genes and cells, and between cells and organisms, what Buss (1987) has called 'somatic ecology', are as important as the more familiar ones between organisms and their environments ('extrasomatic ecology'). An awareness of both is essential if we are to use and, equally, contribute to the ideas and issues that drive evolutionary biology.

## ACKNOWLEDGEMENTS

Many people, knowingly or not, have helped in the preparation of this chapter; I am especially grateful to Joyce Davidson who typed its numerous versions.

## REFERENCES


**Abuzinadah, R.A., Finlay, R.D. & Read, D.J. (1986).** The role of proteins in the nitrogen nutrition of ectomycorrhizal plants. II. Utilization of protein by mycorrhizal plants of *Pinus contorta*. *New Phytologist*, **103**, 495–506.

**Baker, J.M. & Van Bavel, C.H.M. (1986).** Resistance of plant roots to water loss. *Soil Science Society of America Journal*, **78**, 641–644.

**Baldwin, J.P. (1975).** A quantitative analysis of the factors affecting plant nutrient uptake from soils. *Journal of Soil Science*, **26**, 195–206.

**Baldwin, J.P. (1976).** Competition for plant nutrients in the soil; a theoretical approach. *Journal of Agricultural Science, Cambridge*, **87**, 341–356.

**Barraclough, P.B. (1986).** The growth and activity of winter wheat roots in the field: nutrient inflows of high-yielding crops. *Journal of Agricultural Science, Cambridge*, **106**, 53–59.

**Barraclough, P.B. (1989).** Root growth, macro-nutrient uptake dynamics and soil fertility requirements of a high yielding winter oilseed rape crop. *Plant and Soil*, **119**, 59–70.

**Bloom, A.J., Chapin, F.S. & Mooney, H.A. (1985).** Resource limitation in plants — an economic analogy. *Annual Review of Ecology and Systematics*, **16**, 363–392.

**Brownlee, C. & Jennings, D.H. (1981).** The content of soluble carbohydrates and their translocation in mycelium of *Serpula lachrymans*. *Transactions of the British Mycological Society*, **77**, 615–619.

**Buss, L.W. (1987).** *The Evolution of Individuality*. Princeton University Press, Princeton.

**Caldwell, M.M. (1979).** Root structure: the considerable cost of below-ground function. In *Topics in Plant Population Biology* (Ed. by O.T. Solbrig, S. Jain, G.B. Johnson & P.H. Raven), pp. 408–427. Columbia University Press, New York.

**Caldwell, M.M. & Richards, J.H. (1986).** Competing root systems: morphology and models of absorption. In *On the Economy of Plant Form and Function* (Ed. by T.J. Givnish), pp. 251–273. Cambridge University Press, Cambridge.

**Caldwell, M.M., Richards, J.H., Manwaring, J.H. & Eissenstat, D.M. (1987).** Rapid shifts in phosphate acquisition show direct competition between neighbouring plants. *Nature*, **327**, 615–616.

**Calow, P. & Townsend, C.R. (1981).** Energetics, ecology and evolution. In *Physiological Ecology: An Evolutionary Approach to Resource Use* (Ed. by P. Calow & C.R. Townsend), pp. 3–19. Blackwell Scientific Publications, Oxford.

**Chapin, F.S., Van Cleve, K. & Chapin, M.S. (1979).** Soil temperature and nutrient cycling in the tussock growth form of *Eriophorum vaginatum*. *Journal of Ecology*, **62**, 169–189.

**Clarholm, M. (1985).** Possible roles for roots, bacteria, protozoa and fungi in supplying nitrogen to plants. In *Ecological Interactions in Soil* (Ed. by A.H. Fitter, D. Atkinson, D.J. Read & M.B. Usher), pp. 355–365. Blackwell Scientific Publications, Oxford.

**Clarke, A.L. & Barley, K.P. (1968).** The uptake of nitrogen from soils in relation to solute diffusion. *Australian Journal of Soil Research*, **6**, 75–92.

**Clarkson, D.T. (1985).** Factors affecting mineral nutrient acquisition by plants. *Annual Review of Plant Physiology*, **36**, 77–115.

**Coûteaux, M.-M., Faurie, G., Palka, L. & Steinberg, C. (1988).** La relation prédateur-proie (protozoaires-bactéries) dans les sols: rôle dans la régulation des populations et conséquences sur les cycles du carbone et de l'azote. *Revue d'Écologie et de Biologie du Sol*, **25**, 1–31.

**Crawley, M. (1983).** *Herbivory: the Dynamics of Animal–Plant Interactions*. Blackwell Scientific Publications, Oxford.

**Dawkins, R. (1976).** *The Selfish Gene*. Oxford University Press, Oxford.

**Dawkins, R. (1982).** *The Extended Phenotype*. W.H. Freeman, San Francisco.

**Dinkelater, B., Römheld, V. & Marschner, H. (1989).** Citric acid excretion and precipitation of calcium citrate in the rhizosphere of white lupin (*Lupinus albus* L.). *Plant, Cell and Environment*, **12**, 285–292.

**Drew, M.C. & Nye, P.H. (1969).** The supply of nutrient ions by diffusion to plant roots in soil. II. The effect of root hairs on the uptake of potassium by roots of ryegrass (*Lolium multiflorum*). *Plant and Soil*, **31**, 407–424.

**Eissenstat, D.M. & Caldwell, M.M. (1988).** Competitive ability is linked to rates of water extraction. A field study of two aridland tussock grasses. *Oecologia (Berlin)*, **75**, 1–7.

**Eissenstat, D.M. & Caldwell, M.M. (1989).** Invasive root growth into disturbed soil of two tussock grasses that differ in competitive effectiveness. *Functional Ecology*, **3**, 345–353.

**Eissenstat, D.M. & Newman, E.I. (1990).** Seedling establishment near large plants: effects of vesicular–arbuscular mycorrhizas on the intensity of plant competition. *Functional Ecology*, **4**, 95–99.

**Finlay, R.D. & Read, D.J. (1986a).** The structure and function of the vegetative mycelium of ectomycorrhizal plants. I. Translocation of $^{14}C$-labelled carbon between plants interconnected by a common mycelium. *New Phytologist*, **103**, 143–156.

**Finlay, R.D. & Read, D.J. (1986b).** The structure and function of the vegetative mycelium of ectomycorrhizal plants. II. The uptake and distribution of phosphorus by mycelial strands interconnecting host plants. *New Phytologist*, **103**, 157–165.

**Fitter, A.H. (1985).** Functioning of vesicular–arbuscular mycorrhizas in the field. *New Phytologist*, **99**, 257–265.

**Gigon, A. & Rorison, I.H. (1972).** The response of some ecologically distinct plant species to nitrate- and to ammonium-nitrogen. *Journal of Ecology*, **60**, 93–102.

**Grime, J.P. & Hunt, R. (1975).** Relative growth rate: its range and adaptive significance in a local flora. *Journal of Ecology*, **63**, 393–422.

**Grime, J.P., Mackey, J.M., Hillier, S.H. & Read, D.J. (1987).** Floristic diversity in a model system using experimental microcosms. *Nature*, **328**, 420–422.

**Grinsted, M.J., Hedley, M.J., White, R.E. & Nye, P.H. (1982).** Plant-induced changes in the rhizosphere of rape (*Brassica napus* var. Emerald) seedlings. I. pH change and the increase in P concentration in the soil solution. *New Phytologist*, **91**, 19–29.

**Haig, D. (1987).** Kin conflict in seed plants. *Trends in Ecology and Evolution*, **2**, 337–340.

**Hardwick, R.C. (1986).** Physiological consequences of modular growth in plants. *Philosophical Transactions of the Royal Society of London*, **B313**, 161–173.

**Harper, J.L. (1977).** *Population Biology of Plants*. Academic Press, London.

**Hutchings, M.J. (1986).** Plant population biology. In *Methods in Plant Ecology*, 2nd edn (Ed. by P.D. Moore & S.B. Chapman), pp. 377–435. Blackwell Scientific Publications, Oxford.

**Jackson, R.B., Manwaring, J.H. & Caldwell, M.M. (1990).** Rapid physiological adjustment of roots to localized soil enrichment. *Nature*, **344**, 58–60.

**Jarvis, P.G. & McNaughton, K.G. (1986).** Stomatal control of transpiration: scaling up from leaf to region. *Advances in Ecological Research*, **15**, 1–49.

**Jennings, D.H. (1987).** Translocation of solutes in fungi. *Biological Reviews*, **62**, 215–243.

**Keeley, J.E., Osmond, C.B. & Raven, J.A. (1984).** *Stylites*, a vascular land plant without stomata absorbs $CO_2$ via its roots. *Nature*, **310**, 694–695.

**Koide, R. & Elliott, G. (1989).** Cost, benefit and efficiency of the vesicular–arbuscular mycorrhizal symbiosis. *Functional Ecology*, **3**, 252–255.

**Lamont, B. (1982).** Mechanisms for enhancing nutrient uptake in plants, with particular reference to Mediterranean South Africa and Western Australia. *Botanical Review*, **48**, 597–689.

**Loehle, C. & Jones, R.H. (1990).** Adaptive significance of root grafting in trees. *Functional Ecology*, **4**, 268–271.

**Marschner, H. (1986).** *Mineral Nutrition of Higher Plants*. Academic Press, London.

**Maynard Smith, J. (1982).** *Evolution and the Theory of Games*. Cambridge University Press, Cambridge.

**McGonigle, T.P. (1988).** A numerical analysis of published field trials with vesicular–arbuscular mycorrhizal fungi. *Functional Ecology*, **2**, 473–478.

**Medawar, P.B. (1982).** *Pluto's Republic*. Oxford University Press, Oxford.

**Mirmirani, M. & Oster, G. (1978).** Competition, selection and evolutionarily stable strategies. *Theoretical Population Biology*, **13**, 304–339.

**Nadelhoffer, K.J., Aber, J.D. & Melillo, J.M. (1985).** Fine roots, net primary production and soil nitrogen availability: a new hypothesis. *Ecology*, **66**, 1377–1390.

**Newman, E.I. (1988).** Mycorrhizal links between plants: their functioning and ecological significance. *Advances in Ecological Research*, **18**, 243–270.

**Newman, E.I. & Andrews, R.E. (1973).** Uptake of P and K in relation to root growth and root density. *Plant and Soil*, **38**, 49–69.

**Newman, E.I. & Ritz, K. (1986).** Evidence on the pathways of phosphorus transfer between vesicular–

arbuscular mycorrhizal plants. *New Phytologist*, **104**, 72–87.

**Nobel, P.S. (1974).** *Introduction to Biophysical Plant Physiology*. W.H. Freeman, San Francisco.

**Nye, P.H. & Tinker, P.B. (1977).** *Solute Movement in the Soil–Root System*. Blackwell Scientific Publications, Oxford.

**Passioura, J.B. (1963).** A mathematical model for the uptake of ions from the soil solution. *Plant and Soil*, **18**, 225–228.

**Passioura, J.B. (1980).** The transport of water from root to shoot in wheat seedlings. *Journal of Experimental Botany*, **31**, 333–345.

**Pitelka, L.F. & Ashmun, J.W. (1985).** Physiology and integration of ramets in clonal plants. In *Population Biology and Evolution of Clonal Organisms* (Ed. by J.B.C. Jackson, L.W. Buss & R.E. Cook), pp. 399–435. Yale University Press, New Haven.

**Poorter, H. (1989).** Interspecific variation in relative growth rate: on ecological causes and physiological consequences. In *Causes and Consequences of Variation in Growth Rate and Productivity of Higher Plants* (Ed. by H. Lambers, M.L. Cambridge, H. Konings & T.L. Pons), pp. 45–68. SPB Academic Publishing, The Hague.

**Raven, J.A. (1986).** Evolution of plant life forms. In *On the Economy of Plant Form and Function* (Ed. T.J. Givnish), pp. 421–492. Cambridge University Press, Cambridge.

**Read, D.J., Francis, R. & Finlay, R.D. (1985).** Mycorrhizal mycelia and nutrient cycling in plant communities. In *Ecological Interactions in Soil* (Ed. by A.H. Fitter, D. Atkinson, D.J. Read & M.B. Usher), pp. 193–217. Blackwell Scientific Publications, Oxford.

**Robinson, D. (1986).** Limits to nutrient inflow rates in roots and root systems. *Physiologia Plantarum*, **68**, 551–559.

**Robinson, D. (1989).** Can the nutrient demand of a plant be sustained by an increase in local inflow rate? *Journal of Theoretical Biology*, **138**, 551–554.

**Robinson, D. (1990).** Phosphorus availability and cortical senescence in cereal roots. *Journal of Theoretical Biology*, **145**, 257–265.

**Robinson, D. (1991).** Strategies for optimising growth in response to nutrient supply. In *Plant Growth; Interactions with Nutrition and Environment* (Ed. by J.R. Porter & D.W. Lawlor), pp. 177–205. Cambridge University Press, Cambridge.

**Robinson, D. & Rorison, I.H. (1983).** A comparison of the responses of *Lolium perenne* L., *Holcus lanatus* L. and *Deschampsia flexuosa* (L.) Trin. to a localised supply of nitrogen. *New Phytologist*, **94**, 263–273.

**Robinson, D., Griffiths, B., Ritz, K. & Wheatley, R. (1989).** Root-induced nitrogen mineralisation: a theoretical analysis. *Plant and Soil*, **117**, 185–193.

**Robinson, D., Linehan, D.J. & Caul, S. (1991).** What limits nitrate uptake from soil? *Plant, Cell and Environment*, **14**.

**Sackville Hamilton, N.R., Schmid, B. & Harper, J.L. (1986).** Life history concepts and the population biology of clonal plants. *Proceedings of the Royal Society of London*, **B232**, 35–57.

**Sibly, R.M. & Grime, J.P. (1986).** Strategies of resource capture by plants — evidence of adversity selection. *Journal of Theoretical Biology*, **118**, 247–250.

**Smith, F.A. & Smith, S.E. (1986).** Movement across membranes: physiology and biochemistry. In *Proceedings of the First European Symposium on Mycorrhizae* (Ed. by V. Gianinazzi-Pearson & S. Gianinazzi), pp. 75–84. INRA, Paris.

**Stephens, D.W. & Krebs, J.R. (1986).** *Foraging Theory*. Princeton University Press, Princeton.

**Tilman, D. (1982).** *Resource Competition and Community Structure*. Princeton University Press, Princeton.

**Tuomi, J. & Vuorisalo, T. (1989).** What are the units of selection in modular organisms? *Oikos*, **54**, 227–233.

**Vogel, S. (1988).** *Life's Devices*. Princeton University Press, Princeton.

**Watson, M.A. & Casper, B.B. (1984).** Morphogenetic constraints on patterns of carbon distribution in plants. *Annual Review of Ecology and Systematics*, **15**, 233–258.

**Williams, G.C. (1966).** *Adaptation and Natural Selection*. Princeton University Press, Princeton.

# Water and nitrate redistribution in soil as affected by root distribution and absorption

R. HABIB AND F. LAFOLIE*

*Station d'Agronomie, Station de Science du Sol,* INRA Domaine St Paul, BP 91, F-84143 Montfavet Cedex, France*

## SUMMARY

1 The redistribution of water and nitrate in soil as affected by root distribution and the absorption efficiency of the root system is studied using a numerical simulation model. The model simulates one-dimensional (vertical) soil–plant processes.

2 Plant, soil and climate parameters are combined to simulate water flow and nitrate transfer in soil and absorption by a root system. These parameters are (i) root distribution and maximum nitrate-inflow rate, (ii) soil hydraulic properties and nitrate-diffusion coefficient, and (iii) soil temperature and actual transpiration rate.

3 This set of parameters is used in simulating the soil–root interaction processes for annual and perennial plant types. The importance of root system distribution and absorption on water and nitrate redistribution appears to be related to rooting pattern, but this depends on soil type (mainly soil resistance to water flow) and climatic demand.

## INTRODUCTION

In the past few years a considerable effort has been made to study water flow and solute transport in soil through deterministic, mechanistic or stochastic models including diffusion and mass-flow processes (Bresler 1973; van Genuchten, Tang & Guennelon 1984; Addiscott & Wagenet 1985), and even any sink–source interaction term between the salt and the soil matrix (van Genuchten & Wierenga 1976; Bresler 1981).

When considered, water flow and solute transport to plant roots due to absorption by roots were usually modelled with the 'single root model' using cylindrical spatial coordinates. This model assumes that water flow and solute transport take place only radially to the root. Numerous papers based on this approach study simultaneous transport and absorption processes (Passioura 1963; Nye & Spiers 1964; Nye & Marriott 1969). For diffusive transport of ions only and for root systems with regularly or irregularly spaced parallel roots, the uptake of solutes has been described, using an electrical analogue (Sanders, Tinker & Nye 1970; Baldwin, Tinker & Nye 1972). When mass flow was included, numerical solutions to this problem have been found (Passioura & Frere 1967 Nye & Marriott 1969), and applied to different cropping conditions (Phillips *et al.* 1976; Claassen, Syring & Jungk 1986). When soil resistance to water flow and ionic transfer is significant, the spatial configuration of the root system has to be taken into account (Barley 1970; Tardieu 1988). When roots were assumed to be dispersed randomly, an approximate solution of the cylindrical coordinates equation was proposed by Baldwin, Nye & Tinker (1973) to calculate the solute uptake by a root system of increasing density in a finite volume of soil.

More recently an integrated approach has been developed where water uptake by roots was modelled using a macroscopic sink term which is added directly to the transport equation (Feddes, Bresler & Neuman 1974; Aria, Blake & Farrel 1975b; Hillel, Talpaz & van Keulen 1976; Novak 1987). The aim of the present contribution is to combine water flow, nitrate transfer in soil and absorption by roots, using the so-called integrated approach, in order to simulate the effects of root distribution and absorption on water and nitrate redistribution in soil.

## THEORY

### *Modelling water flow in soil and water uptake by roots*

In our analysis it is assumed that water flow obeys Darcy's law for the entire saturated–unsaturated infiltration–redistribution process, that hysteresis in soil hydraulic properties can be neglected and that the soil is non-swelling. In this model, only vertical flow is considered. The combined Darcy's law and continuity equation leads to (symbol definition and units are given in Table 1)

$$ca(\psi)\,\frac{\partial H}{\partial t} = \frac{\partial}{\partial z}\left[K(\psi)\,\frac{\partial H}{\partial z}\right] + S_w(z,\psi,l,t), \qquad (1)$$

where $\psi$ is the soil water pressure head (cm), $H = \psi - z$ is the total head, $ca$ is the soil water capillary capacity ($cm^{-1}$), $\partial\theta/\partial\psi,\theta(\psi)$ is the volumetric water content ($cm^3\,cm^{-3}$), $K$ is the hydraulic conductivity ($cm\,s^{-1}$), $t$ is time (s), $z$ is the distance (cm), considered positive downwards, and $S_w$ is a sink term describing water uptake by the root system ($cm^3\,cm^{-3}\,s^{-1}$).

In addition, it is assumed that the root system is characterized by a root density (root length per soil volume unit, $cm\,cm^{-3}$) function of depth. The absorption capacity per length unit of root for water and nitrate is assumed to be the same whatever part of the root system is under consideration. In the model, water uptake by roots is driven by a climatic constraint, the potential transpiration rate of the crop. A maximum water-uptake rate per unit length of the root system is calculated by dividing the potential transpiration rate by the total length of the root system. This rate is weighted by a function $\alpha$ ($\psi$); the aim of this being to reduce the rate when soil water is at increasingly lower potentials. The sink term is in this case defined as:

$$S_w(z,\psi,l,t) = S_{max}(t)\;\alpha(\psi)\;l(z), \qquad (2)$$

TABLE 1. List of symbols (alphabetic order)

| Symbols | Definition | Units |
|---|---|---|
| $\alpha$ | Weighting function for water uptake | – |
| $\beta$ | Parameter controlling solute exchange between the two phases | $s^{-1}$ |
| $C$ | Nitrate concentration in the soil solution | $mg\ N\ cm^{-3}$ |
| $ca$ | Soil water capillary capacity | $cm^{-1}$ |
| $C_{im}$ | Nitrate concentration in the immobile phase | $mg\ N\ cm^{-3}$ |
| $C_m$ | Nitrate concentration in the mobile phase | $mg\ N\ cm^{-3}$ |
| $D_{iff}$ | Nitrate diffusion coefficient in soil solution | $cm^2\ s^{-1}$ |
| $D_m$ | Diffusion–dispersion coefficient of nitrate in soil solution | $cm^2\ s^{-1}$ |
| $\delta$ | Dispersivity | cm |
| $F_{max}$ | Maximum nitrate uptake rate | $mg\ N\ cm^{-1}\ s^{-1}$ |
| $H$ | Total pressure head | cm |
| $K$ | Hydraulic conductivity | $cm\ s^{-1}$ |
| $k$ | $NO_3^-$ concentration where $S_N = F_{max}/2$ | $mg\ N\ cm^{-3}$ |
| $l$ | Root density | $cm\ cm^{-3}$ |
| $\psi$ | Soil water potential | cm |
| $q$ | Water flux | $cm^3\ cm^{-2}\ s^{-1}$ |
| $S_{max}$ | Potential transpiration rate | $cm^3\ cm^{-1}\ s^{-1}$ |
| $S_N$ | Sink term describing $NO_3^-$ uptake | $mg\ N\ cm^{-3}\ s^{-1}$ |
| $S_w$ | Sink term describing water uptake | $cm^3\ cm^{-3}\ s^{-1}$ |
| $t$ | Time | s |
| $\theta$ | Volumetric water content | $cm^3\ cm^{-3}$ |
| $\theta_{im}$ | Water content of the immobile phase | $cm^3\ cm^{-3}$ |
| $\theta_m$ | Water content of the mobile phase | $cm^3\ cm^{-3}$ |
| $v$ | Average interstitial flow velocity | $cm\ s^{-1}$ |
| $z$ | Depth | cm |

where $\alpha(\psi)$ is the weighting function, $S_{max}$ the potential transpiration rate per unit length of root ($cm^3\,cm^{-1}\,s^{-1}$) and $l(z)$ the root density as function of depth ($cm\,cm^{-3}$). Various formulations have been proposed for $\alpha(\psi)$. In particular, Feddes *et al.* (1978) introduced a piecewise linear function while van Genuchten (1983) has used a smooth S-shaped function. Calculations have been carried out with a piecewise linear form for $\alpha(\psi)$.

### *Modelling nitrate movement in soil and nitrate uptake by roots*

The nitrate movement in the soil profile is modelled with a two-phase first-order physical non-equilibrium model. Equations are as follows:

$$\frac{\partial(\theta_m C_m)}{\partial t} + \frac{\partial(\theta_{im} C_{im})}{\partial t} = \frac{\partial}{\partial z}\left(D_m \theta_m \frac{\partial C_m}{\partial z}\right) - \frac{\partial}{\partial z}(q C_m) - S_N(C,z) \quad (3a)$$

$$\theta_{im}\frac{\partial C_{im}}{\partial t} = \beta(C_m - C_{im}) \quad (3b)$$

$$\theta = \theta_m + \theta_{im}; \qquad D_m = \delta v + D_{iff}; \qquad v = q/\theta_m \quad (3c)$$

$$C = (C_m\theta_m + C_{im}\theta_{im})/\theta, \quad (3d)$$

where $C$ is the nitrate concentration in the soil solution ($mg\,N\,cm^{-3}$) and indexes m and im refer to the mobile and immobile phase, respectively. Flux distributions in the profile, $q$ ($cm^3\,cm^{-2}\,s^{-1}$), are calculated from the solution of Richards' equation and Darcy's law. The dispersion coefficient, $D_m$ ($cm^2\,s^{-1}$), is obtained from the dispersivity, $\delta$ (cm), the velocity, $v$ ($cm\,s^{-1}$), and the molecular diffusion coefficient, $D_{iff}$ ($cm^2\,s^{-1}$) (Bresler 1973). The velocity is evaluated from the flux and the mobile fraction of the soil water content. The mass transfer coefficient, $\beta$ ($s^{-1}$), that controls the rate of solute exchange between the two phases (van Genuchten & Wierenga 1976), has been chosen so as to account roughly for various soil and diffusion coefficients for nitrate (Gaudet 1978). We also considered a certain ranking between the various soil types used in the calculations. For example, the 'clay' soil is simulated with a higher percentage of immobile water than any other soil used, but with a lower value for $\beta$, corresponding to a lower diffusion coefficient. Similarly, higher values of $\delta$ were chosen for the clay soil, to account for a more structured soil, than for the two other soils.

Nitrate uptake by the root system is modelled by Michaelis–Menten type kinetics (Nye & Marriot 1969). The sink term, $S_N$ ($mg\,N\,cm^{-3}\,s^{-1}$), is as follows,

$$S_N(C,z,F_{max}) = F_{max}\frac{C(z)}{k + C(z)}\,l(z), \tag{4}$$

where $C$ is the nitrate concentration in the soil solution ($mg\,N\,cm^{-3}$), $F_{max}$ is the maximum uptake rate per unit time per unit length of the root system ($mg\,N\,cm^{-1}\,s^{-1}$), and $k$ is the value of the concentration where $S = F_{max}/2$. Note that the concentration used in eqn (4) is a weighted mean between mobile and immobile phase concentrations (eqn 3d). Thus, with this expression, nitrate uptake by roots is assumed independent of the water-uptake rate. As reported by Dalton, Raats & Gardner (1975), various opinions can be found in the literature about the possible effects of transpiration upon solute uptake by plant roots. As far as we know, there is no clear evidence for the dependence of nitrate absorption on transpiration flux, and more often, papers conclude on the independence. Grubb (1977) stated that absorption of mineral nitrogen does not depend on transpiration for the range of concentrations which occur in natural soils. Bhat (1982) found that nitrate absorption by apple tree roots in natural conditions was independent of time, occurring at similar rates both during the day and the night. Nitrate uptake by roots of intact seedlings of radish or tomato grown in solution were shown independent of transpiration rate at low solution concentrations, less than 1 mM (Schulze & Bloom 1984). Blondel (1979) observed that the day–night variation of nitrate-uptake rates of wheat seedlings in nutrient solution was much less than that of water uptake, and was more probably related to the day–night variation in nitrate reductase activity than related to the transpiration flux. Bar-Yosef *et al.* (1988) observed no direct relationship between the rate of nitrate uptake of container-grown apple trees and the water evaporation from a pan. Furthermore, they found that nitrate concentration, in the range of 2–11 mM, affected nitrate uptake as predicted by the Michaelis–Menten equation. Hence, it seems reasonable to make the assumption that nitrate uptake by roots does not depend directly upon the water uptake rate. The Richards' equation and the set of equations modelling the displacement of nitrate are approximated by finite differences in space, and the resulting differential equations integrated in time by an implicit scheme. The model is conceived to handle all possible boundary conditions.

## SIMULATION CONDITIONS

Numerical simulations for water flow and nitrate movement in soil have been carried out under various soil–plant–climate conditions. The purpose was to study the distribution of water and nitrate in the profile as influenced by (i) root system distribution (i.e. density profile and pattern), and root system effectiveness (i.e. maximum inflow rates for nitrate and water), (ii) soil types (soil hydraulic properties, dispersivity, nitrate-diffusion coefficient, etc.) and (iii) climatic conditions (i.e. potential transpiration rate). In the simulations, the soil profile was 1 m deep and a unit of ground surface of 1 $m^2$ was considered to normalize applications of water and nitrate. We simulated a 15-day period with two 40 mm irrigations, applied at time $t = 0$ and $t = 180$ h, respectively. Nitrate was applied through the first irrigation water, with 10 g of nitrate-N over 1 $m^2$ corresponding to a 100 kg N $ha^{-1}$ fertilization. The second irrigation was made with nitrate-free water. This set-up is used whatever the soil properties or the root system. Simulations were performed with zero-evaporation conditions at the soil surface in order to stress the role of root uptake within the soil profile, and in particular near the soil surface.

### *Plant parameters*

Simulations have been carried out with three different types of root systems: cereal, forest tree (data from Barley 1970) and fruit tree (unpublished data about peach) as presented in Fig. 1. Thus, static root systems were used during the 15-day simulation period. We know this is an over-simplifying assumption. Nevertheless, this was made in order to stress the particular role of root distribution without interfering with the effects of increasing root densities. The role of the difference in root densities was partly accounted for by the different types of root system. According to published data (Robin 1983; Bar-Yosef *et al.* 1988) the $k$ constant in eqn (4) was set equal to $14 \times 10^{-5}$ mg N $cm^{-3}$. For the tree and cereal root systems, the $F_{max}$ value in eqn (4) was set equal to 0·150 and 0·015 mg N $m^{-1}$ $day^{-1}$, respectively. Thus, higher root density (Fig. 1) and lower maximum inflow rate values were used in simulating a cereal root system. The orders of magnitude for maximum inflow rate are lower than data reviewed by Brewster & Tinker (1972) and Atkinson (1985). They lead to a total simulated nitrate uptake of about 2–3 g during the 15-day period under consideration.

### *Soil and climate parameters*

Simulations have been carried out with three sets of soil hydraulic properties taken from the literature (Beuving 1984). Hereafter, the three corresponding soils are

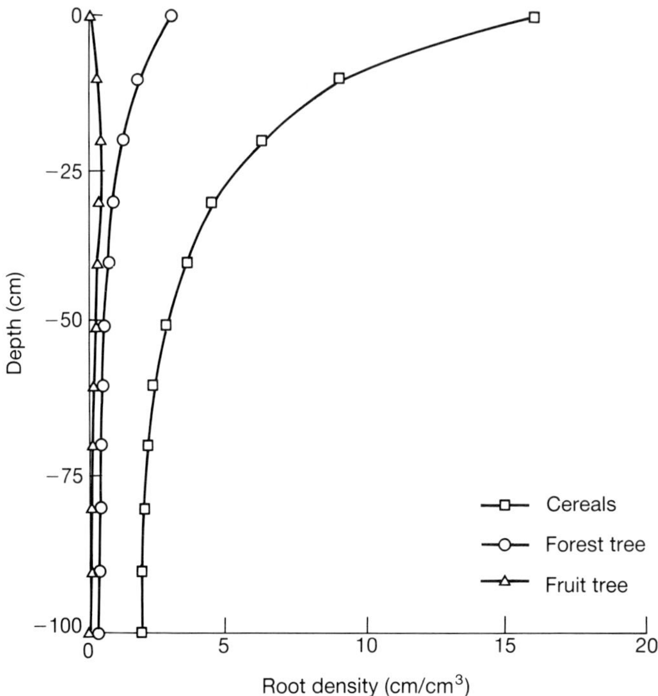


FIG. 1. Root-density profiles used in simulating water and nitrate redistribution in soil.

arbitrarily referred to as 'light' soil (i.e. hydraulic properties close to those of a sandy soil, and a 5% immobile water phase), 'medium' soil (i.e. hydraulic properties close to those of a lime soil, and a 15% immobile water phase), and 'heavy' soil (i.e. soil hydraulic properties close to those of a structured clay soil, and a 25% immobile water phase). Nitrate-diffusion coefficients were $5 \times 10^{-6}$, $4 \times 10^{-6}$ and $3 \times 10^{-6}\,cm^2\,s^{-1}$ for the light, medium and heavy soil, respectively. An initial water potential profile decreasing from −800 cm at the soil surface to −150 cm at the bottom was used for all the simulations. Calculations have been carried out with a high (8 mm $day^{-1}$) and a low (3 mm $day^{-1}$) potential transpiration rate. Soil temperature is known to influence both nitrate diffusion, since the diffusion coefficient increases with temperature, and uptake rate. Three different nitrate-diffusion coefficients were used to assess soil temperature influence at a given uptake rate for the 'medium' soil type only.

## RESULTS AND DISCUSSIONS

### *Typical simulation outputs*

Figure 2 is a typical simulation output. Results are those for the 'medium' soil, with a high density root system (exponential profile) and a high potential transpiration

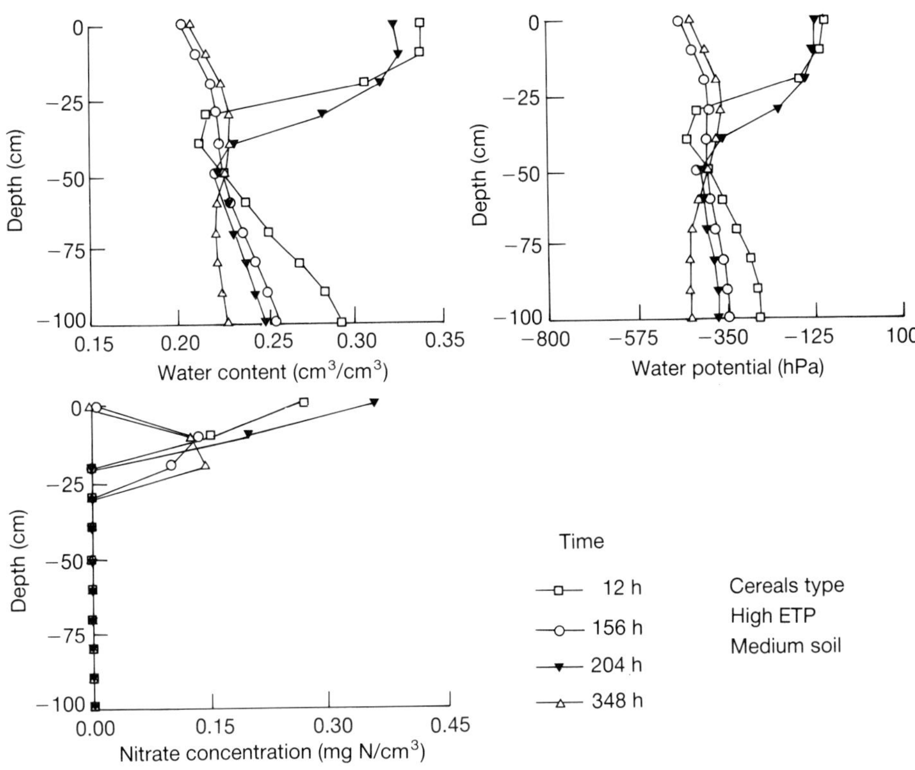


FIG. 2. Simulated water content, water potential and nitrate concentration profiles.

rate. Water content, water potential and nitrate concentration profiles are presented at 12, 156, 204 and 348 h after the beginning of the simulation. Profiles at 12 h illustrate the distributions of water and nitrate just after the first irrigation. Profiles at 156 and 204 h show water and nitrate distributions 24 h before and after the second irrigation which occurs at 180 h. The redistribution of water and nitrate as well as the uptake process can be clearly observed from these profiles. The last profiles, at 348 h, illustrate the situation at the end of the simulation. From these profiles one can observe that deep soil layers are not recharged by irrigation. Nevertheless, they participate largely in supplying water to the crop. First, the roots extract water directly from these layers and secondly because they give water to upper layers. The second irrigation wets the soil profile slightly more deeply since soil layers close to the surface are at lower potentials than those prevailing at the start of the simulation. Note that water content distributions at 156 and 348 h are quite different. In particular, water contents in the upper half of the profile are higher at 348 than at 156 h, while the lower half of the profile is still on drying. If more irrigation–redistribution cycles were simulated, one would probably obtain at the limit a unique water content profile just before each irrigation. In the following, only soil water distributions through water potential profiles are illustrated.

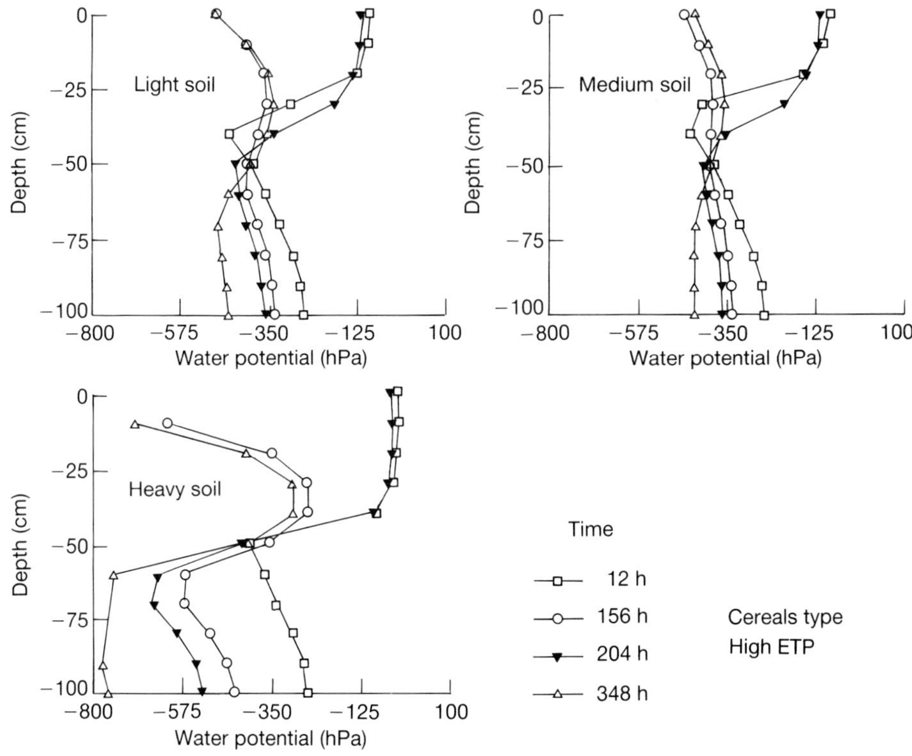


FIG. 3. Water redistribution in soil as affected by soil characteristics.

## *Effect of soil type on redistribution of water and nitrate*

### *Water redistribution*

Figure 3 presents the effects of soil types on water and nitrate distributions for a cereal type root system. Notice that in all cases deep soil layers are not recharged by irrigation. Simulations show that these layers dry out in response to water uptake by roots and upward water flux. The same behaviour was obtained by Hillel, Talpaz & van Keulen (1976) using a macroscopic-scale model of water uptake by a non-uniform root system. A similar trend was also noted by Aria, Blake & Farrel (1975a) in a field experiment. They observed an increasing contribution of lower soil layers to the total water uptake in the presence of growing soya-bean roots. While water potential distributions are very similar for the 'light' and 'medium' soils, the heavy soil shows quite different behaviour. In particular, a trend to dry out in the upper layers to a potential lower than the initial one is noted. Note also a more contrasted potential distribution with depth resulting from a lower hydraulic conductivity. As a consequence the deficits developed are larger on this soil than on any other used in this study. This

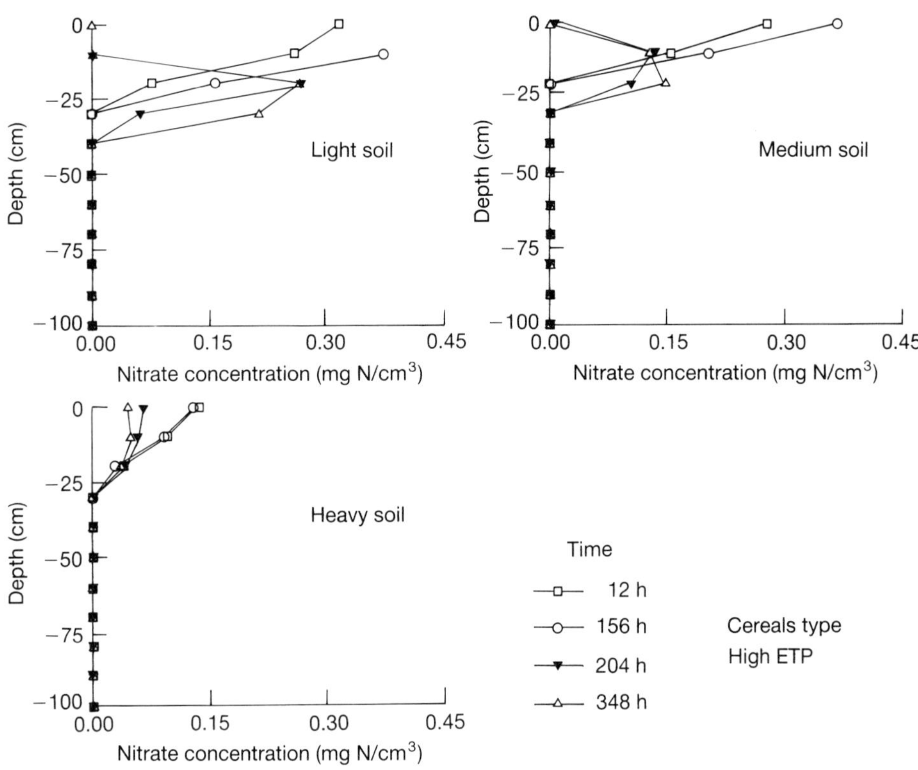


Fig. 4. Nitrate redistribution in soil as affected by soil characteristics.

illustrates the large dependence of water-uptake rates on soil hydraulic properties and root system patterns (Aria, Blake & Farrel 1975a, b).

*Nitrate redistribution*

Figure 4 presents the nitrate distributions corresponding to the previous water potential profiles. For the 'light' and 'medium' soils, a quite usual behaviour is observed (Gaudet 1978). The nitrate pulse, located in the upper layer after the first irrigation, is slowly displaced downwards by the second irrigation. A very small dispersion of the pulse is observed as well, partly due to a low dispersivity, and to the unsaturated conditions. After 15 days, the nitrate pulse is already leached to an average depth of about 25–30 cm for the 'light' soil, while the main part of the root system is located in these first 30 cm. After one or two more irrigations the nitrate would probably be out of reach of the larger part of the root system. This behaviour highlights the need for a smaller nitrate fertilization of this type of soil, not only for groundwater protection but also for a more efficient use of fertilizers. Quite different nitrate profiles are obtained in the 'heavy' soil. Lower nitrate concentrations in this soil are explained by higher initial water contents. Note also that the larger immobile water phase prevents leaching from the upper layer and leads to smoother concentration profiles. The mass of nitrate as a function of depth can be obtained from the water content profile. These three cases illustrate clearly the large dependence of nitrate distributions on water flux and soil structure. It appears that for soils with structural heterogeneities, a good prediction of solute distribution, and in particular of nitrate, will hinge largely upon our ability to predict water flux in such systems.

## *Effects of root system on redistribution of water and nitrate*

### *Water redistribution*

Figure 5 presents the effects of three different root systems on the 'medium' soil for a high potential transpiration rate. At first, the striking similarity between the two patterns obtained may seem surprising. This result highlights the limitations of the water-uptake model used in the simulations. This suggests that at least two modifications should be included in the water-uptake model. First, one should

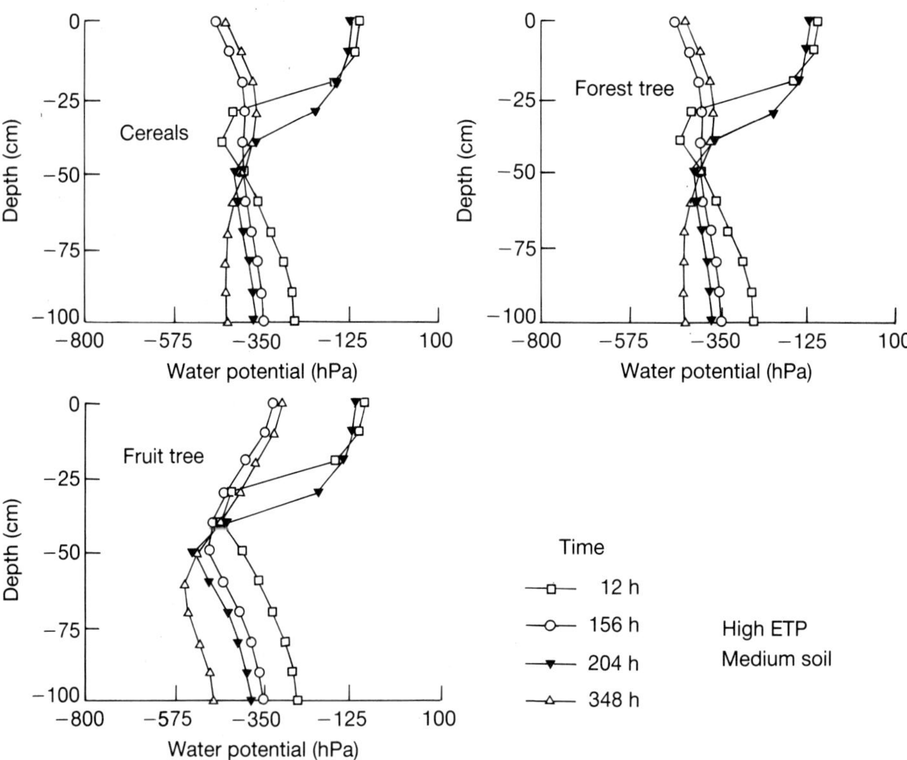


FIG. 5. Water redistribution in soil as affected by root density and pattern.

account for internal root resistance to flow and define a maximum uptake rate per unit root length, mass or surface area. The root water-uptake model does not impose any limitation on the uptake rate per unit root length of root for example, even if there is an upper limit related to the root characteristics. Secondly, the distributed sink term added to Richards' equation, is intuitively better for high density root systems. In a low density system, a soil resistance term should be included to account for higher potential gradients needed to move water towards the surface of the roots (Aria, Blake & Farrel 1975b; Tardieu 1988). The lower the root density and the lower the water potential, the higher the resistance term should be.

The model is very sensitive to the shape of the root system as can be seen when comparing the fruit tree and forest tree types. These differ mainly by the distribution of the roots and not by root densities or total length. In the case of a fruit tree type of root profile, more water is extracted from deep layers. This sensitivity of the model gives some hope for future work aiming at the identification of active parts of the root system by means of mathematical identification techniques (Novak 1987).

### *Nitrate redistribution*

There is a high sensitivity of the nitrate profile to the root system (Fig. 6). It is difficult to separate the effects due to nitrate absorption by different root systems from the indirect effects due to water absorption. Since nitrate moves easily with water in the soil, water-uptake patterns probably influence the distribution of nitrate between two irrigations, at least as much as the nitrate-uptake process by roots itself. A close analysis of water flux, nitrate flux and uptake rates is necessary to separate the various effects. It is remarkable, for example, that for the fruit tree type the relatively low consumption of water in the upper layer leads to a deeper infiltration of water after the second irrigation, and hence to a deeper penetration of nitrate. Similarly, with an exponential type of root system, there is a large reduction in the concentration after the second infiltration, due to a significant consumption of nitrate by the crop. For the fruit tree type of root system, the concentration after the second irrigation is very close to the concentration after the first, in response to a low consumption in the upper layer, since the main part of the root system is located slightly below the depth of penetration of the nitrate pulse.

The main problem with this approach is that nitrate uptake is driven mainly by the maximum uptake rate, $F_{max}$, used in eqn (4). In regions of the soil where nitrate is present the concentrations are much larger than the Michaelis–Menten constant, and thus nitrate uptake occurs at a rate close to $F_{max}$. Total nitrate uptake under the forest tree type of root system as calculated with the model might not be realistic since a maximum inflow rate as estimated from a fruit tree type of root system has been used in the calculations. As noted by Atkinson (1985), it is stressed that a lower root density may be compensated by higher inflow rates. Finally, it may lead to comparable, if not lower, nitrate concentration profiles.

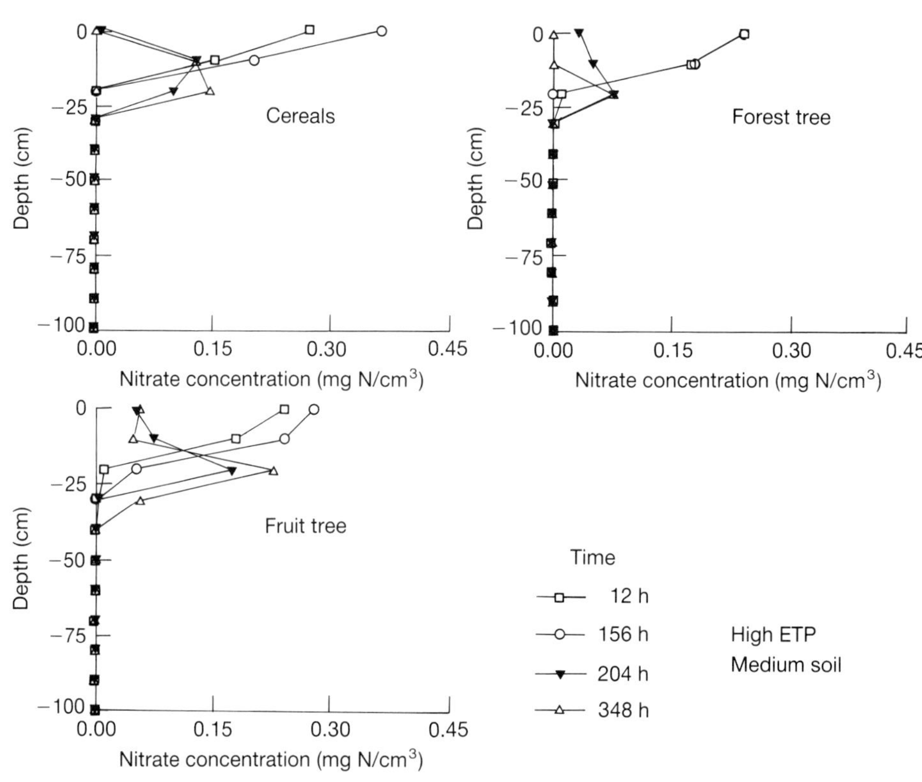


FIG. 6. Nitrate redistribution in soil as affected by root density and pattern.

## *Effects of climatic parameters on redistribution of water and nitrate*

Three different nitrate diffusion coefficients have been used to assess the influence of soil temperature at a given nitrate-uptake rate for the 'medium' soil type. It appears that this temperature effect is very light although the nitrate diffusion coefficient is twice as high at 30 °C than it is at 10 °C. It is worth noting that this difference is of the same order as the one simulated by changing the soil characteristics from 'light' to 'heavy' in one simulation run. The effect of soil temperature through its influence on nitrate-uptake rate (Bhat 1982) should be more important but it has not been considered in the present paper.

Figure 7 presents water and nitrate redistributions at a low transpiration rate for two root system types (cereal and fruit tree) for the 'medium' soil. Water profiles are to be compared with the ones presented in Fig. 5 (high transpiration rate) and nitrate profiles are to be compared with the ones presented in Fig. 6 (high transpiration rate). In particular, the effect of transpiration rate on the distribution of water in the soil is predictably very marked. It is of more interest that the effect of root density and distribution on water redistribution depends on the transpiration rate. Novak (1987) observed that water-extraction patterns of maize roots,

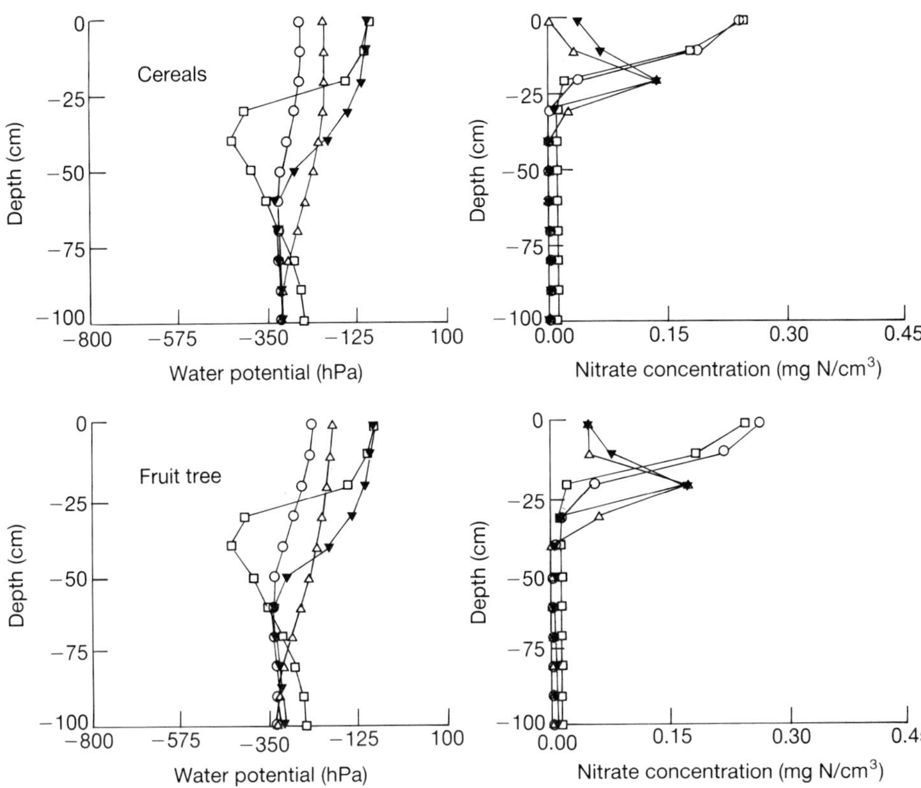


FIG. 7. Water and nitrate redistribution in soil at low transpiration rate, simulated for 'medium' soil characteristics. The symbols are the same as in Fig. 6.

calculated from field data or from the water-uptake model, were dependent on transpiration. It is remarkable that the two sets of water profiles presented in Fig. 7 (low transpiration rate) show only very slight differences in contrast to the water profiles in Fig. 5 (high transpiration rate). Thus, the very large difference in root density and pattern (Fig. 1) is no more noticeable when the climatic demand is low. At low transpiration rate, deep soil layers begin to be recharged by irrigations. This comes from a reduction in water uptake in the soil profile, and consequently higher soil water content and conductivity in the upper profile.

Consequently, the main effects of low transpiration rate on nitrate redistribution in soil are (i) decreasing nitrate concentration by increasing water content, and (ii) increasing nitrate leaching by increasing water flux downwards. These effects are clear from nitrate profiles evolution presented in Figs 6 and 7. Notice that, for a cereal root system and low transpiration rate (Fig. 7), nitrate concentration in the first 20 cm is decreased markedly after the second irrigation. The same general trend, but for deeper soil layer, can be observed from a fruit tree type of root system.

## CONCLUSION

As is well known, water redistribution in soil appears to be very sensitive to soil characteristics such as hydraulic properties and water partitioning between mobile and immobile phases. Nitrate redistribution is dependent largely on water flux and soil structure. The importance of the root system's distribution and absorption on water and nitrate redistribution appears to be the actual root pattern, but this depends on soil type (mainly soil resistance to water flow) and climatic demand.

The integrated approach which has been developed to model water and nitrate movement in soil, including uptake by roots, applies successfully when simulating the effect of root distribution and absorption. Nevertheless, the one-dimensional assumption made in solving the set of equations has some limitations in modelling water uptake by two root systems differing only by their densities. This suggests two modifications which should be included in the water-uptake model, taking into account: (i) root resistance to internal water flow; and (ii) a soil resistance term accounting for higher potential gradients in a soil layer for a low root density. The model appeared to be very sensitive to the distribution of the root system. This highlights that the identification of active parts of the root system by means of mathematical identification techniques should be possible. Under the assumption that we have a correct water-uptake model which relates the soil water potential to the uptake rate, one should be able, with the aid of the model presented in this paper, to determine the distribution of active roots in the soil profile. Another interesting application would be to simulate two-dimensional water and nitrate movement in soil and uptake by roots, thus taking into account the actual spatial distribution of the root system and the soil resistance to water flow in a soil layer.

## REFERENCES

**Addiscott, T.M. & Wagenet, R.J. (1985).** Concepts of solute leaching in soils: a review of modelling approaches. *Journal of Soil Science*, **36**, 411–424.

**Aria, L.M., Blake, G.R. & Farrel, D.A. (1975a).** A field study of soil water depletion patterns in presence of growing soybean roots. II — Effect of plant growth on soil water pressure and water loss patterns. *Soil Science Society of America Proceedings*, **39**, 430–436.

**Aria, L.M., Blake, G.R. & Farrel, D.A. (1975b).** A field study of soil water depletion patterns in presence of growing soybean roots. III — Rooting characteristics and root extraction of soil water. *Soil Science Society of America Proceedings*, **39**, 437–444.

**Atkinson, D. (1985).** The nutrient requirements of fruit trees: Some current considerations. In *Advances in Plant Nutrition* (Ed. by P.B. Tinker & A. Lauchli), Vol. **2**, 93–128. Praeger, New York.

**Baldwin, J.P., Tinker, P.B. & Nye, P.H. (1972).** Uptake of solutes by multiple root systems from soil. II — The theoretical effects of rooting density and pattern on uptake of nutrients from soil. *Plant and Soil*, **36**, 693–708.

**Baldwin, J.P., Nye, P.H. & Tinker, P.B. (1973).** Uptake of solutes by multiple root systems in soil. III — A model for calculating the solute uptake by a randomly dispersed root system developing in a finite volume of soil. *Plant and Soil*, **38**, 621–635.

**Barley, K.P. (1970).** The configuration of the root system in relation to nutrient uptake. *Advances in Agronomy*, **22**, 159–201.

**Bar-Yosef, B., Schwartz, S., Markovich, T., Lucas, B. & Assaf, R. (1988).** Effect of root volume and nitrate solution concentration on growth, fruit yield, and temporal N and water uptake rates by apple trees. *Plant and Soil*, **107**, 49–56.

**Beuving, J. (1984).** Vocht- en doorlatendheidskarakteristieken, dichtheiden samenstelling van bodemprofielen in zand-, zavel, klei- en veengronden. *Rapport* **10**, Instituut voor Cultuurtechniek en Waterhuishouding (ICW), Wageningen, 26 pp.

**Bhat, K.K.S. (1982).** Nutrient inflows into apple roots. II — Nitrate uptake rates measured on intact roots of mature trees under field conditions. *Plant, Cell and Environment*, **5**, 461–469.

**Blondel, A.-M. (1979).** Dynamique comparée de l'absorption des nitrates et de l'eau par des plantules de Blé. *Compte-Rendu de l'Académie des Sciences de Paris*, t. 288, série D, 1545–1548.

**Bresler, E. (1973).** Simultaneous transport of solutes and water under transient unsaturated flow conditions. *Water Resource Research*, **9**, 975–986.

**Bresler, E. (1981).** Models for predicting distribution of chemicals in soil during irrigation. In *Les phénomènes de transport de l'eau et des solutés et l'irrigation*, Coll. Les Colloques de l'INRA, 27–48. INRA, Paris.

**Brewster, J.L. & Tinker, P.B.H. (1972).** Nutrient flow rates into roots. *Soils & Fertilizers*, **35**, 335–359.

**Claassen, N., Syring, K.M. & Jungk, A. (1986).** Verification of a mathematical model by simulating potassium uptake from soil. *Plant and Soil*, **95**, 209–220.

**Dalton, F.N., Raats, P.A.C. & Gardner, W.R. (1975).** Simultaneous uptake of water and solutes by plant roots. *Agronomy Journal*, **67**, 334–339.

**Feddes, R.A., Bresler, E. & Neuman, S.P. (1974).** Field test of a modified numerical model for water uptake by root systems. *Water Resource Research*, **10**, 1199–1206.

**Feddes, R.A., Kowalik, P.J. & Zaradny, H. (1978).** *Simulation of Field Water Use and Crop Yield.* Halsted Press, John Wiley & Sons, New York. 188 pp.

**Gaudet, J.-P. (1978).** *Transferts d'eau et de solutés dans les sols non saturés. Mesures et simulation.* Thèse Doct. d'Etat, Université de Grenoble, 246 p.

**Grubb, P.J. (1977).** Control of forest growth and distribution on wet tropical mountains: with special reference to mineral nutrition. *Annual Review of Ecology and Systematics*, **8**, 83–107.

**Hillel, D., Talpaz, H. & van Keulen, H. (1976).** A macroscopic-scale model of water uptake by a nonuniform root system and of water and salt movement in the soil profile. *Soil Science*, **121**, 242–255.

**Novak, V. (1987).** Estimation of soil-water extraction patterns by roots. *Agricultural Water Management*, **12**, 271–278.

**Nye, P.H. & Spiers, J.A. (1964).** Simultaneous diffusion and mass flow to roots. *8th International Congress on Soil Science*, **3**, 535–542.

**Nye, P.H. & Marriott, F.H.C. (1969).** A theoretical study of the distribution of substances around roots resulting from simultaneous diffusion and mass flow. *Plant and Soil*, **30**, 459–472.

**Passioura, J.B. (1963).** A mathematical model for the uptake of ions from the soil solution. *Plant and Soil*, **18**, 225–238.

**Passioura, J.B. & Frere, M.H. (1967).** Numerical analysis of the diffusion and convection of solutes to roots. *Australian Journal of Soil Research*, **5**, 149–159.

**Phillips, R.E., Nanagara, T., Zartman, R.E. & Leggett, J.E. (1976).** Diffusion and mass flow of nitrate-nitrogen to plant roots. *Agronomy Journal*, **68**, 63–66.

**Robin, P. (1983).** *Contribution à l'étude de la réduction du nitrate chez les plantes cultivées.* Thèse Doc. d'Etat, U.S.T.L., 144 pp.

**Sanders, F.E., Tinker, P.B. & Nye, P.H. (1970).** Uptake of solutes by multiple root systems from soil. I — An electrical analog of diffusion to root systems. *Plant and Soil*, **34**, 453–466.

**Schulze, E.-D. & Bloom, A.J. (1984).** Relationship between mineral nitrogen influx and transpiration in radish and tomato. *Plant Physiology*, **76**, 827–828.

**Tardieu, F. (1988).** Conséquences de la disposition spatiale des racines au champ sur les transferts d'eau sol-plante. In *Etudes sur les Transferts d'Eau dans le System Sol-Plante–Atmosphère* (Ed. by R. Calvet), pp. 181–211. INRA (Paris).

**Van Genuchten, M.Th. (1983).** Analysing crop salt tolerance data: Model description and user's manual. *Research Report No* **120**, U.S. Salinity Laboratory, Riverside, California, 50 pp.

**Van Genuchten, M.Th. & Wierenga, P.J. (1976).** Mass transfer studies in sorbing porous media. I — Analytical solution. *Soil Science Society of America Journal*, **40**, 473–480.

**Van Genuchten, M.Th., Tang, D.H. & Guennelon R. (1984).** Some exact solutions for solute transport through soils containing large cylindrical macropores. *Water Resource Research*, **20**, 335–346.

# Root and shoot activity of two grasses with contrasted growth rates in response to low nutrient availability and temperature

N. KACHI* AND I. H. RORISON
*Unit of Comparative Plant Ecology (NERC), Department of Animal and Plant Sciences, The University, Sheffield S10 2TN, UK*

## SUMMARY

1 Seedlings of two grasses, potentially fast-growing *Holcus lanatus* and potentially slow-growing *Festuca ovina*, were grown for 28 days in a flow culture system with two temperature regimes simulating summer (20°C/15°C diurnal cycle) and autumn (12°C/5°C). Plants were supplied with one nutrient solution of a dilution series containing $NO_3^-$; at 17 mmol $m^{-3}$ (low N), 90 mmol $m^{-3}$ (medium N) or 180 mmol $m^{-3}$ (high N). Dry weight ($W = W_s + W_r$) and total nitrogen accumulation per plant ($N$), specific shoot activity (SSA $= 1/W_s \, dW/dt$), specific root activity (SRA $= 1/W_r \, dN/dt$) and penetration of the root system were measured from weekly harvests, where $W_s$ is the shoot dry weight, $W_r$ is the root dry weight and $t$ is the time.

2 In medium and high N, *H. lanatus* showed higher SSA and SRA and accumulated more dry matter and nitrogen than *F. ovina* at warm and cool temperatures. In low N, SSA and growth rates of *H. lanatus* resembled those attained by *F. ovina*.

3 In low N, *F. ovina* exhibited higher SRA than *H. lanatus*. *F. ovina* accumulated greater amounts of nitrogen than *H. lanatus* particularly at cool temperatures.

4 There was a positive correlation between shoot nitrogen concentrations and SSA in *H. lanatus*, but not in *F. ovina*. This may have resulted from the greater dependency upon nitrogen supply in *H. lanatus*.

5 SRA increased with increasing external concentration of $NO_3^-$ in the medium. In both species, it was saturated at about 40 mmol $m^{-3}$, which is lower than reported for crop plants.

6 In low N, *H. lanatus* produced a longer root system with fewer, finer roots than *F. ovina*. This morphological response may result in a reduction of physical toughness of the root system, and could contribute to seedling mortality following frost heaving of the soil profile during winter.

---

* Present address: Global Environmental Research Division, National Institute for Environmental Studies, Tsukuba, Ibaraki 305, Japan.

## INTRODUCTION

Among plant species native to calcareous grasslands in the British Isles, a clear gradient of potential growth rates has been recognized (Grime & Hunt 1975). With increasing soil fertility, the relative abundance of slow-growing species is decreased (Hodgson 1986a, b).

Nitrogen together with phosphorus is the nutrient which primarily limits plant growth in infertile, semi-natural grasslands (Chapin 1980). Several workers have reported ecophysiological aspects of dry matter growth and nitrogen accumulation for ecologically contrasted species in relation to nitrogen availability and/or temperature (Bradshaw *et al.* 1964; Peterkin 1981; Rorison, Peterkin & Clarkson 1983; Robinson & Rorison 1987). However, even the lowest level of nitrogen supplied by previous workers was still higher than that available under some infertile field conditions. All the nitrogen concentrations found in laboratory-grown plants have been much higher than those observed in field-grown plants (Rorison 1971; Rorison, Gupta & Hunt 1986), even though plant growth itself was reduced compared with control plants supplied with sufficient nutrients. In order to obtain ecologically meaningful information about plant nutrition, it is necessary, therefore, to investigate plant responses at nitrogen availabilities comparable with those in the field.

In many grass species, seedling establishment is the main source of population expansion in a semi-natural grassland (Hillier 1984). In the field, germination occurs mainly in late summer or in early autumn (Grime 1979; Grime, Hodgson & Hunt 1988) and therefore, seedlings grow at suboptimal temperatures. The seedling stage is most critical for the establishment of populations, because seedlings are usually more susceptible to environmental hazards than adult plants (Harper 1977; Rorison 1960). Seedlings germinated in autumn may die in winter through solifluction and frost-heaving of surface soils (Davison 1964). In this case, seedling growth before the onset of winter will be important for survival, because larger seedlings with a well-developed root system are more likely to be resistant to such physical disturbances.

The objectives of this study are to compare growth responses of potentially fast- and slow-growing species to low nutrient availabilities at optimal (summer) and suboptimal (autumn) temperatures and to relate these characteristics to seedling growth and survival in the field.

## SPECIES

Two grasses, *Holcus lanatus* L. and *Festuca ovina* L.* were chosen for the study. *H. lanatus* is a potentially fast-growing species of wide ecological distribution, but occurs most frequently on relatively fertile soils, while *F. ovina* occurs predomi-

* Nomenclature follows Clapham, Tutin & Warburg (1981).

nantly on soils of infertile grasslands and has an inherently low growth rate (Grime & Hunt 1975; Grime, Hodgson & Hunt 1988). Both species are common in unmanaged, calcareous grasslands.

Seeds of both species were obtained from populations grown on a south-facing slope of semi-natural, calcareous grassland in Lathkill Dale, Derbyshire (53°11′N, 1°44′W) (Rorison, Sutton & Hunt 1986; Rorison, Gupta & Hunt 1986). They were air-dried and stored at room temperatures until used. Mean seed weight was 0·32 mg for *H. lanatus* and 0·33 mg for *F. ovina*.

## FLOW-CULTURE EXPERIMENT

### *Culture solutions*

The nutrient solution was based on that of Rorison (in Hewitt 1966, Table 30c) and diluted with deionized water to obtain three nitrate concentrations; 17 mmol $m^{-3}$ (low N), 90 mmol $m^{-3}$ (medium N) and 180 mmol $m^{-3}$ (high N). In order to compare plant growth under nutrient conditions similar to those achieved along a gradient of soil fertility in the field, and also to avoid possible adverse effects of imbalance of other nutrient elements relative to nitrogen, we used the diluted solutions rather than those in which concentrations of all elements except for nitrogen are kept constant. Although in our experiment, we cannot assess the relative contribution of different nutritional elements to plant growth quantitatively, nitrogen is the most likely limiting element in the dilution series, because they contain enough P, K, Ca, Mg and micro-nutrients, relative to nitrogen, demanded for plant growth (the ratios of molar concentrations of P, K, Ca and Mg to N in culture solutions were 0·25, 0·5, 0·5 and 0·25, while the corresponding figures in grasses from calcareous sites are 0·04, 0·4, 0·11 and 0·05, Rorison 1971). pH in the reservoir of culture solution was adjusted initially to 4·5 by adding 0·5 M $H_2SO_4$, but the pH around the roots rose to 6·0 as the solution became depleted. However, this is within a range of pH at which both species show normal growth (Grime, Hodgson & Hunt 1988).

### *Flow-culture system*

Plants were grown in a continuous-flow culture system with a flow rate maintained at 29 ml $h^{-1}$ $pot^{-1}$ by a peristaltic pump. Daily rates of nitrate supply per pot were controlled at 12, 63 and 125 μmol for the three respective nutrient treatments.

### *Determination of effective nitrate concentration*

A preliminary experiment revealed that considerable depletion of nitrogen occurred at concentrations used in the present experiment. It was due to absorption not only by plant roots, but also by micro-organisms including algae which inevitably grew in the pots. In this case, a theoretical estimation of solution depletion in the medium produces an underestimate, because it is based on an assumption that solution depletion is caused only by plant uptake (Edwards & Asher 1974).

Therefore, we determined directly an effective concentration by sampling and analysis. Solution flowing out of each pot was sampled at weekly intervals and nitrate concentration was analysed by a distillation method with Devarda's alloy. The effective nitrate concentration was estimated by the geometrical mean of concentrations of the out-flowing and reservoir solution, assuming that root systems and micro-organisms were distributed homogeneously in a pot and they absorbed nitrate at a rate proportional to the external concentration. No depletion in reservoir solutions was detected during the experiment.

### *Environmental conditions*

The experiment was conducted in a controlled-environment growth room (Rorison 1964). Experimental conditions involved photosynthetically active irradiance of 110–130 μmol photon $m^{-2} s^{-1}$ at plant level over a day length of 16 h, with the maximum temperature during a day period of 20 °C or 12 °C and the minimum temperature during a night period of 15 °C or 5 °C. These respective temperature regimes approximately correspond to mean summer and autumn conditions in Lathkill Dale (Rorison, Sutton & Hunt 1986).

### *Procedure*

Seeds were germinated on floats of Alkathene beads in trays in the growth room at 20 °C day/15 °C night. Each tray contained a culture solution of the same strength as was intended to be used for the subsequent flow culture. Germination started in 3 days in both species and on the seventh day after seed-sowing, thirteen seedlings were transferred to each 1000 ml pot. Another twenty seedlings from the three nutrient treatments were harvested to determine initial dry weights and nitrogen contents. Five replications of pots were used. The time of seedling transfer was designated as day 0, and five, three, three and two plants were taken from each pot on days 7, 14, 21 and 28, respectively. After root penetration (length of the longest root) was measured, the plants were oven-dried at 80 °C for 48 h, weighed and analysed for total nitrogen content using the micro-Kjeldahl digest and an automated colorimetric technique (Allen 1974).

### *Data analyses*

Log–quadratic growth curves were fitted to dry weights and total nitrogen contents of shoots and roots for all plant data by the method of least squares (Hunt & Parsons 1974). In all cases, observed means were within the 95% confidence limits of the fitted curves. The obtained growth curves were used to calculate specific shoot activity (SSA = $1/W_s \, dW/dt$) and specific root activity (SRA = $1/W_r \, dN/dt$), where $W$ is the total dry weight, $W_s$ is the shoot dry weight, $W_r$ is the root dry weight, $N$ is the total N content per plant and $t$ is the time. SSA is a measure of the shoot activity for dry matter accumulation and SRA is a measure of the root activity for nitrogen absorption (Hunt 1982).

## RESULTS AND DISCUSSION

### *Dry weight growth*

In high N at warm temperatures, *H. lanatus* showed the greatest growth (Fig. 1). The mean value of relative growth rate (RGR) averaged over 5 weeks was similar to that of maximal RGR reported by Grime & Hunt (1975) and Robinson & Rorison (1987). In high and medium N, lowering temperatures reduced growth rates in both species, but *H. lanatus* was more sensitive than *F. ovina* (Fig. 1). Such differences between fast- and slow-growing species have also been reported by Rorison (1980) and Peterkin (1981). In low N, however, there was no significant effect of temperature on plant growth even in *H. lanatus*. This means that in low N, plant growth was primarily limited by the nutrient supply.

In medium and high N, *H. lanatus* accumulated dry matter at higher rates than *F. ovina* with both temperature regimes (Fig. 1). In low N, growth rates of *H.*

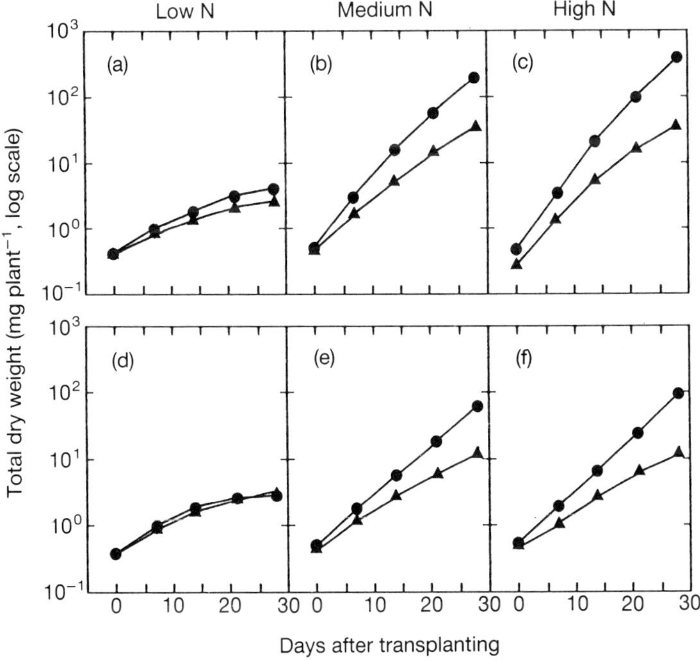


FIG. 1. Fitted curves of total dry weight growth of *H. lanatus* ( ● ) and *F. ovina* ( ▲ ) grown in a flow culture system at three nutrient levels with warm (20 °C day/15 °C night) (a, b, c) and cool (12 °C day/5 °C night) temperature regimes (d, e, f). In all cases, 95% confidence limits lie within the marked point of the mean.

*lanatus* were reduced to a level comparable to that attained by *F. ovina*. Consequently, both species showed similar growth. This result is in accordance with the general trend that the growth of fast-growing species is more sensitive to differential nutrient availabilities (Rorison 1968; Robinson 1983; Wild *et al.* 1974).

### *Specific shoot activity (SSA)*

Generally, trends of time-courses of SSA were approximately parallel to those of RGR. Higher SSAs were obtained under higher nitrogen and/or higher temperature conditions (Fig. 2). Again *H. lanatus* was more sensitive in reducing SSA at lower nutrient availabilities and at lower temperatures. On an absolute basis, *H. lanatus* showed even lower SSA than *F. ovina*, when they were grown in low N at low temperatures.

### *Total nitrogen accumulation*

Nitrogen accumulation by plants showed approximately similar trends to those of dry matter growth. In medium and high N, *H. lanatus* accumulated nitrogen at

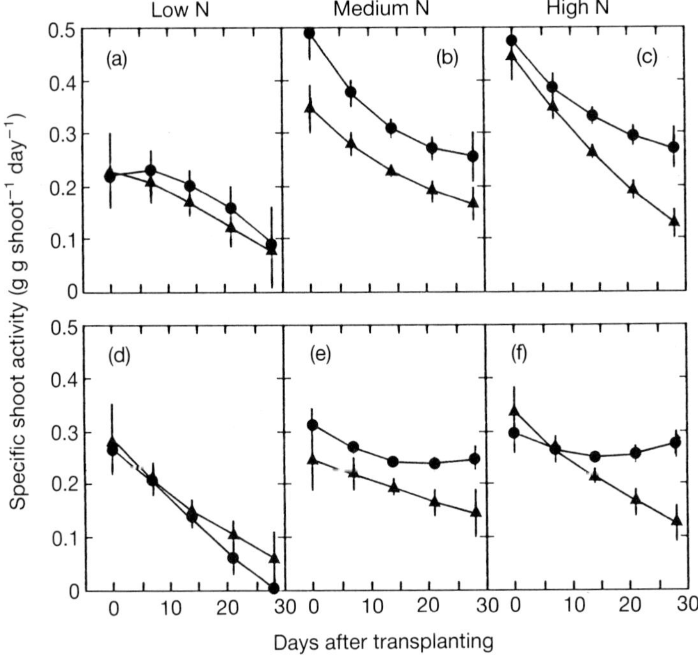


FIG. 2. Fitted progress curves of specific shoot activity for dry matter accumulation in *H. lanatus* ( ● ) and *F. ovina* ( ▲ ). Details of the experimental conditions as for Fig. 1. Each symbol is the fitted mean (±95% confidence limits) of five replicates.

higher rates than did *F. ovina*. In low N, accumulation rates relative to those of *F. ovina* were much reduced (Fig. 3). Particularly at cool temperatures, *F. ovina* accumulated more nitrogen than *H. lanatus* even on an absolute basis.

In both species, when they were grown in low N, nitrogen concentrations in shoots and roots were reduced to 10–20 mg $g^{-1}$, a comparable level observed in field-grown plants (Rorison 1971; Rorison, Gupta & Hunt 1986). Therefore, it can be said that the nutrient availability in the low N treatment of the present experiment approximates to that achieved in the field.

### *Specific root activity (SRA)*

In medium and high N, *H. lanatus* showed, on average, higher SRA than *F. ovina* (Fig. 4). In contrast, in low N, the opposite was observed.

Under higher nitrogen conditions, the SRA of *H. lanatus* tended to decline with time. This is probably because of the considerable depletion of the external medium by this fast-growing species (see also Fig. 6).

When plants were grown in low N, it appeared that SRA increased with time in spite of the solution depletion. This was more remarkable in *F. ovina* (Fig. 4). These increases in SRA might be caused by morphological adaptations to a low

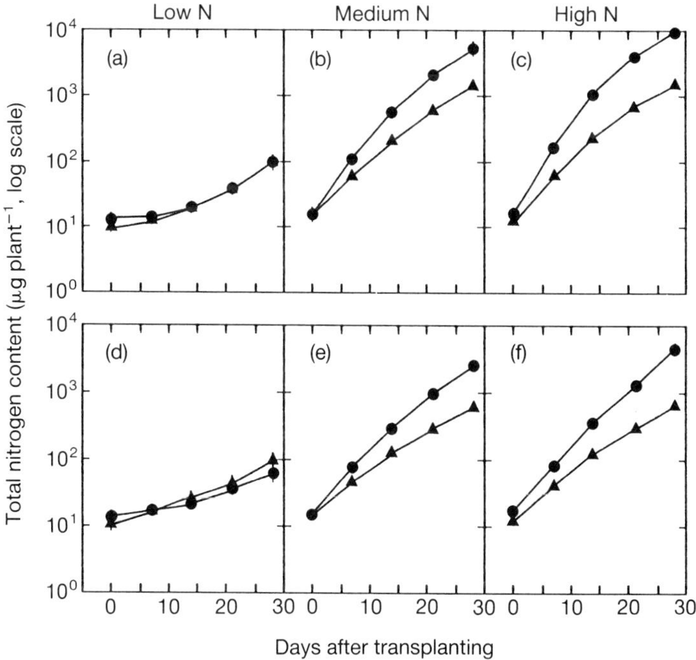


FIG. 3. Fitted progress curves of total nitrogen content per plant in *H. lanatus* ( ● ) and *F. ovina* ( ▲ ). Details of the experimental conditions as for Fig. 1. Each symbol is the fitted mean (± 95% confidence limits) of five replicates.

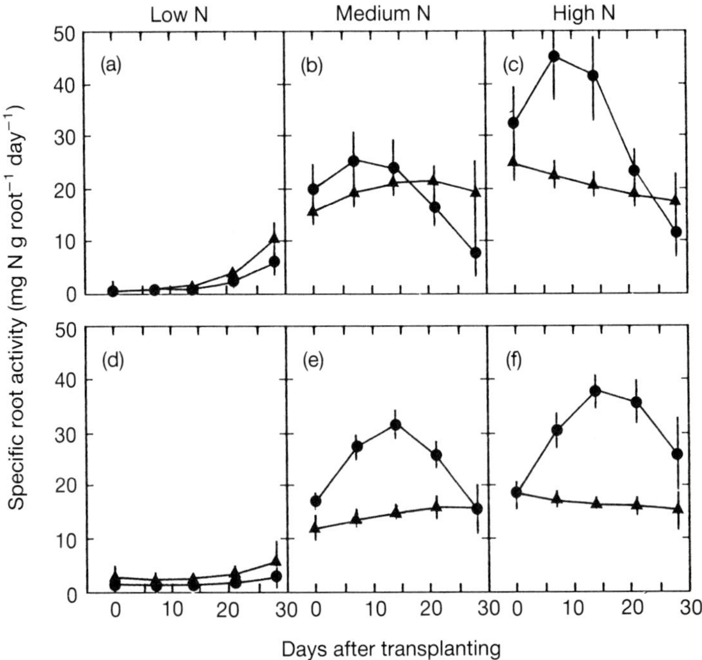


FIG. 4. Fitted progress curves of specific root activity for nitrogen accumulation in *H. lanatus* ( ● ) and *F. ovina* ( ▲ ). Details of the experimental conditions as for Fig. 1. Each symbol is the fitted mean (± 95% confidence limits) of five replicates.

nitrogen supply through production of more fine roots and/or root hairs (Robinson and Rorison 1987).

### *SSA in relation to shoot nitrogen concentration*

Under nutrient-limiting conditions, SSA is related to the nitrogen concentration in the shoot (Hirose & Kitajima 1986). SSA has a physiological basis in leaf photosynthesis and the photosynthetic activity depends largely on the concentration of the primary carboxylating enzyme (Rubisco), which is usually highly correlated with leaf nitrogen concentration (Field & Mooney 1986).

There were positive correlations between shoot nitrogen concentration and SSA in *H. lanatus* ($r$ = 0·82, $P$<0·05 at 20°C/15°C and $r$ = 0·79, $P$<0·05 at 12°C/5°C), but not in *F. ovina* ($r$ = 0·55 and 0·34, for respective temperature regimes) (Fig. 5). This result may reflect that the former species depends more on nitrogen supply for dry matter growth than the latter. When the nitrogen concentration was equal, *H. lanatus* showed higher SSA than *F. ovina*. This suggests that compared with *H. lanatus*, *F. ovina* has a tendency to accumulate nitrogen as photosynthetically-inactive storage and/or defensive substances rather than as metabolically-active enzymes (Bernays 1983; Millard 1988).

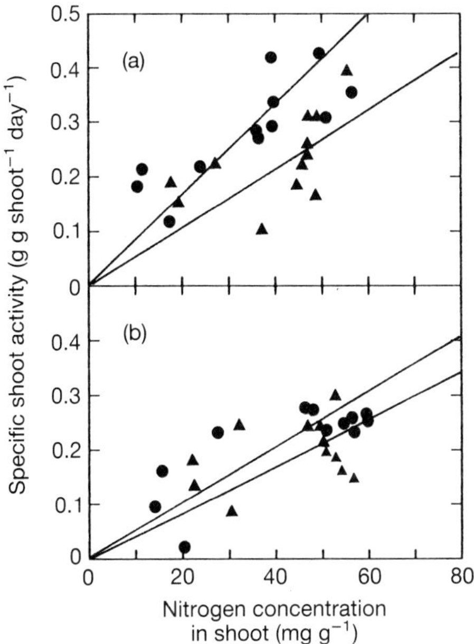


FIG. 5. Specific shoot activity for dry matter accumulation in relation to shoot nitrogen concentration in *H. lanatus* ( ● ) and *F. ovina* ( ▲ ) grown at summer (a) and autumn (b) temperatures. Details of the experimental conditions as for Fig. 1. Calculated regression lines are shown.

## *SRA in relation to external nitrate concentration*

During the experiment, nitrogen was depleted significantly at all levels of supply. Accordingly, to compare SRA between the two species properly, SRA was plotted against the effective external concentration of nitrate in the culture solution (Fig. 6). In both species, dependency of SRA on the external concentration was clear. The efficiency of nitrogen uptake was higher in *H. lanatus* at concentrations more than 40 mmol m$^{-3}$, and at that concentration, SRA appeared to be nearly saturated in both species. This concentration is actually a one hundredth of the full strength solution normally applied (Hewitt 1966). Burns (1980) concluded from a literature survey that near-maximal rates of nitrate uptake or of dry matter production occur at external concentrations of nitrate between 80 and 200 mmol m$^{-3}$. As his conclusion was mainly based on crop plants, our result suggests that nitrogen uptake of native plants grown on infertile soils is saturated at even lower concentrations. This may reflect their lower demand of nitrogen associated with lower growth rates than crop plants (Robinson & Rorison 1985).

## *Penetration of root systems*

In general, *H. lanatus* developed a longer root system than *F. ovina* (Fig. 7). This morphological character of the fast-growing species produces a relative advantage

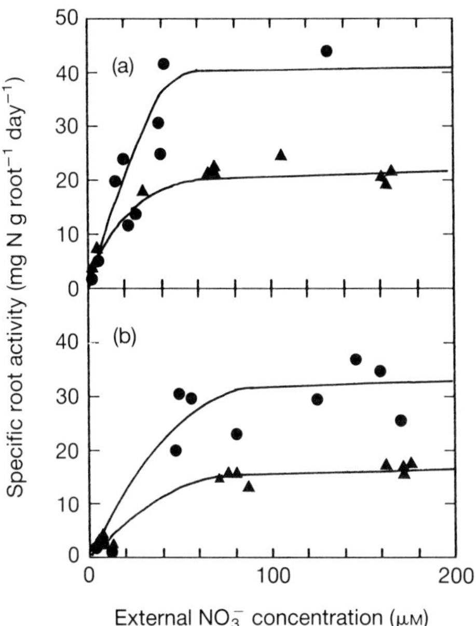


FIG. 6. Specific root activity for nitrogen accumulation in relation to external concentrations of nitrate in *H. lanatus* (●) and *F. ovina* (▲) grown at summer (a) and autumn (b) temperatures. Details of the experimental conditions as for Fig. 1. Curves are fitted by eye.

over the slow-growing species in fertile environments, because continual uptake of nutrients under such conditions depends on the extent to which the root system penetrates to the unexploited region of the soil (Crick & Grime 1987) as well as on the mobility of nutrients (Nye & Tinker 1977).

An increase in root penetration can be achieved by reducing root number and/or a root radius. In fact, *H. lanatus* developed a longer root system with fewer and finer roots than *F. ovina*, when they were grown in low N and both achieved similar root dry weight. This is in accordance with the field observation that *F. ovina* is shallow-rooted compared with *H. lanatus*. Robinson & Rorison (1983) also reported that *H. lanatus* exhibited higher specific root length than *Deschampsia flexuosa*, another slow-growing species, especially in roots subjected to a N-depleted nutrient solution. These differences in root morphology between the two species might be critical in affecting differential seedling survival. A root system with fewer and/or finer roots is expected to be less resistant to disturbance of a soil surface such as frost heaving and solifluction, which is a major cause of seedling mortality during winter (Davison 1964). However, more experimental measurements will be needed to test this hypothesis.

## CONCLUSIONS

In medium and high N, *H. lanatus* maintained its advantage in growth and nitrogen uptake over *F. ovina* at autumn temperatures as well as at summer temperatures.

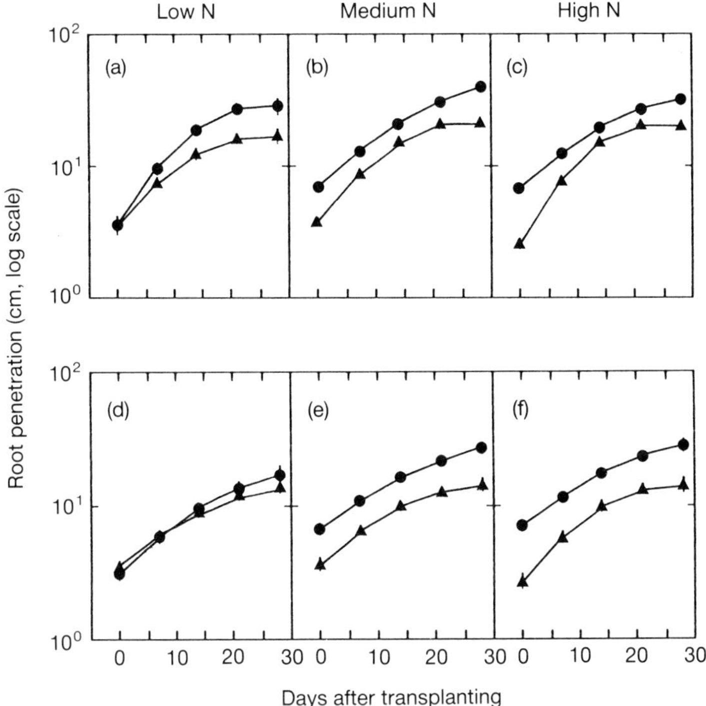


FIG. 7. Fitted progress curves of root penetration in *H. lanatus* ( ● ) and *F. ovina* ( ▲ ). Details of the experimental conditions as for Fig. 1. Each symbol is the fitted mean (± 95% confidence limits) of five replicates.

On the other hand, the relative advantage of *F. ovina* in low N was more marked at autumn temperatures than at summer temperatures. This result suggests that in infertile grasslands, seedlings of *F. ovina*, which germinate in early autumn, may attain greater growth and nitrogen accumulation, even in absolute terms, than the potentially fast-growing *H. lanatus*.

Under fertile conditions, aggressive root penetration in the fast-growing species will be advantageous in absorbing nutrients. However, under infertile conditions, a longer root system is associated with finer, fewer roots. This might result in a reduction of physical toughness of the root system against soil disturbance and consequently of seedling survival in winter.

## ACKNOWLEDGEMENTS

We are grateful to Dr P.L. Gupta and Mrs R.E. Spencer for their technical advice and assistance. The experiment was carried out while N. Kachi was on leave from the National Institute for Environmental Studies, Environment Agency of Japan and with financial support from the Science and Technology Agency of Japan. This study was supported by the Natural Environment Research Council.

## REFERENCES


**Allen, S.E. (Ed.) (1974).** *Chemical Analysis of Ecological Materials.* Blackwell Scientific Publications, Oxford.

**Bernays, E.A. (1983).** Nitrogen in defence against insects. *Nitrogen as an Ecological Factor* (Ed. by J.A. Lee, S. McNeill & I.H. Rorison), pp. 321–344. Blackwell Scientific Publications, Oxford.

**Bradshaw, A.D., Chadwick, M.J., Jowett, D. & Snaydon, R.W.(1964).** Experimental investigations into the mineral nutrition of several grass species. IV. Nitrogen level. *Journal of Ecology*, **52**, 665–676.

**Burns, I.G. (1980).** Influence of the spatial distribution of nitrate on the uptake of N by plants: a review and a model for rooting depth. *Journal of Soil Science*, **31**, 155–173.

**Chapin, F.S., III (1980).** The mineral nutrition of wild plants. *Annual Review of Ecology and Systematics*, **11**, 233–260.

**Clapham, A.R., Tutin, T.G. & Warburg, E.F. (1981).** *Excursion Flora of the British Isles*, 3rd edn. Cambridge University Press, Cambridge.

**Crick, J.C. & Grime, J.P. (1987).** Morphological plasticity and mineral nutrient capture in two herbaceous species of contrasted ecology. *New Phytologist*, **107**, 403–414.

**Davison, A.W. (1964).** *Some factors affecting seedling establishment on calcareous soils.* Ph.D. thesis, University of Sheffield.

**Edwards, D.G. & Asher, C.J. (1974).** The significance of solution flow rate in flowing culture experiments. *Plant and Soil*, **41**, 161–175.

**Field, C. & Mooney, H.A. (1986).** The photosynthesis–nitrogen relationship in wild plants. In *On the Economy of Plant Form and Function* (Ed. by T.J. Givnish), pp. 25–55. Cambridge University Press, Cambridge.

**Grime, J.P. (1979).** *Plant Strategies and Vegetation Processes.* John Wiley, Chichester.

**Grime, J.P. & Hunt, R. (1975).** Relative growth rate: its range and adaptive significance in a local flora. *Journal of Ecology*, **63**, 393–422.

**Grime, J.P., Hodgson, J.G. & Hunt, R. (1988).** *Comparative Plant Ecology: A Functional Approach to Common British Species.* Unwin & Hyman, London.

**Harper, J.L. (1977).** *Population Biology of Plants.* Academic Press, London.

**Hewitt, E.J. (1966).** *Sand and Water Culture Methods Used in the Study of Plant Nutrition*, 2nd edn. Commonwealth Agricultural Bureaux, Farnham Royal.

**Hillier, S.H. (1984).** *A quantitative study of gap recolonization in two contrasted limestone grasslands.* Ph.D. thesis, University of Sheffield.

**Hirose, T. & Kitajima, K. (1986).** Nitrogen uptake and plant growth. I. Effect of nitrogen removal on growth of *Polygonum cuspidatum*. *Annals of Botany*, **58**, 479–486.

**Hodgson, J.G. (1986a).** Commonness and rarity in plants with special reference to the Sheffield flora. Part I: The identity, distribution and habitat characteristics of common and rare species. *Biological Conservation*, **36**, 199–252.

**Hodgson, J.G. (1986b).** Commonness and rarity in plants with special reference to the Sheffield flora. Part II: The relative importance of climate, soils and land use. *Biological Conservation*, **36**, 253–274.

**Hunt, R. (1982).** *Plant Growth Curves: the Functional Approach to Plant Growth Analysis.* Edward Arnold, London.

**Hunt, R. & Parsons, I.T. (1974).** A computer program for deriving growth-functions in plant growth-analysis. *Journal of Applied Ecology*, **11**, 297–307.

**Millard, P. (1988).** The accumulation and storage of nitrogen by herbaceous plants. *Plant, Cell and Environment*, **11**, 1–8.

**Nye, P.H. & Tinker, P.B. (1977).** *Solute Movement in the Soil–Root System.* Blackwell Scientific Publications, Oxford.

**Peterkin, J.H. (1981).** *Plant growth and nitrogen nutrition in relation to temperature.* Ph.D. thesis, University of Sheffield.

**Robinson, D. (1983).** *Plant efficiency in relation to nitrogen supply and utilisation.* Ph.D. thesis, University of Sheffield.

**Robinson, D. & Rorison, I.H. (1983).** A comparison of the responses of *Lolium perenne* L., *Holcus* *lanatus* L. and *Deschampsia flexuosa* (L. ) Trin. to a localized supply of nitrogen. *New Phytologist*, **94**, 263–273.

**Robinson, D. & Rorison, I.H. (1985).** A quantitative analysis of the relationships between root distribution and nitrogen uptake from soil by two grass species. *Journal of Soil Science*, **36**, 71–85.

**Robinson, D. & Rorison, I.H. (1987).** Root hairs and plant growth at low nitrogen availabilities. *New Phytologist*, **107**, 681–693.

**Rorison, I.H. (1960).** Some experimental aspects of the calcicole–calcifuge problem. I. The effects of competition and mineral nutrition upon seedling growth in the field. *Journal of Ecology*. **48**, 585–599.

**Rorison, I.H. (1964).** A double shell plant growth cabinet. *New Phytologist*, **63**, 358–362.

**Rorison, I.H. (1968).** The response to phosphorus of some ecologically distinct plant species. I. Growth rates and phosphorus absorption. *New Phytologist*, **67**, 913–923.

**Rorison, I.H. (1971).** The use of nutrients in the control of the floristic composition of grassland. In *The Scientific Management of Animal and Plant Communities for Conservation* (Ed. by E. Duffey & A.S. Watt), pp. 65–77. Blackwell Scientific Publications, Oxford.

**Rorison, I.H. (1980).** Plant growth in response to variations in temperature: field and laboratory studies. In *Plants and Their Atmospheric Environment* (Ed. by J. Grace, E.D. Ford & P.G. Jarvis), pp. 313–332. Blackwell Scientific Publications, Oxford.

**Rorison, I.H., Peterkin, J.H. & Clarkson, D.T. (1983).** Nitrogen source, temperature and the growth of herbaceous plants. In *Nitrogen as an Ecological Factor* (Ed. by J.A. Lee, S. McNeill & I.H. Rorison), pp. 189–209. Blackwell Scientific Publications, Oxford.

**Rorison, I.H., Gupta, P.L. & Hunt, R. (1986).** Local climate, topography and plant growth in Lathkill Dale NNR. II. Growth and nutrient uptake within a single season. *Plant Cell and Environment*, **9**, 49–56.

**Rorison, I.H., Sutton, F. & Hunt, R. (1986).** Local climate, topography and plant growth in Lathkill Dale NNR. I. A twelve-year summary of solar radiation and temperature. *Plant, Cell and Environment*, **9**, 49–56.

**Wild, A., Skarlou, V., Clement, C.R. & Snaydon, R.W. (1974).** Comparison of potassium uptake by four plant species grown in sand and flowing solution culture. *Journal of Applied Ecology*, **11**, 801–812.

**Note added in proof**

The submission of this paper predates work described in the following two references:

**Kachi, N. & Rorison, I.H. (1989).** Optimal partitioning between root and shoot in plants with contrasted growth rates in response to nitrogen availability and temperature. *Functional Ecology*, **3**, 549–559.

**Kachi, N. & Rorison, I.H. (1990).** Effect of nutrient depletion on growth and on the capacity of roots to absorb nitrogen in *Holcus lanatus* and *Festuca ovina* at warm and cool temperatures. *New Phytologist*, **115**, 531–537.

# The influence of nitrogen availability on growth parameters of fast- and slow-growing perennial grasses

R.G.A. BOOT AND M. MENSINK
*Department of Plant Ecology, University of Utrecht, Lange Nieuwstraat 106, 3512 PN Utrecht, The Netherlands*

## SUMMARY

The present paper discusses the influence of nitrogen availability on the relative growth rate (RGR) and growth parameters of five perennial grasses.

1 The RGR of all species decreased with decreasing nitrogen availability. However, even with low nitrogen supply fast-growing species grew better or as well as slow-growing species.

2 Results suggested that fast-growing species are characterized by a low leaf weight ratio (LWR) and a high specific leaf area (SLA), whereas slow-growing species have a high LWR and low SLA. The between-species differences in the leaf area ratio (LAR) were mainly the result of differences in the SLA, whereas differences in LAR within-species grown at different levels of nitrogen supply were the result of a change in LWR.

3 There was no obvious relationship found between RGR and net assimilation rate (NAR). The two fastest growing species showed a strong reduction in NAR at low nitrogen supply when compared to the high nitrogen treatment. This reduction coincided with low values of nitrogen concentration in the leaves (<1%).

## INTRODUCTION

Species from contrasting habitats differ in their relative growth rate (RGR) under optimal growing conditions. Disturbed and productive habitats are dominated by inherently fast-growing plant species, whereas slow-growing species are often associated with more stable and less productive sites (Grime & Hunt 1975). Plant species also differ in the relative importance of their leaf area ratio (LAR, the amount of leaf area per unit of total dry weight), and net assimilation rate (NAR, the dry weight increment per unit leaf area per time), of the RGR (Corré 1983; Waring *et al.* 1985; Lambers & Dijkstra 1987). Differences in RGR of species from contrasting habitats under suboptimal growing conditions have received less attention than differences under more optimal conditions. In short-term growth experiments, however, species from infertile sites tend to grow more slowly than those from more fertile environments, even at very low nutrient availability (Chapin 1980).

What causes the reduction in relative growth rate when plants are grown at low nutrient supply? Is it the result of a decrease in LAR or NAR, respectively, the morphological or physiological component of the RGR?

Brouwer (1963, 1983) showed that when nitrogen is limiting growth both the dry matter distribution within the plant and the nitrogen concentration in the plant is affected. The effect of a low nitrogen supply on dry matter distribution has been documented (Bradshaw *et al.* 1964; Chapin 1980); it caused an increase in the root : shoot ratio and a reduction in the LAR (Waring *et al.* 1985). The low nitrogen concentration in the plant, on the other hand, might influence NAR, the overall balance between photosynthesis and respiration. In a comparison of RGR and NAR of *Salix aguatica*, receiving either a low or high nitrogen supply, Waring *et al.* (1985) found that the difference in NAR between the nitrogen treatments was only modest whereas the difference in RGR was significant. Hirose (1988) also found the NAR to be constant over a wide range of nitrogen concentrations in the leaf. Only when the nitrogen concentration in the leaf was reduced to $<1\%$, did NAR decrease sharply.

The present study addresses the following questions:

**1** Does the relative importance of the morphological and physiological component of the RGR differ between plant species of contrasting habitats?

**2** Does the RGR of inherently fast- and slow-growing species differ under sub-optimal nutrient conditions?

**3** How are the morphological and physiological component of the RGR affected by the availability of nitrogen?

## MATERIAL, METHODS AND MEASURED PARAMETERS

Five perennial grass species were grown in a sand culture experiment at two levels of nitrogen. The species used, listed in order of increasing nutrient availability in their natural habitats, were: *Deschampsia flexuosa* (L.) Trin., *Festuca ovina* L., *Molina caerulea* (L.) Moench., *Festuca rubra* L. and *Holcus lanatus* L. The experiment was carried out in a greenhouse in late summer. Plants were grown from seeds in 0·6 l plastic containers. Nitrogen treatment, applied as $NH_4NO_3$, was started for all species after they had been grown for 2–4 weeks in sand supplied with a full strength nutrient solution. In the high N treatment plants received, three times a week, 10 ml of 1·0 mM $NH_4NO_3$; the low N plants received 2% of this amount. Nutrients other than nitrogen were provided by a full strength solution of the following composition: 0·27 mM $SO_4^{2-}$, 0·19 mM $H_2PO_4^-$, 0·98 mM $K^+$, 0·60 mM $Ca^{2+}$, 0·27 mM $Mg^{2+}$, 2 μM $Mn^{2+}$, 0·85 μM $Zn^{2+}$, 0·15 μM $Cu^{2+}$, 0·02 mM $H_2BO_3^-$, 0·25 μM $MoO^{2+}$, 0·04 mM $Fe^{2+}$.

For each species and treatment six plants were harvested every 2 weeks over a total period of 10 weeks. At each harvest the plants were separated into leaves, support tissue (stalk and stem), and roots. Fresh and dry weights (48 h drying at 70 °C) were measured. Prior to weight measurement leaf area and root length were recorded. Leaf area, for the two broad-leaved grasses (*H. lanatus* and *M. caerulea*), was measured with a leaf area meter (TFDL-Wageningen). The leaf areas for the three fine-leaved grasses (*D. flexuosa*, *F. ovina* and *F. rubra*) were calculated morphometrically using measurements of leaf length and diameter. Leaf area for these species was calculated as half of the outer surface of the leaf cylinder. Root length was assessed using the line-intersect method (Tennant 1975). From these primary data the following parameters were derived: leaf area ratio (LAR, leaf area / total plant dry weight, $m^2 kg^{-1}$), leaf weight ratio (LWR, leaf dry weight / total plant dry weight, $g g^{-1}$), root weight ratio (RWR, root dry weight / total plant dry weight, $g g^{-1}$), specific leaf area (SLA, leaf area / leaf dry weight, $m^2 kg^{-1}$), specific root length (SRL, root length / root dry weight, $m g^{-1}$). The mean relative growth rate (RGR, in $mg\ g^{-1}$ (total plant) $day^{-1}$) of each species in each treatment was determined from the linear regression of the natural logarithms of total dry weight on time (Hunt 1982). The physiological component of the RGR, the net assimilation rate (NAR, $g m^{-2}$ (leaf area) $day^{-1}$) was calculated as: NAR = $[(W_{T_2} - W_{T_1})/(T_2 - T_1)] \times [(\ln A_{T_2} - \ln A_{T_1})/(A_{T_2} - A_{T_1})]$, where $W$ is the total dry weight of the plant, $A$ is leaf area in $m^2$ and $T$ is time in days.

## RESULTS

### *Relative growth rate*

The relative growth rate of the plant species under high nitrogen conditions corresponded to the nutrient availability in the natural habitat (Table 1). The fast-growing species were those from fertile sites, while the slow-growing species originated from unproductive sites. The relative growth rate was lower when the availability of nitrogen was low, although even with low nitrogen supply, plant species from fertile sites grew better or as well as those from unproductive sites.

### *Leaf area ratio*

Table 2 presents the morphological component of the relative growth rate (LAR), the relative investment of dry weight in the leaves (LWR) and the specific leaf area (SLA). The leaf area ratio at low N tended to increase in the same sequence as the

TABLE 1. The relative growth rate of five perennial grasses at low and high N availability. R (reduction) is the difference in RGR between low and high N expressed as a percentage of the RGR scored in the high N treatment

| | RGR ($mg\ g^{-1}\ day^{-1}$) | | |
|---|---|---|---|
| | Low N | High N | R (%) |
| *H. lanatus* | 42 | 65 | 35 |
| *F. rubra* | 49 | 63 | 22 |
| *M. caerulea* | 54 | 62 | 13 |
| *F. ovina* | 47 | 55 | 15 |
| *D. flexuosa* | 31 | 36 | 14 |

TABLE 2. The influence of nitrogen availability on the leaf area ratio (LAR), the leaf weight ratio (LWR) and the specific leaf area (SLA) of five perennial grasses. Mean values over a time interval of 10 weeks (for *H. lanatus* 8 weeks). R (reduction) is the difference between low and high N expressed as a percentage scored in the high N treatment

| | LAR ($m^2$ $kg^{-1}$) | | | LWR ($g$ $g^{-1}$) | | | SLA ($m^2$ $kg^{-1}$) | | |
|---|---|---|---|---|---|---|---|---|---|
| | Low N | High N | R% | Low N | High N | R% | Low N | High N | R% |
| *H. lanatus* | 20·0 | 23·9 | 16 | ·39 | ·41 | 5 | 56 | 56 | 0 |
| *F. rubra* | 13·6 | 14·5 | 6 | ·38 | ·43 | 12 | 37 | 35 | −6 |
| *M. caerulea* | 11·0 | 12·5 | 12 | ·40 | ·45 | 11 | 29 | 28 | −4 |
| *F. ovina* | 10·5 | 11·2 | 6 | ·50 | ·56 | 11 | 20 | 20 | 0 |
| *D. flexuosa* | 10·2 | 12·6 | 19 | ·49 | ·55 | 11 | 21 | 23 | 9 |

relative growth rate at high N. The LAR of all species at high N was higher than the LAR at low N.

The LAR itself is the product of the LWR and the SLA. The LWR of fast-growing species tended to be lower than that of slow-growing species, although the absolute difference between the fastest and slowest growing species was small (Table 2). Consequently, the SLA showed the same pattern as the LAR. Fast-growing species are characterized by a high SLA, with the absolute difference between *H. lanatus* and *D. flexuosa* being large. Interspecific differences in the LAR result mainly from differences in the SLA. In this experiment the SLA of these species more than compensated for differences in LWR.

The increase in LWR with increased nitrogen availability ranged between 5 and 12%. Fast- and slow-growing species responded equally well in terms of their LWR. The response of the SLA to nitrogen availability was not significant for any species.

In summary the data suggest that inherently slow-growing species tend to have a relatively high LWR but a low SLA. Fast-growing species invest less dry weight in leaves but tend to have a high SLA.

### *Net assimilation rate*

There was no obvious relationship between the RGR and NAR either at low or high nitrogen availability (Table 3). An increase in the availability of nitrogen

TABLE 3. The net assimilation rate of five perennial grasses at low and high N availability. R (reduction) is the difference between low and high N expressed as a percentage scored in the high N treatment

| | NAR ($g$ $m^{-2}$ $day^{-1}$) | | |
|---|---|---|---|
| | Low N | High N | R (%) |
| *H. lanatus* | 2·1 | 2·6 | 19 |
| *F. rubra* | 3·6 | 4·3 | 16 |
| *M. caerulea* | 4·9 | 5·0 | 2 |
| *F. ovina* | 4·5 | 4·9 | 8 |
| *D. flexuosa* | 3·0 | 2·9 | −3 |

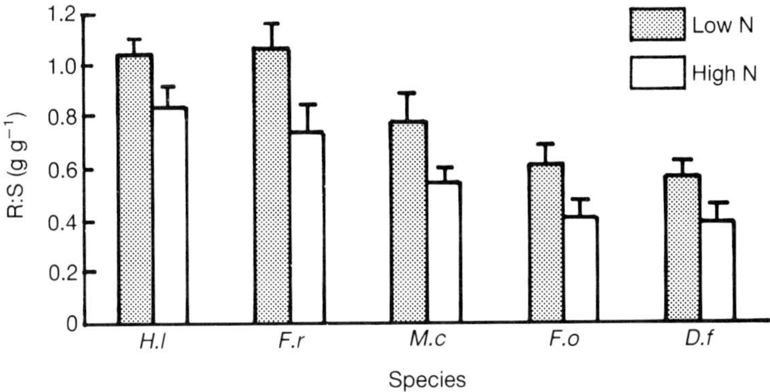


FIG. 1. The influence of nitrogen availability on the root : shoot (R : S) ratio (dry weight) of five perennial grasses. Values are the mean (and S.D.) of five harvests (six replicates each) over a period of 10 weeks. Species are ranked from left to right in order of decreasing RGR at high nutrient supply. H.l = *Holcus lanatus*, F.r = *Festuca rubra*, M.c = *Molinia caerulea*, F.o = *Festuca ovina* and D.f = *Deschampsia flexuosa*. On the $P < 0{\cdot}01$ level the root : shoot ratio for low and high N supply is significantly different for all species. (Two-way-ANOVA followed by Fisher PLSD test.)

resulted in *F. ovina* in a rather modest increase in the NAR and in *D. flexuosa* and *M. caerulea* in no change. *H. lanatus* and *F. rubra* on the other hand responded to low nitrogen availability by a reduction in the NAR of 19 and 16% respectively of their values with high nitrogen.

### *Functional ratios*

The root : shoot ratios showed that all species invested relatively more in root biomass with low nitrogen availability (Fig. 1). Secondly, the data suggest that species from fertile habitats invest relatively more dry weight in roots than do species from infertile sites.

Interpreting root : shoot ratios in terms of the functional equilibrium between the above- and below-ground parts of the plant can be misleading, since the dry weight of roots and shoots is a poor measure of the absorptive 'surfaces', the leaf surface area and the root surface area or root length of the plant (Körner & Renhardt 1987).

In Table 4 the functional equilibrium is expressed as leaf area : root length. While there was a positive correlation between root : shoot ratio and RGR at high nitrogen availability, the leaf area : root length ratio was not correlated with RGR. All species, however, invested relatively more in leaf area when grown at high nitrogen availability. In slow-growing species the leaf area : root length ratio changed more markedly between low and high nitrogen supply than it did in fast-growing species: e.g. 21% for *H. lanatus* and 34% for *D. flexuosa*.

TABLE 4. The leaf area : root length ratio of five perennial grasses at low and high N availability. Values are the mean of five harvests and six replicates each. R (Reduction) is the relative difference between low and high N expressed as a percentage of the ratios scored in the high N treatment

| | Leaf area : root length ($cm^2$ $m^{-1}$) | | |
|---|---|---|---|
| | Low N | High N | R (%) |
| *H. lanatus* | 0·81 | 1·02 | 21 |
| *F. rubra* | 0·81 | 0·97 | 16 |
| *M. caerulea* | 1·53 | 1·97 | 22 |
| *F. ovina* | 0·57 | 0·78 | 27 |
| *D. flexuosa* | 0·92 | 1·40 | 34 |

## DISCUSSION AND CONCLUSIONS

Differences in RGR between five perennial grass species at high nitrogen levels corresponded to differences in the nitrogen availability in their natural habitat. Species from fertile sites were fast-growing, whilst slow-growing species originated from unproductive sites. This is in agreement with the findings of Clarkson (1967), Rorison (1968) and Grime & Hunt (1975). As expected, the RGR of all species was lower when the availability of nitrogen was lower but the results showed that inherently fast-growing species still grew better or equally well at low nitrogen conditions than did slow-growing species.

The results suggest that fast-growing species are characterized by a low LWR and a high SLA, whereas slow-growing species have a high LWR and a low SLA. In this set of species between-species differences in the LAR are mainly the result of differences in the SLA, whereas differences in LAR within-species grown at different levels of nitrogen supply is the result of a change in LWR.

The phenotypic response in LWR correlated, not surprisingly, with the root : shoot ratio. With a low nitrogen supply all species invested relatively more dry weight in their roots than with a high nitrogen supply. This phenotypic response, however, is opposite to the relationship between root : shoot ratio and RGR. Fast-growing species from fertile environments were characterized by high root : shoot ratios, whereas slow-growing species from less-productive environments tended to invest less dry weight in their roots. A similar relation was found by Berendse & Elberse (1990).

Chapin (1980 and references therein, p. 241) suggested the opposite relationship between root : shoot ratio and relative growth rate: i.e. that the species of infertile habitats invest more biomass in their roots relative to the shoot than species associated with productive environments. This discrepancy may be due to the fact that rapidly growing species from high-nutrient habitats show considerable phenotypic plasticity in root : shoot ratio. Fast-growing species in general have a higher ratio at low availability and a lower ratio at high availability than slow-growing species (Christie & Moorby 1975). The availability of nitrogen in the high supply treatment in this experiment may have been suboptimal with the result that

TABLE 5. Nitrogen concentration of five perennial grasses at low and high N availability. Mean values over a time interval of 10 weeks

| | [N] in mg $g^{-1}$ | |
|---|---|---|
| | Low N | High N |
| *H. lanatus* | 12·0 | 14·7 |
| *F. rubra* | 16·6 | 20·3 |
| *M. caerulea* | 21·0 | 27·0 |
| *F. ovina* | 19·2 | 25·6 |
| *D. flexuosa* | 16·4 | 21·0 |

the fast-growing species increased their root : shoot ratio as a response to the suboptimal nitrogen supply more than the slow-growing species. The low nitrogen concentration in the plant (Table 5) and the rather low RGR of *H. lanatus* and *D. flexuosa* in comparison with data from Robinson & Rorison (1985) suggest that the high nitrogen supply was not optimal. However, even when nitrogen availability is optimal at the start of the experiment, fast-growing species will encounter nutrient stress in the course of the experiment at an earlier time than slow-growing species. This is a drawback of all fertilization experiments where a fixed amount of nutrients is added over time and no allowance is made of the size of the plant.

Two of the five species, *F. rubra* and *H. lanatus*, showed a strong reduction in their net assimilation rate at low nitrogen supply when compared to the high nitrogen treatment. The fact that the two fastest growing species in the experiment showed a more than modest difference between their NAR at low and high N supply might be the result of the low nitrogen concentration in the leaves at the end of the experiment; 7·9 and 10·3 mg $g^{-1}$ (N) for *H. lanatus* and *F. rubra*, respectively. This is in agreement with Hirose (1988), who showed that the NAR is constant over a wide range of nitrogen concentrations in the leaves, but is reduced when nitrogen concentration in the leaves becomes less than 1%.

The ratios of leaf area to root length showed a different relation with the RGR at high nitrogen supply than the root : shoot ratio did. Considering the relationship between the leaf area : root length ratio and the growth response, the question arises : How should plasticity be measured and interpreted? In this case, e.g. the relative reduction (R% in Table 4) in the leaf area : root length ratio increases with decreasing RGR at high nitrogen supply, whereas, the relative reduction in RGR (R% in Table 1) decreases with decreasing RGR at high nitrogen supply. In other words, there is a marked response in the leaf area : root length ratio of slow-growing species when the nitrogen treatments are compared, whereas for these species the response, expressed in terms of the relative reduction in the RGR (R% in Table 1) is small. For fast-growing species the opposite is true. The leaf area : root length ratio in these species changes modestly when the availability of nutrients drops, whereas their RGR reduces to a much larger extent under low nitrogen supply.

## ACKNOWLEDGEMENTS

The authors wish to thank H. Lambers and M.J.A. Werger for critical comments to earlier versions of the manuscript.

## REFERENCES

**Berendse, F. & Elberse, W. (1990).** Competition and nutrient availability in heathland and grassland ecosystems. In *Perspectives in Plant Competition* (Ed. by J. Grace & D. Tilman), pp. 93–115. Academic Press, London.

**Bradshaw, A.D., Chadwick, M.J., Jowett, D., Lodge, R.W. & Snaydon, R.W. (1964).** Experimental investigations into the mineral nutrition of several grass species. Part IV. Nitrogen level. *Journal of Ecology*, **52**, 665–676.

**Brouwer, R. (1963).** Some aspects of the equilibrium between overground and underground plant parts. *Mededelingen Instituut voor Biologisch en Scheikundig Onderzoek Landbouwgewassen*, **213**, 31–39.

**Brouwer, R. (1983).** Functional equilibrium: sense or nonsense? *Netherlands Journal of Agricultural Sciences*, **31**, 335–348.

**Chapin, F.S. (1980).** The mineral nutrition of wild plants. *Annual Review of Ecology and Systematics*, **11**, 233–260.

**Christie, E.K. & Moorby, J. (1975).** Physiological responses of semi-arid grasses. I. The influence of phosphorus supply on growth and phosphorus absorption. *Australian Journal of Agricultural Research*, **26**, 423–436.

**Clarkson, D.T. (1967).** Phosphorus supply and growth rate in species of *Agrostis* L. *Journal of Ecology*, **55**, 111–118.

**Corré, W.J. (1983).** Growth and morphogenesis of sun and shade plants. III. The combined effect of light intensity and nutrient supply. *Acta Botanica Neerlandica*, **32**, 277–294.

**Grime, J.P. & Hunt, R. (1975).** Relative growth rate: its range and adaptive significance in a local flora. *Journal of Ecology*, **63**, 393–422.

**Hirose, T. (1988).** Modelling the relative growth rate as a function of plant nitrogen concentration. *Physiologia Plantarum*, **72**, 185–189.

**Hunt, R. (1982).** Plant growth curves. *The Functional Approach to Growth Analysis*. Edward Arnold, London.

**Körner, Ch. & Renhardt, U. (1987).** Dry matter partitioning and root length/leaf area ratios in herbaceous perennial plants with diverse altitudinal distribution. *Oecologia*, **74**, 411–418.

**Lambers, H. & Dijkstra, P. (1987).** A physiological analysis of genotypic variation in relative growth rate: Can growth rate confer ecological advantage? In *Disturbance in Grasslands* (Ed. by J.v. Andel, J.P. Bakker & R.W. Snaydon), pp. 237–252. Dr W. Junk Publishers, Dordrecht.

**Robinson, D. & Rorison, I.H. (1985).** A quantitative analysis of the relationships between root distribution and nitrogen uptake from soil by two grass species. *Journal of Soil Science*, **36**, 71–85.

**Rorison, I.H. (1968).** The response to phosphorus of some ecologically distinct plant species. *New Phytologist*, **67**, 913–923.

**Tennant, D. (1975).** A test of a modified line intersect method of estimating root length. *Journal of Ecology*, **63**, 995–1001.

**Waring, R.H., McDonald, A.J.S., Larsson, S., Ericsson, T., Wiren, A., Arwidsson, E., Ericsson, A. & Lohammar, T. (1985).** Differences in chemical composition of plants grown at constant relative growth rates with stable mineral nutrition. *Oecologia*, **66**, 157–160.

# Root Responses to Soil Conditions

# Consequences of the activity of roots on soil

M.J. GOSS
*AFRC Institute of Arable Crops Research, Rothamsted Experimental Station, Harpenden, Hertfordshire AL5 2JQ, UK*

## SUMMARY

1 Plant roots are important for anchorage, exploration of the soil for water and mineral nutrients, and the production of several growth controlling substances.

2 To carry out these functions root systems grow by proliferation of main axes and branches as well as by longitudinal and radial expansion.

3 About 20% of the carbon fixed by a plant passes into the root system and about one-fifth of that leaks directly into the soil during the growth of the plant; the rest contributes to soil organic matter as death and decay occurs. This activity probably has the greatest effect on the soil and frequently promotes structural stability and modifies nutrient availability.

4 Activities such as compression and shearing of soil during root expansion are likely to affect soil by reducing the stability of natural aggregates. The contribution of roots to soil conditions is discussed in relation to soil husbandry.

## INTRODUCTION

Soil conditions can adversely affect root growth and in some circumstances can reduce crop yield, so much attention has been directed towards understanding the mechanisms underlying these responses. This has also led to a greater recognition of the interactions between soil structure and root growth (Reid 1980). In this paper the growth and function of roots are reviewed in relation to their potential to maintain and improve the structure, stability and chemistry of soil. Much of our knowledge comes from agricultural soils rather than those colonized by natural or semi-natural vegetation such as forest.

## ROOT FUNCTIONS

Roots perform a wide variety of functions, summarized in Fig. 1. *Anchorage* is the most obvious and, initially in the life of most plants, a very important function. The need for anchorage is continuous in herbaceous plants subject to grazing. In trees the need for strong attachment increases because the risk of windthrow (blowing over with uprooting) increases as they get taller. It is the ramification of the root system that provides the anchorage. At anthesis the root system of a wheat plant

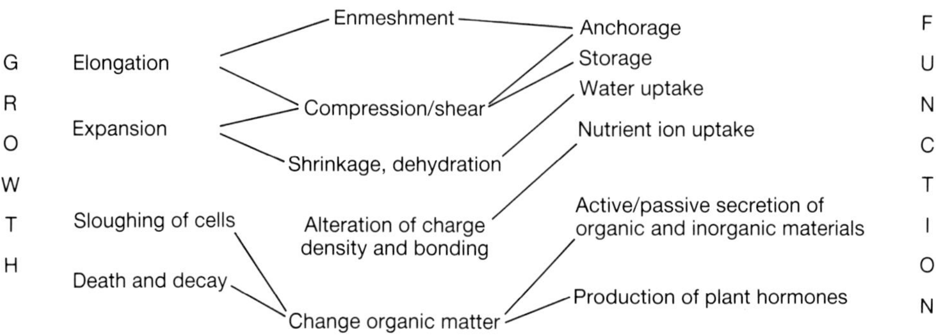


FIG. 1. Some possible interactions between root growth and function and soil conditions.

may have 2000–3000 members totalling 50–100 m in length, some members reaching to a depth of 2 m in well-structured soils. In the surface layers there can be up to 10 cm root cm$^{-3}$. In contrast tree roots can penetrate to a depth of many metres but root length densities are only about 1 cm cm$^{-3}$. Blackwell, Rennolls & Coutts (1990) used a mathematical model to predict the consequences for anchorage of changing the root distribution around trees and the depth of rooting. They concluded that the moment required to overturn a tree was increased by 46% if rooting depth increased from 0.4 to 0.6 m and the angle to the vertical subtended by roots on the windward side was 50° rather than 60°. In addition, Coutts (1986) produced evidence that the density of rooting within the volume of soil explored by roots was more important than depth alone. The tensile strength of roots is also a significant component in anchoring trees (Anderson *et al.* 1989). Anchorage also depends on soil strength (Coutts 1986).

*Storage* of secondary products from carbon fixation is particularly important in biennials and perennials. Starch and sorbitol are commonly stored in roots, and nitrogen can also be stored, often in the form of soluble amino acids (Tromp 1983).

*Uptake of nutrient ions*, both anions and cations, is a crucial function. A wheat crop may take up about 5 kg ha$^{-1}$ day$^{-1}$ of both nitrogen and potassium (Barraclough 1987), and maize about twice this amount (Mengel & Barber 1974). Several ions such as potassium, calcium, iron, magnesium and phosphate are also known to be important for the structure of soils at different levels of organization. At the same time some ions can also be toxic to plants if present at too great a concentration.

*Water absorption* takes place over the whole root surface although fluxes per unit length are greater close to the apex (Sanderson 1983). However in the field the uptake of water per unit length of root is similar at all depths in the profile (Fig. 2). In an environment with a large evaporative demand, the flux across a root is of the order 10 μl cm$^{-1}$ day$^{-1}$, but in the UK fluxes are generally closer to 1 or 2 μl cm$^{-1}$

day$^{-1}$. Nonetheless, over the course of a season roots can remove from 200 to 300 mm water from the soil and in doing so will commonly lower the water potential in the top layers of the soil to about −2 MPa.

*Production of growth controlling substances*, although a very important function for the plant, is of little direct importance for soil structure. However, as some of these substances appear to govern the response of roots to adverse soil conditions, they indirectly determine root distribution. In addition, at least the precursors of growth control substances may also influence the activity of soil microbes so affecting the composition of the rhizosphere. For example, tryptophan released from the roots of legumes attracts rhizobia and they in turn transform it to indole-acetic acid which affects the growth of root hairs before the bacteria invade the roots to form nodules.

*Release of organic and inorganic material* takes place largely as a consequence of roots growing and fulfilling their other functions. The peripheral cells of the root cap are important in the secretion of mucilage. Breakdown of starch in the amyloplasts of these cells is metabolized in the golgi apparatus and passes into the vesicles

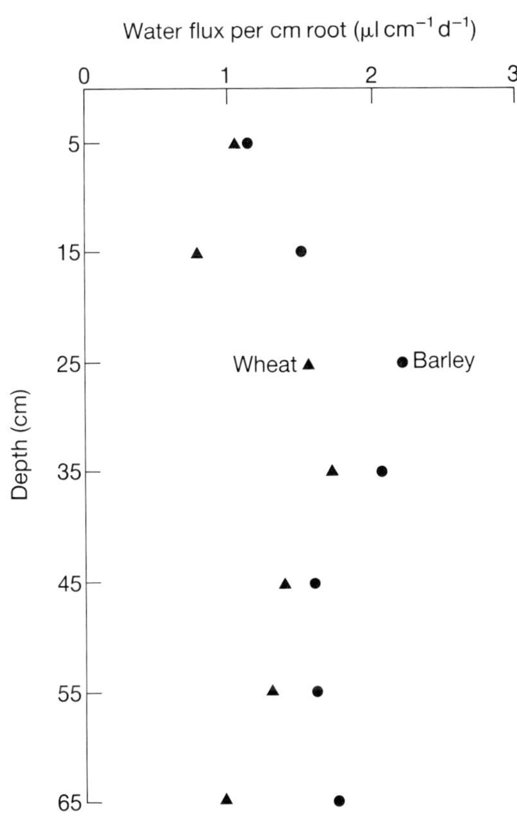


FIG. 2. The uptake of water per unit length of root at different depths in a clay soil, Lawford Series.

which migrate towards the margins of the cells where they fuse with the plasmalemma. The mucilage accumulates preferentially between the plasmalemma and the outer tangential wall of the cells before passing through the wall to the outside of the root (Juniper & Pask 1973). The mucilage is polysaccharide in nature containing glucose, galactose, galacturonate and xylose units (Juniper & Roberts 1966; Mollenhauer & Morré 1966; Harris & Northcote 1970; Kirkby & Roberts 1971). Similar mucilages are also secreted by some epidermal cells and from the tips of root hairs.

In addition to mucilage, the materials released from roots include growth substances, amino acids, simple sugars and organic acid anions. The release of these organic materials is probably the most recognized and important role of roots in the development and maintenance of soil structure.

In the soil the root apex is covered by mucigel, a product of the entire root–soil–microbial complex. It comprises natural and modified plant mucilages, bacterial cells and their metabolic products as well as soil colloids, minerals and organic matter (Rovira 1979). Further back along the root the degree of hydration of the mucigel decreases.

## ROOT GROWTH

To develop an understanding of the potential influence of roots on soil structure, stability and biological activity it is also important to consider the components of root growth. It also has to be recognized that in general the influence of roots on soil structural properties is essentially passive.

*Elongation* includes increases in the length of individual roots as well as proliferation of axes and branches. Elongation of root systems is vital for anchorage, exploration for nutrients and water, while the production of new axes and branches creates root meristems, the site of synthesis of growth control substances. Maximum rates of elongation are about 30 mm $day^{-1}$, but physical and chemical factors in the soil can considerably reduce these rates. In poorly structured, compact soils mechanical resistance may impair elongation. An applied pressure of 20 kPa reduced the rate of extension of barley roots by 50% and other crop plants were also restricted by small confining pressures (Goss 1977). In barley, lateral roots showed a sensitivity similar to the main root axes. Reducing soil aeration also affects the rate of root extension. Blackwell & Wells (1983) showed that elongation of oat roots was slowed at dissolved oxygen concentration less than 0·15 (v/v) which gave a limiting oxygen flux of 56 ng $cm^{-2}$ $min^{-1}$ at 10° C and growth stopped as the flux became zero. The critical oxygen flux for wheat was also between 50 and 60 ng $cm^{-2}$ $min^{-1}$ and an oxygen concentration of 0.05 (v/v) reduced extension by 50% (Powell 1989).

*Radial expansion* occurs to some extent in all roots but increases in diameter can be very large in woody species and is greatest in roots with a storage function. The diameter of the primary roots of most crop plants falls into the range 0·5–2 mm, but the diameter of the root tip is generally about one-third that of the mature root.

Apices of tree roots are of similar size. Mechanical resistance to root extension can increase radial growth by up to 30% in cereals, but in plants with storage roots it can also cause anomalous secondary thickening to take place in branches that would normally remain fibrous (Goss & Russell 1980). Root hairs also increase the effective radius of roots for the uptake of less mobile nutrients such as phosphate (Baldwin & Nye 1974).

*Sloughing of cells* occurs as roots grow through the soil. Cells of the root cap are constantly being lost in this way but in some species epidermal cells may also be shed. The loss of epidermal and cortical cells can also result from attacks by grazing soil fauna and by invasion by soil microbes.

*Death and decay of roots* occurs throughout the season just as some leaves and stem branches die. In cereal plants the maximum total root length is generally registered at anthesis (flowering). The death and decay of a root system can occur over several months and the more lignified tissue may take years before it is unrecognizable.

The sloughing of cells and death and decay of roots make a major contribution to the release of organic materials into the soil.

## POTENTIAL CHANGES IN SOILS CONSEQUENT ON ROOT GROWTH AND FUNCTION

As indicated in Fig. 1 there are a number of ways in which the growth and function of roots can affect the soil. Anchorage is enhanced by root elongation and results in enmeshment of particles and aggregates. Root elongation and radial expansion can produce new channels in the soil and these processes, together with the uptake of water, can cause compression, shearing, shrinkage and dehydration. The uptake of ions can alter charge density whilst the release of polysaccharides can change particle bonding and increase the content of organic matter, so potentially modifying the stability of channels and aggregates within the soil.

### *Enmeshment*

Enmeshment is a function of root extension and branching together with the formation of root hairs. The structure of soil can be rearranged or created since the root system can provide a mechanical support for the soil matrix. The effectiveness of a root system for stabilizing structure depends on the extent to which the movement of particles or aggregates under the erosive influence of wind or water can be restricted. In soils where mycorrhizas can form, fungal hyphae associated with the roots may also impart some degree of stability by enmeshment.

Stevenson & White (1941) assessed the stability of grassland soils in terms of the mechanical strength imparted by the roots. They found that even by taking into account the quantity, distribution, linear dimensions and tensile strength of the roots, the stability of soil still varied under different grass species. Reid & Goss (1981) could find no significant correlation between root length density (with or without mycorrhizas) and aggregate stability. Enmeshment may therefore impart only limited improvement in soil structure over the longer term but it may play a vital role after ploughing in leys, etc., on the establishment of following crops by protecting lighter textured soil from effects of erosion or compaction.

The presence of roots can greatly enhance shear strength and this can be important for increasing the stability of steep slopes (O'Loughlin 1974). Waldron, Dakessian & Nemson (1983) demonstrated the greater shear resistance induced by the woody root systems of yellow pine compared with lucerne.

### *Compression and shear*

Compression and shear can result from root elongation coupled with radial expansion or water extraction. In many soils roots tend to exploit existing pores and channels (Fig. 3), but they will attempt to penetrate through the soil matrix. Barley, Greacen and co-workers have shown that as a root penetrates the soil the main movement of particles is radial about the root. This resulted in significant increases in the bulk density of soil around the roots (Table 1). For example, Greacen (1968), as reported by Nye & Tinker (1977), observed a 7% increase in bulk density of a fine sand, an increase largely from reorientation of particles caused by shear. Dexter (1987) has produced a simplified model to describe changes in porosity and density around a root as it expands. In a clay soil Cockroft, Barley & Greacen (1969) reported the development of crack planes, but it was unclear whether these resulted from shear or shrinkage due to water uptake.

Pfeffer (1893) and Taylor and his co-workers (see for example Eavis, Ratliff & Taylor 1969) obtained values up to 2 MPa for maximum stresses that roots can exert on the soil. However, results from soil tests suggest that compression may not enhance soil stability. Rogowski & Kirkham (1962) found that the stability of an extruded soil to wet sieving increased after it was subjected to pressure. Nonetheless, the application of a pressure of 6.9 MPa failed to restore stability to that of natural aggregates taken from the field. Beacher & Strickland (1955) also found that the water stability of aggregates that had been pressurized in the laboratory was less than that of aggregates of field soils, and Vomocil & Flocker (1965) reported that compaction led to a decrease in stability to excess water.

The stability of soils that have been subjected to pressure or shear tends to increase with time even in the absence of water loss (Arya & Blake 1972; Schweikle, Blake & Arya 1974). The process, known as age-hardening, is due to natural changes in the orientation of water molecules and clay particles. However, whether the stability of sheared soil can recover to that of untreated soil is unclear. It therefore seems probable that the compression and shear caused by roots is more likely to result in reduced soil stability. However, the loss of porosity may increase soil resistance to shear because of the confining pressure exerted by the pore water if the number of air-filled pores is reduced. Furthermore, the greater contact

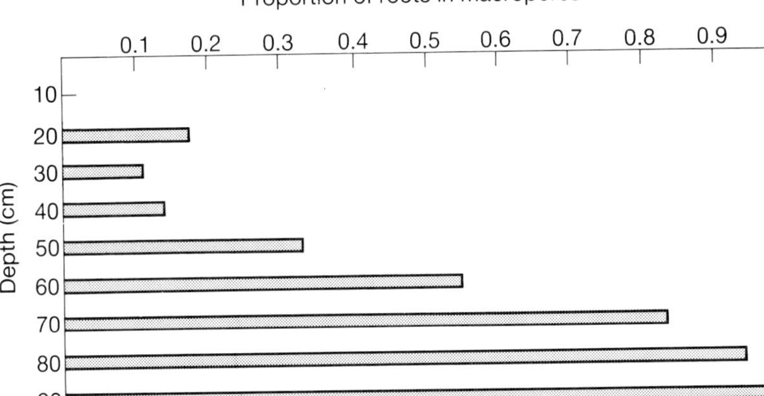


Fig. 3. The importance of large pores (>1mm) for root exploration in a podzolic red earth soil (Goss & Fischer, unpublished).

Table 1. Changes in soil bulk density immediately around roots

| Soil | Plant roots | Initial bulk density (g cm$^{-3}$) | Increase around roots (g cm$^{-3}$) |
|---|---|---|---|
| Fine sand* | Wheat | 1·40 | 0·10 |
| Loam† | Pea | 1·50 | 0·03 |
| Clay‡ | Pea | 1·21 | 0·05 |

* Greacen (1968) (see Nye & Tinker 1977).
† Greacen, Farrell & Cockroft (1968).
‡ From Cockroft, Barley & Greacen (1969).

between particles following compression may increase the effectiveness of other processes that can enhance the stabilization of soils such as the formation of organic matter bridges.

The compression caused by roots also forms channels in the soil which may be of considerable significance for the long-term improvement in soil structure. The pores produced by roots are likely to have more continuity and less tortuosity than intra- or inter-aggregate pores (Oades, private communication). Furthermore, because roots are most likely to follow planes of weakness, the resulting pores are more likely to link with the natural structural voids. Hence such pores may well drain better and so provide a better environment for subsequent roots to exploit. Vertically orientated tubular pores are more resistant to disruption by compaction but the longevity of the pores formed by roots will also depend on the relative contributions to aggregate stability from other processes such as exudation and ion uptake which are considered later.

*Shrinkage and dehydration*

Crops can extract as much as 4 mm water per day from the soil. Under rainfed or irrigated crops this commonly results in the soil, especially topsoil, wetting and drying several times during a season. Initially the cycles may cover only a small range of water potentials but they can eventually have an amplitude from near 0 to −1·5 MPa. Kolodny & Neal (1941) considered that these cycles of wetting and drying under a rye crop were important for the stable aggregation that they observed. Reid & Goss (1982a) used maize to dry the soil which was taken from −8 kPa to −0·1 or −1·5 MPa and back again. This increased the stability of a sandy loam soil by 12 and 13%, respectively, compared with soil maintained at −8 kPa. Cycles of wetting and drying have been shown to increase the aggregate stability of a severely puddled fine sandy loam (Richardson 1976) and of sand and bentonite mixtures but in the latter the water stable aggregation decreased after twenty cycles (McHenry & Russell, 1943). Hofman & De Leenher (1975) showed that repeated wetting and drying of silty soils decreased their stability to wet sieving. Tisdall, Cockroft & Uren (1978) found that there was a decrease in the stability of soils subjected to wetting and drying cycles but effects were smaller or absent in non-sterile soils. The evidence suggests that aggregate stability is improved by wetting and drying cycles only in soils of poor structure. This may be due to the orientation of water molecules (McHenry & Russell 1943) or shrinkage allowing the formation of new bonds with organic matter (Baver, Gardner & Gardner 1972). In dry sodic soils, rewetting may cause dispersion as soluble salts diffuse out. Organic and inorganic cements may become more effective on dehydration (Harris, Chesters & Allen 1966; Baver, Gardner & Gardner 1972). For example, iron oxide resulting from the dehydration of an iron gel was correlated with the aggregation of laterite soils (Lutz 1936). However, there is no indication that these stabilizing agents can be dehydrated at water potentials above −2 MPa.

Probably the most important aspect of the removal of water by plant roots is shrinkage leading to cracking. This can be important in disrupting compact layers and in the formation of natural tilth. The cracking can develop both on drying and on rewetting due to unequal wetting. The rewetting of dry soils can also disrupt aggregates due to the explosion of entrapped air (Panabokke & Quirk 1957).

*Nutrient uptake and effects on charge density*

Plants take up significant amounts of nutrient ions from the soil (Table 2). Uptake leads to a fall in the concentration of most major and minor nutrients around the root (Nye & Tinker 1977) though the zone of depletion varies depending upon the buffering power of the soil. However, some ions such as $Ca^{2+}$ may accumulate in the rhizosphere because of rapid movement towards the root surface by mass flow and the rate may exceed that at which it is absorbed (Barber & Ozanne 1970; Callot 1984). The ionic balance following differential uptake is maintained by the release of $H^+$ or $HCO_3^-$ ions into soil, so changing pH in the rhizosphere (Kirkby 1969).

TABLE 2. Uptake of nitrogen, phosphorus and potassium by some field crops

| Crop | Nutrient (kg $ha^{-1}$) | | |
|---|---|---|---|
| | N | P | K |
| Summer cabbage* | 286 | 39 | 379 |
| Broad beans* | 223 | 22 | 493 |
| Winter wheat† | 224 | 38 | 285 |
| Maize‡ | 315 | 43 | 329 |
| Grass-clover§ | 388 | 45 | 463 |
| Sugar beet‖ | 252 | 28 | 231 |
| Red beet‖ | 146 | 13 | 190 |
| Carrots‖ | 202 | 29 | 157 |

* Costigan, Greenwood & McBurney (1983).
† Barraclough (1987).
‡ Mengel & Barber (1974).
§ McEwen & Johnston (1984).
‖ Johnston, experiments at Woburn (private communication 1988).

Once uptake ceases as a result of changing demands of the shoot, either as a diurnal fluctuation or as a more permanent phenomenon such as senescence, the rhizosphere content of mobile ions will be buffered by diffusion and will return to that in the bulk soil at a rate dependent on the concentration gradient. Replacement of less mobile nutrients depends on the speed of replacement on ion exchange sites, so overall effects of nutrient uptake could be permanent or transient.

The consequences of these changes on soil structural properties of field soils is unclear, and there may well be differences between well-fertilized agricultural soils and semi-natural habitats. The nature of the exchangeable cations greatly affects soils. The swelling of clays varies according to the adsorbed cation with potassium ions decreasing swelling. Sodium-saturated soils usually have a large plastic index, whereas potassium-saturated soils have a small index. The ion concentration in the surrounding solution can affect the flocculation of a clay, but depletion brought about by nutrient uptake is unlikely to result in significant deflocculation. The importance of these changes could well be different for lighter textured soils compared with heavier clays. However, some reductions in soil stability could be expected in poorly buffered soils with appreciable exchangeable sodium. Reid, Goss & Robertson (1982) investigated the effects of maize roots on the stability of a silt loam and a sandy loam. Only in the sandy loam soil did pretreatment of aggregates with a range of sodium chloride solutions increase stability relative to the uncropped control soil. This suggests that ion uptake by the maize roots from the coarser textured soil caused a reduction in stability. These authors also investigated the 14% difference in stability between cropped and uncropped soil that could not be accounted for by nutrient uptake. Aggregates from soils both cropped and uncropped were extracted with 0·5 M acetylacetone to remove iron and aluminium (Giovannini & Sequi 1976) before stability was tested using a turbidimetric technique (Douglas & Goss 1982). Stability in both uncropped soils was significantly correlated with the extractable iron and aluminium, but there was no correlation in the cropped silty loam soil, and the variance accounted for by the correlation was reduced from 84% to only 39% in the sandy soil. It was concluded that the growth of maize roots had destroyed linkages between organic matter, iron or aluminium, and mineral particles. It is known that chelating agents such as citric acid and lactic acid can be present in the rhizosphere of maize (Matsumoto, Okada & Takahashi 1979). Furthermore, they have stability constants that allow them to compete strongly with fulvic acids for iron. Whether these chelating agents are formed only by the maize roots or also by micro-organisms within the rhizosphere is unknown, but the benefits of removing iron in this way could be beneficial for plant nutrition. However, in the pot experiments of Reid, Goss & Robertson (1982) the total uptake of iron was only 0·01% of the acetylacetone-soluble iron. Phosphate, commonly associated with iron and aluminium complexes, may have been released with the metal ions, whilst the breakdown of aggregates released potassium and magnesium ions into solution so that the reduced soil stability could also have improved the chemical fertility. These and other exudates from roots have been shown to chelate phosphate from sparingly soluble compounds (see Barber 1968 and Berthelin *et al.*, pp. 191–197) and the availability of iron can also be modified in this way (Mengel & Kirkby 1987). For example, Brown, Holmes & Tiffin (1961) reported that soya-beans, which were not susceptible to iron chlorosis, increased the production of exudates as the iron supply declined. Root exudates can apparently release nutrient ions from soil without chelating them (Pojasok & Kay 1990), and this can result in greater soil structural stability. It is unclear whether uptake of the released nutrients by plants would nullify the effect on stability.

Root exudates also modify the composition of the rhizosphere flora. Rovira (1979) produced evidence that 5–10% of the root surface was covered in micro-organisms. These free-living rhizosphere microbes will also affect nutrient availability. Zinc deficiency in fruit trees and manganese deficiency in oats can be induced by soil microbes but increased manganese uptake due to bacterially produced substances has also been reported (see Barber 1968). Uptake of molybdenum into higher plants can also be reduced by glucuronic acids derived from rhizosphere bacteria (Tan & Loutit 1977).

### *Changes in soil organic matter*

Best estimates suggest that some 20% of the carbon fixed by a plant passes into its root system. In grass and cereal crops about 75% of roots are found within the topsoil. For a wheat crop this would result in a maximum of 0.6 kg m$^{-3}$ organic carbon into this layer with no account being taken for root respiration. The organic carbon content of most arable soils is in the range 10–30 kg m$^{-3}$, so the annual input from plant roots is generally much less than 10% of the total. During the growth of a cereal root system between 20 and 80% of the root dry weight can pass into the soil through secretion, leakage, autolysis and sloughing of cells (Whipps 1984). The extent of the loss varies according to plant nutrient status (Newman *et al.* 1979) and conditions in the rooting environment but the presence of micro-organisms (Barber & Gunn 1974; Martin 1977) and a fall in soil water potential (Reid & Mexal 1977) stimulate the loss. More material, especially carbohydrate, is lost from roots growing in a particulate medium than in solution culture (Barber & Gunn 1974).

Polysaccharides are the most important compounds stabilizing aggregates in arable soils that contain little organic matter (Greenland, Lindstrom & Quirk 1962). The quantity of polysaccharide in soil is increased in the presence of living roots and mycorrhizas and this can promote the stability of aggregates in the rhizosphere (Tisdall & Oades 1979; Reid & Goss 1981, 1982; Baldock & Kay 1987). The results of Reid & Goss (1982a) indicate that the efficacy of these compounds is enhanced by the drying of the soil.

The release of organic materials from living roots also appears to protect older soil organic matter from breakdown. Thus Reid & Goss (1982b) showed that the release of $^{14}C$ from previously incorporated roots of barley plants was reduced by more than 50% in the presence of maize and ryegrass roots compared with fallowed soil. Similar evidence was obtained by Sparling, Cheshire & Mundie (1982). Effects of living tree roots are not so large possibly because of the smaller rooting density. Roots also suppressed the transformations of soil organic matter by inhibiting the removal of water-soluble organic matter. Maize roots suppressed the processes that transferred humic materials into the fulvic acid fraction (Reid & Goss 1983). These authors also showed a significant uptake of $^{14}C$ from soil organic matter into the living roots of maize and ryegrass.

The effect of roots is thus to contribute to the soil organic fraction, protect older organic matter from breakdown and enhance stability of aggregates around the roots. The consequences of the release of organic molecules from roots on nutrient uptake was discussed above.

## CONSEQUENCES FOR SOIL MANAGEMENT

Harris, Chesters & Allen (1966) concluded that the potentialities for plant roots to form and destroy aggregates was important in the dynamics of soil aggregation. Roots and micro-organisms, especially fungi, are considered to be mainly responsible for the maintenance of aggregates greater than 250 μm (Tisdall & Oades 1982). But roots are equally important for creating larger structural units and the spaces between them (Fig. 4). To improve soil structure, especially below the normal depth of cultivation, the formation of continuous vertically orientated pores are necessary. This will serve to improve drainage and encourage further root development. For drainage only a few pores $>1$ mm $m^{-2}$ would be adequate as long as they linked with natural structural voids deeper in the profile. Rapid growth of strong tap roots when the soil is moist, and hence weak, would be required. To encourage root development a larger number of finer pores, such as those produced by a fibrous root system, would be preferable. Granulation is also likely to be encouraged by a fibrous root system. The rate that roots decompose appears to be important for granulation with roots containing more nitrogen being broken down more readily (see Harris, Chesters & Allen 1966). Cooke and Williams (1972)

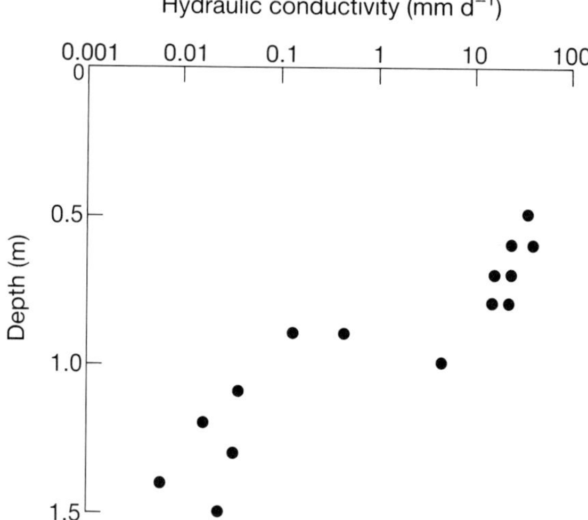


FIG. 4. Variation in hydraulic conductivity with depth in a clay soil (Evesham Series). Horizontal piezometer measurements. The decrease in conductivity around 0·9 m depth is associated with a considerable reduction in the amount of water extracted by roots (89% is extracted above 1 m), although roots do take some water from 1·5 m in most seasons.

showed that roots of lucerne contained almost 2% N and 25% soluble carbohydrates. The main axes grew strongly to depth and the total porosity of soil at 2·5 cm was 42% compared with 39% under grass. Reid & Goss (1981) and Angers & Mehuys (1988) have reported improved aggregate stability due to lucerne.

The fact that roots can be important for forming drainage channels may also be advantageous in minimizing the leaching of soil contaminants. Turner & Steele (1988) showed that the organic matter lining root channels had higher sorptivity for cadmium and manganese than did the bulk soil, particularly in acid conditions.

Growing roots of graminaceous plants can form distinct morphological features in the soil (Wullstein & Pratt 1981; Kane & Mott 1983; McCulley & Canny 1985). The formation of root sheaths (rhizosheaths) is associated with the copious production of root hairs. In a desert grass Wullstein & Pratt (1981) found that the rhizosheath was normally between 0·2–0·5 mm thick, but in maize the sheath was about 1 mm thick (Vermeer & McCulley 1982). The soil particles appear to be bonded to each other and to the root hairs by mucilage with properties similar to that formed by the root cap (Vermeer & McCulley 1982). Within the mucilage are found root-cap cells, bacteria and fungal mycelia. The physical properties of these rhizosheaths can be quite different from the bulk soil (Buckley 1982) and there is evidence for nitrogen fixation within the sheath by *Bacillus polymyxa*-like bacteria (Wullstein, Bruening & Bollen 1979). However, there is no evidence that any of this fixed nitrogen is transferred to the plant. As yet the reasons for differences in the radial extent of root sheaths between species are unknown. The length of root hairs might be important or there may be variety or species differences in the effectiveness of exudates. The practical importance of root sheaths is a major gap in knowledge.

## CONCLUSIONS

Roots can modify soil structure and stability as well as affects its chemical behaviour. Such changes are effected by living roots as well as by their death and decay. Not all changes in structure or stability are beneficial to the long-term maintenance of soil physical conditions, and when soils are subjected to tillage or damage from wheeled traffic, plant roots may not rapidly effect improvement. However, the properties of some plants when grown in mixed stands or in carefully chosen rotation could help to keep soils available for long-term arable cultivation. The properties of the rhizosphere soil may help to reduce the risk of environmental contamination from agricultural land. At the same time, improved knowledge of soil properties in the rhizosphere of certain plants may help the development of low-input systems making maximum use of nitrogen fixation by 'free-living' organisms.

## ACKNOWLEDGEMENTS

I wish to thank J.B. Reid for many stimulating discussions of the ideas expressed above, the Department of Land Resource Science, University of Guelph for support during a visit when much of the literature was surveyed. I am grateful to M.P. Coutts for helpful comments on the contents and to Mrs B. Moseley for preparing the typescript.

## REFERENCES

**Anderson, C.J., Coutts, M.P., Ritchie, R.M. & Campbell, D.J. (1989).** Root extraction force measurement for Sitka spruce. *Forestry*, **62**, 127–137.

**Angers, D.A. & Mehuys, G.R. (1988).** Effects of cropping on macro-aggregation of a marine clay soil. *Canadian Journal of Soil Science*, **68**, 723–732.

**Arya, L.M. & Blake, G.R. (1972).** Stabilization of newly formed soil aggregates. *Agronomy Journal*, **64**, 177–180.

**Backwell, P.S. & Wells, E.A. (1983).** Limiting oxygen flux densities for oat root extension. *Plant and Soil*, **73**, 129–139.

**Baldock, J.A. & Kay, B.D. (1987).** Influence of cropping history and chemical treatments on the water-stable aggregation of a silt loam soil. *Canadian Journal of Soil Science*, **67**, 501–511.

**Baldwin, J.P. & Nye, P.H. (1974).** A model to calculate the uptake by a developing root system or root hair system of solutes with concentration variable diffusion coefficients. *Plant and Soil*, **40**, 703–706.

**Barber, D.A. (1968).** Microorganisms and the inorganic nutrition of higher plants. *Annual Review of Plant Physiology*, **19**, 71–88.

**Barber, D.A. & Ozanne, P.G. (1970).** Autoradiographic evidence for the differential effect of four plant species in altering the Ca content of the rhizosphere in soil. *Proceedings of the Soil Science Society of America*, **34**, 635–637.

**Barber, D.A. & Gunn, K.B. (1974).** The effect of mechanical forces on the exudation of organic substances by the roots of cereal plants grown under sterile conditions. *New Phytologist*, **73**, 39–45.

**Barraclough, P.B. (1987).** Nutrient fluxes in the rhizosphere of high yielding grain crops. *XIII Congress of the International Society of Soil Science, Hamburg 1986. Transactions*, **5**, 217–224.

**Baver, L.D., Gardner, W.H. & Gardner, W.B. (1972).** *Soil Physics.* John Wiley & Sons, New York.

**Beacher, B.F. & Strickland, E. (1955).** Effect of puddling on water stability and bulk density of aggregates of certain Maryland soils. *Soil Science*, **80**, 363–373.

**Blackwell, P.G., Rennolls, K. & Coutts, M.P. (1990).** A root anchorage model for shallowly rooted sitka spruce. *Forestry*, **63**, 73–91.

**Brown, J.C., Holmes, R.S. & Tiffin, L.O. (1961).** Iron chlorosis in soybeans as related to the genotype of rootstalk. 3. Chlorosis susceptibility and reductive capacity at the root. *Soil Science*, **91**, 127–132.

**Buckley, R. (1982).** Sand rhizosheath of an arid zone grass. *Plant and Soil*, **66**, 417–421.

**Callot, G. (1984).** Structure pédologique et functionnement du sol en relation avec la production végétale. Analyse au niveau parcellaire. *Science du Sol — Bulletin de l'Association Française pour l'Etude du Sol*, **2**, 167–181.

**Cockroft, B., Barley, K.P. & Graecen, E.L. (1969).** The penetration of clays by fine probes and root tips. *Australian Journal of Soil Research*, **7**, 333–348.

**Cooke, G.W. & Williams, R.J.B. (1972).** Problems with cultivations and soil structure at Saxmundham. *Rothamsted Report for 1971, Part 2*, 122–142.

**Costigan, P.A., Greenwood, D.J. & McBurney, T. (1983).** Variation in yield between two similar sandy loam soils. I. Description of soils and measurement of yield differences. *Journal of Soil Science*, **34**, 621–637.

**Coutts, M.P. (1986).** Components of tree stability in Sitka spruce on peaty gley soil. *Forestry*, **59**, 173–197.

**Dexter, A.R. (1987).** Compression of soil around roots. *Plant and Soil*, **97**, 401–406.

**Douglas, J.T. & Goss, M.J. (1982).** Stability and organic matter content of surface soil aggregates under different methods of cultivation and in grassland. *Soil and Tillage Research*, **2**, 155–175.

**Eavis, B.W., Ratliff, L.F. & Taylor, H.M. (1969).** Use of a deadload technique to determine axial root growth pressure. *Agronomy Journal*, **61**, 640–643.

**Giovannini, G. & Sequi, P. (1976).** Iron and aluminium as cementing substances of soil aggregates. I. Acetylacetone in benzene as an extractant of soil iron and aluminium. *Journal of Soil Science*, **27**, 140–147.

**Goss, M.J. (1977).** Effects of mechanical impedance on root growth in barley (*Hordeum vulgare* L.). I. Effects on the elongation and branching of seminal root axes. *Journal of Experimental Botany*, **28**, 96–111.

**Goss, M.J. & Russell, R.S. (1980).** Effects of mechanical impedance on root growth in barley (*Hordeum vulgare* L.). III. Observations on the mechanism of response. *Journal of Experimental Botany*, **31**, 577–588.

**Greacen, E.L., Farrell, D.A. & Cockroft, B. (1968).** Soil resistance to metal probes and plant roots. *Transactions of the 9th International Congress of Soil Science (Adelaide)*, Vol. **1**, 769–779.

**Greenland, D.J., Lindstrom, G.R. & Quirk, J.P. (1962).** Organic materials which stabilize natural soil aggregates. *Proceedings of the Soil Science Society of America*, **26**, 366–371.

**Harris, P.J. & Northcote, D.H. (1970).** Patterns of polysaccharide biosynthesis in differentiating cells of maize root tips. *Biochemistry Journal*, **120**, 479–491.

**Harris, R.F., Chesters, G. & Allen, O.N. (1966).** Dynamics of soil aggregation. *Advances in Agronomy*, **18**, 107–169.

**Hofman, G. & De Leenher, L. (1975).** Influence of soil prewetting on aggregate instability. *Pedologie* **25**, 190–198.

**Juniper, B.E. & Roberts, R.M. (1966).** Polysaccharide synthesis and the fine structure of root cells. *Journal of the Royal Microscopical Society*, **85**, 63–72.

**Juniper, B.E. & Pask, G. (1973).** Directional secretion by the golgi bodies in maize root cells. *Planta (Berlin)*, **109**, 225–231.

**Kane, P. & Mott, C.J.B. (1983).** Non-destructive separation of discrete pedogenic features from soils. *Journal of Soil Science*, **34**, 205–212.

**Kirkby, E.A. (1969).** Ion uptake and ionic balance in plants in relation to the form of nitrogen nutrition. *Ecological Aspects of the Mineral Nutrition of Plants* (Ed. by I. Rorison), pp. 215–235. Blackwell Scientific Publications, Oxford.

**Kirkby, E.G. & Roberts, R.M. (1971).** The localized incorporation of $^3$H-L-glucose into cell-wall polysaccharides of the cap and epidermis of corn roots. *Planta (Berlin)*, **99**, 211–221.

**Kolodny, L. & Neal, O.R. (1941).** The use of micro-aggregation or dispersion measurements for following changes in soil structure. *Proceedings of the Soil Science Society of America*, **6**, 91–95.

**Lutz, J.F. (1936).** The relation of free iron in the soil to aggregation. *Proceedings of the Soil Science Society of America*, **1**, 43–45.

**Martin, J.K. (1977).** Factors influencing the loss of organic carbon from wheat roots. *Soil Biology and Biochemistry*, **9**, 1–7.

**Matsumoto, H., Okada, K. & Takahashi, E. (1979).** Excretion products of maize roots from seedling to seed development stage. *Plant and Soil*, **53**, 17–26.

**McCully, M.E. & Canny, M.J. (1985).** Localization of translocated $^{14}$C in roots and root exudates of field-grown maize. *Physiologia Plantarum*, **65**, 380–392.

**McEwen, J. & Johnston, A.E. (1984).** Factors affecting the production and composition of mixed grass/clover swards containing modern high-yielding clovers. *18th Colloquium International Potash Institute*, 47–61.

**McHenry, J.R. & Russell, M.B. (1943).** Soil aggregation. *Soil Science Society of America Proceedings*, **8**, 71–78.

**Mengel, D.B. & Barber, S.A. (1974).** Rate of nutrient uptake per unit of corn root under field conditions. *Journal of Agronomy,* **66**, 399–402.

**Mengel, K. & Kirkby, E.A. (1987).** *Principles of Plant Nutrition.* International Potash Institute, Bern.

**Mollenhauer, H.H. & Morré, D.J. (1966).** Golgi apparatus and plant secretion. *Annual Review of Plant Physiology*, **17**, 27–46.

**Newman, E.I., Campbell, R., Christie, P., Heap, A.J. & Lawly, R.A. (1979).** Root microorganisms in mixtures and monocultures of grassland plants. In *The Soil–Root Interface* (Ed. by J.L. Harley & R.S. Russell), pp. 161–173. Academic Press, London.

**Nye, P.H. & Tinker, P.B. (1977).** *Solute Transport and Plant Growth in Soil.* Blackwell Scientific Publications, Oxford.

**O'Loughlin, C. (1974).** The effect of timber removal on the stability of forest soils. *Journal of Hydrology (NZ)*, **13**, 121–134.

**Panabokke, C.R. & Quirk, J.P. (1957).** Effect of initial water content on stability of soil aggregates in water. *Soil Science*, **83**, 185–195.

**Pfeffer, W. (1893).** Druck- und Arbeitsleistung durch wachsende Pflanzen. *Abh. sachs. Akad. Wiss.*, **33**, 235–474.

**Pojasok, T. & Kay, B.D. (1990).** Effect of root exudates from corn and bromegrass on soil structural stability. *Canadian Journal of Soil Science*, **70**, 351–362.

**Powell, B.A. (1989).** *Aspects of root growth in compacted soils.* Ph.D. thesis, Reading University.

**Reid, C.P.P. & Mexal, J.G. (1977).** Water stress effects on root exudation by lodgepole pine. *Soil Biology and Biochemistry*, **9**, 417–421.

**Reid, J.B. (1980).** *The interaction of roots and the soil with special reference to soil structural stability.* Ph.D. thesis, Reading University.

**Reid, J.B. & Goss, M.J. (1981).** Effect of living roots of different plant species on aggregate stability of two arable soils. *Journal of Soil Science*, **32**, 521–541.

**Reid, J.B. & Goss, M.J. (1982a).** Interactions between soil drying due to plant water use and decrease in aggregate stability caused by maize roots. *Journal of Soil Science*, **33**, 47–53.

**Reid, J.B. & Goss, M.J. (1982b).** Suppression of decomposition of $^{14}$C-labelled plant roots in the presence of living roots of maize and perennial ryegrass. *Journal of Soil Science*, **33**, 387–395.

**Reid, J.B. & Goss, M.J. (1983).** Growing crops and transformations of $^{14}$C-labelled soil organic matter. *Soil Biology and Biochemistry*, **15**, 687–691.

**Reid, J.B., Goss, M.J. & Robertson, P.D. (1982).** Relationship between the decrease in soil stability affected by the growth of maize roots and changes in organically bound iron and aluminium. *Journal of Soil Science*, **33**, 397–410.

**Richardson, S.J. (1976).** Effect of artificial weathering cycles on the structural stability of a dispersed silty soil. *Journal of Soil Science*, **27**, 287–294.

**Rogowski, A.S. & Kirkham, D. (1962).** Moisture, pressure and formation of water-stable soil aggregates. *Proceedings of the Soil Science Society of America*, **26**, 213–216.

**Rovira, A.D. (1979).** Biology of the soil–root interface. In *The Soil–Root Interface* (Ed. by J.L. Harley & R.S. Russell), pp. 145–160. Academic Press, London.

**Sanderson, J. (1983).** Water uptake by different regions of the barley root. Pathways of radial flow in relation to development of the endodermis. *Journal of Experimental Botany*, **34**, 240–253.

**Schweikle, V., Blake, G.R. & Arya, L.M. (1974).** Matrix suction and stability changes in sheared soil. *Transactions of the 10th International Congress of Soil Science (St. Paul)*, Vol. 1, 187–193.

**Sparling, G.P., Cheshire, M.V. & Mundie, C.M. (1982).** Effect of barley plants on the decomposition of $^{14}$C-labelled soil organic matter. *Journal of Soil Science*, **33**, 89–100.

**Stevenson, T.M. & White, W.J. (1941).** Root fibre production of some perennial grasses. *Science in Agriculture*, **22**, 108–118.

**Tan, E.L. & Loutit, M.W. (1977).** Effect of extracellular polysaccharides of rhizosphere bacteria on the concentration of molybdenum in plants. *Soil Biology and Biochemistry*, **9**, 411–415.

**Tisdall, J.M. & Oades, J.M. (1979).** Stabilization of soil aggregates by the root system of ryegrass. *Australian Journal of Soil Research* **17**, 429–441.

**Tisdall, J.M. & Oades, J.M. (1982).** Organic matter and water-stable aggregates in soils. *Journal of Soil Science*, **33**, 141–163.

**Tisdall, J.M., Cockroft, B. & Uren, N.C. (1978).** The stability of soil aggregates as affected by organic materials, microbial activity and physical disruption. *Australian Journal of Soil Research*, **16**, 9–17.

**Tromp, J. (1983).** Nutrient reserves in roots of fruit trees, in particular carbohydrates and nitrogen. *Plant and Soil*, **71**, 401–413.

**Turner, R.R. & Steele, K.F. (1988).** Cadmium and manganese sorption by soil macropore linings and fillings. *Soil Science*, **145**, 79–86.

**Vermeer, J. & McCulley, M.E. (1982).** The rhizosphere in *Zea*: new insight into its structure and development. *Planta (Berlin)*, **156**, 45–61.

**Vomocil, J.A. & Flocker, W.J. (1965).** Degradation of structure of Yolo loam by compaction. *Proceedings of the Soil Science Society of America*, **29**, 7–12.

**Waldron, L.J., Dakessian, S. & Nemson, J.A. (1983).** Shear resistance enhancement of 1.22-meter diameter soil cross sections by pine and alfalfa roots. *Soil Science Society of America Journal*, **47**, 9–14.

**Whipps, J.M. (1984).** Environmental factors affecting the loss of carbon from the roots of wheat and barley seedlings. *Journal of Experimental Botany*, **35**, 767–773.

**Wullstein, L.H. & Pratt, S.A. (1981).** Scanning electron microscopy of rhizosheaths of *Oryzopsis hymenoides*. *American Journal of Botany*, **68**, 408–419.

**Wullstein, L.H., Bruening, M.L. & Bollen, W.B. (1979).** Nitrogen fixation associated with sand grain root sheaths (rhizosheaths) of certain xeric grasses. *Physiologia Plantarum*, **46**, 1–4.

# Involvement of roots and rhizosphere microflora in the chemical weathering of soil minerals

J. BERTHELIN, C. LEYVAL, F. LAHEURTE AND P. DE GIUDICI

*Centre de Pédologie Biologique, CNRS 17, rue Notre-Dame-des-Pauvres B.P. 5, 54501 Vandoeuvre-les-Nancy Cédex, France*

## SUMMARY

1 The weathering of soil minerals in the plant rhizosphere has been studied. Results have been obtained by utilizing experimental devices to study the interactions between 'roots–mycorrhizal fungi–bacteria and soil minerals', in the rhizosphere of graminaceous plants (maize and rice) and of trees (pine).

2 Plant roots alone (axenic roots) are able to uptake and solubilize mineral elements (P, K etc.) from rock phosphates and phyllosilicates (biotite, phlogopite). But the rhizospheric microflora (endo- or ectomycorrhiza and rhizobacteria) are able to modify these processes. Such interactions between roots–micro-organisms and minerals, that depend on the plant growth conditions, promote the transformation of minerals such as the mica phlogopite to the clay vermiculite.

3 Weathering can be attributed, as demonstrated in experiments, to the production of acids and complexing compounds by the roots themselves or by their associated micro-organisms. But it can be attributed also to exchange uptake that can be increased by microbial indirect effect on plant growth.

4 In the experimental conditions, bacteria appear mainly as acid-complexing producing agents and mycorrhiza as agents increasing the soil prospection and the exchange surface.

The rhizosphere must be considered as a very active medium and habitat for weathering and mineral concentration that needs further studies due to the limited available information.

## INTRODUCTION

The weathering of soil minerals involves physical and chemical processes that are the result of the interplay of different environmental factors, mainly parent material, climate, organisms and topography, which lead through time to the transformation and evolution of rocks and minerals and the functioning of biogeochemical cycles of elements.

Among these factors, plants and soil organisms have been recognized to be involved in both physical (e.g. mechanical) and chemical (e.g. solubilization of elements) weathering but, also in deposits and neoformation of minerals (Berthelin 1983, 1988; Robert & Berthelin 1986).

These interactions, 'plants–soil organisms and minerals', are mainly occurring in the upper soil horizons and in the rhizosphere. The rhizosphere receives large amounts of available energy as plant material (shoot and root litters, shoot leachates, root exudates) and is where large populations of micro-organisms are present and active. Also, to satisfy plant nutritional requirements, exchange uptake of elements between the minerals, the soil solution and the plants will take place to a large extent in the rhizosphere (e.g. Callot *et al.* 1982; Marschner 1986). Although the mineral nutrition in higher plants has been extensively studied (Mortvedt, Giordano & Lindsay 1972; Callot *et al.* 1982; Marschner 1986) relatively few studies have been made of the weathering of minerals in the rhizosphere either by the root system directly or by rhizospheric (symbiotic or not) micro-organisms, distinct from effects of the roots (e.g. Berthelin, 1983; Berthelin & Leyval, 1982; Robert & Berthelin, 1986).

The main purpose of this paper is not to provide an exhaustive survey but to present different approaches and different aspects of the chemical weathering of soil minerals in the rhizosphere and to show the importance of microbial rhizospheric activity in the weathering processes.

## MATERIALS AND METHODS

As the results presented in this paper have been obtained in different types of experiments, the methods used to investigate and measure rhizospheric and microbial rhizospheric weathering of minerals are not presented here in detail but are summarized and discussed to show the different approaches to this problem.

### *Experimental methods and materials*

Except for studies concerning either the biological recycling of elements (e.g. Juang & Uehara 1968) or attempts to observe the modifications of the mineral properties of rhizospheric soil samples (Grierson & Jenkins 1986), the weathering of minerals by plant root systems and their associated microflora has been studied mainly in greenhouse and laboratory experiments. Experimental devices have been invented and adopted in order to observe the fate of the minerals and of their constitutive elements and to distinguish and determine the factors (roots, microbes, growth conditions, etc.) and the mechanisms involved.

The abilities of root systems and of their associated microflora to weather minerals have been studied generally in culture pots or in similar systems. Seedlings or young transplants can be cultivated in soil or in acid-washed quartz or sand that have received a well-defined mineral such as the mica biotite (Spyridakis, Chester & Wilde 1967; Mojallali & Weed 1978; Berthelin & Leyval 1982), a rock phosphate or a tricalcium phosphate (Azcon, Barea & Hayman 1976; Chakly & Berthelin 1982; Laheurte & Berthelin 1988). Laboratory, greenhouse and *in situ* lysimeters have all been used. Recently, Leyval (1988) and Leyval & Berthelin (1989) have proposed a lysimeter installed in a greenhouse as an aid to study the interactions between micro-organisms, minerals and forest tree roots (Fig. 1).

In order to study microbial weathering in the rhizosphere and to distinguish the role of the roots alone from those of roots infected with mycorrhizal fungi or rhizobacteria or both, experimental devices in which plant roots grow in axenic conditions (without micro-organisms) have been used (Azcon, Barea & Hayman, 1976; Berthelin & Leyval, 1982; Chakly & Berthelin 1982; Laheurte & Berthelin 1988). Such devices are illustrated in Fig. 1. Generally one microbial inoculation at the beginning of the experiment was sufficient. But, in the lysimeter (Leyval & Berthelin 1989) five root inoculations were performed to maintain a high level of the mycorrhizal infection and of the bacterial population during a 2 year experiment.

The devices, using sand instead of soil as a support for plant growth, appear as a simplified model of the 'soil–root system' but they allow easier and better analysis of plant exudates, microbial components, mineral elements and minerals.

In an even more simplified model, seedlings were incubated or cultivated under

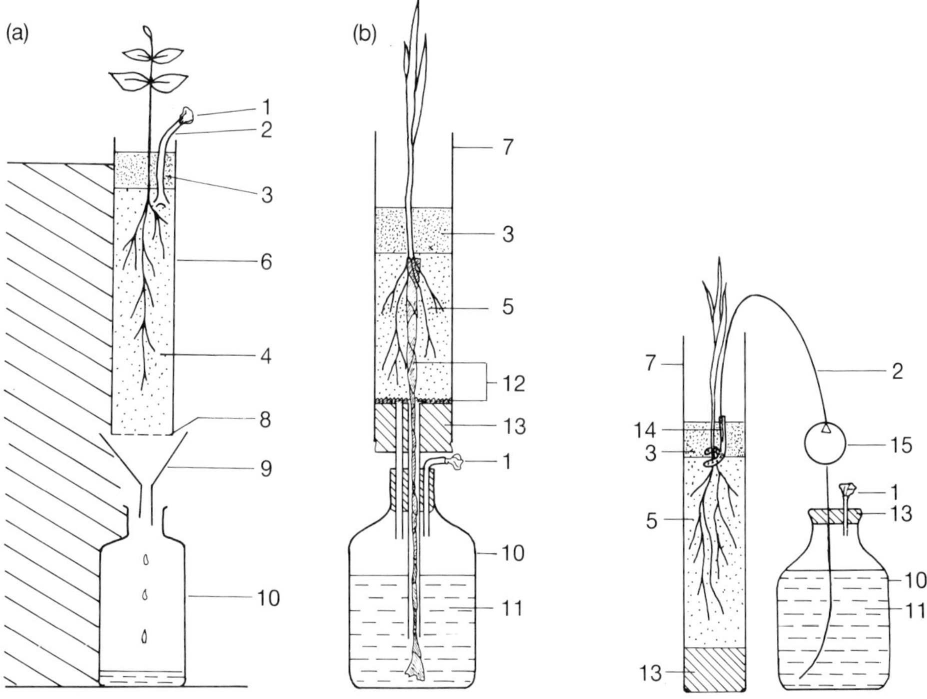


FIG. 1. Experimental devices (a = lysimeter column; b = plant culture system) used to study rhizospheric and microbial rhizospheric weathering of soil minerals (rock phosphate, biotite, phlogopite) (Berthelin & Leyval, 1982; Laheurte & Berthelin 1986, 1988; Leyval 1988; Leyval & Berthelin 1989). 1 = cotton plug; 2 = silicone tube; 3 = hydrophobic sand; 4 = sand + phlogopite + rock phosphate; 5 = sand ± mineral (rock phosphate or mica); 6 = PVC column (30 × 5 cm); 7 = Pyrex tube; 8 = polyamide tissue (20 μm pore size); 9 = funnel; 10 = flask [250 ml in (a) and 1 l in (b)]; 11 = nutrient solution; 12 = glass wool; 13 = rubber stopper; 14 = watering device; 15 = peristaltic pump.

hydroponic conditions (in a liquid medium), in the presence of a suspension of a well-defined mineral, eventually enclosed in a dialysis bag (Conyers & McLean 1968; De Giudici 1985). Where the minerals were placed in dialysis bags, only the direct chemical effects of root and microbial exudates and uptake were observed in the weathering processes.

Plants used in the reported studies included coniferous trees (pine) and grasses. The two main types of minerals were rock phosphates and micas (biotite and phlogopite).

### *Methods used to measure rhizospheric weathering of minerals*

The weathering of the minerals, by comparison with the controls (non-incubated minerals or minerals incubated without plant or microbe), can be determined by measurement of the uptake of mineral elements by the plant, the solubilization of elements from the mineral in the nutrient or perfusing solution and by changes in mineral properties. Three main components have been considered to document the weathering of the minerals and are submitted to different analysis:

1 the plant component;
2 the nutrient solution and water-soluble component; and
3 the non-water-soluble mineral component (or remaining minerals).

Rock phosphate weathering was determined only by plant phosphorus uptake and by phosphorus solubilization in the device. But micas (biotite and phlogopite) weathering was studied by analysis of all three components including the mineral residue.

After different times of growth, at the end of the experiments, the plants were harvested. Nutrient solution or perfusing solution (in lysimetric device) were collected and filtered (0.45 and 0.2 μm) for analysis of water-soluble inorganic and organic compounds. Mycorrhizal infection and bacterial populations were recorded. Dry weights of shoots and roots were determined. Plants were analysed for their mineral composition by atomic absorption and ICP (inductively coupled plasma) after plant tissue mineralization. Phosphorus was also determined using a colorimetric method. In the latest experiments non-rhizospheric minerals were separated from the rhizospheric ones, which remained attached to the root system after hand shaking. The rhizospheric material remaining attached to the roots after this shaking, was recovered by gentle and short agitation of the roots in distilled water. Some small mineral particles can sometimes be more strongly associated, in particular with mycorrhized roots and they are collected with pliers.

Minerals (biotite and phlogopite) were recovered by sieving and eventually a densimetric separation; the mineral particle size was different from that of the sand used as support for plant growth. Mineral transformation has been studied by chemical analysis, CEC (cation exchange capacity) evaluation, X-ray diffraction analysis, scanning electron microscopic observations of root and mineral particles, associated with energy dispersive X-ray analyser or a wavelength dispersive X-ray analyser.

All these methods are described here or in previous papers mentioned in the references.

## RESULTS AND DISCUSSION

The results presented to illustrate the rhizospheric weathering of soil minerals concern the dissolution and weathering of silicates (trioctahedral micas) and of phosphates that have been used as experimental material.

### *Rhizospheric dissolution of insoluble rock phosphate*

The possible improvement of plant growth and phosphorus nutrition, by the release of soluble phosphorus from rock phosphate, under the effect of the roots and their rhizosphere micro-organisms, have suggested (Gerretsen 1948) a number of experiments to study the effect of micro-organisms on the uptake of phosphorus by plants from insoluble phosphates. Two experiments performed using different devices and under different growth conditions have shown that maize growing in axenic conditions (without rhizosphere micro-organisms) can mobilize, over 9 weeks growth, 1·6–2·8 mg of P from insoluble rock phosphate per plant, i.e. 0·4–0·6 mg P $g^{-1}$ of dry plant material (Table 1). The remaining plant phosphorus was provided by the nutrient solutions (containing 2 and 1·5 p.p.m of labile phosphorus, respectively) and by the plants' seed (Laheurte 1985; Laheurte & Berthelin 1986). Chemical analyses have indicated that oxalic, tartaric, oxalo-acetic, isocitric, citric, lactic, fumaric and succinic acids were present in the soluble exudates of these axenic maize roots.

In these experiments, bacterial and endomycorrhizal inoculations did not significantly increase either the phosphorus mobilization from rock phosphate or the plant phosphorus content (Table 1). Inoculation with both micro-organisms (*Glo-*

TABLE 1. Total phosphorus uptake and amount of phosphorus provided by rock phosphate (mg $plant^{-1}$) in two investigations (A and B) of the rhizospheric weathering of rock phosphate in maize

| | Investigation | Plant alone | Plant + endo. | Plant + bact. | Plant + endo. + bact. |
|---|---|---|---|---|---|
| Total P uptake (mg $plant^{-1}$) | A | 4·6 (a) | 4·4 (a) | 4·2 (a) | 4·3 (a) |
| | B | 5·5 (a) | 6·2 (a) | 5·4 (a) | 5·7 (a) |
| P provided by rock phosphate (mg $plant^{-1}$) | A | 1·6 | 1·4 | 1·2 | 1·4 |
| | B | 2·8 | 3·5 | 2·7 | 3·0 |

Two treatments with the same letter on the same line are not significantly different ($P = 0·1$); endo. = plant inoculated by the endomycorrhizal fungi *Glomus mosseae*; bact. = plant inoculated by *Enterobacter agglomerans*.

*mus mosseae* and *Enterobacter agglomerans*), under these experimental conditions, led to the disappearance of citric, isocitric and lactic acids and to the formation of malic acid. This can explain, at least partially, the low mobilization of phosphorus. The phosphate solubilizing ability of the strain *E. agglomerans*, which was demonstrated in pure culture (Laheurte 1985), was not expressed in these experimental conditions in the presence of the plants which provided the energy and carbon source for the bacteria.

Rhizobacteria can also promote plant growth and phosphorus uptake by plant, although there is a significant decrease in the phosphorus concentration in the plant. This suggested, as mentioned by a number of authors, an indirect effect due to the production of plant growth hormone-like substances, which can promote the solubilization of mineral elements by increasing the 'sink effect' of the plant (Laheurte & Berthelin 1988).

Some results have shown (Azcon, Barea & Hayman 1976; Powell & Daniel 1978) that vesicular–arbuscular or endomycorrhizal fungi (*Glomus*) stimulate the uptake of phosphorus from insoluble phosphate. A bacterium able to solubilize rock phosphate *in vitro*, associated with mycorrhizal fungi, enhanced phosphorus uptake (Azcon, Barea & Hayman 1976). Inoculation of calcium phosphate dissolving bacteria of seedlings of *Pinus resinosa* growing in a soil deficient in soluble phosphate enhanced seedling growth as well or better than soluble phosphate fertilizer (Ralston & McBride 1976). In an experiment where *Pinus caribea* was cultivated in a tropical soil (ferrallitic) enriched in tricalcium phosphate and inoculated with an ectomycorrhizal fungi (*Pisolithus tinctorius*) and a bacillus dissolving phosphate, a significant increase in pine growth and phosphorus uptake by the pine shoots was observed after 4 months as a result of the microbial inoculation (Chakly & Berthelin 1982) (Table 2). Shoot phosphorus content was also increased, confirming the ability of mycorrhiza and bacteria to promote the dissolution of insoluble phosphate by the roots.

TABLE 2. Growth (g plant$^{-1}$), phosphorus uptake (mg plant$^{-1}$), phosphorus content (%) of the shoot of *Pinus caribea* inoculated with an ectomycorrhizal fungi (*Pisolithus tinctorius*) and by a phosphate dissolving bacillus, after 4 months cultivation in mixture of ferallitic soil and vermiculite enriched in tricalcium phosphate

| | *Pinus* alone | *Pinus* + *P. tinctorius* | *Pinus* + *Bacillus* | *Pinus* + *P. tinctorius* + *Bacillus* |
|---|---|---|---|---|
| Growth | 0·36 ± 0·16 (a) | 0·87 ± 0·24 (b) | 0·79 ± 0·22 (b) | 1·30 ± 0·35 (b) |
| P uptake | 0·3 ± 0·2 (a) | 2·5 ± 0·9 (b) | 2·4 ± 1·0 (b) | 2·6 ± 1·5 (b) |
| P content | 0·08 (a) | 0·28 (b) | 0·30 (b) | 0·20 (b) |

Two treatments with the same letter on the same line are not significantly different ($P = 0·01$).

TABLE 3. Dissolution of rock phosphate as a result of the production of aliphatic acids by the rhizobacteria of rice during the biodegradation of root exudates as carbon and energy source (after 250 h incubations). * Figures in parentheses indicate maximum content during incubation in p.p.m.

| | Bacteria | | | |
|---|---|---|---|---|
| | *Klebsiella oxytoca* | *Pseudomonas* sp. | *Enterobacter cloacae* | *Azospirillum lipoferum* |
| Phosphorus solubilized (μg P per culture flask) | 222 ± 55 | 193 ± 27 | 347 ± 167 | 23 ± 21 |
| % of mineralization of exudates | 59 | 61 | 26 | 21 |
| Aliphatic acids produced in the medium* | Citric (1·7) | Gluconic | Citric (8·5) | No compound detected |
| | Malic (8·8) | | Malic (14) | |
| | Fumaric | | Fumaric | |
| | One unidentified compound | | Two unidentified compounds | |

The microbial dissolution of rock phosphate by bacteria or fungi has been studied in synthetic or artificial nutrient media containing carbohydrates (mainly glucose) as carbon and energy source (e.g. Louw & Webley 1959; Tardieux-Roche 1966; Bajpai & Sundara Rao 1972; Michoustine 1972; Banik & Dey 1981; Leyval & Berthelin 1986). As a result it has appeared important to determine the ability of the root exudates of axenic plants to act as an efficient nutrient medium to promote the bacterial dissolution of rock phosphates. For this purpose rice root exudates were collected in sterile conditions (De Giudici 1985) and used, as a nutrient medium, in batch cultures to study the dissolution of a natural rock phosphate by different rice rhizobacteria. After 250 h incubation in 20 ml of a liquid nutrient medium, containing 5 mg of organic carbon provided by root exudates, and 20 mg of Togo rock phosphate, the dissolution of rock phosphate by different strains of rice rhizobacteria was observed (Table 3).

High performance liquid chromatography analyses of the medium, and complementary enzyme assays, followed the formation of aliphatic acids by the rhizobacteria (Table 3). Higher solubilization was observed where the bacteria produced large amounts of acids but mineralized small amounts of exudates. During a 250 h incubation the bacteria were able to produce and then use, as nutrients, the same aliphatic acids, e.g. *Klebsiella oxytoca* produced, then utilized malic acid. After 25 h incubation, 2·3 p.p.m. of malic acid were found in the nutrient medium and 6·1 and 8·8 p.p.m. respectively after 45 and 75 h but only 0·8 p.p.m. after 250 h incubation. The same acid was produced by *Enterobacter cloacae*, reaching 14 p.p.m. in the culture flask after 250 h incubation. This may explain the difference in phosphorus dissolution observed with these two strains of rice rhizobacteria.

## *Rhizospheric weathering of silicates (biotite and phlogopite)*

The transformation of biotite to vermiculite by seedlings or growing plants has been observed by Mortland, Lawton & Uehara (1956), Spyridakis, Chester & Wilde (1967). Spyridakis, Chester & Wilde (1967) established an order of effectiveness for the roots of seedlings in producing kaolinite from biotite. Fourteen coniferous and deciduous plants, cultivated for 7–24 months in a greenhouse on acid-washed quartz sand enriched with biotite, were assessed. The formation of kaolinite under the influence of rhizosphere activity with white cedar, hemlock, white pine, white spruce, red oak and hard maple is of doubtful occurrence under natural conditions but was favoured by the experimental conditions. These works, however, did not distinguish between the roles of the rhizosphere micro-organisms and the plants themselves.

Some authors have studied the relative effects of symbiotic and non-symbiotic rhizospheric microflora on the weathering of silicates. Mojallali & Weed (1978) reported that the inoculation of soya-bean plants by the endomycorrhizal fungi *Glomus macrocarpus* increased the uptake of potassium from biotite. However in this study the difference between mycorrhizal v. non-mycorrhizal K uptake was larger in the treatment with soil alone (soil without addition of mica) and complicated the issue. However they observed by X-ray diffraction analysis that the mica had partially weathered to vermiculite. Electron microprobe scans also allowed the detection of the removal of K, and to a lesser extent of Al, from the edges of the biotite flakes.

In an experiment where maize was grown under axenic conditions in laboratory devices (Berthelin & Leyval 1982), in a potassium-deficient medium where biotite was the potassium source, it was observed that maize roots alone (without micro-organisms) were able to use potassium from biotite (Table 4). The inoculation with a mixed rhizosphere bacterial population and the endomycorrhizal fungi *Glomus mosseae* significantly stimulated this potasium uptake (Table 4) and increased the weathering of biotite. However as the shoot potassium concentration was always lower in inoculated plants, compared to non-inoculated ones, this suggests that the

TABLE 4. Uptake of K from biotite by maize shoots during 7 weeks growth (g plant$^{-1}$)

| | Plant alone | Plant + rhizobacteria | Plant + endomycorrhiza | Plant + endomycorrhiza + rhizobacteria |
|---|---|---|---|---|
| Shoot weight | 456 ± 180 (a) | 979 ± 46 (b) | 1050 ± 60 (b) | 1170 ± 334 (b) |
| K uptake | 1·5 ± 0·1 (a) | 2·6 ± 0·1 (b) | 1·9 ± 0·1 (b) | 2·8 ± 0·5 (b) |
| K content (‰) | 3·3 | 2·7 | 1·8 | 2·4 |

K is bought as biotite, but 0·22 and 0·7 mg were bought by the seed and the nutrient solution, respectively. Seed weight is 300 mg.

increase of plant growth by rhizospheric microbial activity modified the chemical equilibrium of potassium solubility with the plant acting as a potassium sink. As in the experiments with phosphates a hormonal-like effect, due to microbes, may be involved in the plant growth stimulation. It may also be that some other nutrient element became limiting in the non-inoculated plants.

The comparison of the results of four experiments performed under different growth conditions has verified that potassium uptake, independently of the microbial root treatments, was well correlated with the production of plant biomass (Fig. 2) (Leyval 1981; Leyval & Berthelin 1982). Significant microbial stimulation of maize growth, and consequently potassium uptake from biotite, was observed (as for phosphorus from insoluble phosphates) under relatively poor culture conditions involving a reduced supply (but not total deficiency) of total potassium and phosphorus in the nutrient solution (Leyval & Berthelin 1982; Laheurte & Berthelin 1988).

A lysimeter placed in a greenhouse was used to study the rhizospheric weathering of phlogopite (Leyval 1988; Leyval & Berthelin 1989). It consisted of lysimeters (Fig. 1) containing a mixture of sand, phlogopite and rock phosphate. Six-month-old pine seedlings (*Pinus sylvestris*) were planted in the lysimeter and inoculated with an acid-producing bacteria (*Agrobacterium* sp.) and/or an ectomycorrhizal fungi (*Laccaria laccata*). A nutrient solution was given to the plants daily to complement the presence of rock phosphate and phlogopite. Phlogopite was the

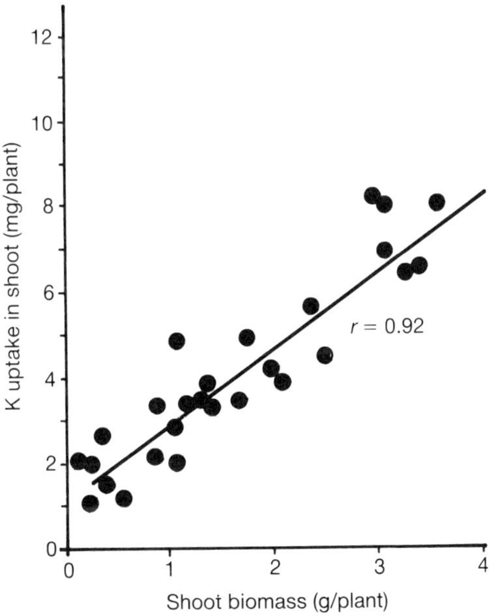


FIG.2. Relation between uptake of K and shoot biomass production during experimental weathering of biotite in the rhizosphere of maize (Leyval 1981; Leyval & Berthelin, 1982).

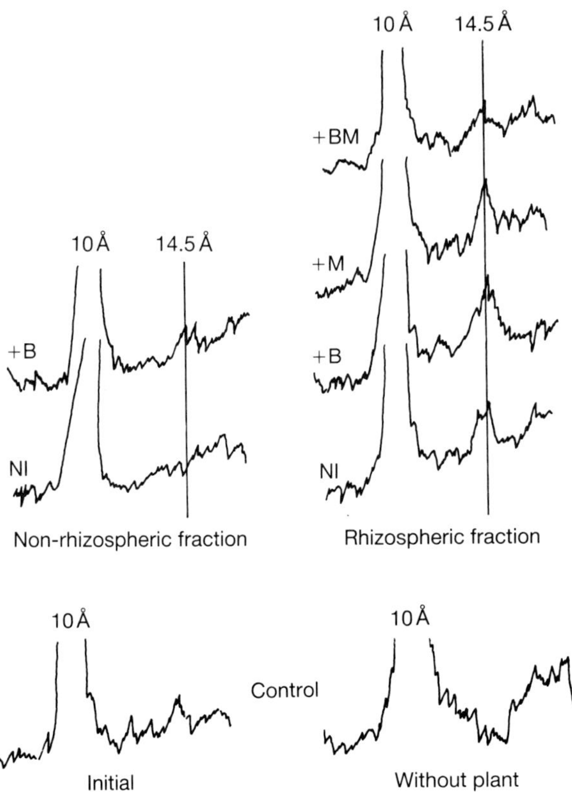


Fig.3. Weathering of phlogopite in the rhizosphere of pine (*Pinus sylvestris*). NI = non-inoculated; + B = inoculated with bacteria; + M = inoculated with ectomycorrhizal fungi; + BM = inoculated with bacteria and ectomycorrhizal fungi (Leyval 1988). X-ray diffraction diagrams.

only source of Mg, K, Fe and Al, but soluble K was added after 1 year when symptoms of deficiency were observed.

After two years cultivation the lysimeters were opened and plants and minerals were collected. Non-rhizosphere, rhizosphere and mineral particles adhering to the roots were recovered as previously described. X-ray diffraction analyses (Fig. 3) were used to detect the mineralogical transformation of phlogopite.

In the pine rhizosphere samples of phlogopite, a 14.5 Å peak appeared beside the 10 Å mica peak (Fig. 3). This peak was not detected in non-rhizospheric samples or in control samples collected from lysimeters without plant. This peak did not change after treatment with ethylene glycol and was attributed to vermiculite. The peak was detected in the pine rhizosphere after 1 year cultivation and was found in the different root treatments, i.e. in inoculated and non-inoculated pine rhizosphere samples. Weathering was however more important where the rhizosphere was inoculated with bacteria as the 14 Å peak is larger but also appeared in the non-rhizosphere samples. The transformation of phlogopite was confirmed by an increase of the cation exchange capacity of the mineral samples and by a decrease in their K content (Leyval 1988). These differences between rhizosphere and non-rhizosphere samples confirm the rhizosphere as an interesting microsite where biosphere, through the roots and their associated micro-organisms, plays a major role in the weathering and evolution of soil minerals. The analyses of water-soluble aliphatic acids, collected by rinsing the residual mineral mixture with distilled water, has shown the presence of relatively large amounts of malic, lactic and gluconic acids and lower amounts of citric and fumaric acids in the rhizosphere of non-inoculated pine. These acids were present in larger amounts in the rhizosphere of pine inoculated with bacteria. A total of 63 μmol per plant of lactic, malic, citric, fumaric, gluconic acids were detected in an inoculated rhizosphere and only of 38 μmol in a non-inoculated one. In mycorrhizal inoculated roots, associated or not with bacteria, these acids were either not detected or were present in very low amounts ($<1$ μmol). These results suggest that depending on the type of root–microbe association, organic acids may be involved in the weathering of minerals. Root–bacteria associations are especially active. With mycorrhizal roots, the greater exploration of the soil volume and the resulting increased exchange-uptake processes, promoting the sink effect of the root systems were more important.

## CONCLUSION

Experimental results have shown that plant roots and their associated symbiotic mycorrhizal and non-symbiotic (rhizobacterial) microflora can weather soil minerals and form clays (vermiculite) from biotite or phlogopite. This mineral weathering is more important in rhizosphere than in the non-rhizosphere materials. Roots alone, in axenic conditions (without micro-organisms) can absorb phosphorus, potassium or iron from rock phosphates of different origin and silicates (e.g. biotite). They can also absorb and solubilize other mineral elements from soil minerals as a consequence of the release of $CO_2$, $H^+$ ions and organic acids which weather minerals. For instance, axenic maize roots can exudate oxalic, tartaric, oxalo-acetic, isocitric, citric, lactic, fumaric and succinic acids that can be involved with carbonic acid and protons in the dissolution and weathering of minerals by reactions such as:

$$M^+(\text{Mineral})^- + H^+R^- \rightarrow H^+(\text{mineral})^- + M^+R^-$$
$$(R^- = HCO_3^-, CH_3COO^-, {}^-O_2C\text{-}CHOH\text{-}CH_2\text{-}CO_2^- \ldots)$$
$$M^+(\text{mineral})^- = \text{silicates, phosphates} \ldots \text{from rocks}$$
$$M^+ = \text{metal ions.}$$

In reactions such as these with organic ligands acting as complexing agents, metal-organic complexes can be formed.

$$M^+(\text{mineral})^- + H^+L^- \rightarrow H^+(\text{mineral})^- + ML$$
$$H^+L^- + LM \rightarrow L_2M + H^+$$
$$L = \text{organic ligands.}$$

Few data are available on the ability of axenic roots to weather soil minerals, but this situation will seldom if ever occur since, in natural conditions, plant roots are always associated with micro-organisms; for example, 90% of plant species form symbiotic associations with mycorrhizal fungi (80% with endomycorrhizal and 10% with ectomycorrhizal fungi). These associations can also be involved in the dissolution of minerals and weathering by the production of acids, including $CO_2$, complexing compounds and protons. During the biodegradation of rice roots exudates, citric, fumaric, malic and gluconic acids were synthesized by rhizobacteria and similar acids have been observed in the pine rhizosphere.

Roots and mycorrhiza can absorb mineral elements and act as sinks in excess of nutritional requirements. Such uptake and exchange uptake of metal ions and of elements such as phosphorus promote the weathering of minerals (Berthelin 1983, 1988; Robert & Berthelin 1986). This exchange uptake is increased by the mycorrhiza hyphae which explore a larger soil volume than do non-mycorrhizal roots. The production of hormone-like substances by the rhizosphere bacteria seems also to be indirectly involved in promoting exchange uptake by increasing plant growth. These different processes are certainly involved in the weathering of different minerals and, probably not just for phosphate and the phyllosilicates which have been observed in the experiments.

Despite the studies reported here, the effects of the plant roots, rhizosphere micro-organisms and weathering mechanisms involved in the rhizosphere are poorly understood and there is a clear need for further studies relating to their pedogenetic and agronomic impacts. Effects related to the activities of symbiotic and non-symbiotic microflora often depend on plant growth conditions, although many unexplained results, with varying types of effect on plant growth and nutrition and mineral weathering, but which may be due to the experimental conditions, remain to be clarified. For example, the observation that plants such as *Festuca sylvatica*, *Molinia caerulea*, *Mercurialis perennis*, *Vaccinium myrtillus* and *Calluna vulgaris* have specific effects on microbial populations, and different activities (iron reduction, phosphate or carbonate dissolution etc.) on the minerals, suggest the need for studies on their role in the weathering (Leyval & Berthelin 1983). It seems important to know if there are specific relations between plant and rhizosphere microflora in relation to pedogenesis and plant nutrition.

Finally the rhizosphere, including the roots with their external mucigel, composed mainly of polysaccharides such as galacturonic acid, the symbiotic and non-symbiotic microflora, and its special environment, must be a very active medium and habitat for weathering and mineral concentration. Further research is needed to augment the limited information available.

## REFERENCES

**Azcon, R., Barea, J.M. & Hayman, D.S. (1976).** Utilization of rock phosphate in alkaline soils by plants inoculated with mycorrhizal fungi and phosphate solubilizing bacteria. *Soil Biology and Biochemistry*, **8**, 135–138.

**Bajpai, P.D. & Sundara Rao, W.V.B. (1972).** Phosphate solubilizing bacteria. Part. I. Solubilization of phosphate in liquid culture by selected bacteria as affected by different pH values. *Soil Science and Plant Nutrition*, **17**, 41–53.

**Banik, S. & Dey, B.K. (1981).** Phosphate solubilizing microorganisms of a lateritic soil. I. Solubilization of inorganic phosphates and production of organic acids by microorganisms, isolated in sucrose calcium phosphate agar plates. *Zentralblatt für Bakteriologie*, **II**, 478–486.

**Berthelin, J. (1983).** Microbial geochemistry. *Microbial Weathering Processes* (Ed. by W.E. Krumbein), pp. 223–262. Blackwell Scientific Publications, Oxford.

**Berthelin, J. (1988).** Physical and chemical weathering in geochemical cycles. *Microbial Weathering Processes in Natural Environments* (Ed. by A. Lerman & M. Meybeck), pp. 33–59. NATO ASI Series, Kluwer Academic Publishers, Dordrecht.

**Berthelin, J. & Leyval, C. (1982).** Ability of symbiotic and non symbiotic rhizospheric microflora of maize (*Zea mays*) to weather micas and to promote plant growth and plant nutrition. *Plant and Soil*, **68**, 369–377.

**Callot, G., Chamayou, H., Maertens, C. & Salsac, L. (1982).** *Les Interactions Sol–Racine, Incidence sur la Nutrition Minérale.* Institut National de la Recherche Agronomique, Paris.

**Chakly, M. & Berthelin, J. (1982).** Les mycorhizes: biologie et utilisation. *Rôle d'une ectomycorhize* 'Pisolithus tinctorius–Pinus caribea' *et d'une bactérie rhizosphérique sur la mobilisation du phosphore de phosphates minéraux et organiques insolubles.* Les Colloques de l'INRA, **13**, 215–220. INRA, Paris.

**Conyers, E. & McLean, E.O. (1968).** Effect of plant weathering of soils clays on plant availability of native and added potassium and on clay mineral structure. *Soil Science Society of America Proceedings*, **32**, 341–345.

**De Giudici, P. (1985).** *Contribution à l'étude de la solubilité microbienne d'un phosphate naturel en modèle rhizosphérique.* Thèse de Doctorat de l'Institut National Polytechnique de Lorraine (ENSAIA), Nancy-Vandoeuvre, France.

**Gerretsen, F.C. (1948).** The influence of microorganisms on the phosphate intake by the plant. *Plant and Soil*, **1**, 51–81.

**Grierson, D. & Jenkins, D.A. (1986).** Modifications to clay minerals in the rhizosphere. *Clay Mineral Group Meeting, Bangor North Wales*, 23–25 March 1986.

**Juang, T.C. & Uehara, G. (1968).** Mica genesis in Hawaïian soils. *Soil Science Society of America Proceedings*, **32**, 31–35.

**Laheurte, F. (1985).** *Recherches sur quelques mécanismes d'interactions rhizosphériques en modèles expérimentaux: Altération microbienne et racinaire, exsudation et humification.* Thèse de Doctorat de l'Université de Nancy I, France.

**Laheurte, F. & Berthelin, J. (1986).** Mycorrhizae: physiology and genetics. *Interactions between Endomycorrhizas and Phosphate Solubilizing Bacteria: Effects on Nutrition and Growth of Maize*, pp. 339–343. 1st SEM Dijon. INRA, Paris.

**Laheurte, F. & Berthelin, J. (1988).** Effect of a phosphate solubilizing bacteria on maize growth and root exudation over four levels of labile phosphorus. *Plant and Soil*, **105**, 11–17.

**Leyval, C. (1981).** *Etude préliminaire de l'altération de minéraux par les microflores rhizosphériques de Graminées et d' Ericacées.* Thèse de Doctorat de 3ème Cycle de l'Université de Nancy I, France.

**Leyval, C. (1988).** *Interactions bactéries-mycorhizes dans la rhizosphère du pin sylvestre et du hêtre: Incidences sur l'exsudation racinaire et l'altération des minéraux.* Thèse de Doctorat d'Etat de l'Université de Nancy I, France.

**Leyval, C. & Berthelin, J. (1982).** Rôle des microflores symbiotiques et non symbiotiques sur l'altération de la biotite et la croissance du maïs (*Zea mays*): influence des conditions du milieu. *Science du Sol*, **1**, 3–12.

**Leyval, C. & Berthelin, J. (1983).** Effets rhizosphériques de plantes indicatrices de grands types de pédogenèse sur quelques groupes bactériens modifiant l'état de minéraux. *Revue d'Ecologie et de Biologie du Sol*, **20**, 191–206.

**Leyval, C. & Berthelin, J. (1986).** Mycorrhizae: Physiology and genetics. *Comparison between the Utilization of Phosphorus from Insoluble Mineral Phosphates by Ectomycorrhizal Fungi and Rhizobacteria*, pp. 340–345 1st SEM, Dijon. INRA, Paris.

**Leyval, C. & Berthelin, J. (1989).** Interactions between *Laccaria laccata*, *Agrobacterium radiobacter* and beech roots: influence on P, K, Mg and Fe mobilization from minerals and plant growth. *Plant and Soil*, **117**, 103–110.

**Louw, H.A. & Webley, D.M. (1959).** A study of soil bacteria dissolving certain mineral phosphate fertilizers and related compounds. *Journal of Applied Bacteriology*, **22**, 227–234.

**Marschner, H. (1986).** *Mineral Nutrition in Higher Plants*. Academic Press, New York.

**Michoustine, E.N. (1972).** Processus microbiologiques mobilisant les composés du phosphore dans le sol. *Revue d'Ecologie et de Biologie du Sol*, **9**, 521–528.

**Mojallali, M. & Weed, S.B. (1978).** Weathering of micas by mycorrhizal soybean plants. *Soil Science Society of America Journal*, **42**, 367–372.

**Mortland, M.M., Lawton, K. & Uehara, G. (1956).** Alteration of biotite to vermiculite by plant growth. *Soil Science*, **82**, 477–481.

**Mortvedt, J.J., Giordano, P.M. & Lindsay, W.L. (1972).** Micronutrients in agriculture. *Soil Science Society of America Inc.*, Madison, USA.

**Powell, C.L. & Daniel, J. (1978).** Growth of white clover in undisturbed soils after inoculation with efficient mycorrhizal fungi. *New Zealand Journal of Agricultural Research*, **21**, 675–681.

**Ralston, D.B. & McBride, R.P. (1976).** Interaction of mineral phosphate dissolving microbes with red pine seedlings. *Plant and Soil*, **45**, 493–507.

**Robert, M. & Berthelin, J. (1986).** Interactions of soil minerals with natural organics and microbes. Role of biological and biochemical factors in soil mineral weathering. *Soil Science Society of America Special Publications*, **17**, 453–495.

**Spyridakis, D.E., Chester, S.G. & Wilde, S.A. (1967).** Kaolinization of biotite as a result of coniferous and deciduous seedling growth. *Soil Science Society of America Proceedings*, **31**, 203–210.

**Tardieux-Roche, A. (1966).** Contribution à l'étude des interactions entre phosphates naturels et microflore du sol. *Annales Agronomiques*, **4**, **5**, 403–471, 479–528.

# Effect of living roots on carbon and nitrogen of the soil microbial biomass

P. BOTTNER, J. CORTEZ AND Z. SALLIH
*CEFE-CNRS, BP 5051, 34033 Montpellier Cedex, France*

## SUMMARY

**1** Two soils were incubated in the laboratory, with $^{14}C$- and $^{15}N$-labelled wheat straw (*Triticum aestivum*). Half of the samples were cropped ten times in succession with spring wheat. After each cropping, the roots were removed from the soil. The other half of the samples were kept bare without plants.

**2** Over the 2 years of the experiment it was sampled eight times for microbial biomass–C and –N determination and the analyses of $NH_4^+$-N, $NO_3^-$-N and N absorbed by the successive croppings.

**3** The presence of living roots induced a gradual increase of the microbial biomass size, but lowered the biomass–$^{14}C$.

**4** An extra amount of N was mobilized by the plants and the rhizospheric microbial biomass from the stable soil native humus compounds. Thus, the successive root systems increased the turnover rates of the microbial biomass–N.

## INTRODUCTION

Among the extreme diversity of microsites in governing the activity and the survival of the micro-organisms, the rhizosphere plays an essential part in modifying the density, the activity and the structure of the microbial communities (Lynch 1982). The aim of this work was to study, in terms of microbial biomass, the role of the rhizosphere in soil nitrogen mobilization and in the transfer of N to the plants.

## MATERIALS AND METHODS

The characteristics and the site locations of the two soils used in this experiment were described in Sallih & Bottner (1988). Soil 1 was collected from the A1 horizon (0–15 cm) of a fersiallitic calcic soil (CPCS 1967) [C = 1·3%; N = 0·12%; clay content = 29%; pH ($H_2O$) = 7·9]. Soil 2 was taken from the A1 horizon (0–15 cm) of a brown soil (C = 2·7%; N = 0·20%; clay content = 11%; pH ($H_2O$) = 6·5). Both soils were collected in southern France from natural grasslands developed under humid Mediterranean climatic conditions.

The detailed description of the experimental part was reported by Sallih & Bottner (1988). Each soil was air-dried, homogenized, subsampled in 18 portions of 800 g dry soil, mixed with 7 g of uniformly $^{14}C$- and $^{15}N$-labelled mature wheat straw (C = 43%; N = 1·0%; specific activity = 2·59 MBq g$^{-1}$ C; $^{15}N$ enrichment =

6·61%), remoistened to 75% water-holding capacity (WHC) and the pots were placed in a growth chamber (daylight, 16 h at 25 °± 4 °C; night, 8 h at 15 °± 3 °C). The experiment lasted over 700 days. Half of the pots were unplanted and the other half were cropped ten times in succession with spring wheat (*Triticum aestivum*; six seedlings to each pot). The wheat was harvested about 1–2 months after sowing, approximately at flowering. After each harvest, the roots were removed from the soil by sieving and hand sampling. They were gently stripped by hand and the fine soil crumbs adhering to the roots were returned to the soil. The next growing period started 2–5 days after harvesting. The soils in the unplanted pots were remixed at the same time. For both treatments with and without plants, soil moisture was maintained to 75% WHC as necessary. Between growing periods 3–4 (days 118–200) and 8–9 (days 456–532) all pots, whether planted or unplanted were kept dry for about 2½ months. During the 700 days of experiment eight samplings were performed on the following days:

1 without plants: days 16, 29, —, 85, 121, 247, 422, 690, and
2 with plants: days —, 29, 64, 85, 121, 247, 422, 690.

At the end of the experiment the two remaining pots were dried and stored for further analyses. At each sampling and for each treatment the whole soil of one pot was used. After the plants were cut and the roots removed from the soil, one portion of the soil (25 g equivalent dry soil, 4 replicates) was immediately $CHCl_3$-fumigated and incubated for the microbial biomass-C (Jenkinson & Powlson 1976) and -N estimation. Another part was immediately incubated without fumigation.

The total C ($MB-C^t$) and labelled C ($MB-C^*$) of microbial biomass were estimated as described in Bottner, Sallih & Billes (1988) using a Kc factor of 0·41. The total N ($MB-N^t$) and labelled N ($MB-N^*$) were estimated, after a 10 day incubation, by measurement of $NH_4^+$-N* and $NO_3^-$-N* extracted from fumigated and unfumigated samples (75 g equivalent dry soil, two replicates) by shaking with 1·0 N $K_2SO_4$ (w/v = 1 : 4) (Shen, Pruden & Jenkinson 1984). $NH_4^+$-N and $NO_3^-$-N were determined by colorimetry. $^{15}N$ enrichments were determined by mass spectrometry using the method of Rittenberg *et al.* (1948).

In estimating the biomass–N the Kn factor was calculated according to the relation of Paul & Voroney (1980): $Kn = 0{\cdot}39 - 0{\cdot}014\ (C^f/N^f)$, where $C^f = CO_2$-C and $N^f = NH_4^+$-N from the fumigated samples after the 10 day incubation. The values obtained from sampling day 16 were not used for the calculation of the microbial biomass, since the $CO_2$ released from the fumigated soils during the early period after incorporation of the plant material was influenced by the high microbial activity and the biomass carbon was underestimated (Martens 1985).

## RESULTS

In Fig. 1, the patterns of the biomass total C ($MB-C^t$), in the treatment without plants are similar for both soils. The biomass, stimulated by the initial amendment increased, reaching maximum values approximately between days 60 and 100. At these levels, $MB-C^t$ was about 5% of the total soil organic C for both soils. Beyond

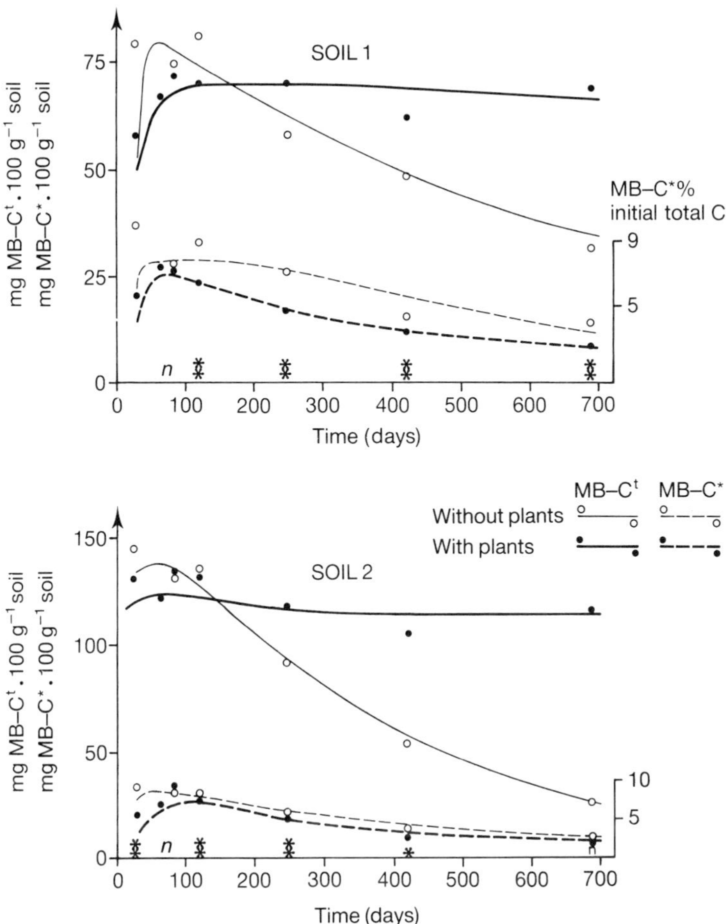


Fig. 1. Total C (MB–C$^t$) and labelled C (MB–C*) of the microbial biomass in mg MB–C$^t$ or MB–C* 100 g$^{-1}$ dry soil. For MB–C*: ** $P = 0{\cdot}01$; *$P = 0{\cdot}05$; ($n$) $P > 0{\cdot}05$.

this time, the biomass progressively decreased. At the end of the experiment MB–C$^t$ accounted for 2% and 0·8% of the total soil C. Thus, in the treatment without plants, after the initial maximum microbial development, in soil 2 with the largest soil C content, the biomass size was twice as high as in soil 1. Nevertheless, at the end of the incubation, both soils tended to reach comparable low values, amounting to between 25 and 35 mg MB–C$^t$ 100 g$^{-1}$ soil.

The biomass was markedly modified by the presence of plants. For both soils, its level remained high throughout the successive wheat cultures. For soil 2, the amounts of MB–C$^t$ measured during the initial 200 days, were comparable in both treatments. For soil 1, during this period, the measured MB–C$^t$ values were a little lower when plants were present, comparatively to the bare soil. Thus, in this soil, the presence of living roots did not contribute further to the increase of the microbial biomass during the initial high activity stages.

For both soils, beyond these initial stages, the biomass progressively became very different in the two treatments. With plants, the biomass size remained approximately at the initial plateau. At the end of the experiment, the amount of $MB-C^t$ in the soils with plants was about 2–4 times larger than in the bare soils.

Concerning the labelled biomass ($MB-C^*$), the question is to know whether its pattern is affected by the successive wheat cultures, since the microbial substrate originated from the roots is unlabelled. The amounts of $MB-C^*$ are also presented in Fig. 1. For both soils, to which the same amount of labelled substrate was added, the $MB-C^*$ curves indicated similar patterns. They levelled off approximately at the same time (between days 60 and 100). In both soils, the presence of plants resulted in lower $MB-C^*$ amounts, except at the end of the experiment for soil 2. In the treatment with plants, $MB-C^*$ accounted for about 70% of the corresponding values in the bare soil. At the maximum levels, $MB-C^*$ was 10–15% of the total labelled $C^*$ remaining in soil; at the end of the experiment, these amounts represented 8–10%. The pattern of total labelled C remaining in the soils throughout the 700 days of the experiment was described by Sallih & Bottner (1988). Thus, the presence of the successive living root systems resulted in a diminution of labelled biomass sizes; this effect was more pronounced in soil 1 than in soil 2.

In the unfumigated soils $NH_4^+$-$N^t$ seldom exceeded 1–2 mg N 100 $g^{-1}$ dry soil. In the unplanted soils, the accumulation of nitrates began around day 100 and $NO_3^-$-$N^t$ increased continuously, amounting at the end of the experiment to 12 mg and 36 mg 100 $g^{-1}$ dry soil in soils 1 and 2. On the contrary, $NO_3^-$-$N^t$ never exceeded 1·5 mg $NO_3^-$-$N^t$ 100 $g^{-1}$ dry soil in the planted soils. The amounts of $NO_3^-$-$N^t$ were not modified after the fumigation and the 10 day incubation, indicating that the $CHCl_3$ inhibited or killed the nitrifiers.

For both treatments and for both soils, the variation of the microbial biomass–N ($MB-N^t$) was found to be almost similar to that of $MB-C^t$ (Fig. 2). Without plants during the first 100 days, i.e. during the active biological periods, $MB-N^t$ was, in soil 2, twice as high as in soil 1; from that time, $MB-N^t$ progressively decreased to reach by the end of the experiment similar low values in both soils (6 mg 100 $g^{-1}$ dry soil). The biomass–N was reduced by a factor of three and five in soils 1 and 2, respectively. The C : N ratios were very close in both soils ranging from 5 to 6.

In the samples with plants, after 100 to 200 days, the $MB-N^t$ level remained relatively constant until the end of the experiment. At that time, $MB-N^t$ was about two and four times higher in planted samples than in unplanted ones, in soil 1 and 2, respectively, and $MB-N^t$ was in soil 2 about twice as high as in soil 1. In this treatment, the accumulated $NO_3^-$-$N^t$ was significantly higher than the decrease in $MB-N^t$. Thus, the $NO_3^-$-$N^t$ production cannot be simply explained by the net decay of the microbial biomass. At the end of the experiment, $NO_3^-$-$N^t$ was three times higher in soil 2 than in soil 1.

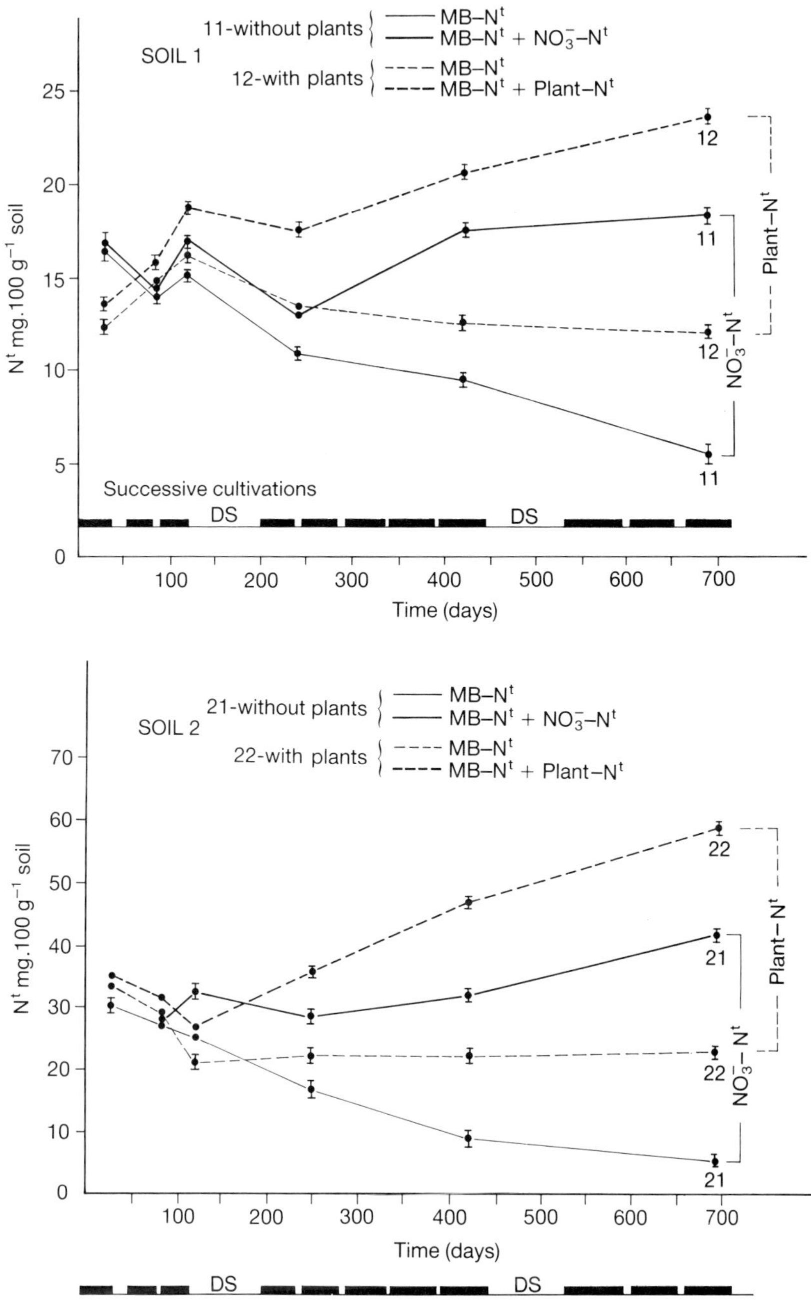


FIG.2. Total N of the microbial biomass (MB–N$^t$), of the nitrates ($NO_3^-$–N$^t$) and of the plants (plant–N$^t$). Plant–N$^t$ represents the cumulative N absorbed by the successive wheat cultivations and corresponding to 100 g dry soil. DS = periods of dry soil (see Methods). Bars: means ±S.E. ($P$ = 0·05).

In Fig. 2, plant–$N^t$ represents the total amount of N accumulated by the successive wheat cultivations (roots and shoots); the N content of the six seeds per 800 g soil has been subtracted. In this figure, the curve (MB–$N^t$ + plant–$N^t$) represents N mobilized by the plants and the micro-organisms for the treatment with plants. The curve (MB–$N^t$ + $NO_3^-$-$N^t$) corresponds, for the unplanted treatment, to N mobilized by the micro-organisms plus the nitrates resulting from the mineralization of N from dead micro-organisms and microbial metabolites after energy exhaustion. For the calculation of the curves the small amounts of $NO_3^-$-$N^t$ which were found in the planted system (never exceeding 1·5 mg 100 g$^{-1}$ dry soil) have been included in plant–$N^t$. If we compare the two curves, it is noteworthy that the presence of plants mobilized more N in the cropped system than in the unplanted one. Beyond the first 200 days, the extra amount of N mobilized in the planted system accounted in both soils for about 25–30% more than in unplanted soils. For the first 200 days, in soil 2 the curves were not significantly different. In soil 1, during the first wheat cultivation, the unplanted samples appeared to mobilize more N than the planted soil. Figure 2 also shows that the amount of plant–$N^t$ assimilated by the successive wheat cultivations and the amount of $NO_3^-$-$N^t$ progressively accumulated in the unplanted soil over the 700 days of experiment were not significantly different. This observation is true for both soils. However in soil 2, more than twice the amount of $NO_3^-$-$N^t$ was accumulated in the unplanted samples and was used by the plants, compared with soil 1. Thus, in both soils, the differences between the curves (MB–$N^t$ + $NO_3^-$-$N^t$) for the bare soil and (MB–$N^t$ + plant–$N^t$) for the planted soils are only explained by the differences in MB–$N^t$.

Table 1 shows the labelled N, expressed as % of total N in the microbial biomass (MB–N*%$N^t$), in the nitrates ($NO_3^-$-N*%$N^t$) and in the plants (plant–N*%$N^t$). In soil 2, the total N content was about twice as high as in soil 1 (N = 0·12 and 0·2% for soils 1 and 2, respectively) and for both soils the same amount of N* had been added; thus the N*%$N^t$ values were lower in soil 2 than in soil 1. In the unplanted samples and for both soils, MB–N*%$N^t$ was very stable over the 700 days of experiment, ranging from 22 to 23% and from 11 to 12% for soils 1 and 2, respectively. After only 29 days of incubation, the maximum percentage was reached, indicating a rapid metabolization of the labelled N. The stability of MB–N*%$N^t$ observed in the unplanted samples for both soils is explained as follows: the decomposition of the labelled plant material with a high initial biological activity involved a fast metabolization of the labelled N. After that initial active stage, the microbial activity slowly decreased because of the available C exhaustion. A part of microbial biomass consequently died. The turnover rates of the surviving organisms became progressively low, resulting in stable MB–N*%$N^t$ values. In the unplanted soils, the $NO_3^-$-N*%$N^t$ values were comparable to those of the MB–N*%$N^t$ indicating that the dead micro-organisms and the microbial metabolites were the essential source of nitrates.

During the first 200–300 days, the MB–N*%$N^t$ values were similar in the planted and unplanted soils ranging from 22 to 24% and from 11 to 13% in soils 1

Table 1. Labelled N expressed in % total N of the microbial biomass (MB–N*%N[t]), of the nitrates $NO_3^-$-N*%N[t]) and of the plant N (Pl–N*%N[t]). a = measured values; b = S.D. for $n = 3–15$

| | Soil 1 | | | | Soil 2 | | | |
|---|---|---|---|---|---|---|---|---|
| | Without plants | | With plants | | Without plants | | With plants | |
| Time (days) | MB–N*%N[t] | $NO_3^-$-N*%N[t] | MB–N*%N[t] | Pl–N*%N[t] | MB–N*%N[t] | $NO_3^-$-N*%N[t] | MB–N*%N[t] | Pl–N*%N[t] |
| | (a) | (a) | (a) | (b) | (a) | (a) | (a) | (b) |
| 29 | 23·1 (23·2; 23) | — | 29·9 (22·1; 23·6) | 18·0 (1·04) | 11·0 (11·2; 11·2) | — | 12·3 (12·3; 12·4) | 13·0 (1·5) |
| 64 | — | — | 23·3 (23·6; 22·9) | — | — | — | 12·6 (12·6; 12·6) | — |
| 85 | 22·9 (23; 22·7) | — | 23·2 (23·2; 23·2) | 16·0 (2·5) | 12·4 (12·3; 12·6) | — | 12·1 (12·3; 12) | 12·0 (0·7) |
| 121 | 21·0 (18; 24·1) | 24·3 (24·2; 24·4) | 24·2 (24·2; 24·1) | 22·0 (1·2) | 11·7 (11·7; 11·7) | 14·9 (15; 14·9) | 12·4 (12·3; 12·5) | 12·0 (1·1) |
| 247 | 22·9 (22·9; 22·9) | 24·9 (25·2; 24·8) | 21·7 (21·7; 21·7) | 19·5 (0·9) | 11·2 (11·5; 10·9) | 14·5 (14·5; 14·4) | 10·6 (10·5; 10·8) | 11·0 (0·5) |
| 422 | 22·5 (22·1; 22·9) | 28·0 (28·2; 27·8) | 19·9 (19·9; 19·9) | 20·0 (2·8) | 11·5 (11·4; 11·6) | 11·3 (10·3; 12·3) | 9·0 (9; 9·1) | 9·0 (0·86) |
| 690 | 22·0 (22·4; 21·6) | 27·0 (26·9; 27·2) | 18·0 (17·9; 18·2) | 14·0 (2·4) | 12·0 (12·1; 12) | 10·5 (10·4; 10·5) | 8·5 (8·5; 8·6) | 7·0 (0·64) |

and 2, respectively. After that time and for both soils in the planted samples, MB–$N^{*}\%N^{t}$ sharply decreased. Figures 1 and 2 show that for both soils from day 200 to 300 the size in biomass became very different between the two treatments. Thus, the presence of successive active root systems contributed to maintain an important microbial biomass with high activities (Fig. 1; Bottner, Sallih & Billes 1988) and with higher turnover rates of the microbial biomass–N, resulting in lowering the MB–$N^{*}\%N^{t}$ values: The labelled N was partially replaced by soil native unlabelled N, whereas in the unplanted soil MB–$N^{*}\%N^{t}$ remained constant.

Over 700 days, plant–$N^{*}\%N^{t}$ evolved differently in the two soils. For soil 1, during the first 100 days, plant–$N^{*}\%N^{t}$ was markedly lower than the corresponding MB–$N^{*}\%N^{t}$ value. In this soil with low N content the competition between the plants and the micro-organisms for N could explain this pattern. During the initial high biological stages, the active microflora essentially developed close by the added labelled plant material particles, which represent the essential energy source, and thus metabolized the available N. On the contrary, in the same time, the root systems of the first cultivations tended to develop preferentially around the soil aggregates which supplied less energy to micro-organisms but which liberated some soil inorganic N. In soil 1, from the beginning until approximately day 400, the plant–$N^{*}\%N^{t}$ values sharply increased, indicating an increasing proportion in $N^{*}$ in the nitrogen absorbed by the plants. The roots progressively colonized the soil-labelled particles, which became a labelled inorganic N source. The highly labelled extra rhizospheric micro-organisms progressively died by available C exhaustion. This explanation would be sustained by the fact that in the unplanted samples (soil 1) the maximum accumulation of labelled nitrates corresponds to the 200–400 day period. From days 400 to 700 soil 1 showed again a highly significant decrease in plant–$N^{*}\%N^{t}$, indicating a predominant uptake of soil native N. During this last period, the microflora essentially stimulated by the root exudation and rhizodeposition, mobilized an increasing proportion of stable humus and soil native compounds. The result was a sharp decrease in the $N^{*}\%N^{t}$ ratios for both the plant– and microbial biomass–N. Thus, the plants can stimulate an extra-mobilization of N from stable soil compounds by their rhizospheric activity.

Similar conclusions might be deduced from soil 2 data (Table 1), except that the MB–$N^{*}\%N^{t}$ and Plant–$N^{*}\%N^{t}$ were close at the initial stage, indicating that the plants (and their rhizospheric microflora) and the extra rhizospheric microflora used similar N pools. This soil has a high N content and the competition between micro-organisms and plants was less pronounced.

## DISCUSSION AND CONCLUSION

In this experiment, over the 700 days, the biological activity markedly varied from high activity stages due to the initial addition of labelled plant material to very low activity stages. In addition, the ten successive wheat cultivations modified the pools of available nutrients, especially of N. The aim of this experiment was, in comparing the planted and unplanted soils, to study the effects induced by successive active root systems on the decomposition processes throughout the varying soil biological stages. The present paper focuses on the microbial biomass.

1 During the initial high activity stages of the soil, the presence of the active roots did not increase the microbial biomass size. In soil 1, the biomass was even significantly lower in the treatment with plants. This could be explained by the competition for N between plants and the extra-rhizospheric microflora stimulated by the high energy resources during the first stages. On the contrary, during the low activity stages, the successive root system doubled and quadrupled the microbial biomass in soils 1 and 2, respectively. This stimulating effect is explained by the C input originated from the roots. Two kinds of compounds are released into the soil: the material released during the plant growth (exudation and rhizodeposition) and the fine root particles artificially left in the soil when the roots were extracted. This later source was probably low since the roots were carefully separated from the soil by sieving and hand sampling.

2 In both soils, the presence of roots significantly reduced the labelled microbial biomass (Fig. 1). In addition, Bottner, Sallih & Billes (1988) demonstrated from this experiment that the roots stimulated the labelled biomass activity and increased the turnover rates of the labelled organisms. Thus the root activity increased the decomposition of the labelled plant material. Nevertheless, the stimulating effect of the active roots cannot be explained by an available C input since, in this experiment, the plants were not labelled. Some mechanisms, like microflora predation by protozoa (Anderson, Coleman & Cole 1981; Clarholm 1984), increased rates of organic matter decomposition or stimulation of the overall soil activity, must be taken into account. Thus, Sallih & Bottner (1988) reported that the presence of living roots increased the rate of the labelled plant material decomposition during the low activity stages.

3 In the planted soil, the successive wheat cultures used approximately the same amount of N as the quantity of $NO_3^-$-N progressively accumulated in the unplanted treatment. Nevertheless, the plant–N and the $NO_3^-$-N in the bare soil originated from separate pools, since their isotopic enrichment was very different (Table 1). Thus, an extra amount of N was mineralized in the presence of plants. This additional quantity was partly used by plants but also by the rhizospheric micro-organisms. The extra N mobilization represented from one-third to one-quarter of the N metabolized by microflora in the unplanted soil. This additional N essentially came from soil native N, i.e. from stable humified compounds. The stable N mobilization by the roots especially appeared during the low soil activity stages when the rhizosphere activity was close to the overall soil activity and when the available N was exhausted.

4 The origin of the nitrates in the bare soils cannot entirely be explained. For both soils the isotopic enrichment of $NO_3^-$-N and the one of the microbial biomass were similar. Nevertheless, the amount of $NO_3^-$-N produced over the 700 days was significantly higher than the quantity of N able to be released from the dead micro-organisms. Thus, to explain these data, a possible reuse of microbial metabolites must probably be taken into account.

## REFERENCES

**Anderson, R.V., Coleman, D.C. & Cole, C.V. (1981).** Effects of saprotrophic grazing on net mineralization. *Terrestrial Nitrogen Cycles* (Ed. by F.E. Clark & T. Rosswall), Vol. 33, pp. 49–115. Ecological Bulletin, Stockholm.

**Bottner, P., Sallih, Z. & Billes, G. (1988).** Root activity and carbon metabolism in soils. *Biology and Fertility of Soils*, **7**, 71–78.

**Clarholm, M. (1984).** Heterotrophic, free-living protozoa: neglected microorganisms with an important task in regulating bacterial populations. *Current Perspectives in Microbial Ecology* (Ed. by M.J. Klug & C.A. Reddy), pp. 443–446. Washington.

**Jenkinson, D.S. & Powlson, D.S. (1976).** The effects of biocidal treatments on metabolism in soil: V. A method for measuring soil biomass. *Soil Biology and Biochemistry*, **8**, 209–213.

**Lynch, J.M. (1982).** Interactions between bacteria and plants in the root environment. *Bacteria and Plants* (Ed. by M.E. Rhodes-Roberts & F.A. Skinner), pp. 1–23. Academic Press, New York.

**Martens, R. (1985).** Limitation in the application of the fumigation technique for biomass estimation in amended soils. *Soil Biology and Biochemistry*, **17**, 57–63.

**Paul, E.A. and Voroney, R.P. (1980).** Nutrient and energy flows through soil microbial biomass. *Contemporary Microbial Ecology* (Ed. by D.C. Ellwood, J.N. Heldberg, M.J. Latham, J.M. Lynch & J.M. Slater), pp 215–237. Academic Press, London.

**Rittenberg, D., Willson, D.W., Nier, A.D.C. & Rieman, P.C. (1948).** *Preparation and Measurements of Isotopic Tracers*, pp. 31–42. Edwards, Ann Arbor.

**Sallih, Z. & Bottner, P. (1988).** Effects of wheat (*Triticum aestivum*) roots on mineralization rates of soil organic matter. *Biology and Fertility of Soils*, **7**, 67–70.

**Shen, S.M., Pruden, G. & Jenkinson, D.S. (1984).** Mineralization and immobilization of nitrogen in fumigated soil and the measurement of microbial biomass nitrogen. *Soil Biology and Biochemistry*, **16**, 437–444.

# Effects of toxic concentrations of metals on root growth and development

M.S. DAVIES
*School of Pure and Applied Biology, University of Wales College of Cardiff, PO Box 915, Cardiff CF1 3TL, UK*

## SUMMARY

1 This paper describes some aspects of the effects of toxic metals on root growth.
2 The extent of metal inhibition of root growth is influenced greatly by the chemical background in which the metal is supplied and the form in which nitrogen is supplied. Many studies of metal toxicity are carried out under conditions which are not related to the chemistry of the soil.
3 The growth history of experimental material can also have a large effect on root growth responses to toxic metals. The dangers of using experimental plants raised in soil conditions quite different from those of their native habitats are emphasized. Short-term exposure of root systems to a toxic metal can also result in enhanced tolerance of those roots to that metal.
4 The effects of some metals, particularly zinc, on cellular events, including the mitotic cell cycle, in the root meristem of tolerant and non-tolerant plant populations are discussed.

## INTRODUCTION

A number of metal ions inhibit biological processes at supra-optimal concentrations. Some are essential trace elements at low concentrations while others are not required for plant growth. Nieboer & Richardson (1980) classified metals on the basis of their ligand-forming properties; this classification is more informative than the widely used term 'heavy metal' which only encompasses those metals with a density greater than five (Lapedes 1974) and thereby excludes metals such as aluminium and beryllium which are toxic at comparatively low concentrations. It must also be stressed that most metals, including those which are major nutrients, are inhibitory to plant growth if present in excessive concentration.

Metal toxicity may be expressed in a number of growth parameters and physiological processes but many observed effects are secondary and the primary toxic effects are largely unresolved (Woolhouse 1983). One of the most conspicuous symptoms of metal toxicity is inhibition of root growth which, in turn, has led to the widespread use of this character to assess metal tolerance of plants (Wilkins 1978).

This metal-induced reduction in root growth and consequent impaired ability of the plant to exploit soil reserves may be particularly detrimental in many metal-contaminated soils, which may have low nutrient concentrations, poor soil structure and water retention capacity (Smith & Bradshaw 1979). There are now many documented examples of the evolution of metal tolerant populations within plant species which are better able to maintain root growth in soils containing toxic concentrations of metals (Baker 1987).

This paper examines the effects of some toxic metals on root growth, particularly in relation to interactions with other nutrient elements, and on the fundamental processes of cell division and cell expansion in the root apex. Genetic and physiological aspects of tolerance of plants to toxic metals will not be dealt with specifically here.

## *Sources of toxic metals*

Elevated concentration of metals such as Zn, Cu and Ni may occur naturally in soils where ore bodies are near the surface, e.g. serpentine soil. Other metals such as Al, Fe and Mn are a natural constituent of soils but may be present in largely insoluble form. Al may become available as the trivalent cation in acid soils but also as the aluminate ($AlO_2^-$) anion in very alkaline soils (Woolhouse 1983). Mn and Fe are also more soluble in acidic soils, but severe Fe and Mn toxicity occurs in flooded soils where the reducing conditions, brought about by soil anaerobiosis, result in the release of divalent Fe and Mn which are readily available to plants (Ponnamperuma 1972). Na may occur in toxic concentrations as the result of sea water inundation (Chapman 1939), or in arid soils (Epstein 1985) or as a result of application of de-icing salt to roads (Davison 1971). Mining activity for 'heavy metals' has generated large amounts of waste or 'tailings', some containing up to 10% of the metal (Smith & Bradshaw 1979). Soil contamination by aerial fallout of metals as a result of smelting activity is also well documented (Wu, Bradshaw & Thurman 1975; Coughtrey & Martin 1977; Cox & Hutchinson 1980). Toxic concentrations of Zn may occur under galvanized wire fencing (Harris 1946) and through other extensive use of Zn galvanizing, e.g. electricity pylons and road crash barriers. Motor car exhaust emission may cause Pb deposition in soils and vegetation (Quarles *et al.* 1974).

The degree of metal toxicity is not easy to assess simply from analyses of metal concentrations in soil, since availability will depend upon factors such as soil pH and organic matter content. Simon (1978) showed that high Ca concentrations in soils reduced Pb toxicity and to a lesser extent Zn toxicity, and lime application to toxic spoils has been shown to improve the performance of both natural populations and seeded cultivars (Street & Goodman 1967; Gemmell 1977). Similarly, Ca in solution culture ameliorates Pb toxicity (Wilkins 1957; Jowett 1964), Zn toxicity (Baker 1978) and Ni toxicity (Robertson 1985).

## *The rooting test*

Bradshaw (1952) showed that markedly reduced root production and growth was a feature of the failure of a pasture population of *Agrostis capillaris* to grow on mine spoil. This inhibitory effect of toxic metals on root growth has led to the widespread use of the 'rooting test' as a rapid measure of metal toxicity and plant tolerance (Wilkins 1978). In some cases measures of plant tolerance, estimated in this way, have been shown to correlate with measures based upon longer term growth. Walley, Khan & Bradshaw (1974) showed a close correlation between the height of 4-month-old seedlings of *A. capillaris* on a copper contaminated waste and an index of Cu tolerance of the same individuals measured by the rooting test. Davies & Snaydon (1973) also demonstrated a close relationship between the responses of contrasting populations of *Anthoxanthum odoratum* to Al measured by a short-term rooting test, and a longer term response of dry weight yield in sand culture. There are, however, many variants of the rooting test including the use of 'sequential' or 'parallel' controls, whether the culture solution is aerated, the range of concentrations of metal used and the method by which tolerance is calculated. These have been extensively reviewed by Wilkins (1978) and more recently discussed by Baker (1987) and will not be considered in detail here.

### *Effect of background nutrient solution*

One major variant of the 'rooting test' is the composition of the background solution in which the metal is supplied. Traditionally, a single salt solution of calcium nitrate (generally 0·5 or 1·0 $g\,l^{-1}$) was used with the metal supplied either as the sulphate or nitrate, thus allowing higher concentrations of metal to be used without danger of precipitation. Other workers, worried by likely complications of nutrient deficiency, have used dilute nutrient solutions, sometimes without phosphate to avoid precipitation problems.

Where comparative studies have been made between root growth responses of genotypes to toxic metals supplied in different background solutions some startling differences have been demonstrated. For example, Davies & Snaydon (1973) found that populations of *A. odoratum* from contrasting plots of the Park Grass Experiment showed differential sensitivity of root extension to Al when it was supplied in full nutrient solution, both with and without P; the responses of individual populations were also correlated with long-term yield responses to Al in sand culture. However, no discrimination between populations occurred in a calcium nitrate background, although the Ca concentration was the same in all these solutions. Similarly McCain & Davies (1983a) found that a diluted full nutrient solution was a more suitable background than calcium nitrate for screening for Al tolerance in *Holcus lanatus*.

The toxicity of 'heavy metals' Cu, Zn and Pb to root growth is also affected by background solution (Table 1). Root growth of *Festuca rubra* was less reduced by increasing Zn and Cu concentrations in diluted Rorison's minus P solution, than it was in calcium nitrate; the tolerant cultivar Merlin, originally derived from a natural population growing on the Pb/Zn spoil at the disused Trelogan mine in Clwyd (Smith & Bradshaw 1979), was less sensitive than S59 in both background solutions. However, Pb toxicity was greater in the diluted nutrient solution than in calcium nitrate, a result also reported by Shaw in Baker (1987) for *Festuca ovina*.

TABLE 1. The percentage reduction in the length of the longest root compared with controls (background solution only) of 15-day old seedlings of two cultivars of *F. rubra* (Merlin & S59) following a 7 day exposure to various concentrations of Cu, Pb and Zn supplied in a background of calcium nitrate (0·5 g dm$^{-3}$) or 0.1 strength Rorison's solution −P

| | Background solution | | | | | |
|---|---|---|---|---|---|---|
| | Calcium nitrate | | | 0.1 strength Rorison's −P | | |
| Cu concentration (μg cm$^{-3}$) | 0·1 | 0·2 | 0·5 | 0·1 | 0·2 | 0·5 |
| Merlin | 34·7 | 44·5 | 53·8 | 14·2 | 22·3 | 41·9 |
| S59 | 38·1 | 44·3 | 67·0 | 24·7 | 36·1 | 59·2 |
| Zn concentration (μg cm$^{-3}$) | 3 | 5 | 7 | 3 | 5 | 7 |
| Merlin | 47·8 | 53·0 | 69·3 | 17·0 | 29·0 | 33·1 |
| S59 | 60·9 | 61·0 | 71·9 | 16·0 | 34·0 | 34·0 |
| Pb concentration (μg cm$^{-3}$) | 2 | 7 | 13 | 2 | 7 | 13 |
| Merlin | 18·7 | 39·9 | 56·6 | 14·3 | 56·4 | 69·1 |
| S59 | 13·0 | 41·2 | 67·0 | 35·6 | 69·5 | 70·1 |

Their data also showed that Pb was least toxic in nutrient solution including P, though it was not stated whether this was partly due to removal of Pb from the solution by precipitation with phosphate. Further work between the interaction of the toxicity of individual metals and background solution is obviously required utilizing a much wider range of species.

However, the problems associated with estimating metal toxicity by means of root growth in solution culture are not confined to whether a single salt or complete nutrient solution is used as the background in which the metal is supplied. The form in which nitrogen is supplied also affects the toxicities of some metals. McGrath & Rorison (1982) found that Mn uptake and toxicity to *Bromus erectus* and *Holcus lanatus* was greater when supplied in a background of $NO_3$-N than $NH_4$-N. However, the root and shoot growth of both species, but particularly of the calcicole *B. erectus*, was adversely affected by $NH_4$-N, to the extent that Mn toxicity was masked. Rorison (1985) has shown similar interactions between N form and Al toxicity. The calcifuge *Deschampsia flexuosa* was tolerant of combinations of $NH_4$-N, $NO_3$-N and Al concentration, *Holcus lanatus* showed Al toxicity in the presence of $NO_3$-N but in *B. erectus* the extremely reduced growth in $NH_4$-N masked any toxic effects of Al. McCain & Davies (1983a) showed that root-induced pH changes in solutions in which *H. lanatus* plants were growing were also markedly affected by the form of N. In acidic solution, where Al was supplied as the trivalent cation, nitrate-based solutions increased in pH causing visible precipitation of Al in the solution, though Al toxicity was not reduced. When N was supplied in equivalent amounts of $NH_4$-N and $NO_3$-N, root-induced solution pH fluxes were acidifying thus posing no problem of Al precipitation. In alkaline solutions, however, when Al was supplied as the aluminate anion, root-induced solution pH fluxes were acidifying and were greatest in $NH_4$-N and $NO_3$-N mixtures where visible precipitation of Al occurred. These fluxes were smaller and

$AlO_2^-$ toxicity greatest in $NO_3$-N solutions. They concluded that screening for cationic Al toxicity is better done in solutions where $NH_4$-N and $NO_3$-N is the N source but screening for toxicity of the aluminate anion in alkaline solution is best achieved with $NO_3$-N and that these combinations are more representative of soil conditions.

It has been suggested (Foy, Chaney & White 1978) that differences between wheat varieties in Al tolerance is linked with the ability of plant roots to raise the pH of the unbuffered nutrient solution by preferential uptake of $NO_3$-N, thereby reducing the availability of Al. The evidence of Rorison (1985) and McCain & Davies (1983a) suggests that this is not the case in natural populations, where in mixtures of $NO_3$ and $NH_4$-N pH fluxes were acidifying and Al toxicity was also greatest in $NO_3$ based solutions. Given that $NH_4$-N is the predominant form of N in acid soils, screening for cationic Al tolerance in nitrate-based solutions may be misleading and may overestimate the effects of Al toxicity. There is urgent need for a critical appraisal of the status of inorganic toxins such as Al and Mn in acid soils in relation to the precise conditions in the rooting zone of plants. Much work to date has been carried out in nutrient solutions of inappropriate chemical composition in relation to the soil solution.

### *Nutritional history of experimental material*

In addition to the chemical composition of the test solution affecting metal toxicity, there is now an increasing body of evidence which indicates that the nutritional history of the experimental material can markedly influence its response to toxic metals.

The initial P status of the plant affects its response to P supply. P deficiency results in greatly increased P absorption by roots in short-term experiments (Clarkson & Scattergood 1982; Lee 1982; Lee & Ratcliffe 1983). McCain & Davies (1983b) showed that P-deficient tillers of *A. capillaris* absorbed more P on transfer to P-sufficient supply than plants pretreated with high P. This effect, which has frequently been reported in short-term experiments, persisted 10 weeks after transfer to high-P supply and over many tiller replication cycles. P-sufficient plants were also better able to withstand subsequent P-deficiency. The initial P status of the experimental tillers also affected their response to Al toxicity; tillers of P-deficient plants produced longer roots than those of P-sufficient plants at low Al concentrations but their root extension was more inhibited by increasing Al concentrations than P-sufficient tillers (McCain & Davies 1985), although the degree of Al toxicity was not directly correlated with the actual P concentration in the shoot of each tiller. P-deficient plants also had a four- to six-fold greater root surface acid phosphatase activity than P-sufficient tillers and this activity was more inhibited by Al than that of P-sufficient tillers (McCain & Davies 1984).

Baker *et al.* (1986) provided a further example of the importance of pre-experimental growth conditions; they demonstrated an average fall of 13% in Cd tolerance, estimated by a rooting test, in populations of a range of grass species from Cd-contaminated soils, when plants were cultured for 2 months prior to testing in potting compost, compared with plants cultivated on their native soils. Even non-tolerant clones showed a lower tolerance when pretreated in compost. *Holcus lanatus* was more subject to phenotypic modification than the other grass species tested and it was even possible to induce a degree of Cd tolerance in clones of this species from uncontaminated soils by precultivation in Cd-contaminated soil.

Pretreatment conditions can also influence the inhibition of root extension by toxic metals over a much shorter term than described above. Brown & Martin (1981) showed that roots of both Cd-tolerant and non-tolerant plants of *H. lanatus*, pretreated with low concentrations of Cd for 10 days, had a greater root growth in a test solution of higher Cd concentration than non-pretreated roots; this enhanced tolerance was restricted to the actual roots pretreated, indicating that the responsive mechanism lay within the root. Similarly Waldren, Davies & Etherington (1987) found that roots of *Geum rivale*, pretreated for 14 days with 25 μg Mn $cm^{-3}$, showed less growth inhibition than non-pretreated roots over the subsequent 7 days, at a range of Mn concentrations, but particularly at the concentration at which they were pretreated (Fig. 1). The effect of this pretreatment was more

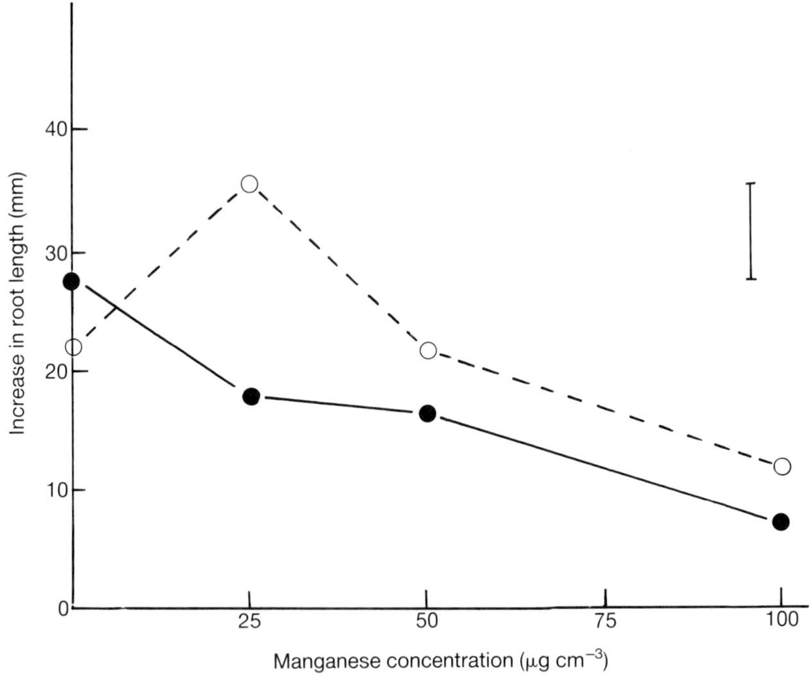


FIG. 1. The increase in the length of the longest root over a 7 day period, at various concentrations of Mn, in *G. rivale* plants pretreated for 14 days with 0·055 μg Mn $cm^{-3}$ (solid line) or 25 μg Mn $cm^{-3}$ (broken line). The vertical bar is the LSD ($P = 0{\cdot}05$) (From Waldren, Davies & Etherington 1987.)

pronounced in the more flooding-tolerant and more Mn-tolerant species, *G. rivale*, than in the flooding-sensitive species *Geum urbanum*. We need to investigate the cellular, physiological and molecular basis of the 'turning on' of the tolerance mechanism.

These findings, particularly the phenotypic modification in metal tolerance produced by pre-experimental soil and nutritional conditions, substantiate the misgivings expressed by Rorison (1969) that the nutritional history of vegetative propagules may drastically influence their response to experimental treatment, and that it is perhaps more ecologically meaningful to raise experimental plants in soil conditions which are similar to those of their native habitats. The short-term reduced sensitivity of individual roots to growth inhibition by a metal, induced by pretreatment with that metal, also has implications for the rooting test, particularly for the 'sequential control' method (Wilkins 1978), where extension of each root in a control solution is compared with its subsequent extension in a solution with the toxic metal.

### *Mycorrhiza*

A further factor which elicits caution in extrapolating root growth response to toxic metals in nutrient solutions to field situations is the role of mycorrhizal infection of roots in enhancing metal tolerance. Mycorrhizal infection of ericaceous plants reduces Cu and Zn toxicity (Bradley, Burt & Read 1982) and Zn toxicity (Brown & Wilkins 1985). Ni and Cu toxicity (Jones & Hutchinson 1986) in *Betula* seedlings is reduced by mycorrhizal infection. Amelioration of toxicity appears to involve a reduction in the amount of metal available to the plant by incorporation into the fungus thereby reducing uptake. The roots of many plant species which may colonize metal-contaminated soils can be infected with mycorrhiza (Harley & Harley 1987). Although there is need for a much more extensive survey of the amelioration of metal toxicity by mycorrhizal infection, it is evident that care must be taken in comparing the performance of plants in sand or solution culture where roots are likely to be non-mycorrhizal, with that in soils where they are probably infected.

## CELLULAR ASPECTS OF ROOT GROWTH

Despite the fact that we have been aware for many years that root growth is particularly sensitive to inhibition by metals, thus accounting for the widespread use of root growth to measure metal toxicity and tolerance, little is known about the physiological, cellular and molecular basis of this metal-induced reduction in root growth.

The growth of roots is an integrated process involving mitotic cell division in the apical meristem, expansion of cells immediately proximal to the meristem, followed by differentiation of these elongated cells into vascular tissue, root-hair cells, etc. Added to this is secondary growth of roots and the formation of lateral roots.

These processes may be perturbed by a variety of physical and chemical factors in the soil (Barlow 1987). Although root extension may occur through cell elongation, this can account for only a finite increase in root length and continued root growth must depend upon the continuous production of new cells in the apical meristem. We know little about the detailed effects of environmental factors on the mitotic cell cycle in the root meristem, particularly those of mineral nutrients and toxic elements.

Several workers have observed that toxic metals such as Zn (Swieboda 1976; Kocik, Wojciechowska & Liguzinskia 1982), Al (Clarkson 1965; Morimura, Takahashi & Matsumoto 1978; Horst, Wagner & Marschner 1983) and Ni (Robertson & Meakin 1980) cause a reduction in the percentage of cells in the root meristem which are in visible stages of mitosis (mitotic index — MI). Horst, Wagner & Marschner (1983) demonstrated a very rapid decline in MI to near zero in roots of *Vigna unguiculata* within 5 h of transfer to solution containing Al, followed by a rise in MI 10 h after transfer which eventually stabilized at a level below that of control roots; this recovery took longer in the Al-sensitive genotype. Figure 2 shows the MI in the root meristem of 6-day-old seedlings of a Zn-tolerant cultivar (Merlin) and a Zn-sensitive cultivar (S59) of *Festuca rubra* at various times after transfer to a solution containing 5 μg Zn cm$^{-3}$ which, in seedlings of this age in the experimental system used, is a very toxic concentration (Powell, Davies & Francis 1986a). The MI in roots of both cultivars in the control solution (0 Zn), which are not presented here, remained relatively constant at *c.* 6·5% throughout the experimental period.

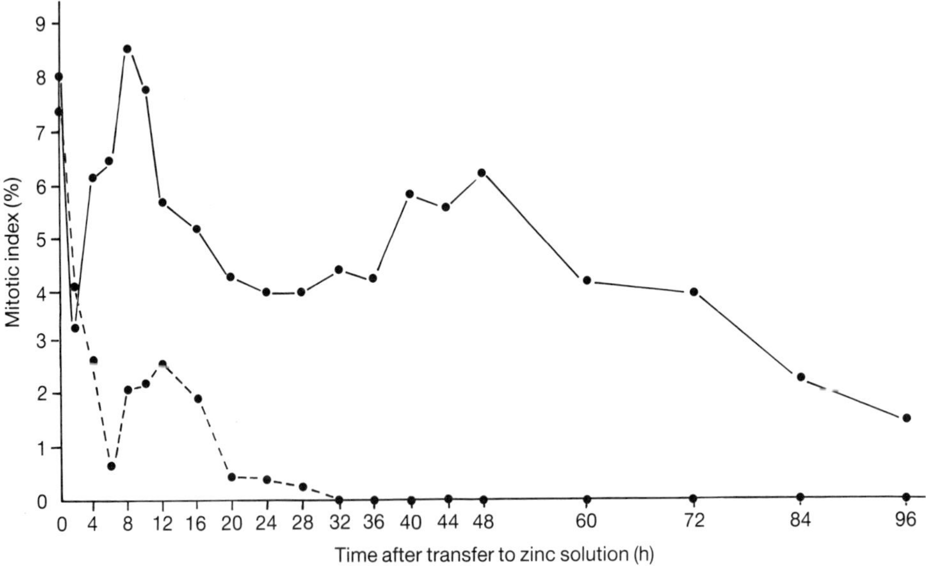


FIG. 2. The mitotic index (%) in root meristems of 7-day-old seedlings of *F. rubra* cvs. Merlin (solid line) and S59 (broken line), sampled various times following exposure to a solution containing 5 μg Zn cm$^{-3}$. (Data of Davies, Francis & Powell, unpublished.)

TABLE 2(a). The length of the longest root as a percentage of that in a control solution (0 mM NaCl) at various concentrations of NaCl in 14-day old seedlings of *F. rubra* cv. Hawk, following a 7 day exposure to the NaCl concentration and the mitotic index of cells in the root meristem (data of Davies & Tiller, unpublished)

| | NaCl concentration (mM) | | |
|---|---|---|---|
| | 0 | 50 | 150 |
| % of control root length | 100 | 87 | 46 |
| Mitotic index (%) | 4·5 | 8.6 | 7·9 |

TABLE 2(b). The percentage of mitotic cells in each stage of mitosis in the root meristem of *F. rubra* cv. Hawk grown at 0 or 50 mM NaCl

| NaCl concentration (mM) | Prophase | Metaphase | Anaphase | Telophase |
|---|---|---|---|---|
| 0 | 41 | 21 | 21 | 17 |
| 50 | 28 | 44 | 15 | 13 |

In the Zn-containing solutions, following an initial fall in MI in both cultivars 2 h after transfer, there was a partial recovery in S59 followed by a further rapid decline culminating in a total absence of cells in mitosis after 30 h. The MI of Merlin was less affected by Zn and a continued decline in MI did not begin until 48 h after transfer.

In these two cultivars, there is a close relationship between the decline in MI in the root meristem at increasing external concentrations of Zn and inhibition of root extension and also with measures of cell doubling time, as estimated by cochicine-induced accumulation of cells in metaphase (Powell, Davies & Francis 1986a). However, great care must be taken in interpreting MI data in relation to the division of cells in the meristem (Lyndon 1967; Clowes 1981). This is dramatically shown by MI data in the root meristem of *F. rubra* grown at various concentrations of sodium chloride (Table 2). Although root growth was inhibited by increasing salt concentration, the MI increased. Examination of cells in mitosis revealed that, in meristems of roots grown at 50 mM NaCl, there was more than a two-fold increase in the proportion of mitotic cells in metaphase compared with meristems of roots grown in the control solutions. These preliminary data suggest that salt may be acting as a partial block on the cell cycle, leading to the accumulation of cells in metaphase, perhaps by perturbing the formation of the mitotic spindle. We are investigating these effects further. In this instance, MI data alone give a false picture of the effects of salt on the activity of meristematic cells. Only when it is known that the duration of the mitosis remains constant and that changes in the duration of the cell cycle are due to alterations in the lengths of other phases, can MI data give an indication of cell doubling time. Unfortunately, most studies of the effects of toxic metals on root meristems have used MI data as the sole measure of mitotic activity and there is little information on the influence of metals on the duration of the cell cycle and its component phases.

## Measurement of the cell cycle

A fundamental property of meristematic cells is their ability to grow and divide. The mitotic cell cycle is divided into a number of distinct phases. Nuclear DNA is replicated in the DNA synthetic (S) phase, preceded by a presynthetic interphase stage (G1) and followed by a postsynthetic interphase (G2) which, in turn, immediately precedes the onset of mitosis (M). Cells in G1 thus have the unreplicated (2C) DNA content, and those in G2 the 4C content. A standard way of measuring the cell cycle components in plant meristems is by the percentage labelled mitosis (PLM) method. Meristems are given a short exposure (pulse) with thymidine, radioactively labelled with tritium ($^3$H-TdR) and then chased with cold (non-radioactive) TdR before being grown on in the treatment solution. Roots are sampled at regular intervals thereafter for a period exceeding the likely duration of the cell cycle and autoradiographs made of squash preparations of the meristem. The relationship between the percentage of cells in mitosis containing radioactive label and the time elapsed since labelling is determined. The method thus monitors the flow of a population of cells from S-phase, through G2 and into and out of mitosis. Generally a double-peaked curve is found (Fig. 3). The distance between the peaks of labelled mitoses gives a measure of the duration of the cell cycle ($T_c$); the width of the first peak at half the maximum height point is equal to the duration of S-phase plus the period of exposure to label; the interval from the start of labelling to 50% of the height of the ascending arm of the first peak is equal to the

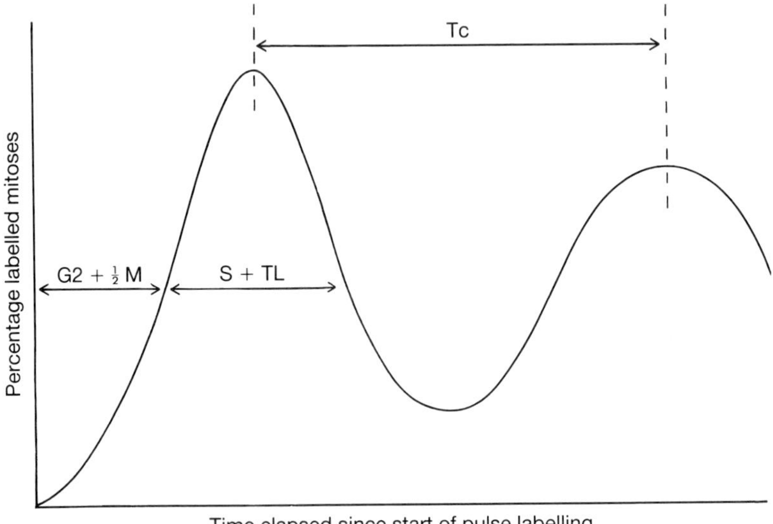


FIG. 3. Hypothetical relationship between the percentage labelled mitoses (PLM) and time elapsed since the start of pulse labelling in populations of meristematic cells in the root apex.

duration of G2 + ½ M; M is from where the initial limb rises to where it plateaus; G1 is obtained by difference (Quastler & Sherman 1959). Scadeng & Macleod (1976) proposed a modification of the method for calculating the duration of M from PLM data.

Powell, Davies & Francis (1986a), using the PLM method and the modification of Scadeng & Macleod (1976), investigated the effect of Zn on the cell cycle in root meristems of a Zn-sensitive (S59) and Zn-tolerant (Merlin) cultivar of *F. rubra*. The concentrations of Zn were chosen to produce only partial (50%) inhibition of root growth in the sensitive cultivar (S59). The duration of the cycle was approximately the same for both cultivars (16 h) in the control treatment (Table 3) but was increased by Zn to a far greater extent in S59 (2·3-fold increase at the higher Zn concentration) than in the tolerant cultivar, Merlin (1.1-fold increase). The major effect of Zn was to increase G1 (8-fold increase in S59, 2·7-fold increase in Merlin). The duration of M was little affected by Zn, which explains why MI was a fortuitous marker of cell-doubling time (see above). Similarly the duration of S phase was little affected by Zn treatment. G2 was slightly shortened in Merlin. The finding that G1 shows the greatest increase is consistent with numerous reports that G1 is the longest phase in extended cell cycles in both plants (Clowes 1976) and animals (Prescott, Liskay & Stancel 1982). Francis & Barlow (1988) review the known effects of temperature on the cell cycle in plants. The drastically lengthened cell cycles in meristems of several plant species at progressively lower temperatures is generally characterized by a disproportionate lengthening of G1, compared with other phases of the cell cycle. They hypothesize that the 2C amount of DNA, characterizing a cell in G1, confers maximal protection of the genome of a somatic cell to low-temperature stress and that cells in extended G1, e.g. in the quiescent centre of root meristems, may be in the best-prepared state to move into S phase when environmental conditions permit. Can this hypothesis be extended to other factors which perturb the cell cycle, e.g. toxic metals? There is clearly a need for much more extensive work on the effect of a range of toxic metals on the cell cycle and its component phases to determine whether G1 extension is a widespread response.

The duration of S phase of the *F. rubra* populations was little affected by Zn, suggesting that nuclei committed to this phase replicate their DNA over a similar

Table 3. Duration (h) of the cell cycle and its component phases, estimated from the percentage labelled mitoses data in cells of the root meristem of *F. rubra* (cvs. Merlin and S59) grown in three concentrations of Zn (from Powell *et al.* 1986a)

| Cultivar | Zn concentration | M | G2 | S | Gl | Tc |
|---|---|---|---|---|---|---|
| Merlin | 0 | 2·5 | 3·4 | 6·6 | 3·5 | 16·0 |
| | 0·1 | 1·4 | 2·3 | 7·0 | 6·2 | 16·9 |
| | 0·2 | 1·5 | 1·7 | 5·9 | 9·5 | 18·6 |
| S59 | 0 | 2·5 | 3·4 | 6·9 | 3·6 | 16·4 |
| | 0·1 | 1·5 | 2·5 | 6·0 | 12·9 | 22·9 |
| | 0·2 | 1·1 | 3·0 | 5·6 | 28·3 | 38·0 |

period, irrespective of Zn concentrations which markedly affect other phases of the cycle. There is evidence that toxic metals reduce DNA synthesis *in vitro* (Sirover & Loeb 1976). Al reduces DNA template activity (Matsumoto & Morimura 1980) and incorporation of $^3$H-TdR into DNA (Wallace & Anderson 1984). Although the low concentrations of Zn used in the *F. rubra* work did not affect the duration of S, it does not follow that higher concentrations would not inhibit DNA replication and perhaps also extend S. Francis & Barlow (1988) conclude that the limited evidence to date on rates of DNA replication, measured by DNA fibre autoradiography, indicates that lowered temperature lowers the rate of replication fork movement, but not replicon size. This technique could also provide valuable data on the effects of toxic metals on DNA-replication rates *in vivo* in root meristems.

### *Cell elongation and differentiation*

Treatment with Zn also markedly reduced the area of cells in the meristem of the sensitive cultivar of *F. rubra* (Powell, Davies & Francis, 1986b). Both cells in mitosis and in interphase were affected. In contrast the cell area of meristematic cells of Merlin was increased by Zn. Inhibition of elongation in cells proximal to the root meristem is another means by which toxic metals reduce root extension; this has been shown in the case of Zn (Wainwright & Woolhouse 1977; Powell 1986; Denny & Wilkins 1987), Pb (Lane, Martin & Garrod 1978) and Al (Wallace & Anderson 1984). Metal toxicity can also result in large perturbations of normal spatial arrangements in roots. Powell, Davies & Francis (1988), using the same cultivars of *F. rubra* and concentrations of Zn described above, showed that increasing Zn concentration reduced the overall length of the root apical meristem by 42% in the sensitive cultivar, compared with 23% in the tolerant cultivar. This was accompanied by a marked reduction in the distance from the root cap boundary to the first visible xylem element and root hair in S59 though, in Merlin, these characteristics were only marginally affected by Zn. Thus Zn in this sensitive cultivar, as well as dramatically lengthening the cell cycle, reduces the number of meristematic cells and elicits precocious cell elongation and differentiation in cells in the elongation zone. Similar disturbances to the normal spatial pattern of development in the root apex have been induced by treatment with salt in this species (Davies & Tiller, unpublished) and also by toxic concentrations of phosphorus which inhibit root growth (Davies & Creber, unpublished). Salt treatment also induces cell vacuolation and the formation of the endodermis much closer to the root apex than in control roots (Hajibagheri, Yeo & Flowers 1985).

### *Cellular organelles*

The nucleolus, the site of ribosome production in the cell, is also sensitive to environmental influences. Plants growing at low temperature have larger nucleoli (Morcillo, Krimer & De La Torre 1978; Jordan *et al.* 1985), mainly as a result of an increase in the granular component. This is accompanied by an increase in the amount of nucleolar RNA polymerase and in the number of ribosomes. These findings are paralleled by the Zn-induced responses in *F.rubra*; Zn results in a reduction in nuclear and nucleolar volume of G2 cells in S59 and also a reduction of cellular RNA content and protein content, determined by cytochemical staining and microdensitometry (Powell, Davies & Francis 1986b). However, in the tolerant cultivar, Merlin, Zn produces a large increase in nuclear and nucleolar volume and in cellular RNA and protein content. The structure of the nucleolus is also sensitive to environmental effects. Heat and cold shocks can result in disruption of nucleolar structure as can anoxia in the root meristem (Barlow 1987), although this latter effect can be reversible. Nucleolar dissolution can also result from Al toxicity (Fiskesjo 1983) and Zn toxicity (Salaj & Hudak 1985). Conspicuous damage to mitochondria in meristematic cells has been induced by treatment with NaCl in a salt-sensitive population of *Agrostis stolonifera* (Smith *et al.* 1982); no damage was apparent in a salt-tolerant population in the same conditions.

### *Intracellular protein*

Increases in the protein content of meristematic cells in roots of the *F. rubra* cultivar Merlin, occur in response to treatment with Zn (Powell, Davies & Francis 1986b) and with Cu and Cd (Davies & Francis, unpublished). The nature and role of this additional protein is not yet known. Meristematic cells have no extensive vacuolar system and therefore are less able to compartmentalize toxic metals in the vacuole, possibly in combination with organic acids, as has been shown for more mature cells in roots (Thurman & Collins 1983). The root apices of metal-tolerant plants contain as much as, and in many cases considerably more of the toxic metal, than non-tolerant plants (Godbold *et al.* 1983; Powell 1986) and some cytoplasmic detoxification system would seem to be necessary. There is recent evidence that low molecular weight metallothionein-like proteins or 'phytochelatins', which are induced by exposure to metals, may be involved in intracellular metal binding (Robinson & Jackson 1986; Robinson & Thurman 1986), although it appears that these may be responsible for binding only a fraction of the metal present. The additional Zn-induced intracellular protein in *F. rubra* may be of a similar nature, or it is possible that it may be a 'stress protein', examples of which have been shown to be synthesized in plant roots in response to heat shock (Cooper & Ho 1983) or anoxic conditions (Sachs, Freeling & Okimoto 1980) and may, as postulated for other shock proteins, serve a protective role in maintaining cellular and membrane integrity. At present we have relatively little insight into the biochemical, cellular and molecular basis of metal tolerance in plant roots. Genetically differentiated populations within species, varying in the degree of tolerance to specific metals, would seem to be ideal experimental tools for these investigations.

## CONCLUSIONS

Toxic metals are a major impediment to root growth in many natural soils and in soils which are artificially contaminated by metals, thereby reducing the capacity of plants to exploit soil reserves which is particularly significant in the generally infertile and/or dry soils in which metal toxicity commonly occurs. However, most research to date on the effects of metals on root growth has been carried out in solution culture, often using short-term measures of root extension. The responses of root growth to toxic metals in these systems may be dramatically altered by the composition of the nutrient solution. Many studies have utilized single salt solutions, chosen for convenience of testing rather than for relevance to the soil solution. Similarly, inappropriate forms of nitrogen in the solution have often been used. We need to know much more about the responses of roots to toxic metals in soils, where mycorrhizal infection may be the norm, and about the nature of the interaction between soil chemistry and metal toxicity in the immediate vicinity of the root.

The findings that the responses of plants to metals are also affected by the growth history of experimental material brings into question the precise ecological interpretations which can be made from much of the work to date, which has generally been carried out on experimental plants raised in high-nutrient compost, in which measured responses may be significantly different from those in the field.

There is little information on the effects of toxic metals on the activity of the 'centre of root growth', the root meristem. Fundamental information is required on the effects of a number of metals on the cell division cycle and the relationships between cell division, cell expansion and differentiation and on the activities of cellular organelles. The effect of metal pretreatment on these attributes may provide further insight into the 'turning on' of tolerance mechanisms.

Genetically-differentiated populations within species are ideal experimental tools for such physiological–ecological studies which are necessary to advance our understanding of the mechanisms of adaptation of plants to extreme soil conditions.

## REFERENCES

**Baker, A.J.M. (1978).** The uptake of zinc and calcium from solution culture by zinc-tolerant *Silene maritima* With. in relation to calcium supply. *New Phytologist*, **81**, 321–330.

**Baker, A.J.M. (1987).** Metal tolerance. *New Phytologist*, **106** (Suppl), 93–111.

**Baker, A.J.M., Grant, C.J., Martin, M.H., Shaw, S.C. & Whitebrook, J. (1986).** Induction and loss of cadmium tolerance in *Holcus lanatus* L. and other grasses. *New Phytologist*, **102**, 575–587.

**Barlow, P.W. (1987).** The cellular organization of roots and its response to the physical environment. In *Root Development and Function* (Ed. by P.J. Gregory, J.V. Lake & D.A. Rose), pp. 1–26. Cambridge University Press, Cambridge.

**Bradley, R., Burt, A.J. & Read, D.J. (1982).** The biology of mycorrhiza in the Ericaceae. VIII. The role of mycorrhizal infection in heavy metal resistance. *New Phytologist*, **91**, 197–209.

**Bradshaw, A.D. (1952).** Populations of *Agrostis tenuis* resistant to lead and zinc poisoning. *Nature*, **169**, 1098.

**Brown, H. & Martin, M.H. (1981).** Pretreatment effects of cadmium on the root growth of *Holcus lanatus* L. *New Phytologist*, **89**, 621–629.

**Brown, M.T. & Wilkins, D.A. (1985).** Zinc tolerance of mycorrhizal *Betula*. *New Phytologist*, **99**, 101–106.

**Chapman, V.J. (1939).** Studies in salt marsh ecology sections IV and V. *Journal of Ecology*, **27**, 160–201.

**Clarkson, D.T. (1965).** The effect of aluminium and other trivalent metal cations on cell division in the root apices of *Allium cepa*. *Annals of Botany*, **29**, 309–315.

**Clarkson, D.T. & Scattergood, C.B. (1982).** Growth and phosphate transport in barley and tomato plants during the development of, and recovery from, phosphate stress. *Journal of Experimental Botany*, **136**, 865–875.

**Clowes, F.A.L. (1976).** The root apex. In *Cell Divisions in Higher Plants* (Ed. by M.M. Yeoman), pp. 253–284. Academic Press, New York.

**Clowes, F.A.L. (1981).** The difference between open and closed meristems. *Annals of Botany*, **48**, 761–767.

**Cooper, P. & Ho, T.H.D. (1983).** Heat shock proteins in maize. *Plant Physiology*, **71**, 215–222.

**Coughtrey, P.J. & Martin, M.H. (1977).** Cadmium tolerance of *Holcus lanatus* from a site contaminated by aerial fallout. *New Phytologist*, **99**, 273–280.

**Cox, R.M. & Hutchinson, T.C. (1980).** Multiple metal tolerance in the grass *Deschampsia caespitosa* (L.) Beauv. from the Sudbury smelting area. *New Phytologist*, **84**, 631–647.

**Davies, M.S. & Snaydon, R.W. (1973).** Physiological differences among populations of *Anthoxanthum odoratum* collected from the park grass experiment. Rothamsted. II. Response to aluminium. *Journal of Applied Ecology*, **10**, 47–55.

**Davison, A.W. (1971).** Effects of de-icing salt on roadside verges. I. Soil and plant analysis. *Journal of Applied Ecology*, **8**, 551–561.

**Denny, H.J. & Wilkins, D.A. (1987).** Zinc tolerance in *Betula* spp. I. Effect of external concentration of zinc on growth and uptake. *New Phytologist*, **106**, 517–524.

**Epstein, E. (1985).** Salt tolerant crops: origins development and prospects of the concept. *Plant and Soil*, **89**, 187–198.

**Fiskesjo, G. (1983).** Nucleolar dissolution induced by aluminium in root cells of *Allium*. *Physiologia Plantarum*, **53**, 508–511.

**Foy, C.D., Chaney, R.L. & White, M.C. (1978).** The physiology of metal toxicity in plants. *Annual Review of Plant Physiology*, **29**, 511–566.

**Francis, D. & Barlow, P.W. (1988).** Temperature and the cell cycle. In *Plants and Temperature*, 42nd Symposium of the Society for Experimental Biology (Ed. by S.P. Long & F.I. Woodward), pp. 181–201. Company of Biologists, Cambridge.

**Gemmell, R.P. (1977).** *Colonisation of Industrial Wasteland*. Studies in Biology, No. 80. Arnold, London.

**Godbold, D.L., Horst, W.J., Marschner, H., Collins, J.C. & Thurman, D.A. (1983).** Root growth and Zn uptake by two ecotypes of *Deschampsia caespitosa* as affected by high Zn concentrations. *Zeitschrift für Pflanzenphysiologie* **112**, 315–324.

**Hajibagheri, M.A., Yeo, A.R. & Flowers, T.J. (1985).** Salt tolerance in *Suaeda maritima* (L.), Dum. Fine structure and ion concentrations in the apical regions of roots. *New Phytologist*, **99**, 331–343.

**Harley, J.L. & Harley, E.L. (1987).** A check list of mycorrhiza in the British flora. *New Phytologist*, **105**, (Supplement), 1–102.

**Harris, T.M. (1946).** Zinc poisoning of wild plants from wire netting. *New Phytologist*, **45**, 50–55.

**Horst, W.J., Wagner, A. & Marschner, H. (1983).** Effects of aluminium on root growth, cell-division rate and mineral element contents in roots of *Vigna unguiculata* genotypes. *Zeitschrift für Pflanzenphysiologie*, **109**, 95–103.

**Jones, M. & Hutchinson, T.C. (1986).** The effect of mycorrhizal infection on the response of *Betula papyrifera* to nickel and copper. *New Phytologist*, **102**, 429–442.

**Jordan, E.C., Cooper, P.J., Martini, G., Bennett, M.D. & Flavell, R.B. (1985).** The effect of temperature and ageing on root apical meristems during seedling growth of *Triticum aestivum*: a specific effect on nucleoli. *Plant, Cell and Environment*, **8**, 325–331.

**Jowett, D, (1964).** Population studies on lead tolerant *Agrostis tenuis*. *Evolution Lancaster*, **18**, 70–80.

**Kocik, H., Wojciechowska, B. & Liguzinskia, A. (1982).** Investigations on the cytotoxic influence of zinc on *Allium cepa* L. roots. *Acta Societatis Botanicorum Poloniae*, **51**, 3–10.

**Lane, S.D., Martin, E.S. & Garrod, J.F. (1978).** Lead toxicity effects on indole-3-acetic acid induced cell elongation. *Planta*, **144**, 79–84.

**Lapedes, D.N. (1974).** *Dictionary of Scientific and Technical Terms*, 1st edn. McGraw-Hill, New York.

**Lee, R.B. (1982).** Selectivity and kinetics of ion uptake by barley plants following nutrient deficiency. *Annals of Botany*, **50**, 429–449.

**Lee R.B. & Ratcliffe. R.G. (1983).** Phosphorus nutrition and the intracellular distribution of inorganic phosphate in pea root tips: a quantitative study using $^{31}$P-NMR. *Journal of Experimental Botany*, **34**, 1222–1244.

**Lyndon, R.F. (1967).** The growth of the nucleus in dividing and non-dividing cells of the pea root. *Annals of Botany*, **31**, 133–146.

**Matsumoto, H.E. & Morimura, S. (1980).** Repressed template activity of chromatin of pea roots treated by aluminium. *Plant Cell Physiology*, **21**, 951–959.

**McCain, S. & Davies, M.S. (1983a).** The influence of background solution on root responses to aluminium in *Holcus lanatus* L. *Plant and Soil*, **73**, 425–430.

**McCain, S. & Davies, M.S. (1983b).** Effects of pretreatment with phosphate in natural populations of *Agrostis capillaris* L. I. Influence on the subsequent long term yield of, and accumulation of phosphorus in, plants grown from tillers. *New Phytologist*, **94**, 367–379.

**McCain, S. & Davies, M.S. (1984).** Effects of pretreatment with phosphate in natural populations of *Agrostis capillaris* L. II. Interactions with aluminium on the acid phosphatase activity and potassium leakage of intact roots. *New Phytologist*, **96**, 589–599.

**McCain, S. & Davies, M.S. (1985).** Effects of pretreatment with phosphate in natural populations of *Agrostis capillaris* L. III. Influences on the inhibition of root extension by aluminium. *New Phytologist*, **99**, 533–543.

**McGrath, S.P. & Rorison, I.H. (1982).** The influence of nitrogen source on the tolerance of *Holcus lanatus* L. and *Bromus erectus* Huds. to manganese. *New Phytologist*, **91**, 433–452.

**Morcillo, G., Krimer, D.B. & De La Torre, C. (1978).** Modifications of nucleolar components by growth temperature in meristems. *Experimental Cell Research*, **115**, 95–102.

**Morimura S., Takahashi, E. & Matsumoto, H. (1978).** Association of aluminium with nuclei and inhibition of cell division in onion (*Allium cepa*) roots. *Zeitschrift für Pflanzenphysiologie*, **88**, 395–401.

**Nieboer, E. & Richardson, D.H.S. (1980).** The replacement of the nondescript term 'heavy metals' by a biological and chemically significant classification of metal ions. *Environmental Pollution Series B*, **1**, 3–26.

**Ponnamperuma, F.N. (1972).** The chemistry of submerged soils. *Advances in Agronomy*, **24**, 29–96.

**Powell, M.J. (1986).** *The influence of zinc on growth, cell division and development in roots of a zinc-tolerant and a non-tolerant cultivar of Festuca rubra L.* Ph.D. thesis, University of Wales.

**Powell, M.J., Davies, M.S. & Francis, D. (1986a).** The influence of zinc on the cell cycle in the root meristem of a zinc-tolerant and a non-tolerant cultivar of *Festuca rubra* L. *New Phytologist*, **102**, 419–428.

**Powell, M.J., Davies, M.S. & Francis, D. (1986b).** Effects of zinc on cell, nuclear and nucleolar size and on RNA and protein content in the root meristem of a zinc-tolerant and a non-tolerant cultivar of *Festuca rubra* L. *New Phytologist* **104**, 671–679.

**Powell, M.J., Davies, M.S. & Francis, D. (1988).** Effects of zinc on meristem size and proximity of root hairs and xylem elements to the root tip in a zinc-tolerant and a non-tolerant cultivar of *Festuca rubra* L. *Annals of Botany*, **61**, 723–726.

**Prescott, D.M., Liskay, R.M. & Stancel, G.M. (1982).** The cell life cycle and the G1 period. In *Cell Growth* (Ed. by C. Nicolini), pp. 305–314. Plenum Press, New York.

**Quarles, III, H.D., Hanawatt, R.B., & Odum, W.E. (1974).** Lead in small mammals, plants and soil at varying distances from a highway. *Journal of Applied Ecology*, **11**, 937–949.

**Quastler, H. & Sherman, F.G. (1959).** Cell population kinetics in the intestinal epithelium of the mouse. *Experimental Cell Research*, **17**, 420–438.

**Robertson, A.I. (1985).** The poisoning of roots of *Zea mays* by nickel ions and the protection afforded by magnesium and calcium. *New Phytologist*, **100**, 173–189.

**Robertson, A.I. & Meakin, M.E.R. (1980).** The effect of nickel on cell division and growth of *Brachystegia speciformis* seedlings. *Kirkia*, **12**, 115–125.

**Robinson, N.J. & Jackson, P.J. (1986).** 'Metallothionein-like' metal complexes in angiosperms; their structure and function. *Physiologia Plantarum*, **67**, 499–506.

**Robinson, N.J. & Thurman, D.A. (1986).** Involvement of metallothionein like copper complex in the mechanisms of copper tolerance in *Mimulus guttatus*. *Proceedings of the Royal Society of London, Series B*, **227**, 493–501.

**Rorison, I.H. (1969).** Ecological inferences from laboratory experiments on mineral nutrition. In *Ecological Aspects of the Mineral Nutrition of Plants* (Ed. by I.H. Rorison), pp. 155–175. Blackwell Scientific Publications, Oxford.

**Rorison, I.H. (1985).** Nitrogen source and the tolerance of *Deschampsia flexuosa*, *Holcus lanatus* and *Bromus erectus* to aluminium during seedling growth. *Journal of Ecology*, **73**, 83–90.

**Sachs, M.M., Freeling, M. & Okimoto, R. (1980).** The anaerobic proteins of maize. *Cell*, **20**, 761–767.

**Salaj, J. & Hudak, J. (1985).** The influence of zinc upon the submicroscopic nucleolus structure in meristematic root cells of *Vicia faba* L. *Biologia (Bratislava)*, **40**, 437–442.

**Scadeng, D.W.F. & Macleod, R.D. (1976).** The effect of sucrose concentration on cell proliferation and quiescence in the apical meristem of excised roots of *Pisum sativum* L. *Annals of Botany*, **40**, 947–955.

**Simon, E. (1978).** Heavy metals in soils, vegetation development and heavy metal tolerance in plant populations from metalliferous areas. *New Phytologist*, **81**, 175–188.

**Sirover, M.A. & Loeb, L.A. (1976).** Infidelity of DNA synthesis *in vitro*: Screening for potential mutagens or carcinogens. *Science*, **194**, 1434–1436.

**Smith, M.M., Hodson, M.J., Opik, H. & Wainwright, S.J. (1982).** Salt-induced ultrastructural damage to mitochondria in root tips of a salt-sensitive ecotype of *Agrostis stolonifera*. *Journal of Experimental Botany*, **33**, 886–895.

**Smith, R.A.H. & Bradshaw, A.D. (1979).** The use of metal tolerant plant populations for the reclamation of metalliferous wastes. *Journal of Applied Ecology*, **16**, 595–612.

**Street, H.E. & Goodman, G.T. (1967).** Revegetation techniques in the lower Swansea valley. In *The Lower Swansea Valley Project* (Ed. by K.J. Hilton). Longmans Green, London.

**Swieboda, M. (1976).** The use of biological tests for establishing the influence of flue dust from lead and zinc on plant development. *Acta Societatis Botanicorum Poloniae*, **45**, 17–32.

**Thurman, D.A. & Collins, J.C. (1983).** Metal tolerance mechanisms in higher plants — a review. *Proceedings of the International Conference of Heavy Metals in the Environment, Heidelberg,* Vol. 1, pp. 298–304. C.E.P. Consultants, Edinburgh.

**Wainwright, S.J. & Woolhouse, H.W. (1977).** Some physiological aspects of copper and zinc tolerance in *Agrostis tenuis* sibth.: cell elongation and membrane damage. *Journal of Experimental Botany*, **28**, 1029–1036.

**Waldren, S., Davies, M.S. & Etherington, J.R. (1987).** The effect of manganese on root extension of *Geum rivale* L., *G. urbanum* L. and their hybrids. *New Phytologist*, **106**, 679–688.

**Wallace, S.V. & Anderson, I.C. (1984).** Aluminium toxicity and DNA synthesis in wheat roots. *Agronomy Journal*, **76**, 5–8.

**Walley, K.A., Khan, M.S. & Bradshaw, A.D. (1974).** The potential for evolution of heavy metal tolerance in plants. I. Copper and zinc tolerance in *Agrostis tenuis*. *Heredity*, **32**, 309–319.

**Wilkins, D.A. (1957).** A technique for the measurement of lead tolerance in plants. *Nature*, **180**, 37.

**Wilkins, D.A. (1978).** The measurement of tolerance to edaphic factors by means of root growth. *New Phytologist*, **80**, 623–633.

**Woolhouse, H.W. (1983).** Toxicity and tolerance in the response of plants to metals. In *Encyclopedia of Plant Physiology, New Series,* Vol. 12C. *Physiological Plant Ecology III. Chemical and Biological Environment* (Ed. by O.L. Lange, P.S. Nobel, C.B. Osmond & H. Zeigler), pp. 245–300. Springer Verlag, Berlin.

**Wu, L., Bradshaw, A.D. & Thurman, D.A. (1975).** The potential for evolution of heavy metal tolerance in plants. III. The rapid evolution of copper tolerance in *Agrostis stolonifera*. *Heredity*, **34**, 165–187.

# The ecological significance of root system architecture: an economic approach

A.H. FITTER
*Department of Biology, University of York, York YO1 5DD, UK*

## SUMMARY

1 Variation in root system architecture among plants and its ecological correlates are discussed. The need to quantify costs and benefits of different root system architectures is emphasized.
2 The components of root system architecture are shown to be magnitude, topology, angle of branching and radial angle, link lengths and link radii. The costs of root systems are affected by all of these angles.
3 Root systems perform two primary functions for plants: anchorage and resource acquisition. Both are shown to depend on architectural characteristics.
4 Evidence is presented suggesting that architecture is a component of the underlying morphological plan or strategy of many plants, and plasticity in architecture is discussed. The problems of obtaining data from field-grown plants are discussed, and the implications of such data explored.

## INTRODUCTION

Root systems vary greatly in form. Diagrams of the root systems of plants *in situ* reveal wide variation in the distribution of root length in space, the dominance of the main axis, the relative dimensions of the members of root systems and the geometry of the branching pattern (Fig. 1). In addition rooting patterns vary through time, though such diagrams give no information on the dynamics of the system. In this paper an attempt is made to assess the ecological significance of these variations.

Faced with variation, ecologists generally seek an explanation involving adaptation. In some cases it is intuitively obvious that the characteristics of root systems are appropriate to habitat conditions, or at least an apparently satisfactory *post hoc* explanation can be produced. For example, in deserts and other arid regions it is common to find that many plants have extensive, shallow root systems which appear to be appropriate for the absorption of water following rain; in sites such as river flood-plains in the same areas, deep-rooted plants such as *Prosopis velutina* may occur, with main roots penetrating as deep as 8 m (Cannon 1911). Annuals in these ecosystems produce rudimentary laterals which only develop fully when soil

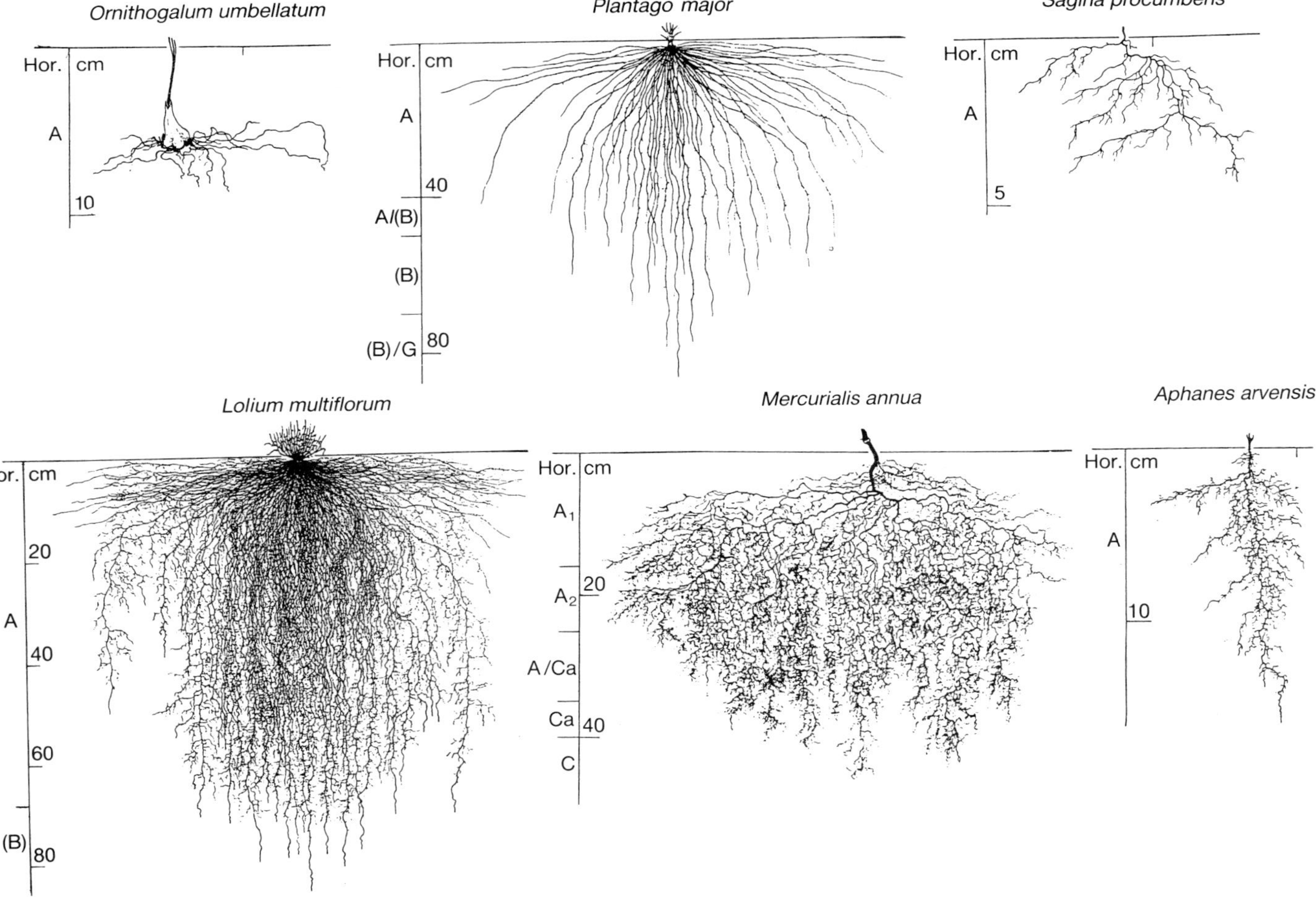


FIG. 1. Examples of contrasting root systems. The top row represent little-branched, the bottom row more strongly-branched systems; in each case there is a trend to increasing dominance of a single main axis from left to right. Diagrams reproduced with permission from Kutschera (1960).

moisture levels are high (Cannon 1911); such rain roots are also found in desert perennials such as *Agave deserti* (Hunt, Zakir & Nobel 1987), emphasizing the importance of dynamics in determining form. Similarly prairie plants growing in an environment where there is often relatively abundant water at depth may have very deep root systems (Weaver 1919). The distribution of roots may be related to many environmental factors other than water. The shallow root systems of trees such as birch (*Betula pubescens*) colonizing wet peat can be explained by anoxia in deeper soil layers, and of conifers such as pines and spruces in boreal forests (or of many trees in tropical rain-forests) by the relative abundance of mineral nutrients in the organic surface layers of the soil, and possibly by the needs of their mycorrhizal partners.

Some of this variation is clearly genetically determined, and there are some underlying construction plans that characterize large groups of plants, in some cases related taxonomically, in others ecologically. The distinctive grass root system is a good example of a taxonomic group, while the root systems of many forest trees (Köstler, Brückner & Bibelriether 1968) and the distinction observed by Cannon (1911) between winter and summer desert annuals (the latter being more profusely branched) represent ecological groups. At an even finer scale, subtle differences in root form exist. Etherington (1987) showed that a clone of *Dactylis glomerata* originating from a shallow soil was better able to develop roots in dry soil than a clone from a deeper soil, while Katyal & Subbiah (1971) demonstrated large differences in the distribution of roots with depth between wheat cultivars. Such differences may determine performance, especially in arid regions (Hurd & Spratt 1975).

Root systems vary in form, therefore, at all scales, and both genetically determined and environmentally induced variation is common. A fundamental question that remains unanswered in any quantitative sense, however, is the way in which form and function are related in root systems. To answer this, a formal quantification of the costs and benefits of particular root morphologies is necessary.

## AN ECONOMIC ANALYSIS OF ROOT SYSTEMS

Roots perform numerous functions in the life of a plant, but two of these can be regarded as primary, namely anchorage and the acquisition of soil-based resources (water and most mineral nutrients). Other functions are secondary, since they arise either as a consequence of the performance of the primary function (e.g. transport) or because root systems, once existent, may be preferred sites of certain activities, notably storage of resources and some metabolic processes. Therefore, only the primary functions will be covered here.

The costs of root systems are relatively simple to quantify, once an appropriate currency has been decided. The construction cost is represented by the quantity of the currency material involved in producing the system. If carbon is selected as currency, $CO_2$ lost in respiration and organic materials lost to the soil during growth must be included, but these are generally well understood (Lambers 1987). Alternatively a mineral element or water may be used as currency (see Hunt, Zakir & Nobel 1987, for an example of this). In addition, there is a maintenance cost which may be very high in carbon terms (Lambers 1987), and as a result the longevity of individual members of the root system will be important, though little studied (Garwood 1967; Coleman, Reid & Cole 1983). It should therefore, be possible to determine the costs and benefits of root systems differing in form so as to determine whether particular types of root system are more likely to be selected in plants growing in particular environments or possessing certain ecological characteristics. Essentially this is an optimization problem, involving the development *a priori* of a hypothesis as to the types of root systems most suitable for given conditions, and the testing of this hypothesis by observation and experiment. Such an approach, however, runs up against a severe problem, namely that it is extremely difficult to define quantitatively the architecture of a root system using the traditional developmental model. For this reason, I have developed a topological approach to the quantification of root system architecture (Fitter 1985, 1986, 1987; Fitter, Nichols & Harvey 1988).

## *The components of architecture*

The overall form or architecture of a root system can be decomposed into five components.

### *Size*

The size or *magnitude* of the system is given by the number of links. A link is the basic unit in this topological system and represents an internode, a segment of the root system either between two branching points (an interior link) or between a branch and a meristem (an exterior link; Fig. 2). Root systems are, mathematically, *trivalent rooted trees*, and as such always have $n$ exterior links and $n - 1$ interior links.

### *Topology*

Topology refers to the distribution of branches within the system. All possible topologies lie between two extremes, one being the herringbone which comprises (in developmental terms) simply a main axis and primary laterals, and the other being a dichotomous pattern, which is in practice rarely achieved. The rhizome axes of primitive plants such as *Psilotum*, the rootlets of *Lycopodium* and the mycorrhizal root clusters of species of *Pinus* are examples of topologically dichotomous patterns. A herringbone topology arises if branch generation is confined to the main axis, whereas a dichotomous system develops when branching occurs with equal probability on all exterior links. Random growth occurs when branching is equiprobable on all links. Simple topological indices can be derived by calculating the exterior pathlength, which is the number of links in the shortest path from an

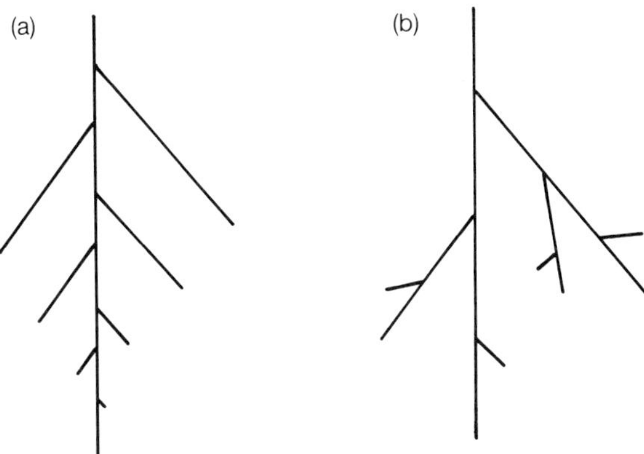


FIG. 2. Extreme topological types of trivalent rooted trees of magnitude 8; (a) a herringbone system and (b) a dichotomous system. Each has eight exterior links and seven interior links.

exterior link to the basal link. The total exterior pathlength ($P_e$) for a system is the sum of the pathlengths for all exterior links, and the altitude ($a$) is the number of links in the longest single path. For magnitude 8, a herringbone system has $P_e = 43$ and $a = 8$ (Fig. 2a), a dichotomous system $P_e = 32$ and $a = 4$ (Fig. 2b). All other systems have values between these two extremes. Expressions for determining minimum, maximum and expected (assuming random growth) values of $P_e$ and $a$ for any magnitude are given by Werner & Smart (1973) and Fitter (1985, 1989). Regressions of log $a$ or log $P_e$ on log magnitude produce linear fits if growth rules are maintained, and the slope of such a regression can be used to characterize the topology of a set of root systems (Fitter 1986; Fitter, Nichols & Harvey 1988). For a single root system appropriate indices are the values of the quotients $a$/E($a$) and $P_e$/E($P_e$), where E($a$) and E($P_e$) are the expected altitude and total exterior pathlength, assuming random growth.

### *Link lengths*

Link lengths are the distances between branching points. Interior link lengths correspond to internodes and exterior link lengths to the length of the unbranched root tips. There is an extensive literature on this, especially for crop plants (e.g. Hackett 1968, 1972; Crossett, Campbell & Stewart 1975) and trees (Lyford 1975, 1980; Kozlowski 1971), and interior link length has been shown to be significant in determining architecture in computer simulations (Lungley 1973). Nevertheless, most data are recorded on the basis of the developmental model, in which lengths of entire laterals rather than strict link lengths are noted.

### *Angles*

Two angles are required to characterize root system architecture: the azimuth (radial) angle corresponds to the origin of lateral roots radially around the root, while the branching angle is the rate at which the lateral departs from the parent. Radial angles depend upon the number of xylem poles and the rules governing the origin of branches (Charlton 1975; Fitter 1987; Pulgarin *et al.* 1988); surprisingly they have been little studied, even to the extent that it is uncertain whether distinct patterns of emergence of radial roots occur under controlled conditions, let alone in the field. Most probably the appearance of a lateral in one protoxylem-based rank reduces the possibility that the next lateral will appear in the same rank. It is obvious that radial angles must play a role in determining the efficiency with which a root system explores space. Branching angles are also important in this respect, since a too steep or too shallow angle will delay the time at which a daughter root escapes from the depletion zone around its parent. The optimum angle of growth for a lateral root after emerging from the epidermis of the parent root should be determined by the distance to the outer shell of the depletion zone for some limiting resource. This distance depends upon the age of the parent root at the point of lateral initiation and the parent growth rate on the one hand, and the diffusivity of the resource on the other. These define two sides of a right-angled triangle (Fig. 3), such that

$$l_1 = g \cdot t$$

$$l_2 = 2\ (D \cdot t)^{0.5},$$

where $l_1$ is the distance from the point of lateral initiation to the tip of the parent, $l_2$ is the radius of the depletion zone, $g$ is parent root growth rate (cm $s^{-1}$), $t$ time (s) and $D$ the diffusion coefficient of the resource ($cm^2\ s^{-1}$). The shortest distance to the outer shell is then:

$$l_3 = l_2 \cdot \sin \Phi$$

and the branching angle, $\theta$, is therefore:

$$\theta = \Phi = \tan^{-1}\ (l_1/l_2).$$

This simple model gives an overestimate of the true angle, since it does not take into account the growth rate of the branch. In the absence of data on branching angles, however, this approach offers testable hypotheses about optimum angles, such as that the angle $\theta$ should be more acute when the ratio of resource diffusion coefficient to root growth rate is greater.

### *Link radius*

Although it is known that the radius (assuming links to be simple cylinders) of members of root systems varies between species, there is little systematic data available on this point. Terminal (exterior) link radius may be as little as 35 μm in many grasses and even some trees (25 μm seems to be an effective absolute minimum, allowing for the need for xylem, phloem, endodermis, cortex and epidermis; Fitter 1987) or greater than 500 μm (e.g. *Podocarpus totara*, Baylis

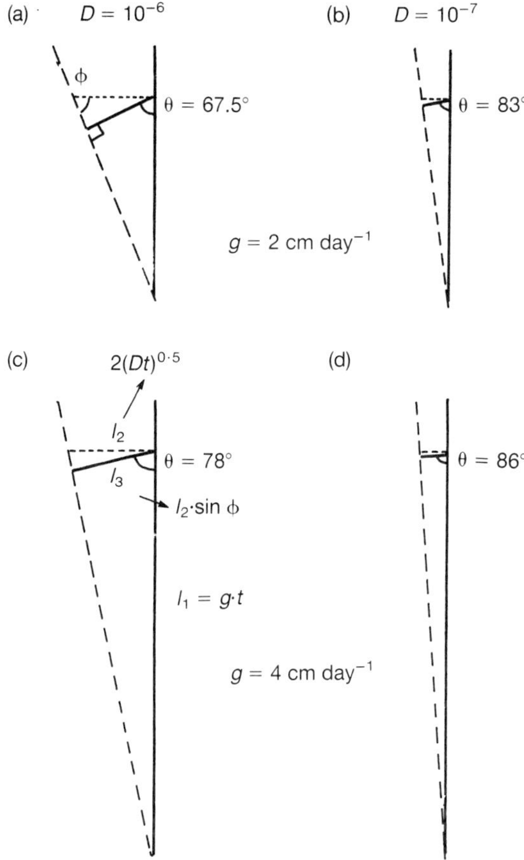


FIG. 3. Schematic representation of optimum branching angles for two different root growth rates ($g$, cm day$^{-1}$) and two different resource diffusitivies ($D$, cm$^2$ s$^{-1}$). (a) $g = 2$, $D = 10^{-6}$; (b) $g = 2$, $D = 10^{-7}$; (c) $g = 4$, $D = 10^{-6}$; (d) $g = 4$, $D = 10^{-7}$.

1975; *Glycyrrhiza lepidota*, Weaver 1919). Link radius is an important variable in determining the exploitation of soil by roots (Nye 1973; Baldwin 1975; Robinson & Rorison 1987), and also in the control of tissue volume and hence the resource cost of the root system. In particular, the relationship between radius and magnitude, or the rate at which interior links become fatter as they serve more exterior links is critical to determining costs (see below), but little studied.

### *Cost of root systems*

Although it may not always be the most appropriate currency for determining costs, carbon is certainly the easiest to measure. Below-ground production figures for a wide range of plant communities show that roots represent a very considerable cost, generally over half and often as much as 70–85% of total production going below-ground [see figures reviewed in Fogel (1985) and Caldwell & Richards (1986)]. Below-ground production figures almost always exceed biomass figures by a large margin, suggesting that turnover and other losses are greater for roots than for aerial organs. The operation of the alternative respiratory pathway (Lambers 1987) and losses by exudation, secretion, excretion and herbivory all contribute to this excess, and taken together mean that any attempt to estimate root system costs from biomass estimates must be treated with caution. Nevertheless maintenance and construction costs are both related to biomass (Amthor 1984), and so that will be used as an indicator of costs in the following discussion.

At a first approximation, construction cost will be a function of root volume. However, xylem cross-sectional area increases as approximately the 3/2 power of root cross-sectional area [data of Dittmer (1969) and Miller (1981) presented in Fig. 8 of Fitter (1987)], and since the construction and maintenance costs of stelar and cortical tissue are likely to differ, it may well be that some similar power law relationship connects cost to tissue volume. Nevertheless, length and, especially, radius of links will determine tissue volume and hence cost.

Exterior links are always longer than interior links, but there is wide variation between species, some, such as *Arrhenatherum elatius*, having relatively long exterior and interior links (Fig. 4). It is, however, the distribution of lengths of link of different radius that contributes significantly to total tissue volume. Interior links have larger radii than exterior links, and interior links of high magnitude (i.e. those serving most exterior links) will be thickest. The more high magnitude links in a system, therefore, the greater will be the total tissue volume. Since herringbone systems have the most high magnitude links at a given magnitude (Fitter 1991), this means that they are more expensive to construct (i.e. have a greater tissue volume),

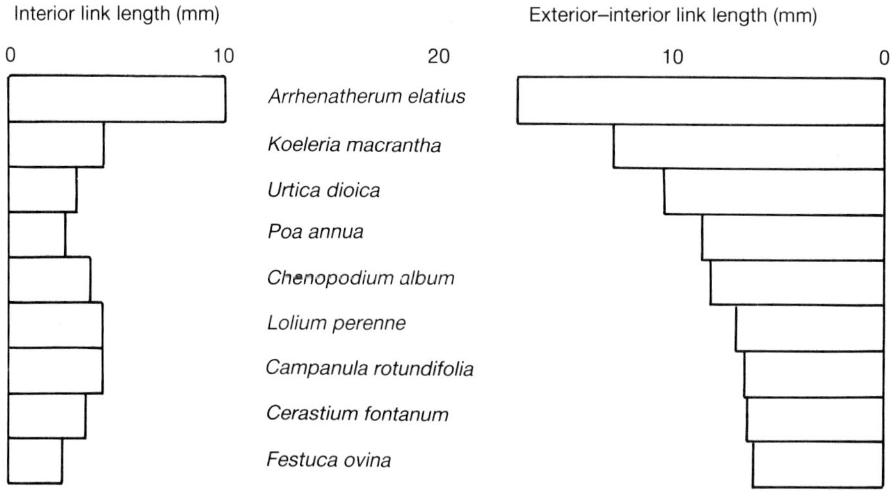


FIG. 4. Mean lengths of interior and exterior–interior links of root systems of nine species grown under uniform high nutrient-supply conditions. (For details of experimental design see Grime pp. 384–388.) Exterior–interior links are exterior links that join interior links.

and that the most important variable is not exterior link radius (since all root systems of equal magnitude have the same number of exterior links) but the rate at which link radius increases with magnitude. This parameter does differ between species (Fitter 1991), but little information is available on it.

## *Benefits*

I shall concentrate here on the two primary root functions, namely acquisition of soil resources and anchorage. Their relative importance will vary greatly, with anchorage becoming progressively more important as the above-ground part of the plant gets larger. In neither case is it yet possible to define formally the components of architecture that determine effective performance of the function.

### *Anchorage*

The stability of the aerial parts of plants is an obvious root system function, best appreciated when failure occurs on a massive scale, as after major windstorms. Conditions such as those of 16 October 1987 in south-east England are clearly outside the range of tolerance of the root systems of many trees, although it is interesting to note that both uprooting and stem failures occur with similar frequency, suggesting that their relative strengths are not dissimilar. Because of its obvious economic significance (at least £3 million per year in direct losses in British forestry, for example), the strength of woody plant root systems has been most studied. Coutts (1986) has demonstrated how roots around Sitka spruce (*Picea sitchensis*) trunks occur mainly in a well-defined zone called the root–soil plate. In high winds, roots on the windward side of the trunk are stressed under tension and failure occurs in the soil under this plate, since this is less elastic than the roots. The strength of roots under tension is then critical, and though the strength of individual roots has been studied to some small extent, their distribution relative to one another and their insertion in soil, both of which are determined by branching pattern, must make a large contribution to the strength of the root–soil system.

In herbaceous plants, lateral forces on the stem are more likely to result in stem failure than in uprooting. Upwards forces can, however, be imposed by grazing animals (and gardeners) and resistance to these is certainly a function of the adhesion of the root system to soil. When such a pull is exerted, failure typically occurs at the stem–root junction, except in very light and dry soils, suggesting that the anchorage function is possibly overfulfilled (Fitter & Ennos 1989). It may be that the evolution of a root system capable of acquiring resources led automatically to satisfactory anchorage. Investigation of plants with reduced root systems, such as hemiparasites, might yield insights here.

### *Resource acquisition*

The recognition of the importance of ion mobility in determining nutrient acquisition from soils, starting with Bray (1954) and culminating in the detailed mathematical explanation of Nye & Tinker (1977), has perhaps obscured the equally fundamental importance of root system geometry. For any given set of soil conditions and plant demands, there must be an optimum distribution of roots in soil, and that distribution will be achieved by a particular architectural configuration. The problem is, in effect, to determine the minimum *tree* that will distribute roots to the most resource-rich parts of the soil volume. This is a problem in optimization for which no analytical solution is currently available. It can, however, be tackled by a simulation model (Fitter 1987), in which a root system is allowed to develop on a screen following defined branching rules which result in different topologies, and depletion zones develop around the roots according to a specified resource diffusivity. The effectiveness of a particular simulated root system can then be determined by measuring the area of the screen 'depleted' after a given time, and this can be expressed as an efficiency by comparing it to the tissue volume of the root system, as an index of cost.

This model suggests that herringbone root systems are the most effective at exploring and exploiting soil, especially for resources of high diffusivity ($D > 10^{-7}\,cm^2\,s^{-1}$) such as nitrate ions and water. This is because dichotomous systems have many high order (secondary and tertiary) laterals which typically have low growth rates and fail to extend beyond the depletion zone of their parent. If the diffusivity is low, however, then even the low growth rates of these roots allows them to exploit fresh soil and the benefits of the herringbone topology are lost.

This leads to the prediction that the herringbone topology, which is both more expensive to construct (see above) and more efficient in the acquisition of mobile resources will be favoured in situations where such resources are limiting growth, as will be the case in many infertile soils, whereas plants on more favourable soils, and those for which the acquisition of soil-based resources is less important, will tend to have roots with a more nearly dichotomous topology.

One test of this prediction is to compare plants differing in life history. There are a number of concepts of plant strategy, such as the one-dimensional *r*–*K* continuum (MacArthur & Wilson 1967) or the two-dimensional triangular model (Grime 1977); in all of them growth rate is a fundamental variable. Both ruderal (*sensu* Grime) and *r*-selected species are fast-growing and are often found in resource-rich environments; generally such species are short-lived, often annual. One can therefore derive secondary predictions that both annual and fast-growing species will tend to have more dichotomous root system topology, than either perennials or slow-growing species, resulting in root systems that are cheaper to construct but, at least in terms of the acquisition of mobile resources, less efficient. The data in Fitter, Nichols & Harvey (1988) demonstrate that annuals do tend to conform to this relationship. A more detailed analysis of these and related data, based on growth rates, is in preparation.

That such relationships can be discerned suggests that root system topology may be a conservative characteristic in plant evolution, correlating well with other variables which together determine a plant's overall growth pattern or strategy. Certainly, Fitter, Nichols & Harvey (1988) found that differences in topology between species were greater than those within a species induced by altering nutrient supply rates. Nevertheless, topological indices were increased (i.e. root systems became more herringbone-like) by low nutrient supply, which is again in conformity with the hypothesis. There was also evidence in this work that responses to differences in nutrient supply rates varied between species, suggesting a degree of genotype–environment interaction.

### *Plasticity*

Root systems have a high degree of morphological plasticity: experiments such as those of Drew, Saker & Ashley (1973) demonstrate the ability of roots to proliferate in nutrient-rich zones. It is not surprising, therefore, that root system topology is sensitive to environmental variation. The analyses described here, however, make it possible to examine the nature of such plasticity in more detail. For example, roots of *Trifolium pratense* grown in the presence of 10 μM $Al^{3+}$ are visibly stunted, but the differences are principally in the lengths of exterior links and not in either interior link lengths or topology (Fig. 5; Fitter, Nichols & Harvey 1988). In other words branching pattern and inter-branch distance are much less affected by aluminium toxicity than is the distance behind the meristem at which branching first occurs. The only topological effect was that, whereas Al-free root systems were able to respond to low nutrient supply by becoming more herringbone-like, Al-treated roots showed no such response.

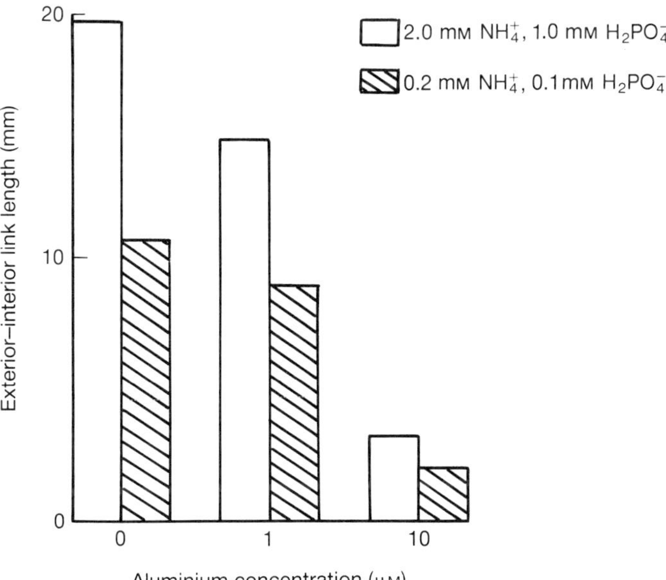


FIG. 5. Lengths of exterior–interior links of *Trifolium pratense* root systems grown at two levels of N and P supply and three levels of $Al^{3+}$. Data from Fitter, Nichols & Harvey (1988).

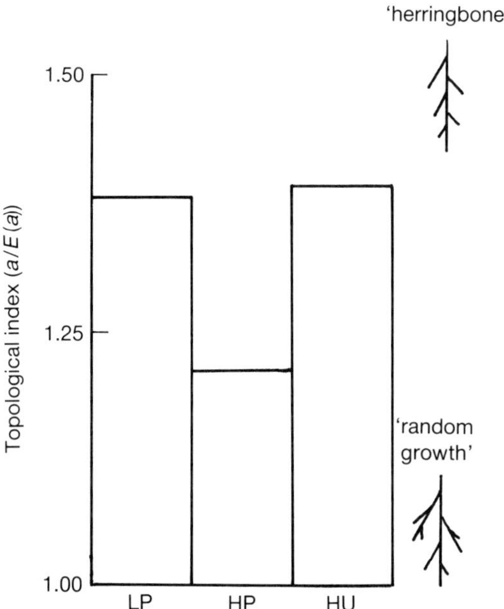


FIG. 6. Mean values of the topological index $a/E(a)$ for nine species grown under either uniform high nutrient supply or with patches of high and low supply. HU = uniform high supply; HP = high supply patch; LP = low supply patch. (See Grime pp. 384–388 for experimental details.)

The ecological significance of variation in link length is, as yet, obscure, and so such responses are difficult to interpret, but they demonstrate that it is possible to quantify plasticity. The same approach can be used for responses within a root system. Grime *et al.* (pp. 384–388) describe experiments in which roots are allowed to explore patches differing in nutrient supply. We have examined root systems from these experiments and found that the parts of individual root systems grown in patches of high nutrient-supply rates are topologically distinct from whole root systems grown at identical, but uniform rates of supply (Fig. 6). Roots from nutrient-rich patches are less herringbone-like than those from uniform high nutrient treatments. No data are available for root systems from uniformly low nutrient-supply rates, but other experiments suggest that these would have even higher topological indices, implying more herringbone structures. It appears that heterogeneity in itself promotes a change in root system topology. Again these findings demonstrate the power of topological analyses to reveal morphological responses of root systems.

## ROOT SYSTEM ARCHITECTURE IN NATURAL COMMUNITIES

The title of this section has been chosen to emphasize a lack of knowledge rather than to reveal any. Although much of the impetus for this work has come from the realization that field-grown roots vary greatly in architecture, and excavations such as those of Cannon (1911) and Weaver (1919) vividly indicate the extent of this variation, it is as yet impossible to make generalizations. It is difficult even to develop well-founded hypotheses; for example, should plants from the same habitat exhibit a common architectural pattern, because architecture is a fundamental component of their ecological make-up, or should they differ widely, because selection for the ability to coexist will have led them to exploit distinct niches and architecture permits them to occupy these?

It has previously been shown by the present author (Fitter 1987) that temporally coexisting plant species may be more similar in terms of exterior link radius than species growing at a later time in the same habitat. Whether the same is true of other features of root system architecture remains to be discovered. One technical problem might be in extracting entire root systems from field soils for analysis. Fortunately Van Pelt & Verwer (1984) have demonstrated that 'cut trees' (i.e. those in which extremities have been removed at random) retain the topological characteristics of the intact system. Sampling of root systems of two widespread plant species, white clover (*Trifolium repens*) and ribwort plantain (*Plantago lanceolata*) from eleven contrasting field sites in Yorkshire and north Wales over two years (Fitter, Stickland & Harvey, unpublished data) has revealed strong site effects on both topology and link lengths.

## CONCLUSIONS

Since roots account for a major part of the biomass and production of most plants growing under natural conditions, a better understanding of their function and development is crucial to understanding their role and behaviour in natural communities. Topological analysis of root systems, as outlined here, permits the quantitative description of various aspects of the architecture of root systems. From preliminary results, it is apparent that there are patterns in the distribution of architectural characteristics among species that are correlated with other important ecological attributes. This approach seems likely to shed light on several important questions about root function and root interactions.

## ACKNOWLEDGEMENTS

M. Harvey, T.R. Stickland and J. Mackey were responsible for the collection of some of the data given here and J.P. Grime designed the plasticity experiments. A.R. Ennos and M.P. Coutts offered valuable comments on anchorage.

## REFERENCES

**Amthor, J.S. (1984).** The role of maintenance respiration in plant growth. *Plant, Cell and Environment*, **7**, 561–569.

**Baldwin, J.P. (1975).** A quantitative analysis of the factors affecting plant nutrient uptake from some soils. *Journal of Soil Science*, **26**, 195–206.

**Baylis, G.T.S. (1975).** The magnolioid mycorrhiza and mycotrophy in root systems derived from it. In *Endomycorrhizas* (Ed. by F.E. Sanders, B. Mosse & T.B. Tinker), pp. 373–390. Academic Press, London.

**Bray, R.H. (1954).** A nutrient mobility concept of soil–plant relationships. *Soil Science*, **78**, 9–22.

**Caldwell, M.M. & Richards, J.H. (1986).** Competing root systems: morphology and models of absorption. In *On the Economy of Plant Form and Function* (Ed. by T.J. Givnish), pp. 251–273. Cambridge University Press, Cambridge.

**Cannon, W.A. (1911).** *Root Habits of Desert Plants.* Carnegie Institute of Washington Publications No. 131.

**Charlton, W.A. (1975).** Distribution of lateral roots and pattern of lateral initiation in *Pontaderia cordata* L. *Botanical Gazette*, **136**, 225–235.

**Coleman, D.C., Reid, C.P.P. & Cole, C.V. (1983).** Biological strategies of nutrient cycling in soil systems. *Advances in Ecological Research*, **13**, 1–55.

**Coutts, M.P. (1986).** Components of tree stability in Sitka spruce on peaty gley soil. *Forestry*, **59**, 173–197.

**Crossett, R.N., Campbell, D.J. & Stewart, H.E. (1975).** Compensating growth in cereal root systems. *Plant and Soil*, **42**, 673–683.

**Drew, M.C., Saker, L.R. & Ashley, T.W. (1973).** Nutrient supply and the growth of the seminal root system in barley. I. *Journal of Experimental Botany*, **24**, 1189–1202.

**Dittmer, H.J. (1969).** Characteristics of the roots of some xerophytes. In *Physiological Systems in Semiarid–Arid Environments* (Ed. by C.C. Hoff & M.L. Riedesel), pp. 231–238. University of New Mexico Press, Albuquerque.

**Etherington, J.R. (1987).** Penetration of dry soil by roots of *Dactylis glomerata* L. Clones derived from well-drained and poorly-drained soils. *Functional Ecology*, **1**, 19–24.

**Fitter, A.H. (1985).** Functional significance of root morphology and root system architecture. In *Ecological Interactions in Soil* (Ed. by A.H. Fitter, D. Atkinson, D.J. Read & M.B. Usher), pp. 87–106. *Special Publication of the British Ecological Society, No. 4.* Blackwell Scientific Publications, Oxford.

**Fitter, A.H. (1986).** The topology and geometry of plant root systems: influence of watering rate on root system topology in *Trifolium pratense*. *Annals of Botany*, **57**, 91–101.

**Fitter, A.H. (1987).** An architectural approach to the comparative ecology of plant root systems. *New Phytologist*, **106**(Suppl), 61–77.

**Fitter, A.H. (1991).** The characteristics and functions of root systems. In *Plant Roots: The Hidden Half* (Ed. by Y. Waisel, U. Kafkafi and A. Eshel), in press. Marcel Dekker, New York.

**Fitter, A.H. & Ennos, A.R. (1989)** Architectural constraints to root system function. In *Roots and the Soil Environment* (Ed. by D.Robinson) *Aspects of Applied Biology*, **22**, 15–22.

**Fitter, A.H., Nichols, R. & Harvey, M.L. (1988).** Root system architecture in relation to life history and nutrient supply. *Functional Ecology*, **2**, 345–351.

**Fogel, R. (1985).** Roots as primary producers in below-ground ecosystems. In *Ecological Interactions in Soil* (Ed. by A.H. Fitter, D. Atkinson, D.J. Read & M.B. Usher), pp. 23–36. *Special Publication of the British Ecological Society, No. 4.* Blackwell Scientific Publications, Oxford.

**Garwood, E.A. (1967).** Seasonal variation in appearance and growth of grass roots. *Journal of the British Grassland Society*, **22**, 121–130.

**Grime, J.P. (1977).** Evidence for the existence of three primary strategies in plants and its relevance to ecological and evolutionary theory. *American Naturalist*, **111**, 1169–1194.

**Grime, J.P. & Hunt, R. (1975).** Relative growth rate: its range and adaptive significance in a local flora. *Journal of Ecology*, **63**, 393–422.

**Hackett, C. (1968).** A study of the root system of barley. I. Effects of nutrition in two varieties. *New Phytologist*, **67**, 287–298.

**Hackett, C. (1972).** A method of applying nutrients locally to roots under controlled conditions and some morphological effects of locally applied nitrate on the branching of wheat roots. *Australian Journal of Biological Science*, **25**, 1169–1180.

**Hunt, E.R., Zakir, N.J.D. & Nobel, P.S. (1987).** Water costs and water revenues for established and rain-induced roots of *Agave deserti*. *Functional Ecology*, **1**, 125–130.

**Hurd, E.A. & Spratt, E.D. (1975).** Root patterns in crops as related to water and nutrient uptake. In

*Physiological Aspects of Dryland Farming* (Ed. by U.S. Gupta), pp. 167–235. Oxford and IBH Publishing Company, New Delhi.

**Katyal, J.C. & Subbiah, B.V. (1971).** Root distribution pattern of some wheat varieties. *Indian Journal of Agricultural Science*, **41**, 788–790.

**Köstler, J.N., Brückner, E. & Bibelriether, H. (1968).** *Die Würzeln der Waldbäume*. Parey, Hamburg.

**Kozlowski, T.T. (1971).** *Growth and Development of Trees*. Chapter 5, Root growth. Academic Press, London.

**Kutschera L. (1960).** Wurzelatlas mitteleuropäischer Ackerunkräuter and Kulturpflanzen. DLG Verlag, Frankfurt-am-Main.

**Lambers, H. (1987).** The fate of carbon translocated to roots. In *Root Development and Function* (Ed. by P.J. Gregory, T.V. Lake & D.A. Rose), *Society for Experimental Biology Seminar Series*. Cambridge University Press, Cambridge.

**Lungley, D.R. (1973).** The growth of root systems — a numerical computer simulation model. *Plant and Soil*, **38**, 145–159.

**Lyford, W.H. (1975).** Rhizography of non-woody roots of trees in the forest floor. In *The Development and Function of Roots* (Ed. by J.G. Torrey & D.T. Clarkson), pp. 179–196. Academic Press, New York.

**Lyford, W.H. (1980).** Development of the root system of northern red oak (*Quercus rubra* L). *Harvard Forest Paper* No. 21, Harvard University, Petersham.

**MacArthur, R.H. & Wilson, E.O. (1967).** *The Theory of Island Biogeography*. Princeton University Press, Princeton.

**Miller, D.M. (1981).** Studies of root function in *Zea mays*. Dimensions of the root system. *Canadian Journal of Botany*, **59**, 811–818.

**Nye, P.H. (1973).** The relation between the radius of a root and its nutrient absorbing power. *Journal of Experimental Botany*, **24**, 783–786.

**Nye, P.H. & Tinker, P.B. (1977).** *Solute Movement in the Soil–Root System*. Blackwell Scientific Publications, Oxford.

**Pulgarin, A., Navascués, J., Casero, P.J. & Lloret, P.G. (1988).** Branching pattern in onion adventitious roots. *American Journal of Botany*, **75**, 425–432.

**Robinson, E. & Rorison, I.H. (1987).** Root hairs and plant growth at low nitrogen availabilities. *New Phytologist*, **107**, 681–693.

**Van Pelt, J. & Verwer, R.W.H. (1984).** Cut trees in the topological analysis of branching patterns. *Bulletin of Mathematical Biology*, **45**, 269–285.

**Weaver, J.E. (1919).** *The Ecological Relations of Roots*. Carnegie Institute Publ. 286, Washington.

**Werner, C. & Smart, J.S. (1973).** Some new methods of topologic classification of channel networks. *Geographical Analysis*, **5**, 271–295.

**Note added in proof**

Since this paper was written, we have been able to extend the simulation model (p. 238) to three dimensions and to use it to reveal the significance of link length (Fitter *et al.*, 1991), and to test the predictions derived from these simulations (Fitter & Stickland, 1991).

**Fitter, A.H. & Stickland, T.R. (1991).** Architectural analysis of plant root systems II. Influence of nutrient supply on architecture in contrasting plant species. *New Phytologist* (in press).

**Fitter, A.H., Stickland, T.R., Harvey, M.L. & Wilson, G.W. (1991).** Architectural analysis of plant root systems I. Architectural correlates of exploitation efficiency. *New Phytologist* (in press).

# The Root System of Different Vegetation Types

# Root production of four arable crops in Sweden and its effect on abundance of soil organisms

A.-C. HANSSON, O. ANDRÉN AND E. STEEN

*Department of Ecology and Environmental Research, Swedish University of Agricultural Sciences, Box 7072, S-750 07 Uppsala, Sweden*

## SUMMARY

1 Production of spring barley roots after 1 and 5 years without nitrogen fertilizer were measured and compared with barley fertilized with 120 kg N $ha^{-1}$ $year^{-1}$. The effects of nitrogen fertilization on absolute and relative allocation of growth above and below ground are discussed.

2 Annual organic matter input to the soil including above-ground harvest residues and stubble, root production and rhizodeposition, was 324 and 402 g $m^{-2}$ containing 4·5 and 7·8 g N $m^{-2}$ in unfertilized and fertilized barley, respectively.

3 Root growth, death and production in a meadow fescue ley (*Festuca pratensis* Huds.) were measured during its second year of growth. About 60% of the annual root production occurred before the second harvest in early August. Since roots produced in preceding years died during the same period, the root biomass did not change. In this crop, 584 g organic matter $m^{-2}$ and 11·5 g N $m^{-2}$ was supplied to the soil annually. Corresponding figures for a lucerne ley (*Medicago sativa* L.) were 591 g $m^{-2}$ and 17·4 g N $m^{-2}$ during its third year of growth.

4 Compared with the unfertilized barley, barley fertilized with nitrogen had a deeper and denser root system. The meadow fescue ley had the most superficial root system of the crops, with 77% of the root biomass in the top 10 cm. In spite of its morphologically different root system, lucerne root biomass distribution was fairly similar to that of the meadow fescue ley, but the root biomass below 27 cm (ploughing depth) was higher in lucerne than in meadow fescue.

5 Soil organism abundance dynamics (bacteria, fungi, nematodes, microarthropods and enchytraeids) in the four different cropping systems are briefly discussed in relation to total input of organic matter, root distribution and water uptake.

## INTRODUCTION

Roots are a major determinant in most ecosystems. In agroecosystems, roots sometimes are the only source of organic matter input to the soil, since low litterfall, high export and in extreme cases even straw burning reduce other inputs.

The interactions between plant roots and other soil organisms have been discussed in several contexts (Mitchell & Nakas 1986) and always the lack of quantitative field data is a problem. Sampling problems and the variability of the results of root samplings are commonly reported although the problems are often reduced with arable crops, where the number of species present and the length of the growing period are controlled factors.

The systems discussed here represent important choices within modern Swedish agriculture, i.e. a spring-sown annual cereal versus perennial leys, and fertilized grass ley versus a nitrogen fixing legume ley. Apart from having different growth and allocation patterns, these crops also imply different agricultural practices, i.e. they are sown and cut at different times and cultivated with different intensities.

The results presented in this paper originate from studies performed within the integrated ecosystem project 'Ecology of Arable Land' aiming at quantifying the role of organisms in carbon and nitrogen cycling in cropping systems representing some typical characteristics of contemporary Swedish agriculture (Andrén 1988; Andrén *et al.* 1989). The primary production studies were methodologically concentrated on estimating plant biomass and its nitrogen concentration, both above and below ground.

## MATERIALS AND METHODS

### *Site description*

The field experiment was located at Kjettslinge, approximately 40 km north of Uppsala, Sweden (60°10′ N, 17°38′E). The regional climate is cool temperate. The mean annual temperature during 1980–85 was +5 °C, with the highest (monthly mean) temperature in July (17 °C) and the lowest in February (−5 °C). Annual precipitation averaged 576 mm, and the highest monthly precipitation occurred in August (85 mm) (Alvenäs, Johnsson & Jansson 1986). The ground is normally covered with snow for 100 days, 10 December–20 March (Rodskjer & Tuvesson 1975). Potential mean annual evapotranspiration for the region is around 500 mm (Wallén 1966).

Plots were laid out in a randomized block design with four blocks and four treatments, giving sixteen plots of 40 × 14 m in size (Fig. 1).

In May 1980 four cropping systems were established: barley (*Hordeum distichum* L.) fertilized with 120 kg N $ha^{-1}$ $year^{-1}$ (B120), barley without N fertilization (B0), a grass ley (*Festuca pratensis* Huds.) fertilized with 120 + 80 kg N $ha^{-1}$ $year^{-1}$ (GL) and nitrogen fixing lucerne (*Medicago sativa* L.) with no N fertilization (LL). The leys were sown with barley as a nurse crop, but because of severe winter damage the leys were resown in a pure stand in May 1981.

Barley was fertilized with nitrogen [$Ca(NO_3)_2$] at sowing, around the second week of May. The grass ley was fertilized in May (120 kg N $ha^{-1}$), harvested around the last week of June and fertilized after harvest with 80 kg N $ha^{-1}$. The leys were harvested a second time, generally in the middle of August. Barley was harvested in early September and the straw (except a 12 cm stubble) was removed from the field. All plots were fertilized with P and K (90 and 168 kg $ha^{-1}$, respectively) in

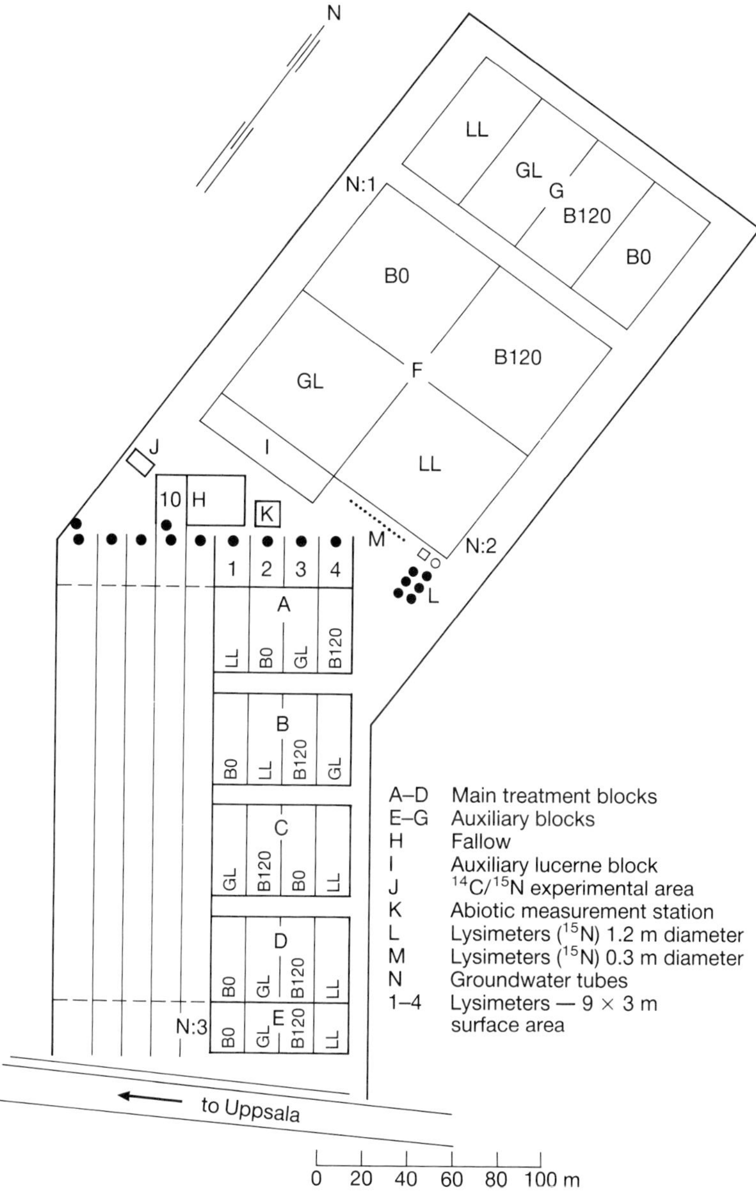


FIG. 1. Layout of the experimental field at Kjettslinge.

1979 when the field site was set up, and again in 1982 and 1983, to maintain standard fertility levels.

The topsoil is classified as a mixed, frigid *typic Haplaquoll* (USDA Soil Survey Staff 1975) or *Mollic Gleysol* (FAO-UNESCO 1974) with loam texture (15–20% clay) overlying a coarser loamy sand layer of variable thickness and a clay loam, with a transition to a clay subsoil at about 1 m depth. At the start of the experiment, the topsoil had a $pH_{(H_2O)}$ of 6·3, the C content varied between 1·6 and 3·1% and the N content was 0·17–0·31% with an average of 2·2% C and 0·23% N for the whole field.

A more complete description of the field site is given by Steen, Jansson & Persson (1984).

## *Methods*

The results concerning root growth and production presented in this paper are based on samplings performed during 1980–85 using three different field methods.

Firstly, below-ground biomass was sampled using a soil corer (7 cm diameter) according to Welbank, Taylor & Williams (1974). In general, six soil cores (0·5 m deep) were taken immediately after sampling of above-ground parts in each plot. Three cores were taken in the plant rows and the rest within the rows. Cores from the same plot were pooled.

The soil cores were stored at −20 °C in plastic bags until they were processed. The organic material was separated from the soil with water over sieves with a mesh size of 0·5 mm according to Cahoon & Morton (1960) and separated into biomass and other organic material. Below-ground biomass (LR) included both living roots and living below-ground stem bases. Living roots were distinguished by their high flexibility and light colour. The remaining material consisted of dead roots from both the current year and preceding years, and harvest residues ploughed under before the ley was sown. This latter was referred to as dead organic material (DOM) and the unseparated samples of below-ground organic material as macro-organic material (MOM).

All samples were dried for 3 days at 50 °C, then weighed and ground. Nitrogen concentration was determined after semi-micro-Kjeldahl digestion (Digestion System 40, Kjeltec Auto 1030 Analyser, Tecator, Landskrona, Sweden). Ash content was determined by combustion for 6 h at 600 °C. All results are presented as ash-free dry mass (organic matter).

Production and uptake of nitrogen in below-ground plant parts were calculated as the sum of significant increases in biomass or of nitrogen in LR (Hansson & Steen 1984; Hansson 1987; Hansson, Pettersson & Paustian 1987; Hansson & Pettersson 1989).

Secondly, during 1980 four mini-rhizotrons (Böhm 1979), two within and two between rows (7 cm diameter, 1 m depth), were installed and used in each treatment (one in each plot) (Hansson & Andrén 1987). Number of roots at each 5 cm depth level were counted three times a week during the main part of the growing season. This year barley was sown in all treatments and meadow fescue and lucerne, respectively, were insown in two of the treatments.

Thirdly, ingrowth mesh bags, i.e. mesh bags with 1 cm mesh and 7 cm diameter, were put into holes in the soil and filled with sieved and root free soil (Steen 1985).

These bags were used during 1 year to follow growth of roots produced the current year in the perennial grass ley (Hansson & Andrén 1986).

$^{15}$N-labelled fertilizer was applied in small enclosed plots (0·3 m diameter, 0·45 m depth) adjacent to the field experiment (Hansson & Pettersson 1989). Below-ground samples were treated according to the description above, and nitrogen isotopic composition of the samples was determined on a mass spectrometer (Micromass 602). Uptake of fertilizer nitrogen was calculated according to Vose (1980).

In parallel with the field studies, barley and meadow fescue were grown in a gas-tight growth chamber supplied with $^{14}$C-labelled $CO_2$ and $^{15}$N-labelled $Ca(NO_3)_2$. The soil atmosphere was separated from the above-ground atmosphere with a gas-tight sealant (Johansson, unpublished). Complete plant carbon and nitrogen budgets, including root respiration and rhizodeposition, were compiled from these experiments. The proportional relationships between the different plant fractions obtained in the growth chamber were used when extrapolating these results to the field situation in order to include them in the ecosystem budgets (Paustian *et al.* 1990).

Above-ground crop in the field experiment was sampled weekly or bi-weekly. The vegetation was cut at ground level with scissors and litter was swept up. The samples were dried and sampled into living and dead parts. After weighing, the samples were ground and analysed for nitrogen and ash content (Pettersson *et al.* 1986; Pettersson 1987).

Detailed methods for sampling and analysing soil micro-organisms and soil fauna can be found in Schnürer, Clarholm & Rosswall (1986), Sohlenius, Boström & Sandor (1987) and Lagerlöf (1987).

The field samplings of the different organisms were coordinated, i.e. monthly samplings were performed for all organisms during the growing season of 1982 in B120 and GL. Furthermore, yearly September samplings in all treatments were performed, to monitor long-term dynamics of the organisms.

Abiotic factors such as soil temperature, wind speed and precipitation were automatically monitored; soil water content, water discharge and nutrient leaching were measured semi-automatically (Alvenäs, Johnsson & Jansson 1986; Bergström 1987).

## RESULTS AND DISCUSSION

### *Root production and depth distribution*

The two barley treatments in our experiment represent two extremes considering nitrogen availability, since one (B120) received 120 kg N $ha^{-1}$ $year^{-1}$ and the other (B0) did not receive any nitrogen fertilizer during 6 years. The results presented here originate from the second (1981) and sixth year (1985). Some of the results from 1981 were previously presented by Hansson, Pettersson & Paustian (1987).

Nitrogen influences the amount of light intercepted, i.e. production is regulated by the growth rate of the leaf area, with good nitrogen availability promoting a high rate of leaf area growth. Under such circumstances the supply of assimilates to the tiller buds stimulate tillers to develop, contributing to a higher total leaf area (Gallagher & Biscoe 1978).

Compared with the unfertilized barley, the fertilized barley had a higher and denser canopy, i.e. there were more stems and the ears had more grains. The higher stem density and higher green area per stem resulted in a higher green area index in B120, resulting in a doubling of total production (Hansson, Pettersson & Paustian 1987; Pettersson 1987; Pettersson 1989) (Table 1).

A reduced nitrogen availability in the soil reduces leaf area growth rate which gives a lower photosynthesis in later stages of plant growth. However, the photosynthetic rate of existing leaves is little affected (Gregory, Marshall & Biscoe 1981), and carbohydrates not used for leaf growth may be allocated to the roots, i.e. an increased proportion of total production is used for root production. If nitrogen availability is very low, also the photosynthetic rate is reduced since it is a function of the nitrogen concentration in the plant, which is lower at nitrogen deficiency (van Keulen & Seligman 1987).

The change of dry matter allocation between above- and below-ground parts of the plant, due to reduced nitrogen availability, has been demonstrated in several laboratory studies (Brouwer 1962; Troughton 1962). In the growth chamber experiment here a lower proportion of the total assimilation was translocated below ground in the fertilized barley, 25%, compared with 29% in the unfertilized barley (Johansson, unpublished), while in our field experiment, the below-ground proportion was higher in B0 than in B120 in 1981, but the same in 1985 (Table 1).

In Fig. 2 the change of the root/shoot ratio over time in B0 and B120 in 1981 and 1985 is compared with other published field estimates for spring barley. In general, the unfertilized treatments showed a higher ratio than the fertilized. However, the influence of other factors (climate, soil, cultivar) on allocation between above- and below-ground parts is clearly illustrated by the different curves. During 1985 in our experiment, maximum green leaf area index and its duration was unusually low in B120 which resulted in lower total production than preceding years (Pettersson 1989). In addition, the higher total nitrogen uptake in

TABLE 1. Primary production, above and below ground, in unfertilized (B0) and nitrogen fertilized barley (B120). OM = organic matter (g $m^{-2}$), N = nitrogen (g $m^{-2}$), R : S = root : shoot ratio

| | 1981 | | | | 1985 | | | |
|---|---|---|---|---|---|---|---|---|
| | B0 | | B120 | | B0 | | B120 | |
| | OM | N | OM | N | OM | N | OM | N |
| Above ground | 430 | 4·2 | 844 | 11·8 | 377 | 4·1 | 678 | 14·3 |
| Below ground | 128 | 1·6 | 160 | 3·2 | 80 | 1·5 | 145 | 3·4 |
| Total | 558 | 5·8 | 1004 | 15 | 457 | 5·6 | 823 | 17·7 |
| R : S | 0·30 | 0·38 | 0·19 | 0·27 | 0·21 | 0·37 | 0·21 | 0·24 |

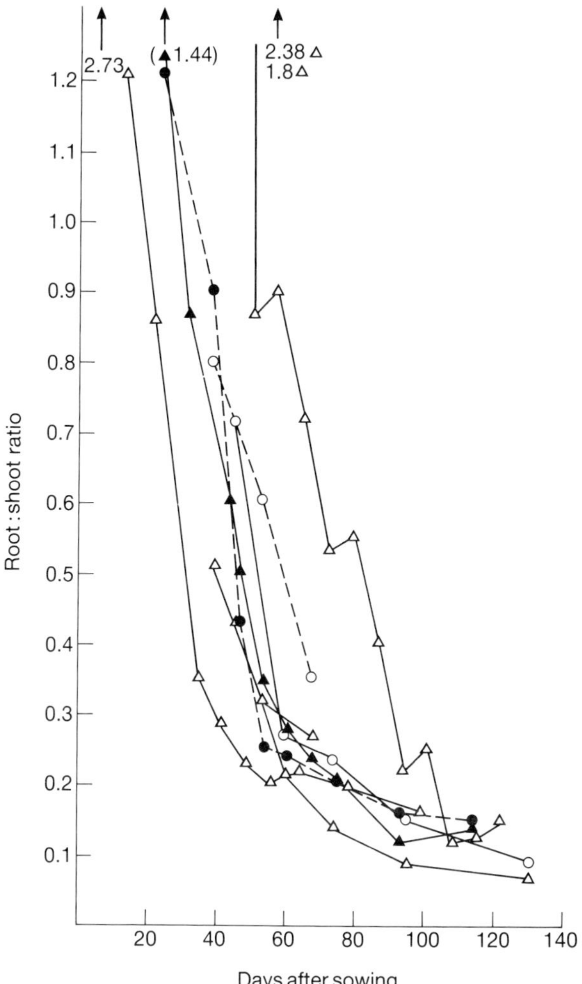


FIG. 2. Ratio between standing crop of biomass of root and shoot in spring barley. Circles stand for unfertilized crops and triangles for fertilized. Redrawn from Biscoe, Scott & Monteith (1975); Nielsen (1980); Welbank & Williams (1968); Hansson, Pettersson & Paustian (1987); Hansson, unpublished.

B120 in 1985 compared with 1981 indicated that growth was being restricted by a factor other than nitrogen (Table 1), and as precipitation in June was only half the normal value, and little rain fell until the last week of July (Alvenäs, unpublished), it is likely that lack of water impeded growth in both treatments during the summer of 1985.

The nitrogen was more efficiently used in B0 than in B120, i.e. more biomass was produced per unit nitrogen and time (nitrogen productivity, cf. Ågren 1985), both when comparing maximum values, and over the whole season (Hansson, unpublished). When comparing 1981 and 1985, the nitrogen productivity was lower in 1985 than 1981 for both treatments. The lower efficiency in B120 compared with B0 can probably be explained by higher competition for light in the fertilized crop, whereas the lower efficiency during 1985 was due to lack of water.

The increase in root biomass ceased earlier and at a lower level in 1985 compared to 1981 in both treatments, although it was more apparent in B0 than in B120 (Fig. 3). The low precipitation during June and July, discussed above, probably caused an early decrease in root production, or increased root death. After flowering root biomass decreased due to significant root death. Maximum relative growth rate (RGR) occurred during the first 2 weeks in both treatments. In B120, how-

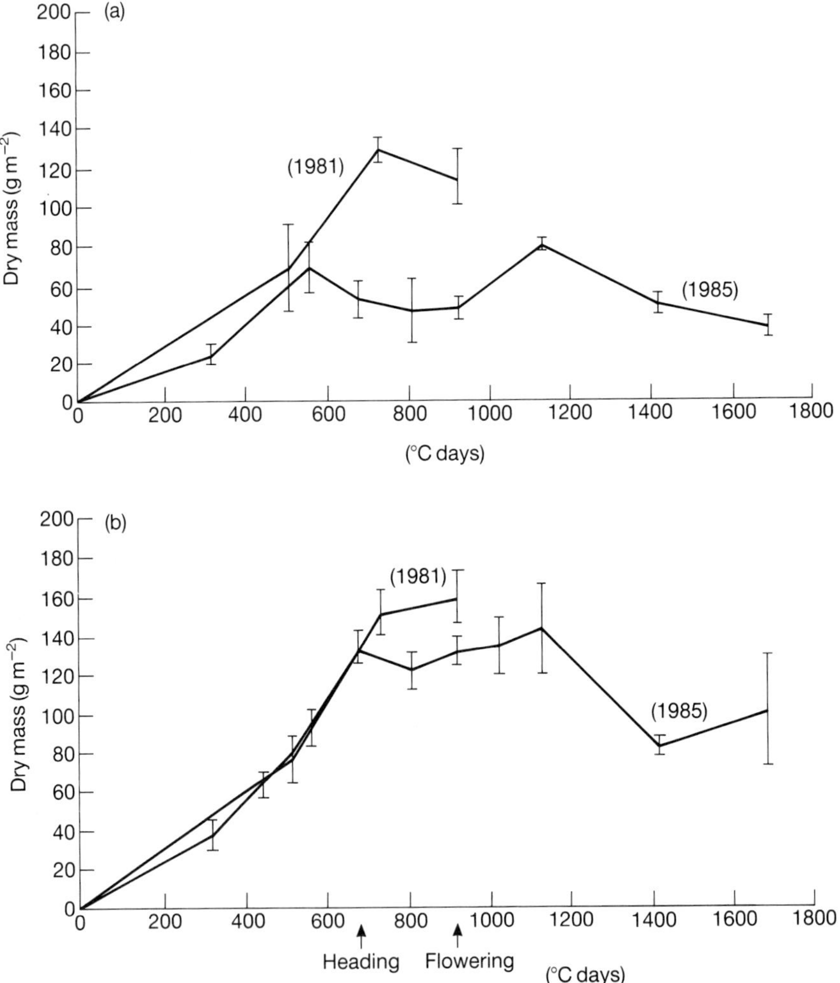


FIG. 3. Root biomass in barley (a) without (B0) and (b) with (B120) nitrogen fertilization, g m$^{-2}$. The bars indicate 1 S.E., $n = 4$.

ever, RGR remained at the same high level 1 week longer than in B0 (Hansson, unpublished).

The growth chamber experiments indicated, for both fertilized and unfertilized barley, that about a third of the carbon translocated below-ground was immediately lost via root respiration, 20–30% delivered to the soil as rhizodeposition while the rest was found as root biomass at harvest (Johansson, unpublished). The results from other experiments using $^{14}C$-labelling have been summarized in Table 2. In most cases the distribution corresponds with Johansson's results. The carbon input to the soil expressed as a proportion of the total assimilation was 29 and 25% in B0 and B120 respectively (Johansson, unpublished). Keith, Oades & Martin (1986) estimated it to be 15·5% and Sauerbeck & Johnen (1977) 33%. If the relative carbon allocation found in the laboratory studies are extrapolated to the field situation then the root production in barley calculated from the field samplings (cf. Table 1) would represent an underestimation of 45% due to rhizodeposition. However, this extrapolation of laboratory results to the field presents some difficulties. $^{14}C$-labelling experiments have shown that rhizodeposition does occur and that it can be significant although it is not known whether the measured proportions are extreme or normal values or how environmental conditions affect rhizodeposition.

The ratio between below- and above-ground biomass at the end of the experiment, after 7 weeks, was considerably lower in the growth chamber experiment (0·14) than in the field experiment (0·4). This could be the result of a faster phenological development in the laboratory experiment due to higher temperature and different light conditions.

Apart from roots, harvest residues and stubble were incorporated into the soil each year at ploughing when cultivating annual cereals. Since the straw was removed and the stubble low (12 cm), these parts did not represent major inputs; only 156 and 111 g $m^{-2}$ containing 1·8 and 1·2 g N $m^{-2}$ in B120 and B0, respectively (Pettersson 1989).

Total input of organic matter and nitrogen to the soil from roots and above-ground harvest residues were considerably lower in B0, than in B120 (324 and 402 g $m^{-2}$, and 4·5 and 7·8 g N $m^{-2}$, respectively). The organic material supplied to the soil cultivated with B120 had a higher mean nitrogen concentration than in B0 (Table 3) (Paustian *et al.* 1990).

Below-ground biomass in the top 10 cm was about two times higher in B120 than in B0 (Fig. 4). A higher proportion of the below-ground biomass consisted of stem bases in B120 than in B0, since the number of tillers was higher in B120 (Pettersson 1989). However, the results from the mini-rhizotron studies showed a higher number of roots and a larger root depth in B120 (Fig. 5) (Hansson & Andrén 1987). This is well in line with the observation by Weaver (1926) that fertilization promotes root development in cereals and the results of Fehrenbacher, Ray & Alexander (1969), who showed that rooting depth increased when nitrogen was supplied to soils with low natural fertility.

Since crops are sown in rows, crop below-ground biomass could be expected not to be horizontally uniformly distributed over a field. The mini-rhizotron results

TABLE 2. Distribution of $^{14}C$ in barley and wheat plants cultivated in field (F) and laboratory (L). $R_R = CO_2$ produced as root respiration; Rhizodep = $CO_2$ produced from decomposing roots and rhizodeposition; $CO_2$ = total amount of $CO_2$ produced, i.e., $R_R$ + Rhizodep

| Crop | Plant age (days) | Growth stage | Site | % Distribution of $^{14}C$ | | | | | | Reference |
|---|---|---|---|---|---|---|---|---|---|---|
| | | | | Root | Soil | Root + Soil | $CO_2$ | $R_R$ | Rhizodep | |
| B0 | 53 | Late tillering | L | 34 | 14 | 48 | 52 | 35 | 17 | Johansson (unpublished) |
| B120 | 53 | Late tillering | L | 40 | 12 | 52 | 48 | 40 | 8 | Johansson (unpublished) |
| Wheat | 21 | Seedling | L | 60 | 19 | 79 | 21 | — | — | Barber & Martin (1976) |
| Wheat | 153 | Maturity | L | 20 | 4 | 24 | 76 | 15 | 61 | Sauerbeck & Johnen (1977) |
| Wheat | 140 | Maturity | F | 25 | 25 | 50 | 50 | — | — | Keith, Oades & Martin (1986) |
| Wheat | 49 | Early tillering | F | | | 35 | 65 | — | — | Martin & Kemp (1986) |
| Wheat | 70 | Late tillering | F | | | 55 | 45 | — | — | Martin & Kemp (1986) |

TABLE 3. Input of organic matter, and its nitrogen content, to the soil including above-ground parts, roots and rhizodeposition. Based on results from 1981 and 1985 for barley, and 1982 and 1983 for the leys. OM = organic matter (g m$^{-2}$); N = nitrogen (g m$^{-2}$)

| | OM | N | %N | C:N ratio |
|---|---|---|---|---|
| B0 | 324 | 4·5 | 1·4 | 72 |
| B120 | 402 | 7·8 | 1·9 | 52 |
| GL | 584 | 11·5 | 2·0 | 51 |
| LL | 591 | 17·4 | 2·9 | 34 |

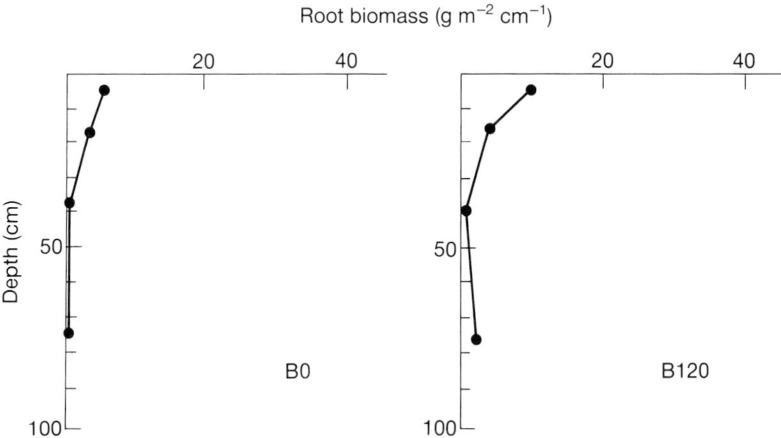


FIG. 4. Depth distribution of root biomass in barley at anthesis. B0 = barley without nitrogen fertilization; B120 = barley with nitrogen fertilization. From Hansson (1987).

from the first half of the growing season, when the insown ley crops still were suppressed by the barley, were used for comparing the number of roots within and between rows. Until mid July, larger numbers of roots were recorded within barley rows than between, but due to high variance the differences were not statistically significant (Fig. 6). According to Welbank, Taylor & Williams (1974) and Andersen (1986) root length density did not differ in core samples taken within and between rows, whereas root biomasses in rows were two times higher than between rows in a study by Groth (1987). These contradictory results may be explained if stem bases were included in the latter investigation, contributing to root biomass.

In managed perennial leys on arable soil the standing crop of root biomass increases during the first years. The results discussed here originate from a meadow fescue ley, GL, during its second growing season and a lucerne ley, LL, during its third growing season. Below-ground biomass increased in both leys during the growing season, although the increase was slow in GL and rapid in LL (Fig. 7). In GL parallel studies with soil coring, decomposition experiments and mesh

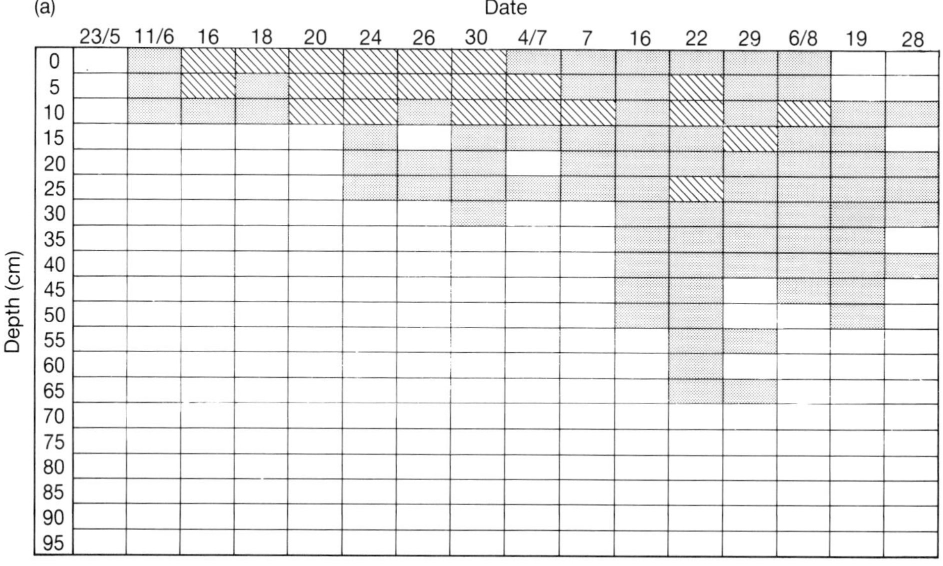


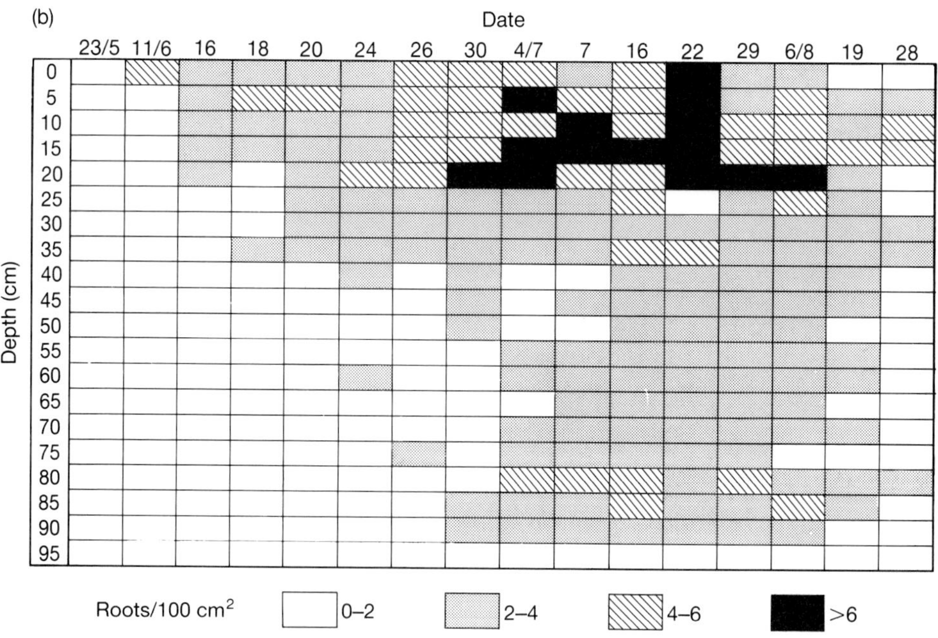


FIG. 5. Number of roots in a 5 × 20 cm square at sixteen dates and twenty depth levels down to 1 m in 1980. Mean of four rhizotrons. (a) B0 = barley without nitrogen fertilizer; (b) B120 = barley with nitrogen fertilizer. From Hansson & Andrén (1987).

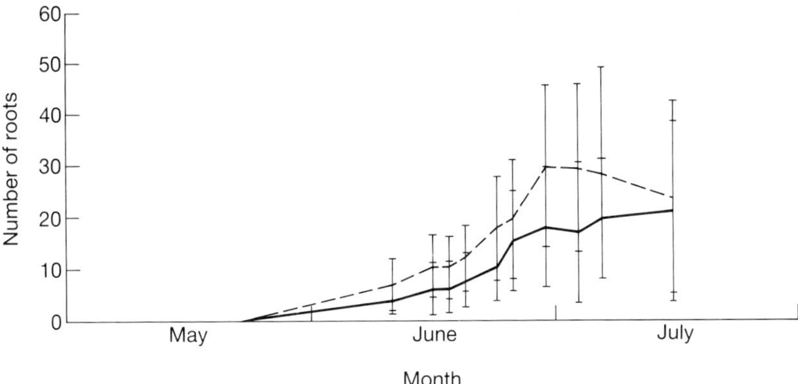


FIG. 6. Number of roots in row (broken line) and between rows of barley (solid line) in a 5 × 20 cm square sown to 1 m in 1980. Mean of eight rhizotrons. Recalculated from Hansson & Andrén (1987).

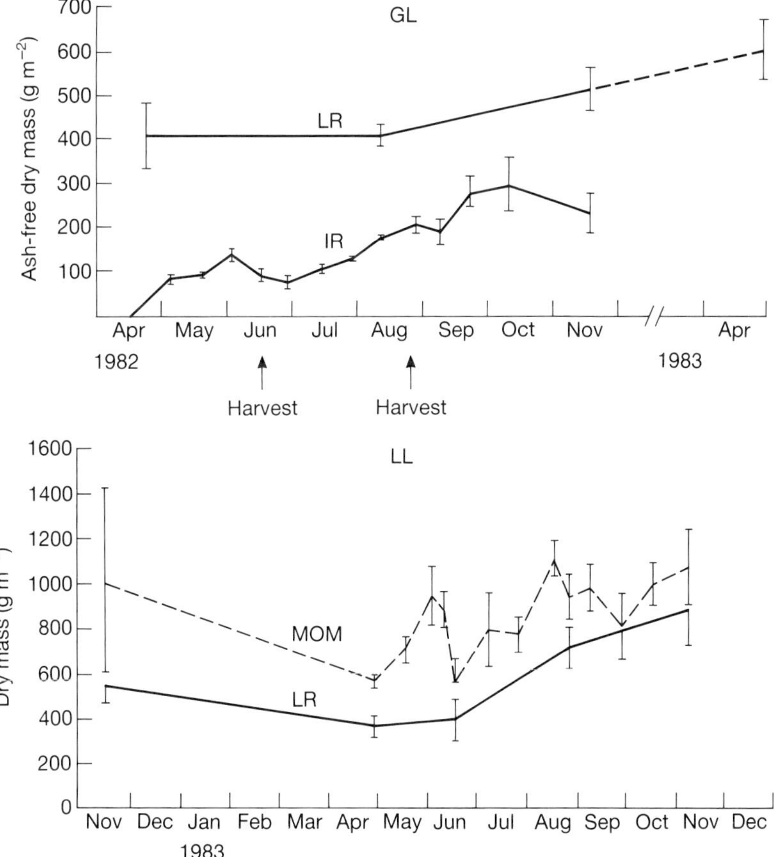


FIG. 7. Root biomass (LR), roots grown into mesh bags (IR), and unsorted organic matter (MOM), consisting of dead organic material and living roots. GL = meadow fescue ley; LL = lucerne ley.

bags showed that the slow increase of root biomass was due to a simultaneous growth and death of roots (Fig. 7) (Hansson & Andrén 1986). From the results of the $^{15}$N-labelling experiment it was concluded that the roots dying during the first half of the growing season, i.e. before the second harvest, were mainly those produced in preceding years, while the roots dying during the autumn were produced in the current year (Fig. 8).

According to estimates based on field samplings, root production in GL ranged between 190 and 460 g m$^{-2}$ of which the latter was considered to be the more realistic estimate (Table 4) (Hansson & Andrén 1986). Results from the $^{15}$N-labelling experiment corroborates root production estimates of the order of 450–500 g m$^{-2}$ (Hansson & Pettersson 1989). The root death during the growing season amounted to 270 g m$^{-2}$. The growth chamber experiments showed that additional 172 g m$^{-2}$ was delivered to the soil as rhizodeposition (Johansson, unpublished). Input to the soil of above-ground harvest residues and litter were of the same magnitude as in the barley treatments. Summing the different flows about 584 g m$^{-2}$ and 11·5 g N m$^{-2}$ was supplied annually to the soil in GL (Table 3) (Paustian *et al.* 1990).

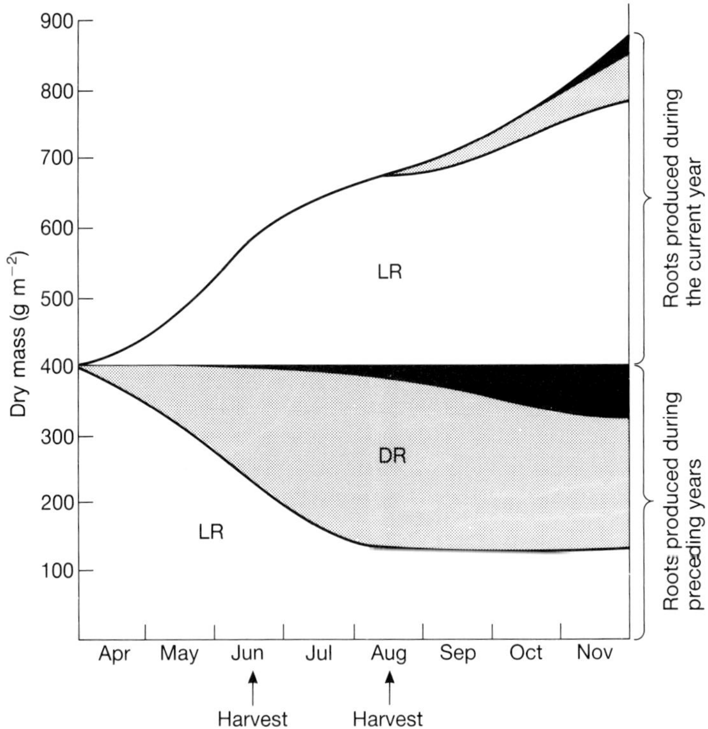


FIG. 8. Hypothetical dynamics of root biomass (LR) and dead roots (DR) produced during preceding years and during the current year in meadow fescue. Black areas represent decomposed material. From Hansson (1987).

TABLE 4. Primary production above and below ground in meadow fescue ley (GL) and lucerne ley (LL). OM = organic matter (g $m^{-2}$); N = nitrogen (g $m^{-2}$); R : S = root : shoot ratio

| | GL | | LL | |
|---|---|---|---|---|
| | OM | N | OM | N |
| Above ground | 1008 | 29·8 | 1104 | 40 |
| Below ground | 460 | 7·8 | 480 | 14 |
| Total | 1468 | 37·6 | 1584 | 54 |
| R : S | 0·46 | 0·21 | 0·43 | 0·35 |

The root studies in LL only utilized one method, i.e. the soil coring method. Due to the lack of supplementary information on growth dynamics and decomposition rates, estimates of root death could not be achieved. Thus, an estimate of root production, 480 g $m^{-2}$, was based on dynamics in the frequently sampled macro-organic material (MOM), consisting of living and dead roots, and other kinds of dead organic matter (Pettersson *et al.* 1986). Neither were growth chamber experiments performed with lucerne. Thus, the proportion of rhizodeposition for GL was used when calculating total input to the soil cultivated with lucerne. Including an input of 256 g $m^{-2}$ as above-ground litter (Pettersson & Hansson 1990), total input of organic matter was estimated to 591 g $m^{-2}$ and 17·4 g N $m^{-2}$ (Table 3) (Paustian *et al.* 1990).

Root biomass in GL was concentrated in the topsoil, i.e. 77% of the below-ground biomass was found in the top 10 cm (Fig. 9). According to the mini-rhizotron studies the mean root depth was only around 20 cm. In spite of this shallow root system, 5% of the below-ground biomass was found in the subsoil, below 27 cm depth.

The morphology of lucerne root system, dominated by a thick tap root, differs radically from the grasses. In spite of this, the relative depth distribution of root

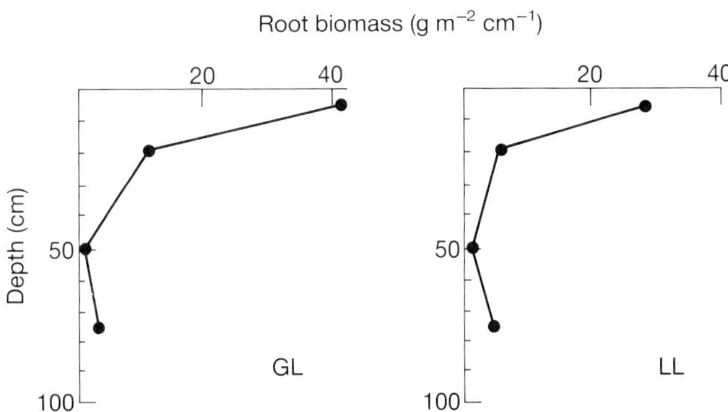


FIG. 9. Depth distribution of root biomass in meadow fescue ley (GL) during its second year of growth, and in lucerne (LL) during its third year of growth.

biomass did not differ significantly between GL and LL, but the amounts found in the subsoil were greater in LL than in GL (Fig. 9).

The quality, e.g. nitrogen concentration, of the organic matter supplied to the soil is considered to have an influence on the characteristics in the associated fauna (Usher 1985; Heal & Dighton 1985). Assuming no major translocations of nitrogen immediately before death, nitrogen concentration in living roots gives an indication of differences in quality of organic matter from the different treatments. Nitrogen concentrations in living roots in B120 were significantly higher than in the other treatments (Fig. 10). Further, the concentration differed considerably between all treatments at the beginning and end of the growing season, and when considering the total input of roots and above-ground crop residues, the mean nitrogen concentration was higher in B120 and LL than in B0 and GL (Table 3).

### *Effects on soil organisms*

The root system can be expected to generally affect soil organisms as a result of the quantity and quality of the input of organic matter and the uptake of nutrients. In addition, the spatial distribution of the roots can affect the distribution of the organisms. The productivity of a crop has indirect effects on the soil environment as transpiration and thus water uptake are positively correlated with the biomass. Thus, the soil in our experimental field usually was much drier under the meadow

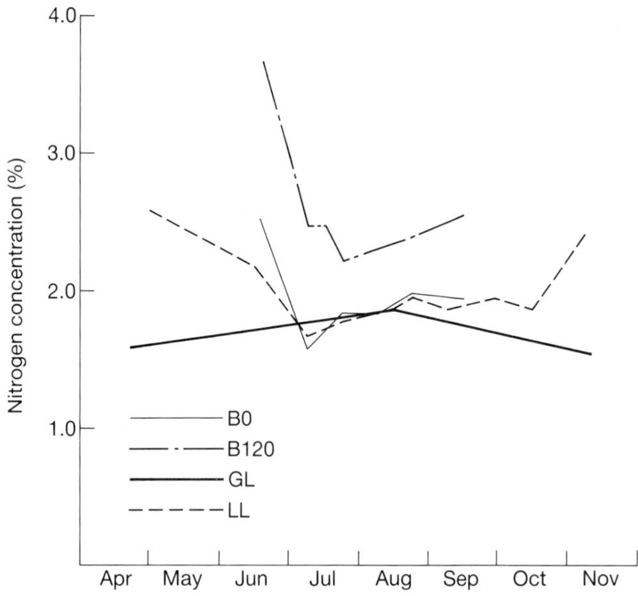


FIG. 10. Nitrogen concentration (%) of ashfree dry mass in living roots. B0 = barley without nitrogen fertilizer, B120 = barley with nitrogen fertilizer, GL = meadow fescue ley, LL = lucerne ley.

fescue than under the fertilized barley, which was drier than under the unfertilized barley (Alvenäs, Johnsson & Jansson 1986). Further, the length of the growing season is of importance both for the soil water content and the supply of organic matter. Finally, different managements result from the production of annual and perennial crops, e.g. annual crops are sown, harvested and the soil ploughed the same growing season, whereas the below-ground standing crop in the perennials increases during the first years, and the soil is not cultivated.

The production of bacteria and fungi is mainly dependent on the input of organic matter from the crop, soil temperature and soil moisture. Determinations of total microbial biomass with the chloroform fumigation method (Jenkinson & Powlson 1976) showed a fairly stable soil biomass, ranging from 390 to 510 $\mu g\,C\,g^{-1}$ dry soil, during the growing season of fertilized barley, B120 (Schnürer 1985). Normally C input from the crop limits microbial growth in arable soils in temperate regions. The average generation time for the microbial biomass in soils cropped with cereals in England, Canada and Sweden has been estimated to range between 1 and 3 years (Jenkinson, Ladd & Rayner 1981; Voroney 1983; Schnürer 1985; Schnürer *et al.* 1985).

The population densities of micro-arthropods in B120 and the grass ley were similar during the summer months, but during the autumn months the density decreased in B120, whereas they remained about the same in GL (Lagerlöf 1987).

Although most Collembola and Acari became more abundant in the leys than in the barley treatments, the species number remained similar during the 5 year period. The lower densities found under barley compared with the leys were probably a result of the more intense soil cultivation in the annual crops (Lagerlöf 1987).

The greatest part of the Collembola fauna in B120 was found in the deeper part of the profile, while in GL it was concentrated near the surface (Lagerlöf & Andrén, in press). Acari were superficially distributed in GL, while they were more uniformly distributed in B120 (Lagerlöf & Andrén 1988). The uniform depth distribution in the barley treatments was mainly a result of the ploughing each autumn, when the crop residues were incorporated to the soil and soil density reduced.

Vertical distribution of enchytraeids was even throughout the sampling depth in B120 but concentrated to the uppermost part of the profile in GL (Lagerlöf, Andrén & Paustian 1989). The concentration of enchytraeids in the top layer in GL may be due to root distribution, which was more superficial in GL than in the other crops (cf. Figs 4 and 9). Since they are sensitive to desiccation, the shallow depth distribution in the leys, especially in GL, left them more exposed to drought than the deep dwelling enchytraeids in the barley crops. During the warm and dry summers of 1982 and 1983 low enchytraeid abundance was observed in GL (Lagerlöf, Andrén & Paustian 1989). Further, the lower soil moisture in B120 than in B0 (Alvenäs, Johnsson & Jansson 1986) may explain the lower enchytraeid abundance in B120 towards the end of the experiment.

## CONCLUDING REMARKS

In general the management connected with each crop, i.e. fertilization and soil cultivation had a more significant effect on soil animal abundance and distribution than the size and distribution of the root system. Further, indirect effects, such as low soil water content due to high transpiration, had significant effects on the organisms.

The crops were selected to represent a wide range of carbon inputs, which would cause significant differences in microbial and faunal biomass and activity. However, the input of organic matter, as long as the leys remained intact, differed much less than could be expected from the net total production. Greater differences occur when the leys are ploughed, and a substantial input consisting of below-ground biomass and above-ground stubble will be supplied.

## ACKNOWLEDGEMENTS

The authors wish to thank their colleagues in the project for access to unpublished data. The project was financed by the Swedish Council for Planning and Coordination of Research, the Swedish Council for Forestry and Agricultural Research, the Swedish Natural Science Research Council and the Swedish Environmental Protection Board.

## REFERENCES

**Ågren, G.I. (1985).** Theory of growth of plants derived from the nitrogen productivity concept. *Physiologia Plantarum*, **64**, 17–28.

**Alvenäs, G., Johnsson, H. & Jansson, P.-E. (1986).** Meteorological conditions and soil climate of four cropping systems: Measurements and simulations from the project 'Ecology of Arable Land'. Department of Ecology and Environmental Research, Report 24. Swedish University of Agricultural Sciences, Uppsala.

**Andersen, A. (1986).** Rodvækst i forskellige jordtyper. *Tidsskrift for Planteavls Specialserie*, S **1827**.

**Andrén, O. (1988).** Ecology of Arable Land — an integrated project. *Ecological Bulletins*, **39**, 131–133.

**Andrén, O., Lindberg, T., Paustian, K. & Rosswall, T. (1989).** Ecology of Arable Land. Organisms, carbon and nitrogen cycling. *Ecological Bulletins*, **40**. Copenhagen, in press.

**Barber, D.A. & Martin, J.K. (1976).** The release of organic substances by cereal roots into soil. *New Phytologist*, **76**, 69–80.

**Bergström, L. (1987).** Leaching of nitrate and drainage from annual and perennial crops in tile-drained plots and lysimeters. *Journal of Environmental Quality*, **16**, 11–18.

**Biscoe, P.V., Scott, R.K. & Monteith, J.L. (1975).** Barley and its environment. III. Carbon budget of the stand. *Journal of Applied Ecology* **12**, 269–293.

**Böhm, W. (1979).** Methods of studying root systems. *Ecological Studies*, Vol. 33. Springer Verlag, Berlin.

**Brouwer, R. (1962).** Nutritive influences on the distribution of the dry matter in the plant. *Netherlands Journal of Agricultural Sciences*, **10**, 399–408.

**Cahoon, G.A. & Morton, E.S. (1960).** An apparatus for the quantitative separation of plant roots from soil. *Proceedings from the American Society for Horticultural Science*, **78**, 593–596.

**FAO-UNESCO (1974).** *Soil map of the world*, Volume I.

**Fehrenbacher, J.B., Ray, B.W. & Alexander, J.D. (1969).** How soils affect root growth. *Crop and Soil*, **21**, 14–18.

**Gallagher, J.N. & Biscoe, P.V. (1978).** A physiological analysis of cereal yield. II. Partitioning of dry matter. *Agricultural Progress*, **53**, 51–70.

**Gregory, P.J., Marshall, B & Biscoe, P.V. (1981).** Nutrient relations of winter wheat. III. Nitrogen uptake, photosynthesis of flag leaves and translocation of nitrogen to grain. *Journal of Agricultural Science*, **96**, 539–547.

**Groth, R. (1987).** Messungen zum wurzelwachstum von Winterweizen auf Tieflehm-Fahlerde. *Archiv für Acker- und Pflanzenbau und Bodenkunde*, **31**, 197–204.

**Hansson, A.-C. (1987).** Roots of arable crops: Production, growth dynamics and nitrogen content. Ph.D. thesis, Department of Ecology and Environmental Research, Report 28. Swedish University of Agricultural Sciences, Uppsala.

**Hansson, A.-C. & Steen, E. (1984).** Methods of calculating root production and nitrogen uptake in an annual crop. *Swedish Journal of Agricultural Research*, **14**, 191–200.

**Hansson, A.-C. & Andrén, O. (1986).** Below-ground plant production in a perennial grass ley (*Festuca pratensis* Huds.) assessed with different methods. *Journal of Applied Ecology*, **23**, 657–666.

**Hansson, A.-C. & Andrén, O. (1987).** Root dynamics in barley, lucerne and meadow fescue investigated with a mini-rhizotron technique. *Plant and Soil*, **103**, 33–38.

**Hansson, A.-C. & Pettersson, R. (1989).** Uptake and above- and below-ground allocation of soil mineral-N and fertilizer-$^{15}$N in a perennial grass ley (*Festuca pratensis* Huds.). *Journal of Applied Ecology*, **26**, 259–271.

**Hansson, A.-C., Pettersson, R. & Paustian, K. (1987).** Shoot and root production and nitrogen uptake in barley, with and without nitrogen fertilization. *Zeitschrift für Acker- und Pflanzenbau*, **158**, 163–171.

**Heal, O.W. & Dighton, J. (1985).** Resourse quality and trophic structure in the soil system. *Ecological Interactions in Soil. Plants, Microbes and Animals* (Ed. by A.H. Fitter, D. Atkinson, D.J. Read, & M.B. Usher), pp. 339–354. *Special Publications Series of the British Ecological Society, No 4*. Blackwell Scientific Publications, Oxford.

**Jenkinson, D.S. & Powlson, D.S. (1976).** The effects of biocidal treatments on metabolism in soil. — V. A method for measuring soil biomass. *Soil Biology and Biochemistry*, **8**, 209–213.

**Jenkinson, D.S., Ladd, J.N. & Rayner, J.H. (1981).** Microbial biomass in soil — measurement and turnover. *Soil Biochemistry* (Ed. by E.A. Paul & J.N. Ladd) Vol. 5, pp. 415–471. Marcel Dekker, New York.

**Keith, H., Oades, J.M. & Martin, J.K. (1986).** Input of carbon to soil from wheat plants. *Soil Biology and Biochemistry*, **18**, 445–449.

**Keulen, H. van & Seligman, N.G. (1987).** *Simulation of water use, nitrogen nutrition and growth of a spring wheat crop*. 310 pp. PUDOC, Wageningen.

**Lagerlöf, J. (1987).** *Ecology of soil fauna in arable land. Dynamics and activity of microarthropods and enchytraeids in four cropping systems*. Ph.D. thesis, Department of Ecology and Environmental Research, Report 29. Swedish University of Agricultural Sciences, Uppsala.

**Lagerlöf, J. & Andrén, O. (1988).** Abundance and activity of soil mites (Acari) in four cropping systems. *Pedobiologia*, **32**, 129–145.

**Lagerlöf, J. & Andrén, O. (1991).** Abundance and activity of Collembola, Protura and Diplura (Insecta, Apterygota) under four arable crops. *Pedobiologia*, in press.

**Lagerlöf, J., Andrén, O. & Paustian, K. (1989).** Dynamics and contribution to carbon flows of Enchytraeidae (Oligochaeta) under four cropping systems. *Journal of Applied Ecology*, **26**, 183–199.

**Martin, J.K. & Kemp, J.R. (1986).** The measurement of C transfers within the rhizosphere of wheat grown in field plots. *Soil Biology and Biochemistry*, **18**, 103–107.

**Mitchell, M.J. & Nakas, J.P.** (Eds.) **(1986).** *Microfloral and Faunal Interactions in Natural and Agro-ecosystems*. Martinus Nijhoff/Dr W. Junk Publisher, Dordrecht.

**Nielsen, N.E. (1980).** Forløbet af rodudvikling, næringsstofoptagelse og stofproduktion hos byg, dyrket på frugtbart morænelerjord. *Planternes Ernæring, Kgl. Veterinær- og Landbohøjskole*, Report No. 1119, København.

**Paustian, K., Andrén, O., Clarholm, M., Hansson, A.-C., Johansson, G., Lindberg, T., Pettersson, R. & Sohlenius, B. (1990).** Carbon and nitrogen budgets of four agroecosystems with annuals and perennial crops, with and without N fertilization. *Journal of Applied Ecology*, **27**, 60–84.

**Pettersson, R. (1987).** *Primary production in arable crops: Above-ground growth dynamics, net production and nitrogen uptake.* Ph.D. thesis, Department of Ecology and Environmental Research, Report 31. Swedish University of Agricultural Sciences, Uppsala.

**Pettersson, R. (1989).** Above-ground growth dynamics and net production of spring barley in relation to nitrogen fertilization. *Swedish Journal of Agricultural Research*, **19**, 135–145.

**Pettersson, R. & Hansson, A.-C. (1990).** Net primary production of a perennial grass ley (*Festuca pratensis*) assessed with differential methods and compared with a lucerne ley (*Medicago sativa*). *Journal of Applied Ecology*, **27**, 788–802.

**Pettersson, R., Hansson, A.-C., Andrén, O. & Steen, E. (1986).** Above- and below-ground production and nitrogen uptake in Lucerne (*Medicago sativa*). *Swedish Journal of Agricultural Research*, **16**, 167–177.

**Rodskjer, N. & Tuvesson, M. (1975).** Observations of temperature in winter wheat at Ultuna, Sweden, 1968–72. *Swedish Journal of Agricultural Research*, **5**, 223–234.

**Sauerbeck, D. & Johnen, B. (1977).** Root formation decomposition during plant growth. *Symposium on Soil Organic Matter Studies*, pp. 141–148.

**Schnürer, J. (1985).** Fungi in arable soil — Their role in carbon and nitrogen cycling, Ph.D. thesis, Department of Microbiology, Report 29. Swedish University of Agricultural Sciences, Uppsala.

**Schnürer, J., Clarholm, M. & Rosswall, T. (1985).** Microbial biomass and activity in an agricultural soil with different organic matter contents. *Soil Biology and Biochemistry*, **17**, 611–618.

**Schnürer, J., Clarholm, M. & Rosswall, T. (1986).** Fungi, bacteria and protozoa in soil from four arable cropping systems. *Biology and Fertility of Soils*, **2**, 119–126.

**Sohlenius, B., Boström, S. & Sandor, A. (1987).** Long-term dynamics of nematode communities in arable soil under four cropping systems. *Journal of Applied Ecology*, **24**, 131–144.

**Steen, E. (1985).** Root and rhizome dynamics in a perennial grass crop during an annual growth cycle. *Swedish Journal of Agricultural Research*, **15**, 25–30.

**Steen, E., Jansson, P.-E. & Persson, J. (1984).** Experimental site of the 'Ecology of Arable Land' project. *Acta Agriculturae Scandinavica*, **34**, 153–166.

**Troughton, A. (1962).** The roots of temperate cereals (wheat, barley, oats and rye). *Commonwealth Bureau of Pasture and Crops*, mimeographed publications No. 2. Hurley, Berkshire.

**USDA Soil Survey Staff (1975).** *Soil Taxonomy.* A basic system of soil classification for making and interpreting soil surveys. *Agriculture Handbook* No 436. Soil Conservation Service, US Department of Agriculture.

**Usher, M.B (1985).** Population and community dynamics in the soil ecosystem. *Ecological Interactions in Soil. Plants, Microbes and Animals* (Ed. by A.H. Fitter, D. Atkinson, D.J. Read & M.B. Usher), pp. 339–354. *Special Publications Series of the British Ecological Society, No 4.* Blackwell Scientific Publications, Oxford.

**Wallén, C.C. (1966).** Global solar radiation and potential evapotranspiration in Sweden. *Tellus*, **18**, 786–800.

**Weaver, J.E. (1926).** *Root Development in Field Crops.* McGraw-Hill, New York.

**Welbank, P.J. & Williams, E.D. (1968).** Root growth of a barley crop estimated by sampling with a portable powered soil-coring equipment. *Journal of Applied Ecology* **5**, 477–481.

**Welbank, P.J., Taylor, P.J. & Williams, E.D. (1974).** Root growth of cereal crops. *Report from Rothamsted Experimental Station*, **2**, 236–241.

**Voroney, R.P (1983).** *Decomposition of Crop Residues.* Ph.D. thesis. University of Saskatchewan, Oxford.

**Vose, P.B. (1980).** *Introduction to Nuclear Techniques in Agronomy and Plant Biology*, pp. 391. Pergamon Press, New York.

# Root turnover and nitrogen cycling in low input grass/clover ecosystems

P.J. GOODMAN
*AFRC Institute for Grassland and Animal Production, Welsh Plant Breeding Station, Plas Gogerddan, Aberystwyth, SY23 3EB, Dyfed, UK*

## SUMMARY

**1** White clover is often used to improve native pasture, but no adequate explanation is available for the mechanism of improvements.

**2** Evidence suggests that nitrogen is released by clover, but the pathway of this nitrogen from clover into the ecosystem is incompletely understood. Several possibilities exist, including excretion of N by clover roots, nodule shedding, root decomposition, and cycling through animals. The relative importance of these processes is not known.

**3** Four experiments were designed to trace the transfer pathway of clover nitrogen.

**4** In the first experiment, which lasted 3 years on low input upland pastures, herbage accumulation and nitrogen parameters were found to be very variable, with the exception of $^{15}N$ recovery. Nitrogen was actually lost by plants in old sward in the dry season of 1984, a process that was unlikely to be sustained.

**5** A second experiment with regular destructive harvests showed that in grass/clover mixtures, grass stubble plus root was greater in early summer but then returned to a lower, apparently steady, level. Loss of grass material to the soil was therefore not abnormal. Clover stolon plus root was greatest in mid-season, but declined in autumn. Grass and clover thus contributed to the soil organic matter at different times of the year.

**6** In the third experiment, a replacement series of perennial ryegrass and red clover showed that in the first year negligible nitrogen transfer occurred. There was strong competition between grass and clover for soil nitrogen. Grass increased in the presence of clover, but clover was little affected by grass. First results from a 14-month-long white clover replacement series showed that transfer of N from clover to grass occurred after the first winter.

**7** In the fourth experiment, using $^{15}N$-labelled urea to simulate urine deposition by sheep in the field, grass dry matter production was increased for about 100 days after treatment. Clover herbage accumulation decreased to a minimum at 100 days, but recovered after 140 days.

**8** Competition between grass and clover for soil nitrogen (released from soil organic matter) is suggested as an important feature of low input pasture. Flow rates of nitrogen now need to be determined, and work begun on manipulating flows by changes in management.

## INTRODUCTION

Native pastures often contain legumes as well as grasses. This association has been exploited in many parts of the world by deliberately introducing forage legumes into marginal grazing land. Productivity of the grass pasture is thereby increased, using the nitrogen fixing legume as a substitute for nitrogen fertilizer.

Among the most successful of the forage legumes are the annual and perennial *Trifolium* species (Sprent 1979), which are usually grown in mixtures with grasses, forming a distinct agricultural ecosystem. Annual species of legume, such as *Trifolium subterraneum*, die after setting and maturing the seed. Runners and roots then decompose, and nitrogen is made available to the companion grass before the next generation of clover plants is established (Simpson 1976).

The mechanism and timing of return of nitrogen to the soil are much less obvious in perennial forage legumes such as the European white clover, *T. repens*, because the plant overwinters, and so retains at least part of its nitrogen. How and when does fixed nitrogen from *T. repens* become available to the ecosystem?

Our work set out to discover the destiny of nitrogen after fixation by the clover plant. In the absence of grazing animals, several possible pathways exist for transferring clover nitrogen to grass: e.g. Walker, Orchiston & Adams (1954) suggested that legumes 'excrete' nitrogen to companion grasses, but we have found no evidence of this. According to Simpson (1965) such a mechanism is probably restricted to specific environmental conditions. Indeed, when we have detected transfer of fixed nitrogen from white clover to companion grass (Goodman & Collison 1986), the amounts have been variable, usually small, and as yet, quite unpredictable. When inorganic $^{15}$N-labelled tracers are added to the system, only about 25% is recovered by plants, so direct mineral pathways appear to be unimportant. Most of the applied mineral nitrogen does not move immediately into plants.

Lack of simple pathways for nitrogen from clover to grass suggests that transfer involves more complex substances, probably organic material which is decomposed and recycled. Wilson (1942) suggested that nodules were shed, but we have not observed such an organized process.

Our present explanation of transfer inclines to processes of root senility and death which passively include nodules. Haystead & Marriott (1979) reached similar conclusions from pot experiments, and regarded defoliation as the likely trigger for root death. Another possibility is that white clover plants die back in winter, either as a result of cold temperatures or lack of solar radiation (Ollerenshaw 1983). This view is supported by the work of Simpson (1965) who suggested that white clover competed with grass in the (Australian) summer but 'donated' nitrogen to grass during winter. Dead clover nitrogen from winter could thus be mineralized and become available to both grass and clover plants in the following spring. If the pathway of nitrogen transfer is root turnover in the field in winter, accurate measurement of biomass change and decomposition could be difficult.

This paper details our observations on the grass/clover system starting in the field, using the stable isotope $^{15}N$ to discover pathways of fixed nitrogen from the legume to its grass companion. Upland pastures have been deliberately included because in Britain these marginal areas have been most successfully improved by introduction of clover. After making observations in the field, two biological models were tested, one simulating competition between grass and clover for nitrogen; the other simulating return of urine nitrogen by the grazing animal. Both models were tested using $^{15}N$.

## *Field observations on herbage production, nitrogen fixation, uptake and transfer in low input grass/clover swards (experiment 1)*

Although the potential production of grassland in Britain is as large as 25 000 kg ha$^{-1}$ year$^{-1}$, average dry matter production may be only 7000 kg ha$^{-1}$ year$^{-1}$ (Cooper & Breese 1971). Production in low-nitrogen-input permanent grassland, particularly in upland areas, is usually smaller than average, 5000 kg ha$^{-1}$ year$^{-1}$ or less, but may be substantially increased by the introduction of white clover to around 6000 kg ha$^{-1}$ year$^{-1}$ (Hopkins *et al.* 1985). Nitrogen supply is usually considered to be limiting.

In our field experiments, we measured herbage production and nitrogen turnover in 3 years at three different sites in Wales: a lowland (altitude 30 m), and an upland area (300–350 m), both on brown earth soils of the Denbigh series; and another upland area (300–350 m) on a peaty gley soil of the Ynys series (Goodman & Collison 1986).

Herbage accumulation by clover plus grass at the different sites in the different years ranged from 1160 (upland peat in 1985) to 8520 kg ha$^{-1}$ year$^{-1}$ (lowland soil in 1983) (Table 1). Clover plus grass in lowland swards generally exceeded upland

TABLE 1. Sowing dates and dry weights (g m$^{-2}$) of accumulated herbage in grass/clover mixtures on different soils in different years

| | Lowland soil | Upland soil | Upland peat | S.E. |
|---|---|---|---|---|
| 1983 | | | | |
| Sowing date | 1982 | 1982 | 1982 | |
| Grass dry wt | 620 | 712 | 596 | N.S. |
| Clover dry wt | 232 | 39 | 82 | 14 |
| 1984 | | | | |
| Sowing date | 1982 | 1967 | 1967 | |
| Grass dry wt | 157 | 76 | 140 | 19 |
| Clover dry wt | 60 | 76 | 75 | N.S. |
| 1985 | | | | |
| Sowing date | 1983 | 1980 | 1982 | |
| Grass dry wt | 127 | 173 | 87 | 19 |
| Clover dry wt | 215 | 87 | 29 | 16 |

swards, but sward age was less important than season. In 1984, a drier than average year, herbage accumulation was small at all sites.

Nitrogen fixation in herbage and in stubble plus roots was measured by isotope dilution using $^{15}N$ (Goodman & Collison 1986). About 5 kg $^{15}N$ ha$^{-1}$ year$^{-1}$ was applied within open-ended stainless steel cylinders of 28 cm diameter and 5 cm depth. The results were very variable, in the range from almost 0 to 260 kg N ha$^{-1}$ year$^{-1}$ (Table 2). In earlier experiments a similar but slightly narrower range (15–104 kg N ha$^{-1}$ year$^{-1}$) was found in pastures with a known history of improvement (Goodman & Collison 1984). These variations were not apparently correlated with soil type or age of sward but, except in the dry season of 1984, nitrogen fixation at the lowland site was greater than at the upland sites.

Recovery of $^{15}N$ applied to the sward was much more uniform than either herbage production or nitrogen fixation (Table 3). Uptake of the isotope by grass was several times greater than uptake by clover. Similar results were obtained in earlier experiments using $^{32}P$ labelling in the soil (Goodman & Collison 1982). Phosphate labelling demonstrated the strong competition for uptake by grass

TABLE 2. Nitrogen fixation (g m$^{-2}$) by clover mixed with grass, on different soils in different years

| | Lowland soil | Upland soil | Upland peat | S.E. |
|---|---|---|---|---|
| 1983 | | | | |
| Leaves | 5·4 | 0·4 | 2·3 | 1·0 |
| Stubble + root | 5·6 | 0·1 | 0·9 | 0·5 |
| 1984 | | | | |
| Leaves | 2·2 | 0·7 | 0·1 | N.S. |
| Stubble + root | n.a. | 0·1 | 26·0 | 1·5 |
| 1985 | | | | |
| Leaves | 7·4 | 3·0 | 1·2 | 1·3 |
| Stubble + root | 4·9 | 1·5 | 1·5 | 0·4 |

TABLE 3. Recovery of applied $^{15}N$ (%) in grass and clover components of mixtures, on different soils in different years

| | Lowland soil | Upland soil | Upland peat | S.E. |
|---|---|---|---|---|
| 1983 | | | | |
| Grass | 16·4 | 21·5 | 16·4 | 1·6 |
| Clover | 6·2 | 1·6 | 1·6 | 0·3 |
| 1984 | | | | |
| Grass | 6·2 | 7·2 | 10·3 | 0·3 |
| Clover | 1·0 | 0·3 | 0·6 | N.S. |
| 1985 | | | | |
| Grass | 13·8 | 16·1 | 12·7 | N.S. |
| Clover | 4·4 | 2·7 | 1·5 | 0·8 |

in mixtures with clover. Similar competition must occur for nitrogen, and is apparently stronger than the effects of site, season or sward age. Recoveries ranged between 0.3 and 6.2% of the applied $^{15}N$ for clover and between 10·3 and 21·5% for grass. Although these were wide ranges, they did not overlap. More than 75% of the $^{15}N$ applied must have been lost by leaching, in gaseous form, or incorporation into the soil organic matter. Recovery values were much smaller than those recorded by Ryden (1984) for fertilized perennial ryegrass. We cannot, as yet, explain this large difference between recovery values on fertilized English downland pasture and our results on unfertilized, mainly upland Welsh pasture. Possible explanations include nitrogen incorporation into soil organic matter which is known to be rapid, even in lowland (Bristow, Ryden & Whitehead 1987); or loss by leaching in the higher rainfall of the western uplands.

Uptake of nitrogen from the soil by grass and clover plants, calculated from isotope dilution, when the only added fertilizer was 5 kg $^{15}N$ ha$^{-1}$ year$^{-1}$, was small and even included negative values (Table 4). These negative values show that, in old swards, plants can lose nitrogen to the soil. This loss was probably caused by leaf litter fall or death of roots, in the dry season of 1984. No clear pattern of nitrogen transfer from clover to grass could be established (Table 5). In these experiments the only clue that was found which might explain transfer was the loss of nitrogen from plants in the old sward. Could a similar loss occur in an organized way during sward growth, but at a time not noticed in previous sampling, for example, in winter? Were there losses from clover as well as from grass?

### *Stubble and root turnover in low input grass/clover swards (experiment 2)*

An investigation of possible losses (or gains) in stubble plus root dry weight and nitrogen content in grass and clover in mixed sward was begun on a lowland site in 1987. Six destructive harvests were taken, from April to November. Herbage was removed before and after each measurement interval (Figs 1 and 2). In the

TABLE 4. Uptake of soil nitrogen (g m$^{-2}$) by grass and clover components of mixtures, on different soils in different years

| | Lowland soil | Upland soil | Upland peat | S.E. |
|---|---|---|---|---|
| 1983 | | | | |
| Grass | 2·0 | 11 | 5 | 0·5 |
| Clover | 0·6 | 0·4 | 0·3 | N.S. |
| 1984 | | | | |
| Grass | 6 | −11 | −6 | 1 |
| Clover | 2·6 | −1·0 | −0·3 | N.S. |
| 1985 | | | | |
| Grass | 5·1 | 9·5 | 8·8 | 1·1 |
| Clover | 1·4 | 1·9 | 0·9 | 0·3 |

TABLE 5. Nitrogen transfer (g $m^{-2}$) from clover to grass, in different soils in different years

| | Lowland soil | Upland soil | Upland peat | S.E. |
|---|---|---|---|---|
| 1983 | −0·9 | 13·2 | 0·1 | N.S. |
| 1984 | 3·7 | 1·1 | 25·8 | 1·4 |
| 1985 | −0·2 | 2·3 | −1·6 | 0·9 |

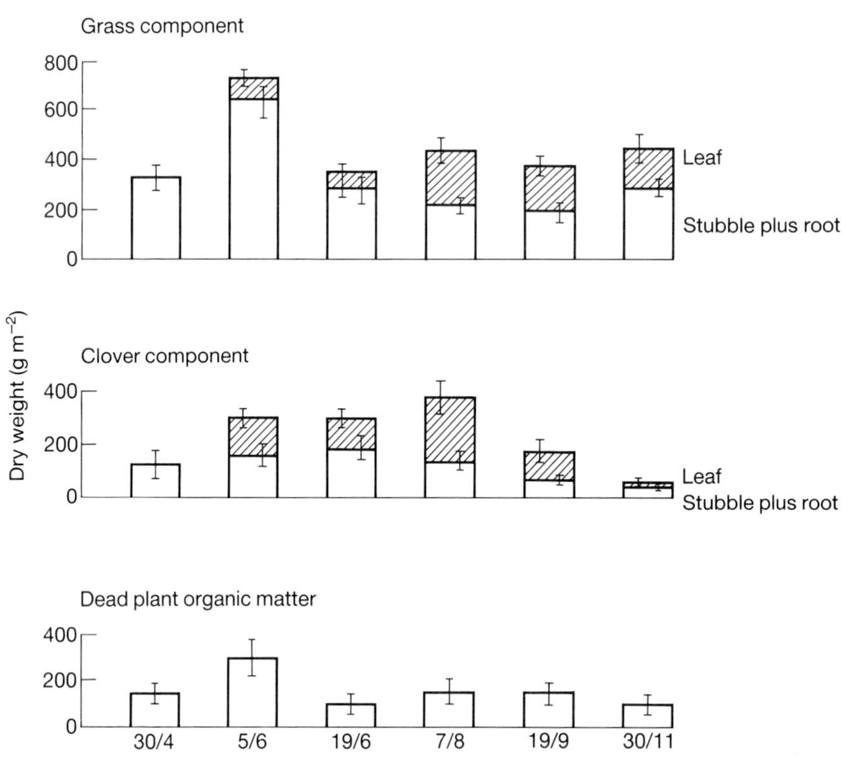


FIG. 1. Accumulated herbage; instantaneous stubble plus root, and recognizable but dead plant organic matter in soil (dry weight g $m^{-2}$) for six harvests on lowland soil in 1987 (L.S.D. = 5%).

grass component of the mixed sward, stubble and root were greatest in dry weight and in nitrogen content (g $m^{-2}$) in early June, followed by a sudden decline to a 'steady state' after mid-June. Clover stubble and root showed a different seasonal pattern: there was a gradual build-up of stolon and root until mid-June, followed by a decline to very small values at the end of the season.

Dead plant material, which was recognizable but not easily distinguished as clover or grass, accumulated at a time which coincided with the decline in grass

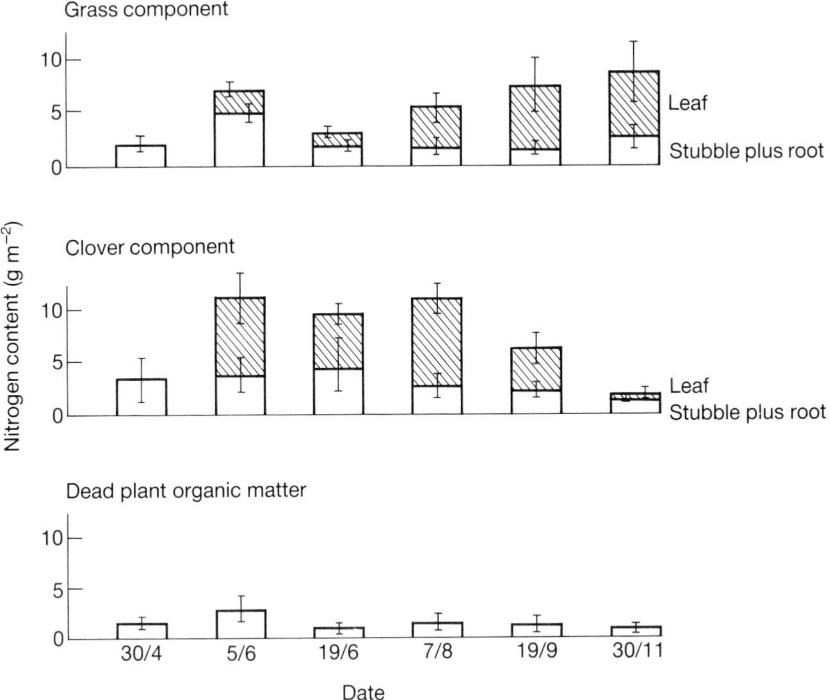


FIG. 2. Accumulated herbage; instantaneous stubble plus root, and recognizable but dead plant organic matter in soil (nitrogen, g m$^{-2}$) for six harvests on lowland soil in 1987 (L.S.D. = 5%).

stubble plus root in early June. After June less dead plant material was found in the soil, both as dry weight, and as nitrogen. Presumably warmer temperatures by mid-June brought about the decomposition and mineralization of the earlier recognizable plant material.

Soil organic matter was evidently increased by grass stubble and root death in early summer whereas clover contributed to soil organic matter in autumn (Fig. 1). In another context, Ryden (1984) made the surprising statement that, in grassland, accumulation of nitrogen in the soil organic matter is essentially independent of the input of nitrogen. That author considered that carbon was the limiting factor in soil organic matter which controlled nitrogen accumulation. If this is true, then each year carbon from grass must form a microbiological and biochemical matrix into which grass and clover nitrogen is incorporated, adding to the existing store of soil organic matter.

Soil organic matter undoubtedly accumulates under grass/clover pastures. Garwood *et al.* (1977) reported that, under grazed ryegrass–clover sward, nitrogen in soil organic matter increased asymptotically at an annual rate which decreased from 95 to 28 kg N ha$^{-1}$ year$^{-1}$ over a 15 year period. Could it be that no immediate pathway of transfer of fixed nitrogen exists, but that grass and clover contribute nitrogen to the soil organic matter in early summer and early winter, respectively? In the following spring, both grass and clover would then draw nitrogen from a common mineral pool derived from soil organic matter accumulated over a (possibly long) time period. The growth of grass and clover in spring would then depend on root competition between the two species for nutrients, especially nitrogen, fixed in the previous season.

### *Competition between grass and clover for nitrogen (experiment 3)*

The third experiment in this series concerned competition between grass and clover. Methods for studying competition between species have been discussed in detail by Hall (1978). A highly relevant earlier experiment on competition for

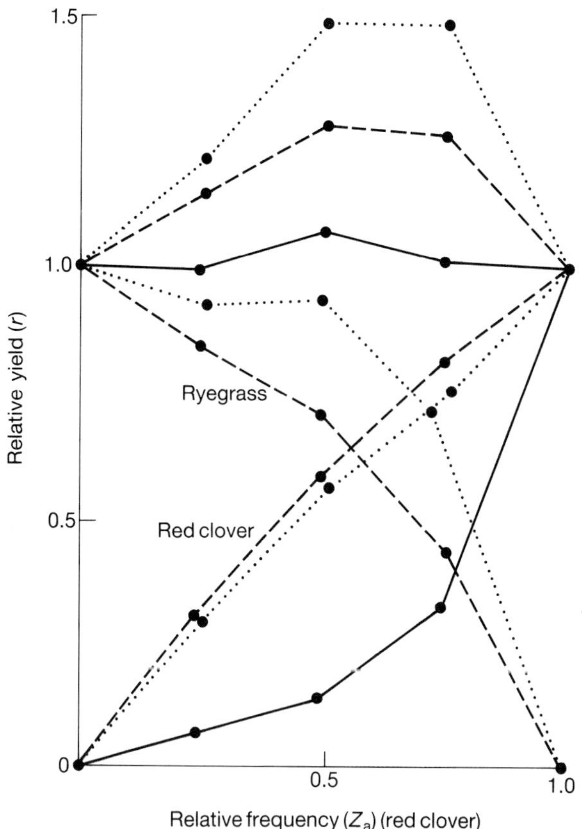


FIG. 3. Replacement diagram in which relative yields ($r_a$ or $r_b$) of dry matter (dashed line), total nitrogen (dotted line) and soil nitrogen (solid line) for red clover (species a) and perennial ryegrass (species b) and relative total yields are plotted against the relative frequency ($Z_a$ or $Z_b$) of each species. Accumulated herbage plus stubble plus roots after 28 weeks.

nitrogen between Rhodes grass (*Chloris gayana* Kunth.) and the Australian forage legume *Stylosanthes humilis* H.B.K. described by Vallis *et al.* (1967) was analysed more completely by Hall (1974), and Begon & Mortimer (1986). The experiment was designed to assess the intensity of competition for soil nitrogen between the two species, in an association where the legume was able to use an alternative resource: symbiotically fixed nitrogen. The authors set up a replacement series in which each pot contained the same number of plants, but the ratio of the two species varied: 1:0, 0·25:0·75, 0·50:0·50, 0·75:0·25, 0:1. This fulfills the requirement of a replacement series that each treatment added up to unity.

Uptake of soil nitrogen was distinguished from symbiotic fixation in the legume by labelling the mineralized soil nitrogen with $^{15}N$. We undertook a similar investigation of competition for nitrogen between perennial ryegrass and *Trifolium* species. For this we used red clover rather than white clover, because plant ratios are easier to maintain in the former than in the latter species. Eleven herbage collections were made from the plants, between March and September, each followed by the addition of 200 μg labelled N, per box containing twenty-four plants. With modern mass spectrometers it is possible to give much less nitrogen than was used by Vallis *et al.* (1967), so that the results are even more typical of low input systems.

Although herbage was cut eleven times during the growing season, transfer of nitrogen from clover to grass was negligible. This result clearly shows that transfer was not triggered by defoliation.

Data for dry matter yields of the two species, including accumulated herbage and stubble plus roots collected in September, are shown in Fig. 3. Yields are expressed relative to monocultures. Relative yields of the two species and total relative yields (of grass plus clover) are plotted against relative frequency of the species in the mixture. Relative crowding coefficients have also been calculated, together with their product (Table 6).

In these very low nutrient conditions, the relative growth of ryegrass was distinctly increased in the presence of red clover. Relative growth of red clover, presumably because of its access to an alternative source of nitrogen was scarcely affected by the presence of ryegrass. Ryegrass competed very strongly for soil nitrogen, taking most, whereas red clover took up much less of this nutrient from the soil. Substantial fixation of nitrogen occurred, supplementing the soil supply available to the clover.

TABLE 6. Relative crowding coefficients for red clover ($k_{ab}$) and perennial ryegrass ($k_{ba}$) and their product ($k_{ab} \times k_{ba}$) derived by using the best squares method of Thomas (1970)

| | Red clover | Perennial ryegrass | |
|---|---|---|---|
| | ($k_{ab}$) | ($k_{ba}$) | ($k_{ab} \times k_{ba}$) |
| Dry matter yield | 1·47 | 2·65 | 3·90 |
| Total nitrogen | 1·41 | 2·27 | 3·20 |
| Soil nitrogen | 0·60 | 2·27 | 1·36 |

The product of the crowding coefficients of the two species, and hence the relative yield total for soil nitrogen is close to unity, suggesting that this resource is potentially available to both species. Where some part of a resource, for example, total nitrogen, is not available to both species, the relative yield of the species with single access to a resource (atmospheric N by fixation in red clover, as here) and the relative yield product may exceed unity. However, the small value of $K_{ab}$ for soil nitrogen shows that strong competition is occurring, and that soil nitrogen would be a limiting factor for clover growth if clover was not able to supplement the nitrogen supply by fixation.

White clover replacement experiments are more difficult to maintain, because growth of stolons changes the proportions of clover to grass. However, in addition to the red clover experiment, a smaller scale but longer term (14 month) experiment was set up with white clover and perennial ryegrass at only three ratios (1 : 0, 0·5 : 0·5, 0 : 1) in a replacement series. First results of this experiment showed that, after the first winter, as much as 50% of grass nitrogen content had come from the clover. This result resembles those of some previous workers. In a shorter (6 month) experiment, which began in winter, de Wit, Tow & Ennik (1966) reported that white clover roots had 'deteriorated' and that transfer from them to ryegrass had occurred. Simpson (1965) reported similar results over an Australian winter.

Two features emerge from these replacement series: there is competitive interference for soil nitrogen between grass and clover. Soil nitrogen could be the limiting factor governing the yields of the two species in mixture. Non-competitive interference also occurs between the species and is evidently sustained by the alternative nitrogen supply from atmospheric nitrogen. Although there is evidently no measurable transfer of nitrogen from red clover to grass during the first season, and defoliation is not the trigger, transfer occurred from white clover after winter. It is not yet clear whether this deviant result shows that a difference exists in transfer between white and red clover or that transfer starts in both species after the first winter. Overwintering apparently causes root death in white clover, followed by decomposition and N uptake by companion grass.

## *Transfer of nitrogen in simulated urine (experiment 4)*

In grazed pastures, transfer of nitrogen from clover to grass takes place through animals. Dung and urine are returned directly to plants and soil. Opinions differ on the importance of this pathway for the transfer of nitrogen from clover to grass plants (Simpson 1965, 1976; Marriott, Smith & Baird 1987).

Estimates by several authors (Doak 1952; Marriott, Smith & Baird. 1987) using different methods, suggest that urine patches from sheep cover from 27 to 40% of a pasture, and that the effects last for about 13 weeks. Urine patches have been found to contain about 300–600 kg N $ha^{-1}$ (Whitehead 1970).

In our experiments carried out in early summer 1986, three treatments were applied to replicated microplots (30 cm × 30 cm) in a field of perennial ryegrass/

white clover in the absence of sheep: 5 kg N $ha^{-1}$ of ammonium sulphate labelled with $^{15}N$; a similar treatment, but with the addition of 500 kg N $ha^{-1}$ of urea; and thirdly, 500 kg N $ha^{-1}$ of urea labelled with $^{15}N$ plus 5 kg N $ha^{-1}$ of unlabelled ammonium sulphate. The main difference between these experiments and those by other authors was in the use of $^{15}N$-isotope labelling.

The differences between treatments are shown in Fig. 4. Grass dry matter production was increased by the presence of urea for about 100 days. There was an increase in total nitrogen and $^{15}N$ content in the grass for a similar period. Although in our experiments, there was no 'scorching' of grass or clover, urea caused clover herbage accumulation to decrease by up to 70 g $m^{-2}$, nitrogen content by 3·5 g $m^{-2}$ and nitrogen fixation by 3·2 g $m^{-2}$ at 100 days. These decreases disappeared after 140 days. In our field experiments of 0·6 ha with animals, we found that visible urine patches covered 40% of the field.

Marriott, Smith & Baird (1987) comment that the results of urine patch experiments are likely to differ if they are measured earlier or later in the year, or in different weather conditions. Despite this, animal urine return is an important pathway for clover nitrogen to reach companion grasses. Recovery of urine N in grass herbage between days 7 and 81 in the experiments of Marriott, Smith & Baird (1987) was only 11·6%. Only 27% was recovered in the soil mineral nitrogen pool, but this is perhaps not surprising in the light of our findings of rapid disappearance

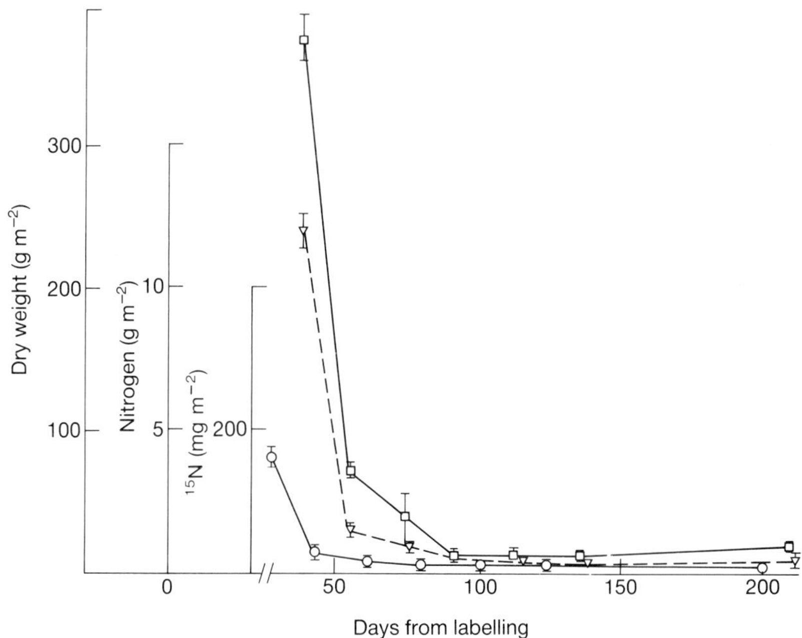


FIG. 4. Effect of simulated urine patches, on grass component of a grass/clover mixture, on dry weight (g $m^{-2}$) (squares); nitrogen (g $m^{-2}$) (triangles) and $^{15}N$ recovery (circles) on lowland soil in 1986 (L.S.D. = 5%).

of $^{15}N$-labelled inorganic salts applied to the soil. Urine N probably behaves like inorganic N except that it will contribute carbon as well as nitrogen to the plant or soil in which it is incorporated.

## DISCUSSION

Field experiments are obviously fundamental in ecological research but often, as here, give complex results because of the many interacting factors involved. This is especially true of marginal upland areas exposed to limiting factors such as climate, soil and nutrition. Measurements of herbage accumulation, dry weight of stubble plus roots, and nitrogen fixation, gave greatly varying results at different sites and in different years. Despite this, one set of observations was consistent and apparently predictable. These were the labelled nitrogen uptakes, in which grass always exceeded clover. Strong competition was taking place.

Analysis of grass/clover competition in a replacement series showed that perennial ryegrass and red clover competed specifically for soil nitrogen. Relative growth of the grass component, in the first year at least, seems to depend on the soil nitrogen supply. Competition for this resource probably determined early clover growth as well. Later though, clover is able to exploit the alternative resource of biologically fixed nitrogen. No doubt because of this alternative resource, relative growth of the legume was able to continue, virtually unaffected by the presence of grass. Relative growth of grass was slightly increased in the presence of clover.

The outcome of grass/clover competition will determine the amount of carbon (or energy, from grass photosynthesis) and nitrogen (from biological fixation) available for animals. If part or all of this carbon and nitrogen is not removed by cutting or grazing, leaf senescence, litter fall, and stubble and root turnover will contribute to soil organic matter. In the field, some grass stubble plus root appears to be lost from plants, and hence passes to soil organic matter in early summer. Part of the clover stolons and roots passes to organic matter in autumn.

The carbon and nitrogen of organic matter are further supplemented by dung and urine from grazing animals: an effect which lasts for about 100 days — a substantial part of the grazing season. Return of urine also creates a patchwork of nitrogen rich areas (up to 40%), among untreated areas with consequent variation in grass and particularly clover growth, at least in the short term. Simpson (1965, 1976) comments that this process is inefficient, because sheep 'camp' in local areas where urea accumulates and ammonia losses are large.

Direct transfer of nitrogen from clover to grass, through soil, does not occur in the first year of growth (at least in red clover) — nor in response to defoliation. It seems to occur after winter in white clover, although our evidence for this is, as yet, slight. Simpson (1965, 1976) showed large differences between legume species in N transfer through soil. Marriott, Smith & Baird (1987) suggest that, under grazing, and presumably with older swards, 50% of nitrogen passing to grass comes by transfer from white clover, 50% from dung and urine.

The role of the soil organic matter pool in this interpretation of cycling is to receive contributions of carbon and nitrogen from senescing plant parts, dung and urine. In time, probably mainly in spring, soil organic matter is mineralized, producing nitrate and ammonium for growth of grass and clover. Competition between roots of grass and clover for uptake of nitrogen will then resume. Both Simpson (1965) and Vallis, Henzel & Evans (1977) have commented on the changing balance between nitrogen competition and 'transfer' with time, depending on the growth cycles of the species in mixtures. This remark about legumes in general has a specific relevance in white clover: competition and the 'donation' of N to soil by death, decay and mineralization occur at different times. Summer competition is based on the products of autumn 'dieback' of clover and spring 'dieback' of grass.

Clover is notoriously difficult to maintain in pastures, but if continued clover presence depends on competition for the soil nitrogen resource, predictions of clover behaviour may be possible. Management of soil organic matter is also conceivable. In the long term, changes in carbon or nitrogen in the soil organic matter resource could be achieved, for example by altering the speed of mineralization by partial cultivation, introducing more oxygen into the soil; so stimulating microbiological activity in spring, or by introducing varieties of clover which differ in their competitiveness for soil nitrogen, or their capacity for donating nitrogen to the soil during winter.

## REFERENCES

**Begon, M. & Mortimer, M. (1986).** *Population Ecology. A Unified Study of Animals and Plants*, 2nd edn. Blackwell Scientific Publications, Oxford.

**Bristow, A.W., Ryden, J.C. & Whitehead, D.C. (1987).** The fate at several time intervals of $^{15}$N-labelled ammonium nitrate applied to an established grass sward. *Journal of Soil Science*, **38**, 245–254.

**Cooper, J.P. & Breese, E.L. (1971).** Plant breeding: forage grasses and legumes. In *Potential Crop Production. A Case Study* (Ed. by P.F. Wareing & J.P. Cooper), pp. 295–318. Heinemann, London.

**de Wit, C.T., Tow, P.G. & Ennik, G.C. (1966).** Competition between legumes and grasses. *Agricultural Research Reports 687*. Centre for Agricultural Publication and Documentation, Wageningen.

**Doak, B.W. (1952).** Some chemical changes in the nitrogenous contents of urine when voided on pasture. *Journal of Agricultural Science, Cambridge*, **42**, 162–171.

**Garwood, E.A., Tyson, K.C. & Clement, C.R. (1977).** A comparison of yield and soil conditions during 20 years of grazed grass and arable cropping. *Technical Report 21*, Hurley: The Grassland Research Institute, Maidenhead.

**Goodman, P.J. & Collison, M. (1982).** Varietal differences in uptake of $^{32}$P labelled phosphate in clover plus ryegrass swards and monocultures. *Annals of Applied Biology*, **100**, 559–565.

**Goodman, P.J. & Collison, M. (1984).** Growth, nitrogen uptake and fixation by ryegrass–white clover mixtures on Welsh upland soils. *Forage Legumes* (Ed. by D.J. Thomson), pp. 199–200. Occasional Symposium No. 16, British Grassland Society, Maidenhead.

**Goodman, P.J. & Collison, M. (1986).** Effect of three clover varieties on growth, $^{15}$N uptake and fixation by ryegrass/white clover mixtures at three sites in Wales. *Grass and Forage Science*, **41**, 191–198.

**Hall, R.L. (1974).** Analysis of the nature of interference between plants of different species. I. Concepts and extension of the de Wit analysis to examine effects. *Australian Journal of Agricultural Research*, **25**, 739–747.

**Hall, R.L. (1978).** The analysis and significance of competitive and non-competitive interference between species. In *Plant Relations in Pastures* (Ed. by J.R. Wilson), pp. 163–174. CSIRO, Melbourne.

**Haystead, A. & Marriott, C. (1979).** Transfer of legume nitrogen to associated grass. *Soil Biology and Biochemistry*, **11**, 99–104.

**Hopkins, A., Dibb, C., Bowling, P.J., Gilbey, J., Murray, P.J. & Wilson, I. (1985).** Production from permanent and reseeded grassland in England and Wales: results from a multi-site cutting trial. *Grass and Forage Science*, **40**, 245–246.

**Marriott, C.A., Smith, M.A. & Baird, M.A. (1987).** The effect of sheep urine on clover performance in a grazed upland sward. *Journal of Agricultural Science, Cambridge*, **109**, 177–185.

**Ollerenshaw, J.H. (1983).** Genetic variation in yield components of *Trifolium repens* at low temperature. In *Temperate Legumes Physiology, Genetics and Nodulation* (Ed. by D.G. Jones & D.R. Davies), pp. 89–101. Pitman, London.

**Ryden, J.C. (1984).** *The Flow of Nitrogen in Grassland*. The Fertiliser Society, London.

**Simpson, J.R. (1965).** The transference of nitrogen from pasture legumes to an associated grass under several systems of management in pot culture. *Australian Journal of Agricultural Research*, **16**, 915–926.

**Simpson, J.R. (1976).** Transfer of nitrogen from three pasture legumes under periodic defoliation in a field environment. *Australian Journal of Experimental Agriculture and Animal Husbandry*, **16**, 863–869.

**Sprent, J.I. (1979).** *The Biology of Nitrogen-fixing Organisms*. McGraw-Hill, London.

**Thomas, V.J. (1970).** A mathematical approach to fitting parameters in a competition model. *Journal of Applied Ecology*, **7**, 487–496.

**Vallis, I., Haydock, K.P., Ross, P.J. & Henzel, E.F. (1967).** Isotopic studies on the uptake of nitrogen by pasture plants. III. The uptake of small additions of $^{15}N$ labelled fertilizer by Rhodes grass and Townsville lucerne. *Australian Journal of Agricultural Research*, **18**, 865–877.

**Vallis, I., Henzel, E.F. & Evans, E. (1977).** Uptake of soil by legumes in mixed swards. *Australian Journal of Agricultural Research*, **28**, 413–425.

**Walker, T.W., Orchiston, H.D. & Adams, A.F.R. (1954).** The nitrogen economy of grass legume associations. *Journal of the British Grassland Society*, **9**, 249–274.

**Whitehead, D.C. (1970).** *The role of nitrogen in grassland productivity. Bulletin 48*. Farnham Royal: Commonwealth Agricultural Bureaux.

**Wilson, J.K. (1942).** The loss of nodules from legume roots and its significance. *Journal of the American Society of Agronomy*, **34**, 460–471.

# Relationship between root production and tiller appearance rates in perennial ryegrass (*Lolium perenne* L.)

C. MATTHEW, J.X. XIA, A.C.P. CHU, A.D. MACKAY*
AND J. HODGSON
*Massey University, Palmerston North, New Zealand; and *DSIR, Grasslands Division, Private Bag, Palmerston North, New Zealand*

## SUMMARY

1 Seasonal variations in root mass and new root production, and the relationship between root production and tiller appearance rates for a ryegrass (*Lolium perenne* L.) sward under two contrasting grazing managements were monitored over a 12 month period.

2 Root mass and new root production were found to be relatively insensitive to grazing management. Seasonal fluctuations in new root production were large in comparison to those produced by contrasting hard and lax grazing regimes.

3 A perception of a tiller as an elongating axis moving upwards towards an intermittently rising soil surface emerges from this study. There is a seasonal peak of tiller appearance from underground tiller axes in early summer. Peak root production coincided with this peak in tiller appearance.

4 Sward manipulation during this post-flowering peak of high tiller and root appearance appears to offer scope for achieving pasture growth rate efficiencies.

## INTRODUCTION

Root systems of grasses are much less frequently studied than shoot systems and there is a disparity in knowledge about the above- and below-ground structure of grasslands. Davidson (1978) referred to root systems as 'the forgotten component of pastures'.

Despite Davidson's (1978) comment, there is a sizeable body of literature on the underground organs of herbage grasses. A review by Troughton (1957) cites approximately 800 references and although there has been no recent review, investigation has continued. What has been lacking in most previous studies of grass swards however, is an integration of data on root dynamics with data on shoot dynamics. Where the relationship between the root and shoot systems of grasses has been addressed at all, consideration has usually been confined to the question of partitioning of mass (e.g. Parsons & Robson 1981; Deinum 1985). Many studies have used young seedlings, potted plants, or even hydroponic culture (Evans 1971;

Hunt & Thomas 1985) to avoid technical difficulties of sampling roots in the field or of determining live : dead ratios in samples taken from established swards. While a useful starting point, data from such studies cannot necessarily be extrapolated to provide a definitive understanding of root–shoot relationships in grazed pasture.

In this paper we present an overview of results from a study conducted at Palmerston North, New Zealand, in which root mass and a root replacement rate index were measured for ryegrass swards under lax or hard grazing managements over a 12 month period. Our aims in this experiment were first to define seasonal variation in root mass and seasonal patterns of root replacement for ryegrass, secondly to see if such variation might be linked to variations in tiller density or tiller appearance rates (TAR) and thirdly to explore the possibility of achieving pasture growth rate advantages if the dynamics of the root system were understood and taken into account when formulating grazing management strategies.

## MATERIALS AND METHODS

### *Site description*

The experiment was conducted at the Pasture and Crop Research Unit, Massey University, on a perennial ryegrass dominant pasture. In July 1986 the site was sprayed with herbicide (active ingredients picloram and 2,4-D) to remove clover. The resulting sward was predominantly ryegrass but contained approximately 500–1000 tillers $m^{-2}$ *Poa trivialis* L. After removal of clover, nitrogen was applied as urea at approximately 15 kg N $ha^{-1}$ every 3 weeks. The soil was derived from greywacke loess, and classified under the soil taxonomy system as a Typic fragiaqualf. Mean annual rainfall is 995 mm and long-term average temperatures range from 8·0 °C (July) to 17·6 °C (February).

### *Experimental design*

In winter 1986 sixteen 100 $m^2$ plots (four replicates of four treatments) were fenced in a randomized complete block design. Measurements commenced in December 1986 and initially were confined to eight plots (four replicates of two treatments — 'hard' or 'lax' grazing) with the remaining eight plots reserved for the introduction of cross-over lax–hard and hard–lax treatments in spring 1987 (Matthew *et al.* 1989b). Target herbage masses were 800–1500 and 2500–4000 kg DM $ha^{-1}$ on 'hard' and 'lax' grazed plots, respectively. These grazing regimes were intended to contrast the extremes of management which pastures might be subjected to in normal farm practice and were also expected to have contrasting tiller densities. Grazing was with fifteen to thirty sheep per 100 $m^2$ plot approximately every 3 weeks, but less frequently in winter, and sometimes less frequently on lax plots. Data from successive harvests were analysed as split plot in time subtreatments of grazing main effects.

## *Methods*

Root mass and root length in intact soil cores (IC) and in 'refilled' (ingrowth) cores (RC) was measured as previously described by Matthew, Mackay & Chu (1986). The RC technique is identical in principle to the mesh bag technique (Bohm 1979; Steen 1983) except that no mesh bag is used. RC data was used as an index of seasonal root replacement and to compare root replacement for the two grazing managements. A hydropneumatic elutriation system (Smucker, McBurney & Srivastava 1982) was used to extract root samples from cores.

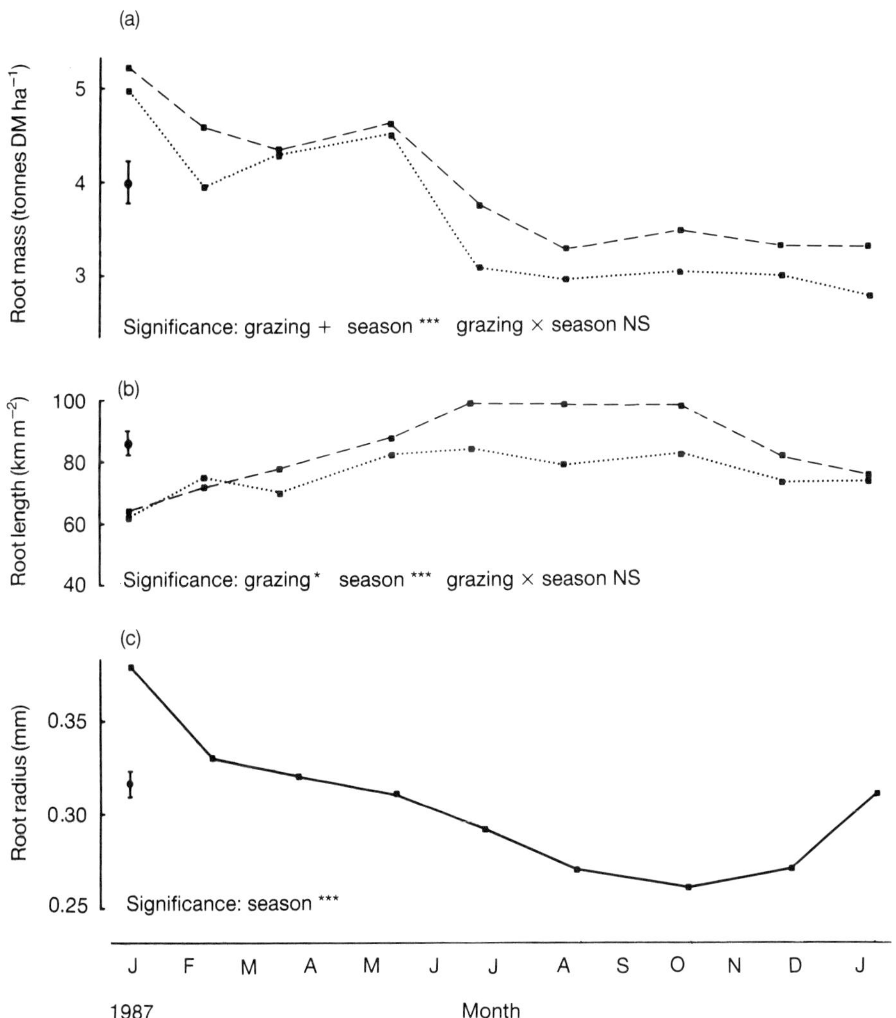


FIG. 1. Intact cores (0–250 mm depth): (a) root mass; (b) root length in hard (dotted line) and lax (dashed line) grazed plots; (c) mean root radius.

Tiller population densities were calculated by counting tillers in thirty 53-mm-diameter plugs per plot and tiller appearance by monitoring tagged tillers in fixed quadrats (Korte 1986). Pasture growth rate was measured by tissue turnover techniques (Davies 1981) from September 1987, and prior to this was estimated using a computer model (Butler, unpublished) which adjusts an arbitrary 'potential growth rate' for rainfall, temperature, water balance, soil fertility, and corrects for reproductive growth.

### *Sampling*

At 6–7 week intervals from December 1986 IC and RC root mass and root length were determined for three soil depths (0–70 mm, 70–250 mm, 250–600 mm). Root data presented are totals for the 0–250 mm soil depth. This was the approximate extent of the A horizon and usually more than 90% of total root harvested was within this horizon. Sward parameters were measured as soon as possible after each root harvest.

## GRAZING MANAGEMENT EFFECTS

### *Intact core root mass and root length*

Grazing management effects were small. Root mass (0–250 mm) was consistently higher on lax than on hard grazed swards (Fig. 1a), but only slightly so, averaging 3·9 and 3·5 tonnes DM $ha^{-1}$ ($P<0{\cdot}10$), respectively, with a parallel difference in root lengths (81 and 71 km root $m^{-2}$, $P<0{\cdot}05$, Fig. 1b).

### *Root replacement index*

Apart from a period when reproductive growth occurred on lax plots between October and December, root replacement for the two treatments was also similar, averaging 9·0 and 7·5 kg DM $ha^{-1}$ $day^{-1}$ (N.S.) on lax and hard grazed plots, respectively (Fig. 2a). Very possibly a technique such as the use of viewing windows to enable day to day evaluation of root elongation might have detected checks to root growth after defoliation similar to those for leaf extension (Fig. 3c). Such effects are well known from pot experiments. Evans (1971) found that defoliation of ryegrass plants to 75 mm reduced root growth in the following week by approximately 40%, while defoliation to 25 mm reduced root growth by almost 90%; and Troughton (1957) reviewed a number of experiments reporting similar reduction of root growth after defoliation.

There was no evidence that the RC technique overestimated new root production. When an endoscope was used to count roots adjacent to viewing tubes in RC and undisturbed soil, root counts in RC were lower than in surrounding soil for the 0–70 mm soil depth, and similar for depths of 70–250 mm.

Two factors probably contribute to the unexpectedly small grazing management

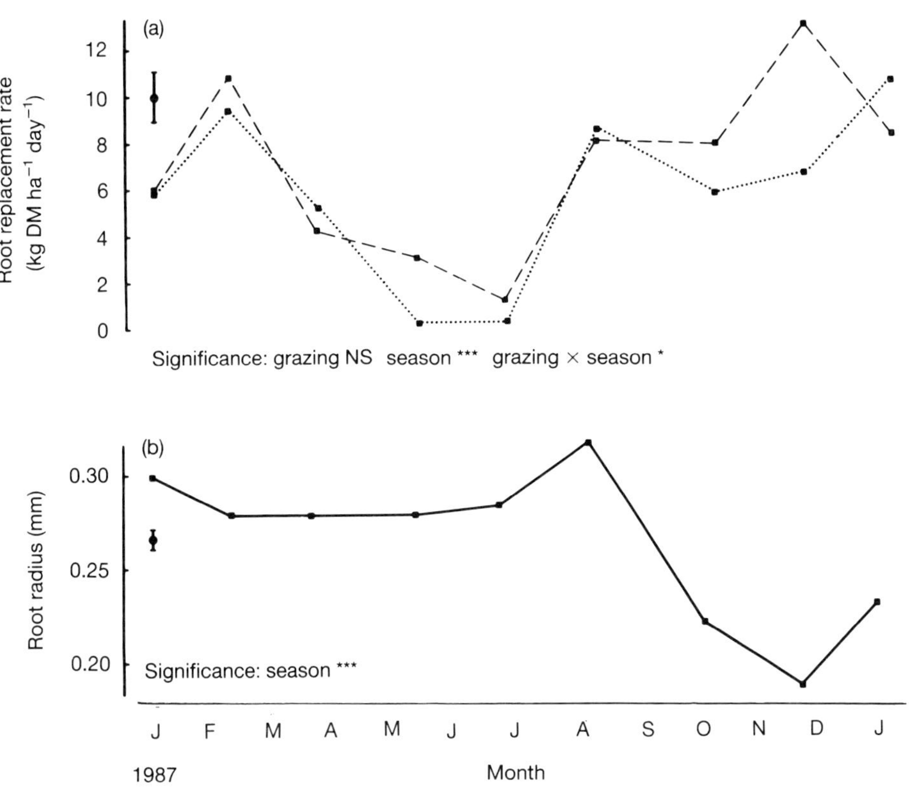


FIG. 2. Refilled cores (0–250 mm depth): (a) root replacement rate in hard (dotted line) and lax (dashed line) grazed plots; (b) mean root radius.

effects observed in this experiment. Firstly, defoliation of established swards, as in our experiment, with several tonnes $ha^{-1}$ root mass already in place, and with 3–6 weeks interval between grazing, would be of much less overall import than in seedling plants studied by Evans (1971), or in work reviewed by Troughton (1957). Secondly, swards adapted rather quickly to the different grazing heights (Fig. 4) resulting in similar amounts of leaf mass in hard and lax grazed swards, despite differences in grazing height.

### *Above-ground measurements*

Tiller population densities responded to grazing management more strongly than root mass or root replacement and increased under hard grazing (Fig. 3a), means for hard and lax plots being 7720 and 5120 tillers $m^{-2}$, respectively. TAR were generally higher in hard grazed swards (Fig. 3b), reflecting higher tiller numbers.

After burial in autumn, by earthworm casting, of quadrats placed in summer, new fixed quadrats were placed in May. Tiller densities in fixed quadrats tended to rise after the first tagging and then stabilize at levels 10–20% above those in the

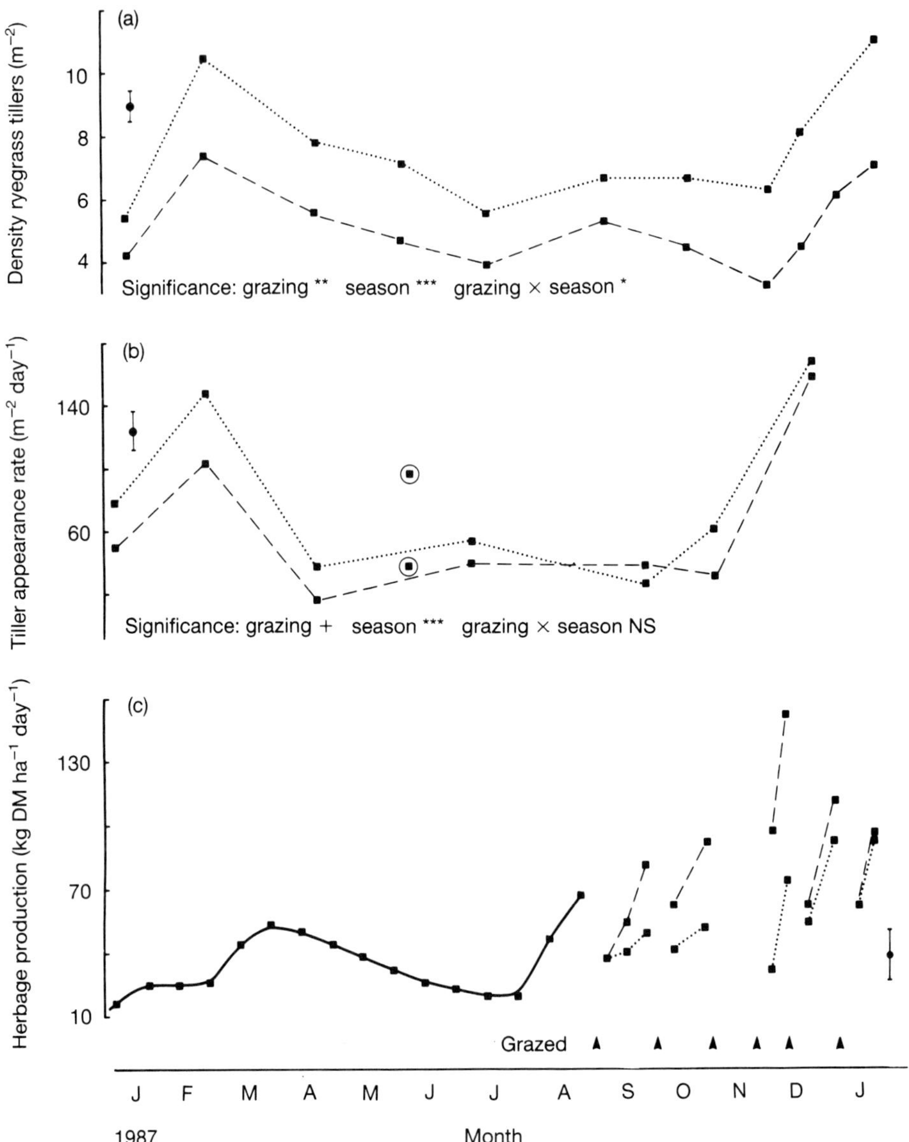


FIG. 3. Sward measurements: (a) tiller population density; (b) tiller appearance rate; (c) herbage production, in hard (dotted line) and lax (dashed line) grazed swards.

surrounding pasture. High tiller appearance rates in fixed quadrats for May (circled, Fig. 3b), when tiller numbers in the surrounding pasture were falling (Fig. 3a) are therefore thought to be an artefact of this 'disturbance' effect on TAR. A similar disturbance effect was also noted by Arosteguy (1982). Pasture growth rate (Fig. 3c) is synthesized from computer predictions of net production (December–

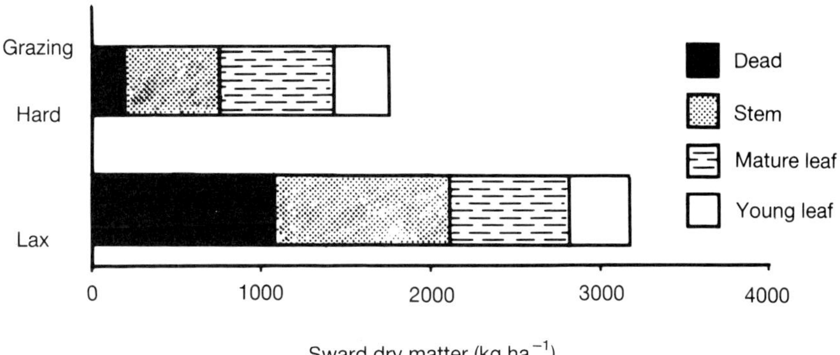


FIG. 4. Herbage mass components for hard and lax grazed swards before grazing, October 1987.

August) and actual accumulation measurements (August–December). The latter data are derived from leaf extension measurements on individual tillers and their daughters and are expressed in kg DM ha$^{-1}$ day$^{-1}$. They show a consistent check to herbage production in the first measurement period after each grazing.

## SEASONAL EFFECTS

### *Intact core root mass and root length*

Seasonal fluctuations in parameters measured were large in comparison to differences induced by grazing management. The only exception was IC root mass which declined from 5·1 to 3·3 tonnes ha$^{-1}$ over the course of the experiment, but showed no clear seasonal variation (Fig. 1a). The decline is unlikely to be due to the decay and disappearance of dead clover roots as experimental measurements did not begin until 5 months after clover had been sprayed, and dead roots should have largely disappeared in that time (Gibbs 1986).

IC root length increased during autumn and winter, against the background trend of falling root mass (Fig. 1b) and mean IC root radius therefore declined (Fig. 1c). This might reflect proliferation and branching of existing roots at a time of low root production (Fig. 2a).

### *Root replacement and tiller appearance rate*

Root replacement, tiller densities and TAR all showed broadly parallel summer peaks and winter minima (Figs 2a and 3a,b), but with root production in spring preceding tiller appearance. Dry conditions in December 1986 appeared to delay the peak in tiller appearance.

Our findings of high summer root production contrast sharply with earlier New Zealand work. Jacques (1956) from unpresented data, described a seasonal pattern of ryegrass root activity in which an annual crop of new roots appears between March and September and grows on down through the soil profile. Over summer, from October to March, Jacques (1956) found little root activity near the soil surface. Caradus & Evans' (1977) counts of numbers of new nodal roots support the pattern described by Jacques.

One possible explanation is that there are two different categories of roots involved. RC roots harvested from July to October tended to be large in diameter (Fig. 2b) and correspond with Jacques' (1956) and Caradus & Evans' (1977) peak of nodal root production. Roots produced from October onwards were finer (Fig. 2b) and coincided with the appearance of new tillers (Fig. 3b) which were observed to begin producing roots of their own within a week of emergence. It may be that the above authors mistakenly classified the nodal roots of young tillers as secondary branches of the larger diameter roots formed by mature tillers in late winter. Alternatively it may be that the ryegrass cultivar used in this experiment (Ellett) has a seasonal pattern of tillering (and hence root appearance) inherently different from the cultivar used by Caradus & Evans (Grasslands Ruanui).

There is evidence that seasonally high rates of tiller and root production in summer may be widespread in grass swards. For ryegrass Colvill & Marshall (1984), Korte (1986) and L'Huillier (1987) have all reported high TAR in summer. For timothy (*Phleum pratense* L.) Atkinson (1984) has reported a summer peak in root production and a period of high TAR in summer is indicated by tiller cohort survivorship patterns (Jewiss 1972).

## TOWARDS INTEGRATION OF ROOT AND SHOOT DYNAMICS

Integrated measurement of tiller and root dynamics highlighted the fact that the tiller does not lie passively at the soil surface. Rather the tiller is an elongating axis moving upwards towards a soil surface intermittently raised by seasonal burial, for example by earthworm activity. Where addition of phytomers (Silsbury 1970) to the axis is insufficient to keep pace with the rate of burial, underground internode elongation occurs. In this experiment up to 150 $m\,m^{-2}$ ground surface of these underground tiller axes could be sorted out from root samples. Harris *et al.* (1979) termed these underground tiller axes stolons and using gel-electrophoresis showed that individual genotypes of ryegrass had, with time, spread over patches up to 1.2 m diameter. A description of the seasonality of burial and stolon formation and the significance for patterns of TAR (Fig. 3b) was presented by Matthew *et al.* (1989a). The tiller axis, then, provides a principle which integrates root and shoot dynamics of a grass sward.

Questions relevant to root dynamics are (i) the number of plastochrons or leaf appearance intervals delay between leaf elongation and root initiation at a given segment (phytomer) on the axis; (ii) the number of plastochrons for which a root remains active on the axis; and (iii) whether the 1 : 1 relationship between root and leaf appearance established by Hunt & Thomas (1985) in hydroponic culture also holds under field conditions. This may not necessarily be the case as Troughton (1981) found that root initiation does not occur in dry soil conditions. We plan a follow up experiment to attempt to clarify some of these points.

For tiller dynamics the central question is the rate of branching of the axis, and in the variety of ryegrass studied there is a distinct seasonal pattern of sward renewal by rapid proliferation of daughters on the axes of post-flowering tillers (Matthew *et al.* 1989a).

Allowing a period of reproductive growth for increased assimilation (Parsons & Robson 1981) just prior to this phase of rapid TAR, appears to enhance daughter tiller growth, and so allow pasture growth rate efficiencies in the following summer and autumn (Matthew *et al.* 1989b).

## CONCLUSIONS

Root dynamics in grass swards are much less sensitive to manipulation by grazing management than might have been predicted from earlier research. The importance of field measurement for an accurate understanding of sward dynamics is highlighted by the perception which emerges from this study of the tiller as a continually elongating axis, and by the discrepancies between our results and those obtained from observations of young plants under artificial conditions. Seasonal patterns of tiller and root appearance are marked and sward manipulation during the period of high tiller and root appearance in early summer does appear to offer scope for achieving pasture growth rate efficiencies.

## ACKNOWLEDGEMENTS

The authors would like to thank Mr T.J. Lynch and his staff for technical assistance, and the Massey University Agricultural Research Foundation for a research grant in support of this study.

## REFERENCES

**Arosteguy, J.C. (1982).** *The dynamics of herbage production and utilisation in swards grazed by cattle and sheep.* Ph.D. thesis. University of Edinburgh.

**Atkinson, D. (1984).** Spatial and temporal aspects of root distribution as indicated by the use of a root observation laboratory. In *Ecological Interactions in Soil* (Ed. by A.H. Fitter, D. Atkinson, D.J. Read & M.B. Usher), pp. 43–65. *Special Publications Series of the British Ecological Society No. 4.* Blackwell Scientific Publications, Oxford.

**Bohm, W. (1979).** *Methods of Studying Root Systems.* Ecological studies 33 (Ed. by W.D. Billings, F. Golley, O.L. Lange & J.S. Olson) Springer Verlag, Berlin, 188pp.

**Caradus, J.R. & Evans, P.S. (1977).** Seasonal root formation of white clover, ryegrass, and cocksfoot in New Zealand. *New Zealand Journal of Agricultural Research*, **20**, 337–342.

**Colvill, K.E. & Marshall, C. (1984).** Tiller dynamics and assimilate partitioning in *Lolium perenne* with particular reference to flowering. *Annals of Applied Botany*, **104**, 543–557.

**Davidson, R.L. (1978).** Root systems — the forgotten component of pastures. In *Plant Relations in Pastures* (Ed. by J.R. Wilson), pp 86–94. CSIRO, Melbourne.

**Davies, A. (1981).** Tissue turnover in the sward. In *Sward Measurement Handbook* (Ed. by J. Hodgson, R.D. Baker, A. Davies, A.S. Laidlaw & J.D. Leaver), pp. 179–208. British Grassland Society, Maidenhead.

**Deinum, B. (1985).** Root mass of grass swards in different grazing systems. *Netherlands Journal of Agricultural Science*, **33**, 377–384.

**Evans, P.S. (1971).** Root growth of *Lolium perenne* L. II. Effects of defoliation and shading. *New Zealand Journal of Agricultural Research*, **14**, 552–562.

**Gibbs, R.J. (1986).** *Changes in soil structure under different cropping systems*. Ph.D. thesis. Lincoln College, Canterbury.

**Harris, W., Pandey, K.K., Gray, Y.S. & Couchman, P.K. (1979).** Observations on the spread of perennial ryegrass by stolons in a lawn. *New Zealand Journal of Agricultural Research*, **22**, 61–68.

**Hunt, W.F. & Thomas, V.J. (1985).** Growth and developmental responses of perennial ryegrass grown at constant temperature II. Influence of light and temperature on leaf, tiller and root appearance. *Australian Journal of Plant Physiology*, **12**, 69–76.

**Jacques, W.A. (1956).** Root development in some common New Zealand pasture plants IX. The root replacement pattern in perennial ryegrass (*Lolium perenne*). *New Zealand Journal of Science and Technology*, **A38**, 160–165.

**Jewiss, O.R. (1972).** Tillering in herbage grasses — its significance and control. *Journal of the British Grassland Society*, **27**, 65–81.

**Korte, C.J. (1986).** Tillering in 'Grasslands Nui' perennial ryegrass swards 2. Seasonal pattern of tillering and age of flowering tillers with two mowing frequencies. *New Zealand Journal of Agricultural Research*, **29**, 629–638.

**L'Huillier, P.J. (1987).** Tiller appearance and death of *Lolium perenne* in mixed swards grazed by cattle at two stocking rates. *New Zealand Journal of Agricultural Research*, **30**, 15–22.

**Matthew, C., Mackay, A.D. & Chu, A.C.P. (1986).** Techniques for measuring the root growth of a perennial ryegrass (*Lolium perenne* L.) dominant pasture under contrasting spring managements. *Proceedings Agronomy Society of New Zealand*, **16**, 59–64.

**Matthew, C., Quilter, S.J., Korte, C.J., Chu, A.C.P. & Mackay, A.D. (1989a).** Stolon formation and significance for sward tiller development in perennial ryegrass (*Lolium perenne* L.). *Proceedings of the New Zealand Grassland Association*, **50**, 255–259.

**Matthew, C., Xia, J.X., Hodgson, J. & Chu, A.C.P. (1989b).** Effect of late spring grazing management on tiller dynamics and autumn pasture growth rates in a perennial ryegrass (*Lolium perenne* L.) sward. *Proceedings XVI International Grassland Congress*, Nice, 521–522.

**Parsons, A.J. & Robson, M.J., (1981).** Seasonal changes in the physiology of S24 perennial ryegrass (*Lolium perenne* L.) III. Partition of assimilates between root and shoot during the transition from vegetative to reproductive growth. *Annals of Botany*, **48**, 733–744.

**Silsbury, J.H. (1970).** Leaf growth in pasture grasses. *Tropical Grasslands*, **4**, 17–36.

**Smucker, A.J.M., McBurney, S.L. & Srivastava, A.K. (1982).** Quantitative separation of roots from compacted soil profiles by the hydropneumatic elutriation system. *Agronomy Journal*, **74**, 500–503.

**Steen, E. (1983).** The net stocking method for studying quantitative and qualitative variation with time of grass roots. In *Proceedings of an International Symposium, Gumpenstein* (Ed. by W. Bohm, L. Kutschera & E. Lichtenegge), pp. 63–74.

**Troughton, A. (1957).** The underground organs of herbage grasses. *Bulletin No. 44*. Commonwealth Bureau of Pastures and Field Crops, Hurley, 163pp.

**Troughton, A. (1981).** Root-shoot relationships in mature grass plants. *Plant and Soil*, **63**, 101–105.

# The development of clover and ryegrass root systems in a pasture and their interactions with the soil fauna

C.A.G. SACKVILLE HAMILTON AND J.M. CHERRETT
*School of Biological Sciences, University College of North Wales, Bangor, Gwynedd LL57 2UW, UK*

## SUMMARY

1 The effects of removing different components of the soil fauna on the establishment of a short-term ley were examined. Shoot dry weight, root length and populations of soil animals were analysed in plots surrounded by mesh to exclude three size-classes of fauna.

2 The yield of dry matter above ground and the length of roots were greater in plots from which animals were excluded; results for the exclusion of (i) earthworms only and (ii) all soil fauna were similar. This was possibly because the movement of animals is limited in the absence of earthworm burrows.

3 Chance local events such as the death of an animal have large effects on total numbers of animals. As a result there were large differences between plots given the same treatment.

4 Physical exclusion alone did not keep plots free from soil fauna. A combination of physical and chemical means (pesticides) must be used in the future.

## INTRODUCTION

Forty to sixty per cent of pasture production occurs below the soil surface and the use of pesticides has produced increases in herbage production of up to 100% (Clements *et al.* 1982). International Biological Programme (I.B.P.) studies in the USA have indicated that both primary production and herbivory were greater below than above ground (French 1979), although data on herbivory are normally obtained by extracting soil fauna and estimating plant consumption from assumed feeding relations. There have been few studies on root herbivory *in situ* (Harding 1968; Carpenter 1985), but recently Brown & Gange (1989), working in the field on the differential effects of above- and below-ground insect herbivory during early plant succession have shown that plant species richness, diversity, vegetation frequency and cover abundance were all increased after treatment with the soil insecticide, Dursban (chlorpyrifos). The present study is part of a programme to examine the effects of the soil fauna on root growth by continuous direct observation in the field.

## METHODS

The rhizotron used in these studies and the treatments have been described by Sackville Hamilton *et al.* (see pp. 49–59). Briefly, on the north side of the rhizotron, fifteen windows were assigned, in a randomized complete block design, to three treatments: (i) all fauna except moles allowed to colonize the one cubic metre cage adjacent to each window; (ii) moles and earthworms excluded; and (iii) as many of the macro- and meso-fauna as possible excluded. Each cage was filled with sterilized soil. Full fauna plots are described as control plots, i.e. representing the normal fauna invading from the surrounding pasture; reduced fauna plots are referred to as treatments. Animals were observed *in situ* through the windows of the rhizotron using a hand lens or a binocular microscope. The roots of plants were mapped onto polyester film 80 cm × 40 cm × 0·1 mm, using a different colour for each cohort, root lengths being measured and counted five times between August 1987 and December 1987. The numbers of animals, divided for convenience into thirty-one taxa were recorded twenty-four times during the establishment year for each window. Particular attention was given to any visible interactions involving soil organisms.

Above-ground plant material in the square metre adjacent to each window was cut to a height of 4 cm and harvested by hand. It was cut once during 1987 in September. Initially plots were harvested monthly during 1988, then fortnightly from 13 May. Plant material was dried and weighed after each harvest.

All plots were fertilized with NPK (20 : 10 : 10) at agricultural rates, 18·83 g $m^{-2}$ at planting and 37·66 g $m^{-2}$ in spring 1988. From 30 April 1988 following approximately 4 weeks with little rain, plots were irrigated regularly to assist in maintaining close contact between the soil and the glass windows. When the soil dries out it contracts away from the glass, making the observation of roots difficult and allowing the soil fauna to move deeper in the soil than they would normally be able to, simply by moving down the gap between soil and glass.

By the end of 1987 earthworms had invaded some of the earthworm exclusion and minimal fauna plots. On 8 January therefore a 0·2% solution of formaldehyde was applied in two aliquots each of 5 l to each plot in which earthworms were to be excluded (Raw 1959). Earthworms driven from their burrows were collected, washed, identified and counted. Enchytraeids and slugs (*Arion* spp. and *Deroceras reticulatum* Müller) were also expelled from the soil by this treatment. The application of formaldehyde was repeated on 9 May 1988.

## RESULTS

### *Root length — per horizon*

Roots were shorter overall in the full fauna control ($P < 0·05$) than in the reduced fauna treated plots although there were no significant differences between treatment within each horizon (Fig. 1). Residual variation between windows was high

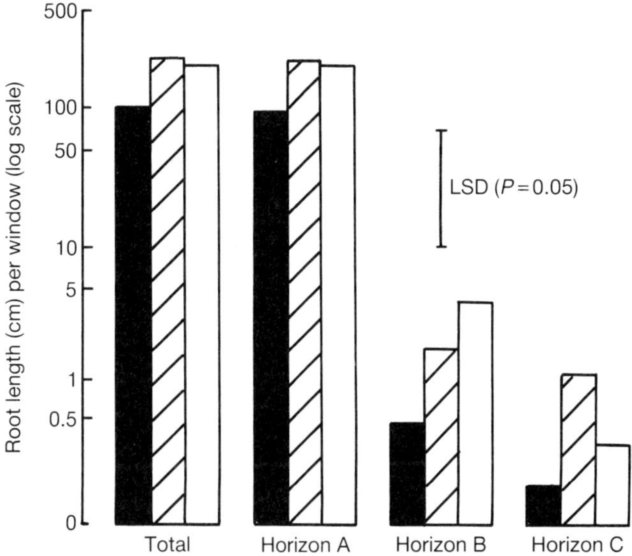


FIG. 1. Mean length of root (cm) in each window per horizon in plots with a complete fauna ( ■ ), earthworms excluded ( ▨ ) or a minimal fauna ( □ ). Data were log transformed and are plotted on a log scale.

(coefficient of variation (cv) = 349%) as was the residual variation between horizons within windows (cv = 160%). There was no interaction between treatments and horizons.

### *Root length — per cohort*

Roots were shorter in the control than in the treated plots when averaged over all times ($P<0{\cdot}05$) (Fig. 2). On the final date there was half the root length in the control compared with the treated plots. The residual variation between windows was again high (cv = 175%), whereas the residual variation between successive observations on the same window was much lower (cv = 39%), i.e. random differences between windows were consistent from one time to the next. There was a significant increase ($P<0{\cdot}001$) in the total length of root with time, the curves being similar in the control and the treated areas.

### *Dry weight of above-ground material*

Overall the production of shoot dry matter in the control plots was less than in the minimal fauna plots ($P<0{\cdot}05$). The standing crop of dry matter in the control was significantly lighter than that in the minimal fauna treatment in the harvests of September 1987, 10 June and 8 August (Fig. 3). Conversely on 27 May the plants in

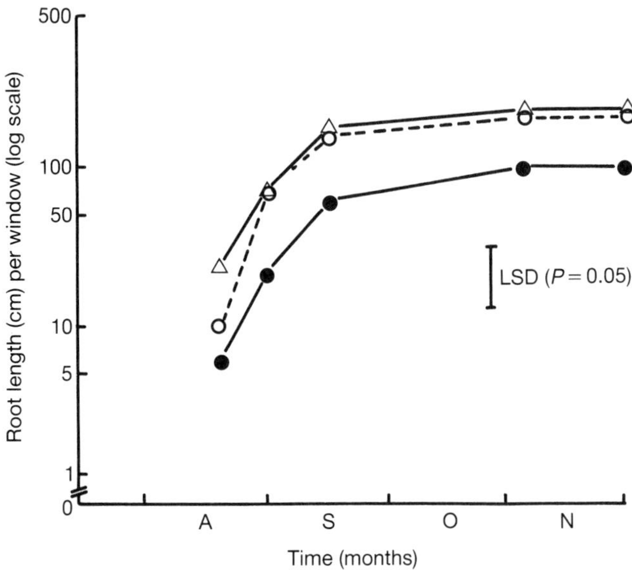


FIG. 2. Mean length of root (cm) in each window for each of five observation dates in the establishment year. Plants were grown in soil with a complete fauna ( ● ), without earthworms ( O ) or with a minimal fauna ( △ ). Data were log transformed and are plotted on a log scale.

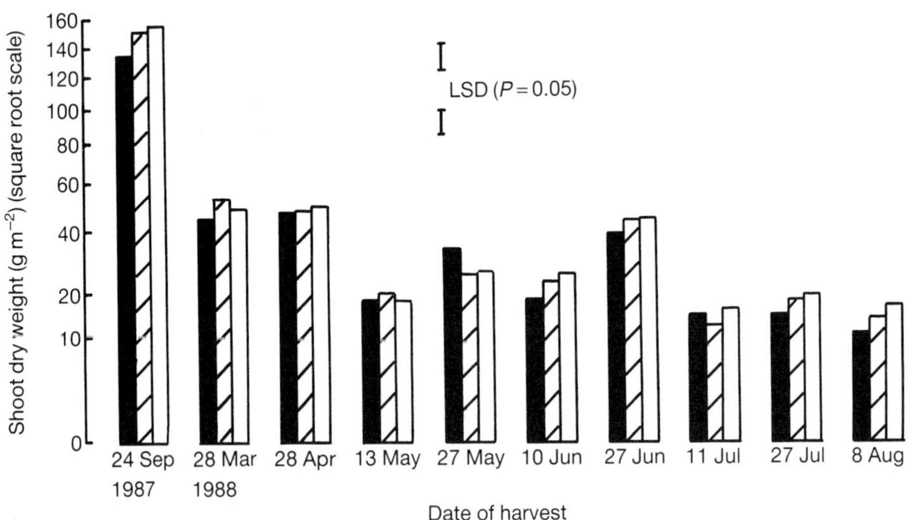


FIG. 3. Mean dry weight ($g\,m^{-2}$) of herbage in plots on the north side of the rhizotron in which plants were grown in soil with a complete fauna ( ■ ), without earthworms ( ▨ ) or with a minimal fauna ( □ ). Data were square root transformed and are plotted on a square root scale.

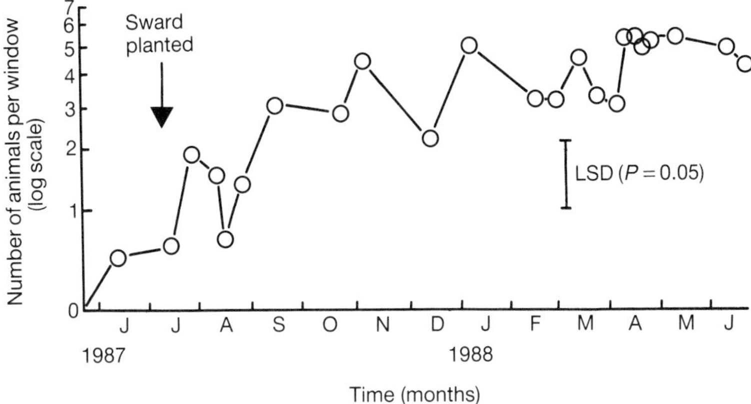


FIG. 4. Mean total number of all animals (except earthworms) per window on the north side of the rhizotron, from the time soil was reinstated until June 1988. Data were averaged over all treatments as there were no significant differences between treatments. Data were log transformed and are plotted on a log scale.

the control were significantly heavier than either of those in the treated plots. On the remaining six dates there were no significant differences. The annual production of herbage was increased by 8·6% in the absence of earthworms and by 12·4% in the absence of all animals. There was a low residual variation between windows averaged over time (cv = 23·8%) and no consistent differences between windows.

## *Total soil fauna*

There was no effect of treatment on the total number of animals ($P = 0{\cdot}175$). The differences between replicates were no greater than random differences within replicates, whether averaged over dates or for each date. The random variation between windows was consistent from date to date. The total population of animals increased with time ($P < 0{\cdot}001$) (Fig. 4).

## *Earthworms*

Overall there were four times as many earthworms in the control as in the treated plots ($P < 0{\cdot}001$) (Fig. 5). There was on average no increase in the number of earthworms with time but there were treatment-specific changes with time ($P < 0{\cdot}05$).

Following the first application of formaldehyde to the treated plots, the number of earthworms in these plots decreased compared with the control ($P < 0{\cdot}05$). The second application of formaldehyde did not have such a clear effect.

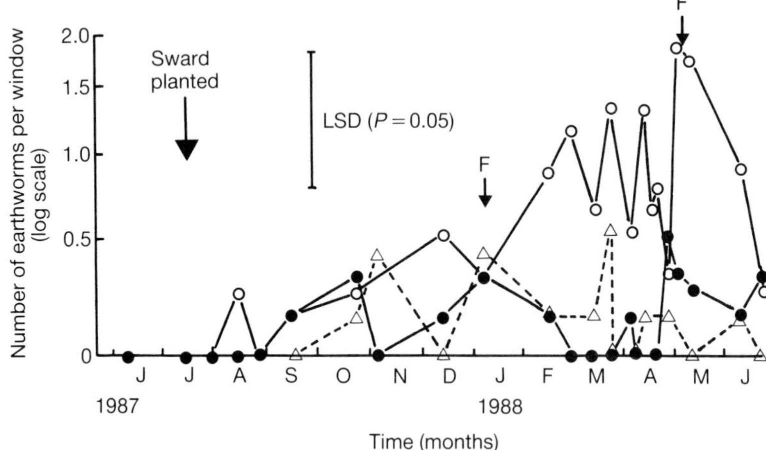


FIG. 5. Mean number of earthworms per window on the north side of the rhizotron in each of three treatments: plots with a complete fauna ( ○ ); earthworms excluded ( ● ); minimal fauna ( △ ). Data are for the time from reinstatement of the soil until June 1988. The arrows marked F show the dates when formaldehyde was applied to the animal exclusion plots. Data were log transformed and are plotted on a log scale. Only data since 24 October 1987 have been analysed.

## *Observations on animal behaviour*

### *Responses to light*

Certain species of soil fauna moved away from bright light. Earthworms and *Lithobius* spp. had to be observed as soon as the shutters were opened or using red light. Earthworms were not as sensitive to light when they were mating, an event that was observed seven times during the year.

### *Feeding*

*Campodea* and *Collembola* were seen eating live plant roots. The former grazed the root *c.* 2 mm behind the root apex just where the zone of root hairs started. The *Collembola* grazed root hairs and mycorrhizae, but did not eat the root proper unless it had been damaged by another species, such as earthworms or *Campodea*.

Leatherjackets, slugs and earthworms all pulled live leaves down into the soil to a maximum depth of 15 cm. When the leaves had been fragmented by these species other animals, especially *Collembola* ate the remains.

Earthworms were observed eating soil and slugs' faeces. *Collembola* were important consumers of faeces. *Collembola, Campodea* and enchytraeidae were all seen grazing on the lining of worm burrows, which support large numbers of micro-organisms (Parle 1963).

Animals are not uniformly distributed. Their local abundance and behaviour is affected by localized events such as the death of another animal. After the

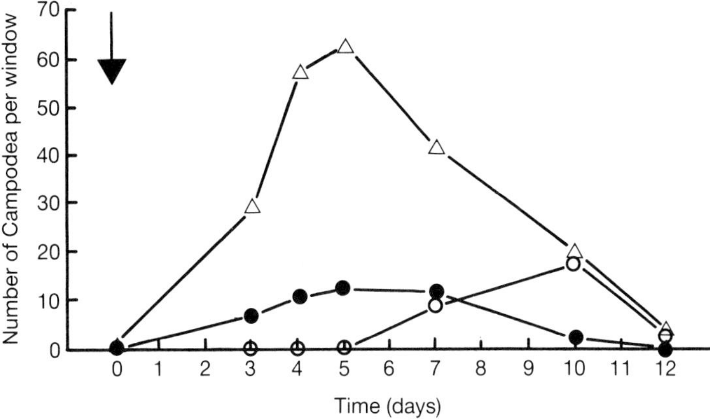


FIG. 6. Number of *Campodea staphylinus* in window number 15 following the death of an individual of *Allolobophora chlorotica* after the application of formaldehyde to remove earthworms from an earthworm exclusion treatment. Total number of *Campodea* in the window ( △ ); number of *Campodea* feeding on the dead worm ( ● ); total number of other animals feeding on the dead worm ( ○ ).

first application of formaldehyde in January 1988 an earthworm (*Allolobophora chlorotica* Savigny) died next to one of the windows. This allowed a study of the breakdown and colonization of the dead animal by the soil fauna (Fig. 6). The day the worm died there were no animals observed in the window, but 5 days later there were sixty-three individuals of *Campodea staphylinus* Westwood. Of these about twelve were feeding on the worm at any one time whilst the others were in a cavern below it. There was a constant movement of *Campodea* from the cavern to feed. Other animals including *Collembola* spp., an enchytraeid and a centipede appeared on or after the fifth day. On the eleventh day fungal hyphae were seen and the worm was flaccid; by the following day there was no sign of where the worm had been.

In spring 1988 an earthworm was observed being caught by a centipede and whilst it was dying other animals (*Campodea*, *Collembola*, mites and enchytraeids) appeared and began feeding on it. Seventeen hours later there was no sign of where the worm had been.

### *Burrowing*

During the winter earthworms burrowed deeper in the soil; in the summer a few earthworms were seen aestivating. Burrowing by earthworms allowed a larger volume of soil to be seen from the window and therefore potentially more roots. The soil lining earthworm burrows consists of faeces and egested matter, has a fine tilth and often has a mottled appearance from the mixing of soils from different areas.

Roots moved in the soil both downwards and to each side. The movements of roots from side to side were caused mainly by animal activity, particularly burrowing by earthworms. In the establishment year roots in the A horizon moved down as much as 23 mm largely because of soil compaction.

All soil fauna used burrows created by earthworms to reach other parts of the soil profile. The only other animals seen burrowing were *Polydesmus* spp. and Enchytraeidae. Leatherjackets and slugs greatly enlarged the earthworm burrows. The slugs formed caverns in the soil just below the surface where they sheltered. When worms defecated next to the glass, roots were obscured in that area by the worm cast.

## DISCUSSION

The removal of any components of the soil fauna resulted in the increase of root length and herbage yield. That the removal of earthworms was as beneficial to growth as the removal of the entire soil fauna was unexpected; earthworms are generally thought to be advantageous to plant growth (Edwards & Lofty 1977). A possible explanation is that earthworms graze living plant roots (Carpenter 1985; Baylis, Cherrett & Ford 1987). In this study however earthworms were not observed eating roots of living plants. A second possibility relates to their fossorial (burrowing) habit. Earthworms are the major fossorial invertebrate and the burrows they create allow non-fossorial animals to explore the soil. In the absence of earthworms there are few channels in the soil except those created by decaying plant roots, enchytraeids, *Polydesmus* spp. and *Boettgerella pallens* Simroth. The exclusion of earthworms therefore, had an effect similar to the removal of all of the soil mesofauna as they could not reach the lower horizons. This will be tested by creating artificial burrows in the earthworm exclusion cages so as to separate the physical and biotic effects of the earthworms.

The total population of all soil fauna increased with time as populations of animals invaded the sterilized soil but the numbers of animals fluctuated (Fig. 4). At least some of these fluctuations were attributable to fluctuations in the food supply; for example the death of a worm adjacent to the window resulted in an increase (Fig. 6).

The data presented are for the establishment year only; there is evidence in the literature that the effects of removing soil fauna on the herbage yield varies between years. Linzell & Madge (1986) found that during the establishment year there were no significant differences in the dry matter yield of grass between pesticide-treated and non-treated plots although there were significant differences the second year. Conversely Clements *et al.* (1982) found that in grassland treated with pesticides more plots showed a yield increase in the first year than in the second year. Brown & Gange (pp. 453–470) found that the effects of foliar and soil insecticides were inconsistent between the first and the second years.

The residual variation between windows was high for root length. This reflects the stochastic event of a root hitting a window. The observation windows in the rhizotron display a vertical slice through the soil (a profile) which is analogous to a line transect above ground.

The soil is a heterogeneous environment. The death of a single individual of *Allolobophora chlorotica* can cause a rapid increase in fauna over a short period (Fig. 6). This type of observation shows the importance of frequent observations and the need for continuous recording techniques such as time-lapse photography.

Plots in which all fauna were present had 50% less total root length (Fig. 2) and yielded 12% less herbage (Fig. 3) than plots in which fauna were removed. Even so above ground there were no symptoms of damage. This is close to the range of 1–10% reduction in herbage suggested by Curry (1987) for estimates of insidious damage by invertebrate herbivores in natural ecosystems. Henderson & Clements (1974) showed increases of 30% in annual dry matter production in grassland after treatment with pesticides even though the swards were apparently healthy before the pesticides were applied and no insect problem was thought to exist. They suggested that the loss of yield was due to the large population of soil fauna normally present in grassland. In a subsequent study of fourteen grassland sites in England it was found that this effect was widespread (Henderson & Clements 1977). The yield increases of 0–32% were attributed to the suppression of aphids, plant-sucking bugs, stem-boring larvae and nematodes.

The effect of formaldehyde on the earthworm populations in the cages suggests that it can be used as a way of removing earthworms from the exclusion cages. However it is not equally effective on all species. Vertical burrowers such as *Lumbricus terrestris* are extracted most efficiently by this method (Edwards & Lofty 1977). Formalin was used in a fescue meadow in the USA to create an artificial habitat containing reduced numbers of earthworms (Malone & Reichle 1973) and it was successful (two worms $m^{-2}$ cf. 108 worms $m^{-2}$ in the control). However it was not as effective as phorate or sodium chlorate which removed all earthworms. Malone & Reichle (1973) found that litter from a sward of fescue initially decomposed more rapidly in the absence of arthropods and earthworms, and after 1 year, there was no effect on litter decomposition. This was attributed to compensation for faunal activity by increased microbial action. The reduction in any type of fauna decreased the rate of decomposition of roots and this was thought to be a reflection of the role of animals in fragmenting plant material before it is decomposed by micro-organisms.

Formaldehyde seems to be phytotoxic (Fig. 3). The yield of herbage in the plots in which no animals were removed was only once significantly greater than the yield in the absence of part or most of the soil fauna and this followed the application of formaldehyde. Fortunately this effect seems to be short lived. Following trials in the greenhouse in which formaldehyde was found to be phytotoxic to both *L. perenne* and *T. repens*, it was decided to use a solution of mustard instead.

## CONCLUSIONS

Overall patterns observed in root growth and animal numbers were dominated by local events such as the death of an animal, or roots immigrating or emigrating from the plane of observation.

Earthworm burrows facilitated the movement of other fauna through the soil so that removing worms had the same effect as removing all soil fauna.

Earthworms moved roots, exposed and obscured roots by burrowing and defecating, formed burrows which allowed the movement of other animals through the soil, ingested soil in one area and defecated it in another and died. All of these will continually change the small-scale patchiness of nutrients in the soil and move roots into different patches.

## ACKNOWLEDGEMENTS

The senior author is grateful to Nicky Legg for her help with collecting these data, to N.R. Sackville Hamilton for his advice on the statistics and to G.R. Sagar and N.R. Sackville Hamilton for their comments and constructive criticisms. The programme is funded by AFRC.

## REFERENCES

**Baylis, J.P., Cherrett, J.M. & Ford, J.B. (1986).** A survey of the invertebrates feeding on living clover roots (*Trifolium repens* L.) using $^{32}$P as a radiotracer. *Pedobiologia*, **29**, 201–208.

**Brown, V.K. & Gange, A.C. (1989).** Differential effects of above- and below-ground insect herbivory during early plant succession. *Oikos*, **54**, 67–76.

**Carpenter, A. (1985).** *Studies on invertebrates in a grassland soil.* Ph.D. thesis, University of Wales.

**Carpenter, A., Cherrett, J.M., Ford, J.B., Thomas, M. & Evans, E. (1985).** An inexpensive rhizotron for research on soil and litter-dwelling organisms. In *Ecological Interactions in the Soil* (Ed. by A.H. Fitter, D. Atkinson, D.J. Read & M.B. Usher), pp. 67–71. Blackwell Scientific Publications, Oxford.

**Clements, R.O., French, N., Guile, C.T., Golightly, W.H., Lewis, S. & Savage, M.J. (1982).** The effect of pesticides on the establishment of grass swards in England and Wales. *Annals of Applied Biology*, **101**, 99–107.

**Curry, J.P. (1987).** The invertebrate fauna of grassland and its influence on productivity. III. Effects on soil fertility and plant growth. *Grass and Forage Science*, **42**, 325–341.

**Edwards, C.A. & Lofty, J.R. (1977).** *Biology of Earthworms.* 2nd edn. Chapman & Hall, London.

**French, N.R.** (Ed.) (1979). *Perspectives in grassland ecology. Ecological Studies*, Vol. 36. Springer Verlag, Berlin.

**Harding, D.J.L. (1968).** A preliminary survey of the relationships between the soil fauna and the roots of fruit trees. *Annual Report of the East Malling Research Station for 1967*, pp. 169–172.

**Henderson, I.F. & Clements, R.O. (1974).** The effect of pesticides on the yield and botanical composition of a newly-sown ryegrass ley and of an old mixed pasture. *Journal of the British Grassland Society*, **29**, 185–190.

**Henderson, I.F. & Clements, R.O. (1977).** Grass growth in different parts of England in relation to invertebrate numbers and pesticide treatment. *Journal of the British Grassland Society*, **32**, 89–98.

**Linzell, B.S. & Madge, D.S. (1986).** Effects of pesticides and fertilizer on invertebrate populations of grass and wheat plots in Kent in relation to productivity and yield. *Grass and Forage Science*, **41**, 159–174.

**Malone, C.R. & Reichle, D.E. (1973).** Chemical manipulation of soil biota in a fescue meadow. *Soil Biology and Biochemistry*, **5**, 629–639.

**Parle, J.N. (1963).** A microbial study of earthworm casts. *Journal of General Microbiology*, **13**, 13–32.

**Raw, F. (1959).** Estimating earthworm populations using formalin. *Nature*, **184**, 1661.

# Below-ground dynamics in a wet grassland ecosystem

M. DUMORTIER
*Laboratory of Plant Ecology, State University Ghent, Coupure Links 653, B-9000 Ghent, Belgium*

## SUMMARY

1 Below-ground dynamics in a wet grassland ecosystem were investigated using four different parameters: total root dry weight; new root dry weight; species root dry weight; and species root activity.

2 In wetter conditions above- and below-ground dry weights were higher than in drier conditions (respectively, 800 and 1409 g $m^{-2}$ *v*. 400 and 1199 g $m^{-2}$). The shoot : root ratio was highest in the wetter conditions. The total root dry weight was highest in September and lowest in December. The difference was used to get a minimum estimate of productivity: 1222 g $m^{-2}$ $year^{-1}$.

3 The growth of new (lateral) roots was lower in wetter conditions (381 g $m^{-2}$ $year^{-1}$ *v*. 596 g $m^{-2}$ $year^{-1}$). Taking into account the higher total root dry weight, this suggests a lower turnover rate in these conditions.

4 The species root dry weights, measured in pot experiments, showed a similar vertical distribution to that found in the field, although the vertical nutrient decrease was absent in the pots. The vertical root distribution in the field and in the pots of *Carex acuta* and *Phalaris arundinacea* was less pronounced when the water table was raised, while the vertical root distribution of *Glyceria maxima* stayed approximately equal.

5 While all root dry weights declined sharply between 5 and 15 cm depth, root activity by vigorous species remained relatively constant. From May to July the absorption intensity of *Carex acuta* increased slightly, while that of *Glyceria maxima* decreased.

## INTRODUCTION

While the above-ground components of the dynamics of wet grassland have often been investigated (Verlinden 1985), below-ground data are scarce. The principal reasons for this lack of information are:

1 the inaccessibility and the variability of root material; this involves much work to obtain significant results;

2 the difficulty (impossibility?) of identifying the roots of different species and separating dead and live materials. Nevertheless, knowledge of below-ground dynamics is important (Böhm 1979; Dickinson & Polwart 1982; Ingham & Detling 1984). Spatial and temporal differences in rooting patterns may enhance the

possibilities of coexistence (Berendse 1979, 1981, 1982; Fitter 1982; Slydes & Grime 1984; Veresoglou & Fitter 1984). They may contribute to the survival mechanisms of species in extreme environmental conditions, e.g. in periods with high water tables (Crawford 1983). Also the regrowth after cutting and grazing merely depends on below-ground responses (Korte & Harris 1987). Ulehlova, Tesarova & Ostry (1970) described the soil as a subecosystem affected, in the course of the growing season, by the development, maturation and degradation of the vegetational cover. Root dynamics are the link between soil and vegetational cover.

Many estimates of below-ground dynamics and production have been based upon total root dry weights. These figures have many restrictions, as they neglect inactivity, mortality and decomposition of roots (Hansson & Andrén 1986). For a better understanding of the below-ground influences on ecosystem composition and structure, several parameters have to be combined. This paper presents a comparative study of four estimates of root dynamics which are useful in semi-natural or natural grassland ecosystems. Total root dry weight, the ingrowth of new roots and the root dry weight of selected species were measured. Root activity was investigated using absorption experiments. The effects of a high water table were assessed.

In these grasslands, the below-ground biomass consists of roots and rhizomes. When talking about root dry weight, rhizomes are also included. The term dry weight was preferred to biomass, because the amount of dead roots is not known.

## STUDY SITE

The investigation was carried out in two hayfields in the nature reserve 'Bourgoyen-Ossemeersen' in the alluvial plains of the river Leie (51°6′N, 3°40′E). The water table varies between −60 and +20 cm in the wetter hayfield, which is flooded during several months each winter, while the water table in the drier hayfield fluctuates between −90 and 0 cm. The soil profiles show no profile development. The humuficious upper layer is restricted to ±10 cm, while the rest of the profile consists of homogeneous rich clay (Table 1). Important species in the wetter hayfield are *Carex acuta* (nomenclature follows De Langhe *et al.*), *Glyceria maxima, Ranunculus repens*, *Phalaris arundinacea*, *Poa trivialis, Cardamine pratensis, Lychnis flos-cuculi, Carex disticha*, etc. As conditions become drier they are less

TABLE 1. Vertical distribution (cm) of chemical characteristics (p.p.m.) and total root dry weight (g $m^{-2}$) in the upper soil layers

| Depth | N | P | K | Ca | Mg | Na | pH | Roots |
|---|---|---|---|---|---|---|---|---|
| 0–5 | 10 243 | 22 | 182 | 4400 | 181 | 52 | 5·03 | 724 |
| 5–10 | 7280 | 12 | 150 | 5367 | 109 | 32 | 5·23 | 207 |
| 10–15 | 4526 | 4 | 103 | 6767 | 83 | 23 | 5·47 | 167 |
| 15–20 | 2683 | 4 | 102 | 7433 | 87 | 23 | 6·22 | 100 |

productive and accompanied by *Holcus lanatus, Anthoxantum odoratum, Festuca rubra*, etc. The hayfields are cut once a year during summer. Above-ground yields vary from ±800 g $m^{-2}$ in the wetter hayfield to ±400 g $m^{-2}$ in the drier hayfield.

## METHODS AND MATERIALS

### *Total root dry weight*

Total root dry weight was measured by taking three (at 2-monthly intervals, in a wet plot, 1984–85) or nine (in a wet and a dry plot, November 1987) random samples with a steel corer with a diameter of 8·2 cm. From each sample above-ground parts were cut at the soil surface. The cores were transported to the laboratory in polyethylene bags, divided into vertical sections of 5–10 cm and stored at 2 °C for a few days. From the November 1987 sample soil was removed for chemical analysis. Homogeneous clay samples were oven-dried (24 h; 90 °C) and then soaked in a 0·27% $Na_4P_2O_7$ solution for the dispersion of clay particles. Soil samples were washed with water and the roots captured on a 450 μm -sieve. Visual separation of living and dead roots was not possible, because of the density of the root network and the presence of a number of different species. Finally the washed roots were oven dried (24 h; 90 °C), weighed, ashed, and dissolved in 1 N HCl. This indicated a residual soil fraction of 16%, which was used as a correction factor.

### *New root dry weight*

To obtain supplementary estimates of the production of new roots during 1 year, root-free soil samples ($\phi$ = 8·2 cm, H = 10 cm) were put into the ground (September 1986). Eighty soil cores were taken *at random* from both the wetter and drier plots, and roots were extracted manually. This seemed to be the best method, as wet sieving washes out the nutrients and dry sieving was too labour intensive. Some sandy soil was added to the remaining soil to obtain the original volume and conserve soil structure. The root-free soil was sterilized (8 h; 110 °C) and put back in the holes. The soil was compacted to a density similar to that of the surrounding soil. No resowing or replanting was done. One year later, in September 1987, the soil was removed and roots washed out and dried (see above).

### *Species root dry weight*

Species root dry weights of *Carex acuta, Glyceria maxima* and *Phalaris arundinacea* were investigated in pot experiments. To evaluate root responses to water saturation three permanent water tables were established (−32, −16 and 0 cm). Eighty-one plants of each of the above species were dug out (±10 cm soil) from the field and transplanted into pots (22·5 × 24 × 48 cm deep) (March–April 1986). Each pot contained nine plants of the same species, so each water table–species combination was repeated three times. The soil was obtained from neighbouring agricultural land, which had not been treated with pesticides during the previous year. Its structure was loamy clay. Roots were extracted manually from the soil, which was sterilized (8 h; 110 °C). After 1 year (March–April 1987) soil from the middle of each pot (15 × 15 cm) was divided into vertical sections 5–10 cm long. Roots were washed out and dried (see above).

### *Species root activity*

Measurements of root dry weight concentrate emphasis on the thick roots, while absorption mainly occurs in the fine roots. To investigate this aspect of root dynamics, a strontium absorption technique was used (Veresoglou & Fitter 1984). A strontium solution (3 mg $ml^{-1}$) was injected in three different 5 × 5 grids at a spacing of 5 cm. Each grid used a different injection depth: 5, 10 or 15 cm. Grids were sited where the species under investigation were abundant. Plants with similar intensities of flowering were chosen as preliminary measurements and showed that flowering shoots accumulated only approximately half of the concentrations of strontium in their leaves. One month after injection, the above-ground biomass was cut at the soil surface, separated by species, oven dried and weighed. After ashing, the strontium concentration was measured by atomic absorption. The experiment was repeated in May, June and July 1986.

## RESULTS

### *Total root dry weight*

The total root dry weight between 0 and 20 cm depth varied from 1199 g $m^{-2}$ in the drier plot to 1409 g $m^{-2}$ in the wetter plot. As above-ground dry weights were 400 g $m^{-2}$ and 800 g $m^{-2}$ respectively the shoot : root ratio varied from 0·3 to 0·6 respectively. As the above-ground biomass is removed each year, while some roots may stay for over 1 year, the actual shoot : root ratio for productivity may be higher. On the other hand, if the amount of decomposed roots is higher than the mean root dry weight then the ratio may be lower. The vertical root distribution is presented in Fig. 1. The root system under wetter conditions was shallower than under drier conditions but still penetrated up to 100 cm depth under conditions where the reduction boundary was at 70 cm depth.

Time variations in total root dry weight are presented in Fig. 2. The quantity of roots showed a peak in September (2383 g $m^{-2}$), while the amount was lowest in January (1161 g $m^{-2}$). This means that at least 1222 g $m^{-2}$ new roots were produced between 0 and 40 cm during 1 year. As the curve is a result of productivity, mortality and decomposition, actual production could be higher.

### *New root dry weight*

In the wet plots, 273 ± 187 g $m^{-2}$ roots were formed between 0 and 5 cm, and 108 ± 89 g $m^{-2}$ roots were produced between 5 and 10 cm depth. In the dry plots,

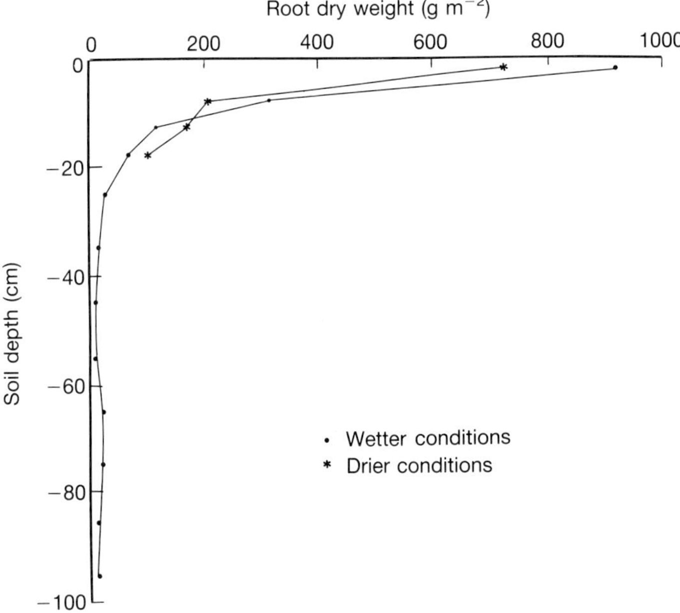


Fig. 1. Vertical root distributions.

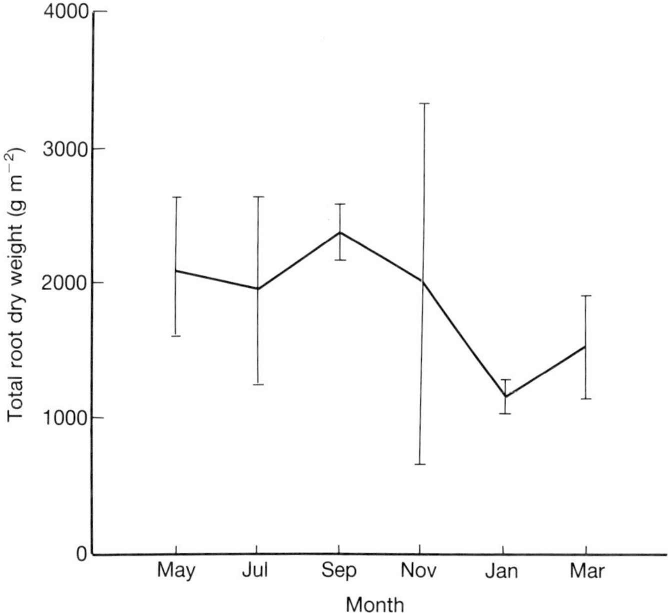


Fig. 2. Seasonal changes in total root dry weight (wet plot).

comparable values were 422 ± 184 and 174 ± 96 g m$^{-2}$. These quantities are much lower than the previous calculations which were considered an underestimate. The cause of this may be that when using root-free soil cores only lateral root growth is measured, while normal soil cores are regularly taken directly under the shoot. Other possible reasons for the lower estimate are that the experiment with the root-free soil cylinders was carried out in the extreme wet summer of 1987, that the soil conditions in the sterilized, slightly sandy clay are not exactly the same as in the surrounding soil, and surrounding roots are cut off or injured during installation which may influence the root production (Hansson & Andrén 1986).

New root production was higher in drier than under wetter conditions while the total root dry weight (between 0 and 10 cm) was lowest in drier conditions. This indicates a higher root-turnover in the drier plots and suggests that the difference in

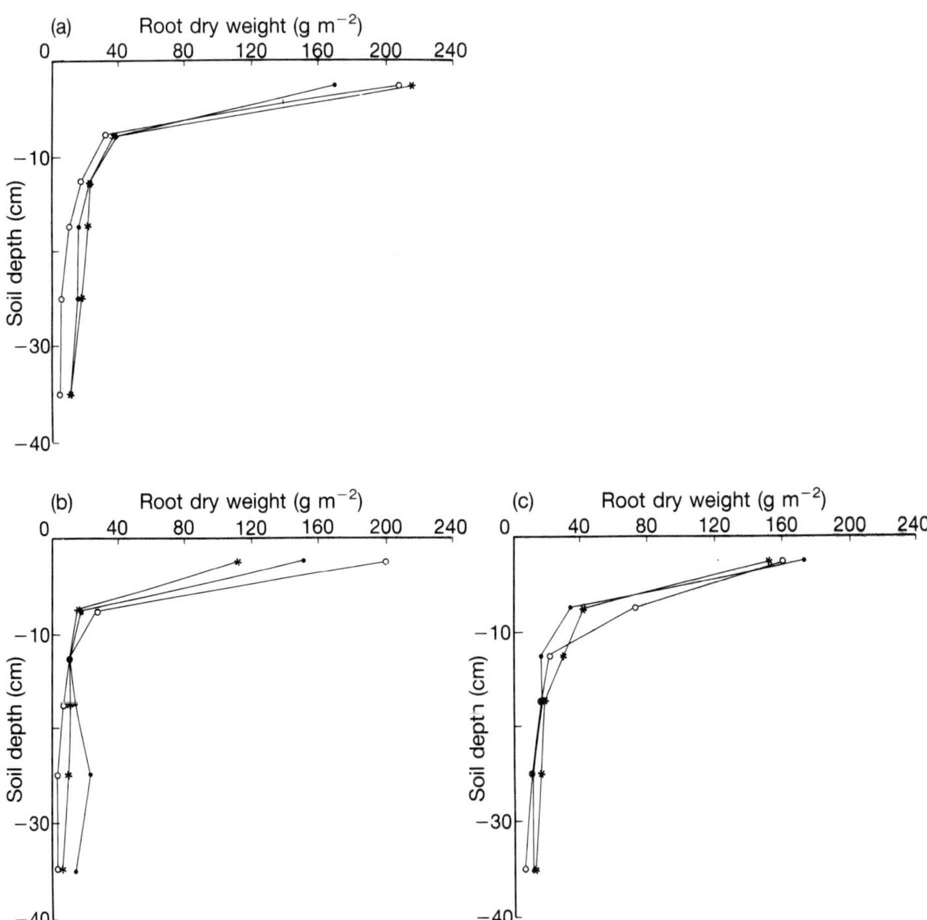


FIG. 3. Species vertical root distribution: (a) *Carex acuta*; (b) *Phalaris arundinacea*; (c) *Glyceria maxima*. ● = water table at −32 cm; * = water table at −16 cm; ○ = water table at 0 cm.

shoot : root ratio, indicated above, should be higher. The variability in root production was higher under wetter conditions.

*Species root dry weight*

Root dry weight between 0 and 40 cm was 295 g m$^{-2}$ for *Carex acuta*, and 265 g m$^{-2}$ for *Glyceria maxima* and 226 g m$^{-2}$ for *Phalaris arundinacea*. This is less than determined with the root-free soil cylinders, although the roots immediately under the shoots were investigated in this experiment. The small quantity of roots recovered may be a consequence of transplantation, the absence of ecosystem interactions, the homogeneity of species composition etc. The vertical root distribution of the three species, in relation to the water table, is presented in Fig. 3. The global vertical root distribution was very similar to the root distribution found in the field, although the potplants got similar amounts of nutrients over all the soil profile. *Phalaris arundinacea* formed more roots with the high water table, although the roots were shallower. *Carex acuta* also formed shallower roots with a high water table although the total root mass stayed the same. *Glyceria maxima* formed slightly less roots with a high water table with little effect on vertical distribution.

*Species root activity*

The strontium absorption of *Carex acuta* is presented in Fig. 4. *Carex acuta* absorbed little strontium, analogous to its low calcium uptake (Soileau 1973). The strontium absorbed by *Glyceria maxima* decreased during the same period. While the quantities of root decreased sharply between 5 and 15 cm depth, activity was relatively constant. This was also the case for *Phalaris arundinacea* but not for all

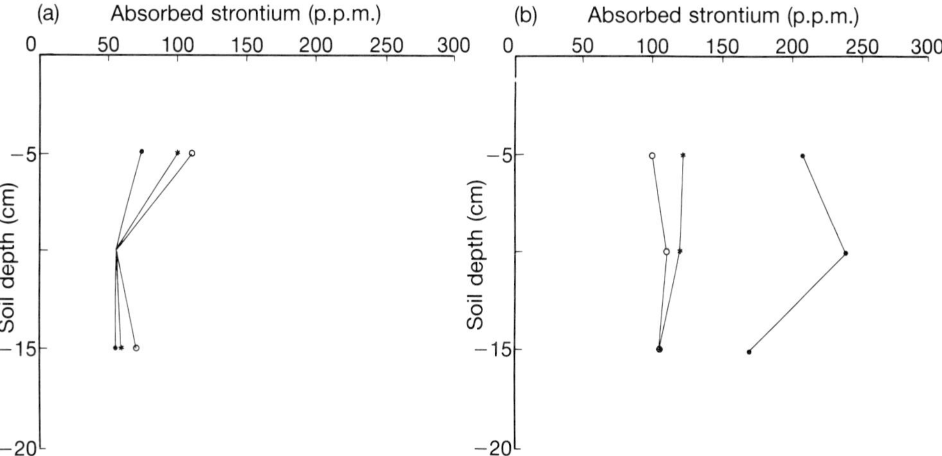


FIG. 4. Strontium absorbed by (a) *Carex acuta* and (b) *Glyceria maxima*. ● = May; * = June; ○ = July.

species. The absorption pattern of *Ranunculus repens* decreased between 5 and 15 cm depth. This also occurred with *Agrostis cappilaris*, *Holcus lanatus* and *Poa pratensis* (Veresoglou & Fitter 1984).

## DISCUSSION

The estimates of below-ground root mass and production correspond with other figures from wet grassland ecosystems (Bradbury & Grace 1983). Marshes and swamps are amongst the most productive ecosystems, which is generally attributed to a plentiful supply of nutrients, due to flushing with nutrient-rich water, and sufficient water availability during most of the year (Bradbury & Grace 1983).

Some obvious differences between the wetter and the drier hayfield were observed. The shoot : root ratio was higher in wetter conditions and the root system was shallower. As the below-ground turnover was assumed to be lower, the percentage of dead root material has to be higher in these conditions. These observations mean that in spite of the stresses due to water-logged soil conditions, the remaining amounts of active roots are able to support higher shoot production. The capacity of some investigated species to survive in these extreme conditions was illustrated in pot experiments. The applied stagnant water table here resulted in a greater level of stress. The fluctuating water table in the field improves oxygen supply and reduces levels of toxic substances by sweeping them away. Some specific adaptations were described in Crawford (1983).

In the pot experiments mean vertical root distribution stayed similar to that found in the field. The homogeneous availability of nutrients over the soil profile did not result in a deeper rooting pattern. This suggests that nutrient availability is not important to root dynamics in these ecosystems.

The discordance between the vertical distribution of root weight and root activity for some of the species was remarkable. A possible explanation is the fact that some roots in the upper layer, which are generally thicker, function for the transport of water and nutrients, and do not (or only slightly) normally absorb. This hypothesis is supported by the fact that the more vigorous species continue to absorb with the same intensity in deeper soil layers. The ratio of 'transporting' roots : 'absorbing' roots declined with depth. Root dry weight, which is mainly dependent on the thick roots, declines sharply. The absorption, by the finer roots, which are relatively more abundant in the deeper layers, remains the same. Species with shallow root systems have little need for transport through the upper soil layer. The root dry weight is therefore mainly dependent on 'absorbing' fine roots, and consequently limited. Root dry weight, and absorption intensity, decline sharply with depth.

This discordance illustrates the necessity to apply different parameter types to understand ecosystem dynamics. This was also suggested by Veresoglou & Fitter (1984), who remarked that vegetative growth did not necessarily coincide with nutrient uptake.

The results of the absorption experiment demonstrated some seasonal changes in root activity. While the activity of *Glyceria maxima* obviously decreased between

May and July, the activity of *Carex acuta* increased slightly. Below-ground activity appeared to be lower during flowering. In addition different mowing dates, intervening in different stages of seasonal cycles, may have different consequences for ecosystem composition and structure. Below-ground reactions under different mowing regimes are currently under investigation.

## ACKNOWLEDGEMENTS

This research is carried out with the financial support of the Institute for the encouragement of Scientific Research in Industry and Agriculture (I.W.O.N.L.). The author gratefully acknowledges Professor J. Schalck for promoting the investigation and A. Verlinden and H. Beeckman who assisted with a critical review of the manuscript.

## REFERENCES

**Berendse, F. (1979).** Competition between plant populations with different rooting depths. I. Theoretical considerations. *Oecologia*, **43**, 19–26.

**Berendse, F. (1981).** Competition between plant populations with different rooting depths. II. Pot experiments. *Oecologia*, **48**, 334–341.

**Berendse, F. (1982).** Competition between plant populations with different rooting depths. III. Field experiments. *Oecologia*, **53**, 50–55.

**Bradbury, I.K. & Grace, J. (1983).** Primary production in wetlands. *Ecosystems of the World. 4A. Mires: Swamp, Bog, Fen and Moor. General Studies* (Ed. by D. Goodall), pp. 285–310. Elsevier, Amsterdam.

**Böhm, W. (1979).** *Methods of studying root systems. Ecological Studies. Vol. 33*. Springer Verlag. Berlin.

**Crawford, R.M.M. (1983).** Root survival in flooded soils. *Ecosystems of the World. 4A. Mires: Swamp, Bog, Fen and Moor. General Studies* (Ed. by D. Goodall), pp. 257–283. Elsevier, Amsterdam.

**De Langhe, J.E., Delvosalle, L., Duvigneaud, J., Lambinon, J. & Van den Berghen, C. (1983).** *Flora van België, het Groothertogdom Luxemburg, Noord-Frankrijk en de aangrenzende Gebieden*. Nationale Plantentuin, Meise.

**Dickinson, N.M. & Polwart, A. (1982).** The effect of mowing regime on an amenity grassland ecosystem: above- and below-ground components. *Journal of Ecology*, **19**, 569–577.

**Fitter, A.H. (1982).** Influence of soil heterogeneity on the coexistence of grassland species. *Journal of Ecology*, **70**, 139–148.

**Hansson, A.-C. & Andrén, O. (1986).** Below-ground plant production in a perennial grass ley (*Festuca pratensis* Huds.) assessed with different methods. *Journal of Ecology*, **23**, 657–666.

**Ingham, R.J. & Detling, J.K. (1984).** Plant–herbivore interactions in a North American mixed-grass prairie. III Soil nematode populations and root biomass on *Cynomys ludovicianus* colonies and adjacent uncolonized areas. *Oecologia*, **63**, 307–313.

**Korte, C.J. & Harris, W. (1987).** Effects of grazing and cutting. *Ecosystems of the World. 17B. Managed Grasslands. Analytical studies* (Ed. by R.W. Snaydon), pp. 71–79. Elsevier, Amsterdam.

**Slydes, C.L. & Grime, J.P. (1984).** A comparative study of root development using a simulated rock crevice. *Journal of Ecology*, **72**, 937–946.

**Soileau, J.M. (1973).** Activity of barley seedling roots as measured by strontium uptake. *Agronomy Journal*, **65**, 625–628.

**Ulehlova, B., Tesarova, M. & Ostry, I. (1970).** Immobilisation of nitrogen in the soils of certain alluvial meadow ecosystems. *Acta Oecologia/Oecologia Plantarum*, **5**(19), 147–178.

**Veresoglou, D.S. & Fitter, A.H. (1984).** Spatial and temporal patterns of growth and nutrient uptake of five coexisting grasses. *Journal of Ecology*, **72**, 250–272.

**Verlinden, A. (1985).** *De dynamiek van kruidachtige vegetaties in funktie van waterhuishouding en beheer van natuurgebieden*. Ph.D. thesis, University of Ghent.

# Root function related to the morphology, life history and ecology of tundra plants

T.V. CALLAGHAN,* A.D. HEADLEY[†] AND J.A. LEE[‡]
* *Institute of Terrestrial Ecology, Merlewood Research Station, Grange-over-Sands, Cumbria LA11 6JU, UK;* [†] *Department of Environmental Sciences, The University, Bradford BD7 1DP, UK; and* [‡] *Department of Environmental Biology, The University, Manchester M13 9PL, UK*

## SUMMARY

1 Tundra areas are extensive and characterized by extreme physical environments related to short, cool growing seasons and long, cold winters. The physical environment of the soil is dominated by low temperatures (rarely above 7 °C) and freeze–thaw cycles which occur in the active soil layer above the permafrost. Soil displacement during freeze–thaw cycles heaves poorly adapted species out of the soil. Survival probabilities will be increased in plants with small roots, elastic roots, strong tap roots, moderately branched root systems, non-vertically orientated root systems or annual roots.

2 Tundra soils may be waterlogged when permafrost prevents drainage, or dry when gravelly on wind-exposed fell fields. Members of the Cyperaceae which pump oxygen to roots are particularly abundant in the moist and wet areas of the tundra whereas plants with xeromorphic features characterize fell-fields and High Arctic polar deserts.

3 Tundra soils are generally cold and infertile and there are many mechanisms which tundra plants display to overcome this.

(a) Tundra plant roots can withstand freezing, some down to −195 °C in winter and −23 °C in summer.

(b) The roots of tundra plants can grow and function at low temperatures, approaching 0 °C while respiration rates are less sensitive to temperature than those of temperate species and apparent activation energies are low.

(c) Nutrient uptake can occur at temperatures as low as 0 °C.

(d) Arctic and alpine plants have the highest below- to above-ground biomass quotients (6–34) but there is variation between regions with polar desert plants resembling those of hot deserts with quotients of 0·5–3.

(e) The roots of many tundra plants are long-lived and economize on the investment of nutrients and carbon. Clonal species predominate and the roots of old leafless generations of sedge modules supply younger generations with nutrients.

(f) Where the soil environment is patchy, rhizomatous and stoloniferous species forage for nutrients and exploring growing points receive subsidies of water, nutrients and energy from older generations within clones.

(g) Nutrients may be conserved by recycling between leaves, between leaf cohorts and between generations of modules within clones. Particularly efficient recycling in some species allows plants to survive with very little biomass allocated to roots.

(h) Many tundra species conserve nutrients by functioning with lower nutrient concentrations in their tissues than temperate species.

(i) Increased root surface area to volume quotients would allow more efficient uptake of nutrients but tundra species appear to have simpler branching patterns than temperate species, while cryptogams without roots are particularly abundant.

(j) An increased absorptive surface area for nutrients would be achieved by the development of mycorrhizae but, as much of the tundra is dominated by members of the Cyperaceae which generally lack them, mycorrhizae must be less important in the tundra than elsewhere.

**4** It was hypothesized that (i) there is no unique combination of options for species to respond to low soil fertility in a particular habitat; (ii) the severity of the stress experienced by a species should be reflected in the number of options exhibited; and (iii) species exhibiting different combinations of options should have an increased chance of coexisting. A survey of options exhibited by a few species showed that no two species adopted identical combinations of options and that none adopted all of the options. Thus none of the species was maximally adapted to low soil fertility whilst the potential for coexistence was considerable. The sample size was extremely small, however, and more quantitative data are required.

## INTRODUCTION

Tundra areas cover 17% of the earth's land surface yet relatively little is known about root structure and function of tundra plants. Tundra areas are characterized by short, cool growing seasons, long, cold winters and primitive impoverished humus soils which often consist of a lower permanently frozen layer — the permafrost — and an upper 'active layer' which thaws in summer. There are, however, great geographical differences in the environment, mainly associated with oceanic/continental gradients. Oceanic South Georgia (54°S 17′E), for example, has a minimum monthly mean temperature in winter of only −2 °C and a maximum monthly mean temperature in summer of only +5 °C with monthly mean temperatures above 0 °C for 8 months of the year (Rosswall & Heal 1975). In contrast, Devon Island in the High Arctic has a minimum monthly mean temperature as low as −36 °C and a maximum monthly mean summer temperature of up to 9 °C with mean monthly temperatures above 0 °C for only 3 months. Precipitation also varies greatly along oceanic/continental gradients and many polar areas — the polar deserts and semi-deserts — receive little precipitation. For example, Devon Island and Point Barrow (Alaska) have an annual precipitation of only 151 and 83 mm, respectively (Barry, Courtin & Labine 1981), although Point Barrow has much surface water due to the thawing of frozen ground in summer.

Within the tundra region there are great differences in environment often associated with relatively small changes in topography which affect the duration of snow-lie, drainage and temperature through insolation related to aspect and shelter. The communities of the various tundra habitats range from closed forest–tundra communities, which represent the ecotone from taiga to tundra with scattered trees of low stature, to open fell-field communities of scattered graminoids, cushion forbs and prostrate dwarf shrubs on exposed hilltops, and snow bed communities with contracted growing seasons in sheltered depressions (Callaghan & Emanuelsson 1985). Moist meadow communities, such as those dominated by *Eriophorum vaginatum*, are particularly extensive ($0{\cdot}9 \times 10^6\,km^2$; Miller 1982) on intermediately drained areas, often with underlying permafrost. Biomass and rates of net primary production are low in the tundra. Biomass ranges from near zero in polar deserts to $4000\,g\,m^{-2}$ in shrub and forest tundra and net primary production in these habitats ranges from $1\,g\,m^{-2}\,year^{-1}$ to $1200\,g\,m^2\,year^{-1}$, respectively (Wielgolaski *et al.* 1981).

The structure, development, function and reproduction of tundra plant species have evolved primarily in response to selection pressures of the physical environment, rather than the interspecific competition which tends to dominate the selection pressures on species in mesic temperate environments (Callaghan & Emanuelsson 1985). In harsh tundra environments strongly deterministic growth forms predominate with growing points physically protected within compact shoots. In the closed vegetation of the less severe tundra environments, more opportunistic growth occurs. This allows growing points, provided with subsidies of nutrients, carbon and water by older established parts of the plant, to be extended beyond the shelter of the main body of the plant (Headley, Callaghan & Lee 1988a, b; Callaghan 1988; Jonsdottir & Callaghan 1990).

In general, the vegetation of the Arctic is young, and true adaptations to arctic environments are few (Savile 1972). The indigenous species tend, therefore, to avoid the harshest conditions rather than show elaborate adaptations. Thus, compact, dwarf and mat-forming growth forms are common and most plant structures — up to 98% of total living biomass (Billings, Peterson & Shaver 1978; Wallen 1986) — are below ground. Between-plant competition below ground is likely, therefore, to be more intense than that above ground.

The rhizospheres of tundra environments share many characteristics with those of other environments, for example, infertility, aridity or anoxia. However, low summer temperatures and the temperature fluctuations associated with daily and seasonal freeze–thaw cycles, and the soil movement associated with these, are unique to the tundra, particularly to the Arctic.

This paper concentrates on aspects of root form and function which are characteristic of tundra plants. As data are scarce, a certain amount of conjecture is unavoidable and gaps in our knowledge, together with priorities for research, are highlighted.

## THE PHYSICAL ENVIRONMENT OF TUNDRA SOILS

The physical environment of tundra soils is dominated by temperature and freeze–thaw cycles. Tundra soils do not experience the wide fluctuations in temperature common in other soils. Winter temperatures are rarely lower than −20 °C under snow cover and surface temperatures rarely exceed 20 °C in summer. In the spotted tundra of western Taimyr, there were 20 days per year with soil temperatures above 5 °C at 5 cm depth, and only 12 days when temperatures exceeded 5 °C at 20 cm depth (Everett *et al.* 1981).

Freeze–thaw cycles, or 'cryoperturbation' processes, occur in the upper layers of the soil throughout the day, particularly at the beginning and end of the growing season. In some areas where water is present, these cycles result in frost heave with annual ground displacement of several centimetres and cyclical differential ground pressures of many kilopascals per square centimetre (French 1987). Geomorphic indicators of frost heave include the upheaval of bedrock blocks and the sorting and migration of soil particles which leads to the formation of solifluction lobes and patterned ground.

The surface layer of the soil where most of the freeze–thaw cycle occurs, and water freezes at 0 °C, is called the 'active layer' and it may be only 20 cm thick in the coldest tundra soils (Everett *et al.* 1981). The predominantly permanently frozen zone below is called the 'permafrost' (Fig. 1). This extends down to 600 m on Devon Island (Brown 1977). Some of the permafrost may undergo freeze–thaw cycles even though the soil temperature is below 0 °C and as much as 40% of soil water may be unfrozen at −1 °C (Williams 1976, 1977; French 1987). Presumably, this unfrozen water is particularly rich in mineral ions. Sometimes the permafrost occurs as ice wedges and these may grow over a period of years and attenuate the active layer by 20–40% (French 1987). In contrast, the longer term thawing of permafrost leads to subsidence, erosion and the development of hummocky thermokarst topography which will assume particular significance in the future as climate changes.

Freeze–thaw cycles in the active layer may be one or two sided (Fig. 1). One-sided freezing and thawing in periglacial areas experiencing seasonal frost, such as some temperate and alpine areas, leads to soil movement which decreases down the profile. In areas of permafrost, however, two-sided freezing and thawing occurs and the formation of ice lenses at the surface of the permafrost leads to lateral variations in the rate of soil movement and the formation of patterned ground in which the soil particles are either sorted (polygons) or not sorted (hummocks). On vegetated slopes with hummocks, soil movement — or 'solifluction' — may occur at 1 cm year$^{-1}$ (Mackay 1981) with the hummocks sliding over water-saturated sediments lying above the permafrost. When there is high hydrostatic pressure in the active layer created by excessive pore water trapped between the permafrost and a rigid carapace, mud boils may form and in some areas these may form exceptionally rapid large-scale mud flows as the active layer slides over the permafrost.

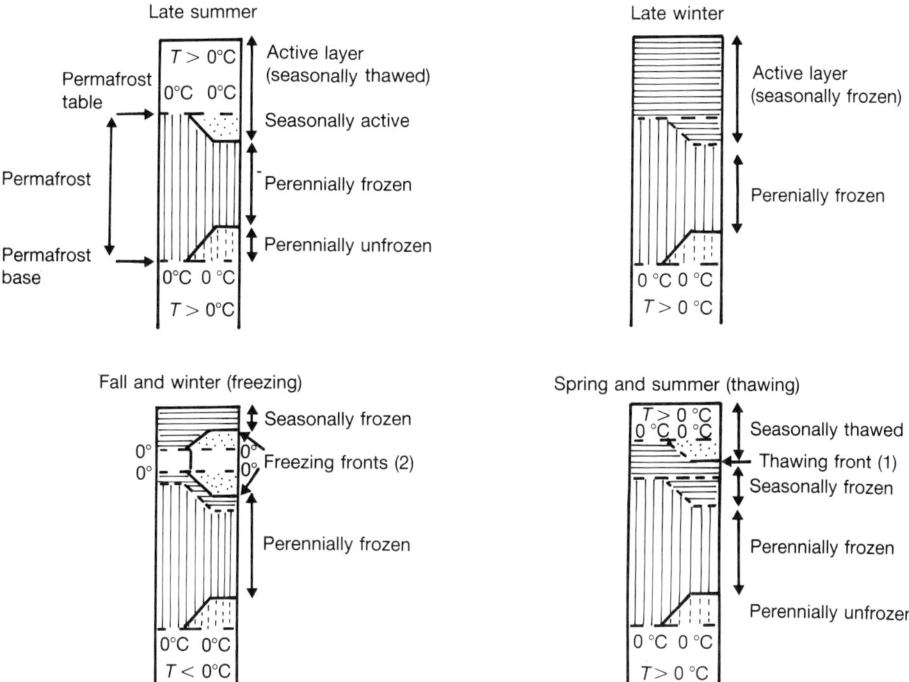


FIG. 1. Diagrammatic representation of the seasonal freeze–thaw cycles in tundra soils with permafrost (from French 1987).

## THE EFFECT OF FROST HEAVE ON PLANTS

As the soil freezes in autumn, needle ice forms and primary frost heave produces minor displacements of the soil. However, the following winter-long formation of ice lenses produces a major soil displacement or secondary frost heave (Perfect, Miller & Burton 1988). This secondary frost heave may produce vertical soil displacements with amplitudes commonly reaching 10–100 mm (James 1971; Fahey 1979). Primary frost heave tends to displace seedlings whereas secondary frost heave can displace mature overwintering plants, particularly crops and transplanted trees (Perfect, Miller & Burton 1988). However, frost heave can also be a major constraint on the survival of native species in the Arctic (Wager 1938).

Recent experiments on alfalfa (*Medicago sativa* L.) at Ithaca, New York, where the major sources of mortality are crown exposure and root breakage, showed that the major displacement took place in midwinter and that this was not totally reversed (Fig. 2; Perfect, Miller & Burton 1988). However, continual primary frost heave events could lead to the 'jacking-up' of plants (Fig. 2) even though the forces of each displacement were relatively small. Attempts to measure

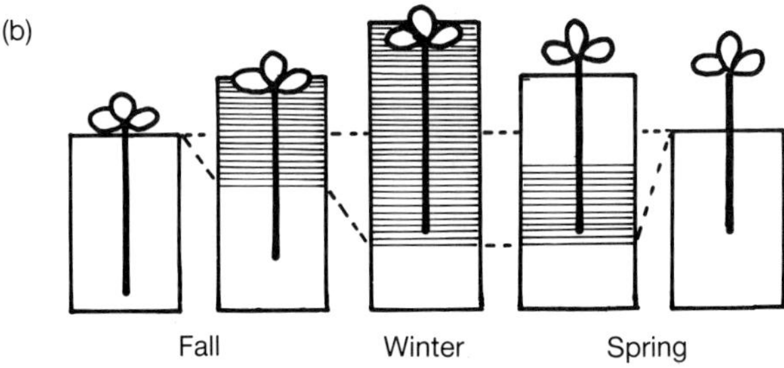


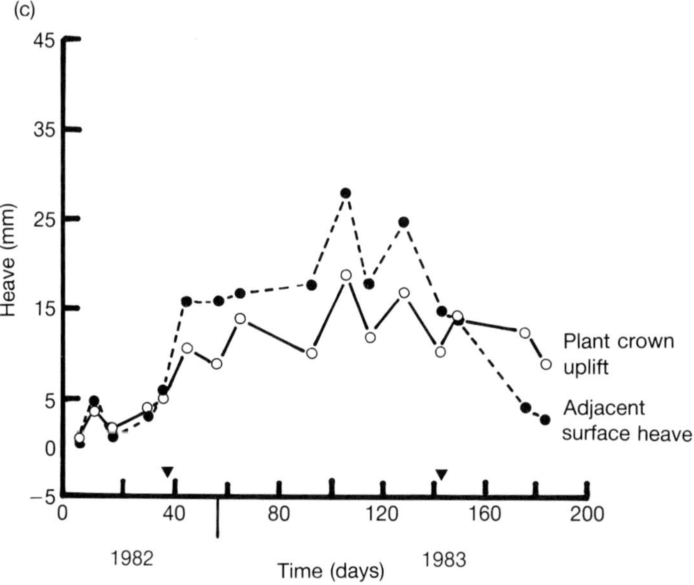


FIG. 2. Seasonal pattern of frost heave and uplift of crowns of 2-year-old alfalfa plants (c) and diagrammatic representation of the primary heave — or jacking up — of plants (a) and secondary winter-long heave (b). Vertical shading denotes needle ice, and horizontal shading denotes ice lenses (from Perfect, Miller & Burton 1988).

the forces responsible for lifting mature alfalfa plants in midwinter were unsuccessful but exceeded 0·5 kN. Calculations predicted values in the range 0·42–1·89 kN per plant.

The final uplift of plants by spring is a function of the amplitude of the seasonal frost heave cycle and biotic factors related to root morphology (Perfect, Miller & Burton 1988). Perhaps the most important aspects of the root morphology of species naturally occurring in frost heave areas are:

**1** small root systems which allow plants to float, e.g. *Pinguicula* spp.;

**2** elastic root systems which allow displacement and recovery without damage, e.g. the coiled roots of *Phippsia algida* (Bell & Bliss 1978) and the zig-zag roots of *Tofieldia pusilla*;

**3** tap roots with high tensile strength, presumably such as those of *Silene acaulis*, etc.;

**4** moderately branched root systems — unbranched root systems would be easily jacked-up whereas finely branched systems would be susceptible to breakage;

**5** roots which are not vertically orientated, e.g. those of many rhizomatous sedges such as *Carex bigelowii*; and

**6** annual roots, such as those of *Eriophorum angustifolium* (Shaver & Billings 1975) in which breakage would be irrelevant.

Detailed analyses of root morphology are now required and these should be related to the position of native species within patterned ground (Jonasson 1986), i.e. the degree of soil movement which they can tolerate. Jonasson (1986) for example, showed a variation in the occurrence of growth forms on patterned ground from rhizomatous species on stable ground, through adventitiously-rooted dwarf shrubs on ground of moderate disturbance to fibrous-rooted species where disturbance (up to 40 mm heave) is greatest. Studies are at present underway to relate the variation in tensile strength and elasticity of species associated with patterned ground to displacement (Callaghan & Jonasson, personal observation).

## THE EFFECT OF WATER STRESS AND WATERLOGGING ON PLANTS

Tundra soils generally experience waterlogging for at least part of the summer because permafrost prevents free drainage of melt water and any summer rainfall. For example, much of the ground at Point Barrow, Alaska, is waterlogged or has surface water even though the annual precipitation is only 83 mm. Depressions have very low oxygen concentrations, particularly at depth. These areas tend to be occupied by meadow or wetland species which are tolerant of anoxic environments in more temperate regions, e.g. *Eriophorum angustifolium*, *E. vaginatum* and *Carex aquatilis*.

The pumping of oxygen through an extensive rhizome system to the finer roots is not confined to arctic plants, but may explain the predominance of members of the Cyperaceae in the arctic tundra and the generally thicker roots (i.e. greater oxygen pumping capacity) found in plants from wet meadows at Point Barrow (Chapin 1977, 1978).

Fell-fields at the other extreme, experience dry soils in both summer and winter. In winter, water is locked up in ice while wind exposure and little insulating snow cover leads to desiccation. In mid- to late summer, the gravelly soils have a deep thaw, are well drained and precipitation may be low. *Saxifraga oppositifolia*, typically found in some of the driest parts of the Arctic, can withstand very low water potentials of −4·4 to −5·5 MPa (Teeri 1973). Many plants of the fell-field habitat have xeromorphic features and deep tap roots similar to plants of hot deserts.

In less extreme habitats, xeromorphic leaves may be more related to nutrient conservation in some species (Small 1972; Thomas & Grigal 1976; Headley, Callaghan & Lee 1985) as few plants with xeromorphic leaves at two study sites in Alaska experience low enough water potentials to reduce growth significantly (Miller *et al.* 1978; Oberbauer & Miller 1982). Severe root pruning had no significant effect on the water potentials of *Betula nana* but significantly reduced those of *Empetrum nigrum* (Oberauer & Miller 1982). This suggests that parts of the large, deep and extensive root system of deciduous *Betula nana* can subsidize parts of the shoots normally receiving water from other parts of the root system. This can also occur in the stoloniferous evergreen cryptogam *Lycopodium annotinum* in which 3- and 4-year-old root systems normally provide most of the water for the growing point of the stolon, but when rooting is impossible, all of the water can be provided by roots up to 11 years old (Headley, Callaghan & Lee 1989).

## PLANT RESPONSES TO INFERTILE SOILS

Generally, tundra soils are infertile, although there are exceptions such as those supporting bird cliff communities (Bliss 1971). There are many ways in which plants have responded to tolerate infertile soils (Table 1). The enumeration of these options allows comparisons to be made between species and habitats. Three generalizations can be made (Callaghan 1987):

1 there is no unique combination of options for a particular habitat;

TABLE 1. Potential responses of tundra plants to cold, infertile soils

| |
|---|
| Ability to function at low temperatures |
| Increased efficiency of nutrient uptake by roots at low temperatures |
| Increased root to shoot biomass |
| Increased root longevity to exploit soil with reduced cost of carbon and nutrients |
| Presence of foraging habit to locate favourable microsites |
| Conservation of nutrients by recycling |
| Increased efficiency of nutrient use in tissues with low nutrient concentrations |
| Increased surface area of roots relative to weight |
| Presence of mycorrhizae to forage for N and P beyond zones depleted by roots |
| Functioning with low growth rates to reduce nutrient requirements |
| Presence of store of non-structural carbohydrates to allow opportunistic growth |

**2** greater degrees of nutrient stress should lead to a greater number of options being exhibited by a species; and

**3** two species in a habitat which have evolved different combinations of options are likely to coexist: two species with similar combinations of options are likely to compete.

The different options shown by various growth forms in different tundra habitats are detailed below using this 'check list' approach.

## *Physiological function at low temperatures*

The roots of arctic plant species have to survive long periods of freezing and take up water and nutrients from cold soils during a short growing season. Prerequisites for the roots of tundra plants are therefore, that they must be resistant to freezing and they must be able to grow and respire efficiently at low temperatures.

### *Freezing resistance*

From the few studies on the freezing tolerance of arctic–alpine species from alpine environments, it would appear that freezing damage to plants is rare (Larcher 1981, 1985). The freezing tolerance of the roots of *Silene acaulis* and *Carex firma* is generally lower in mid-winter (−30 to −195 °C and −30 to −70 °C respectively) than in summer (−11 to −23 °C and −8 to −13 °C respectively) (Larcher 1981). The potential freezing resistance of hardened arctic–alpine plants can be as low as −196 °C in *Saxifraga oppositifolia* and *Silene acaulis* and even after de-hardening, freezing resistance is still as low as −25 to −45 °C (Larcher 1985).

### *Root initiation and growth*

The soils of the arctic in summer would normally halt the growth of temperate plant roots, as the soils rarely rise above 7 °C in the growing season (Klepper 1987). Root initiation can occur at very low temperatures, with *Eriophorum angustifolium* having optimum root initiation at 5 °C (Chapin 1974). The temperature optimum for root elongation rates of *E. angustifolium, Dupontia fischeri* and *Carex aquatilis* in the same study was between 5 and 10 °C (Chapin 1974) whilst the fastest rates of mean root growth observed in the field for *E. angustifolium* and *Carex aquatilis* were at 0 to 0·9 °C and *Dupontia* roots grew fastest at 7–7·9 °C (Billings *et al.* 1977). In fact the roots of these plants were observed to carry out growing down to −0·5 °C under controlled conditions (Billings *et al.* 1977). Root growth in these species (Billings *et al.* 1977, 1978), and in a number of species from the High Arctic (Bell & Bliss 1978), ranges from 0·2 to 2·5 mm $day^{-1}$ below 5 °C.

### *Respiration*

The meristematic growth of roots is partly dependent on respiration rates (Billings *et al.* 1977) but also on the photoperiod experienced by the shoots. The supply of energy from respiration may be particularly important in determining potential rates of nutrient uptake. Root respiration in arctic plant species can operate at about 0·1–0·3 mg $CO_2$ $h^{-1}$ $g^{-1}$ at just above 0 °C (Chapin 1977; Billings *et al.* 1977). *E. angustifolium* roots generally respire faster than *Dupontia fischeri* or *Carex aquatilis* roots at low temperature. The respiration rates of arctic plant roots show less temperature sensitivity than those of temperate species with arctic graminoids having $Q_{10}$ in the order of 1·3–1·4 as opposed to $Q_{10}$ of 2·0–2·4 in temperate species (Chapin 1974, 1977).

The arctic species may achieve the higher respiration rates at lower temperatures by having lower apparent activation energies ($E_{app}$) for root respiration as demonstrated by Earnshaw (1981) in *Carex misandra*, *Dupontia fischeri* and *E. scheuchzeri* from Svalbard ($E_{app}$ 17–29 kJ $mol^{-1}$). However, data calculated from Billings *et al.* (1977) show the $E_{app}$ for *E. angustifolium* (69), *Dupontia* (66–69) and *Carex aquatilis* (65–80 kJ $mol^{-1}$) to be higher and similar to values for temperate species. Changes in the levels of saturation of phospholipids in mitochondrial membranes may alter activation energies of respiration and other membrane-bound processes. This seems the most plausible explanation for the continued respiration and growth of arctic plant roots down to, or even below 0 °C. The lower activation energy of respiration in some arctic plants may be related to membrane fluidity maintained at low temperatures by decreased levels of saturation of phospholipids (Drew 1987). Acclimation at low temperatures has been demonstrated to increase the desaturation of fatty acids, but this is not reversible (Clarkson, Hall & Roberts 1980; Osmond, Wilson & Raper 1982). There is, however, much scope for more detailed work looking at the mechanisms by which arctic species maintain respiration and root growth at very low temperatures.

## *Nutrient uptake*

Problems of nutrient acquisition in cold tundra soils would be overcome to some extent if roots of tundra plants were more efficient in taking up nutrients than their temperate counterparts, particularly at low temperatures.

The uptake of phosphate by arctic graminoids has been thoroughly investigated by Chapin and co-workers, but little work has been carried out on other growth forms or nutrients, particularly nitrogen. The rates of $P_i$ uptake by arctic graminoids can proceed at temperatures as low as 0 °C with rates in *E. vaginatum* being 40 nmol $h^{-1}$ $g^{-1}$ FW at 0·1 °C (Chapin 1974). The rate of phosphate uptake by arctic species or ecotypes is generally less temperature sensitive than in temperate species or ecotypes (Fig. 3; Chapin & Bloom 1976). This again suggests that the activation energy for $P_i$ uptake in arctic species is lower than that for temperate species. The calculated $E_{app}$ value of $P_i$ uptake at 10 μmol $P_i$ by *E. vaginatum* is 64 kJ $mol^{-1}$ which, however, is in a similar range to that for *Zea mays* roots, 40–80 kJ $mol^{-1}$ (Carter & Lathwell 1967; Bravo & Uribe 1981). Arctic species do not however have a low temperature optimum for $P_i$ uptake (20–30 °C: Fig. 3; Chapin & Bloom 1976), and this is much higher than the soil temperatures they

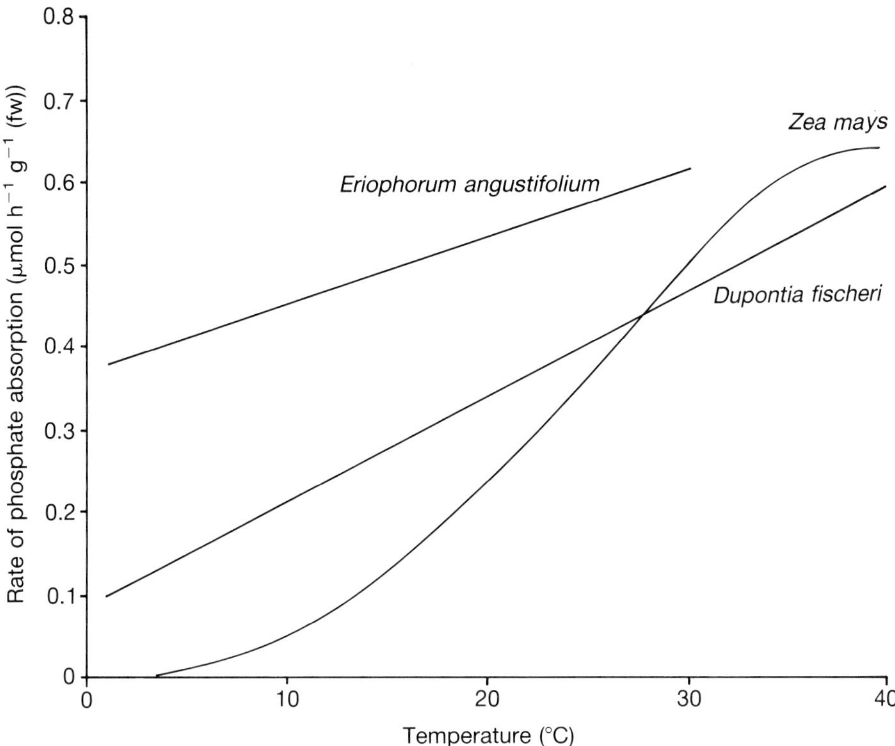


FIG. 3. Effect of temperature on rates of phosphate absorption (μmol $h^{-1}$ $g^{-1}$ fresh weight) by excised roots of *Eriophorum angustifolium*, *Dupontia fischeri* and *Zea mays* from 20 or 25 μM $P_i$ solution (redrawn from Chapin & Bloom 1976 and Bravo & Uribe 1981).

experience. The problem with the interpretation of most laboratory studies is that they have concentrated on $Q_{10}$ values between 10 and 20 °C which the roots of arctic plants hardly ever experience. The rates of $P_i$ uptake at temperatures below 5 °C do not differ as much between species as those at higher temperatures.

The maintenance of phosphate uptake at low temperatures by arctic plants has been suggested to be related to increased levels of de-saturation of membrane lipids (Kedrowski & Chapin 1978). However, the evidence from chilling-resistant temperate crop plants indicates that acclimation to low temperatures is not necessarily linked to lipids or to the number of membrane proteins (Robards & Clarkson 1984; Drew 1987).

On the whole, rates of $P_i$ uptake by excised roots from field-collected material show that deciduous shrubs and graminoids such as *Salix pulchra* and *Carex aquatilis* have higher rates than the evergreen dwarf shrubs *Ledum palustre* and *Vaccinium vitis-idaea* (Chapin & Tryon 1982; 1983). The lateral roots have much higher uptake rates than primary roots of nearly all species.

Some plants may use externally located phosphatase enzymes (Kroehler &

Linkins 1988) to tap soil organic $P_i$ resources. A major uncertainty which cannot at present be addressed is the significance of phosphatases to root systems in arctic soils. In temperate soils, the root surface phosphatase activities are much lower than those in the rhizosphere. However, in the non-mycorrhizal arctic sedges (see below), they could, perhaps, prove to be an important process of P acquisition.

### *Hemiparasites*

The hemi-parasitic habit is another way by which some plants have overcome limited soil nutrient and water supply. Most hemiparasites are members of the Scrophulariaceae and are particularly abundant in the tundra. The hemi-parasite tends to greatly reduce the resource allocation to its own root system once it has become attached to the root of a host plant, particularly in the annual species. These plants accumulate high concentrations of solutes, particularly potassium (13–77 mg $g^{-1}$ dry weight) compared to the host (Rumsey, personal communication). This means that there will be preferential water movement to the parasite from the host root system. This is confirmed by the high transpiration rates observed in the hemi-parasite *Bartsia alpina* and some other hemi-parasites (Rumsey, personal communication).

The occurrence of a wide variety of hemi-parasites in a stressed environment would, at first sight, appear to be a poor strategy. There may, however, be a potential advantage to a host in having a hemi-parasite attached. The rapid transpiration rate may draw more nutrients to the host root system in the soil water by mass flow than would normally occur in an unparasitized root system where the low summer temperatures do not result in rapid transpiration rates.

### *Biomass allocation to roots and shoots*

Nutrient shortage could be overcome to some extent if plants allocated proportionately more biomass to roots than shoots, thereby increasing the relative size of the nutrient acquisition system, or source, and decreasing the relative size of a major nutrient utilizing system, or sink.

A large proportion of the work on arctic plant species or communities has been concentrated on the biomass allocation to above- and below-ground structures (Shaver & Cutter 1979; Dennis 1977; Dennis & Johnson 1970; Miller, Mangan & Kummerow 1982; Webber 1977; Wein & Bliss 1974; Kjelvik & Karenlampi 1975; Wielgolaski 1972; Bell & Bliss 1978; Wielgolaski *et al.* 1981; Wallen 1986). Only a few of these studies give a breakdown of the below-ground components into stems, rhizomes and roots (Bell & Bliss 1978; Shaver & Cutter 1979; Miller, Mangan & Kummerow 1982; Wallen 1986). On the whole, the below : above-ground biomass quotients in the arctic tundra are between 4 and 34 (Bliss 1971). Alpine and arctic tundras have the highest values found in the world (6–17). Within the tundra, polar desert areas tend to have lower values (0·5–3) (Webber 1977; Bell & Bliss 1978) similar to those of hot desert ecosystems (3·5). Temperate and subtropical grasslands have quotients between 0·25 and 2 while temperate and tropical forests have quotients generally less than 0·25 (Webber 1977).

The graminoids such as *Eriophorum vaginatum*, *Carex aquatilis* and *Dupontia fischeri* tend to have a high allocation of biomass to roots (29–80%). The proportion of biomass below ground in this group can be very high, as in *Carex bigelowii* (Table 2). Forbs generally have a similar proportion of biomass allocated to roots (30–40%) but a much higher rhizome and below-ground stem component (Table 2). However, when they are growing in a polar semi-desert environment, much lower proportions of biomass (below 10%) are allocated to roots (Bell & Bliss 1978). The deciduous shrubs have a lower percentage of biomass in roots (2–24%), but they have a high proportion of their stems below ground. Evergreen dwarf shrubs have the lowest overall percentage of biomass allocated to roots (4–19%) and to total below-ground structures (5–76%).

Few studies have been carried out on the nutrition of arctic plants and its effects on root structure and biomass. One study on *E. vaginatum* growing plants in solution culture at three different concentrations of nitrogen (as nitrate) and phosphate showed that there was only a small increase in root biomass when the ions were applied separately, but in combination at high concentrations marked increases in root biomass occurred. The root : shoot biomass quotient was not

TABLE 2. Estimates of the proportion of biomass in the below-ground tissues and roots of tundra plants. * Indirect method of calculation

| Species | Biomass in roots (%) | Biomass below ground (%) | References |
|---|---|---|---|
| Evergreen dwarf shrubs | | | |
| *Lycopodium annotinum* | 5 | 5 | Callaghan *et al.* (1986a) |
| *Huperzia selago* | 10 | 10 | Headley (1986) |
| *Ledum palustre* | 4 | 62 | Shaver & Cutter (1979) |
| *Ledum palustre* | 16 | 42 | Chapin *et al.* (1980) |
| *Ledum palustre* | 5 | 76 | Miller *et al.* (1982) |
| *Ledum palustre* | | 43 | Chapin (1980) |
| *Empetrum nigrum* | 5 | 56 | Miller *et al.* (1982) |
| *Vaccinium vitis-idaea* | 19 | 53 | Miller *et al.* (1982) |
| *Empetrum hermaphroditum* | | 90* | Wallen (1986) |
| *Andromeda polifolia* | | 98* | Wallen (1986) |
| Deciduous shrubs | | | |
| *Salix pulchra* | 24 | 70 | Chapin *et al.* (1980) |
| *Salix pulchra* | 22 | 72 | Miller *et al.* (1982) |
| *Salix pulchra* | | 70 | Chapin (1980) |
| *Betula nana* | 2 | 53 | Shaver & Cutter (1979) |
| *Betula nana* | 24 | 89 | Miller *et al.* (1982) |
| *Betula nana* | | 80 | Chapin (1980) |
| *Vaccinium uliginosum* | 5 | 75 | Shaver & Cutter (1979) |
| *Vaccinium uliginosum* | 6 | 69 | Miller *et al.* (1982) |
| *Vaccinium microcarpus* | 44 | 67 | Miller *et al.* (1982) |

TABLE 2. Cont.

| Species | Biomass in roots (%) | Biomass below ground (%) | References |
|---|---|---|---|
| Graminoids | | | |
| *Alopecurus alpinus* | 58 | 65 | Bell & Bliss (1978) |
| *Phippsia algida* | 35 | | Bell & Bliss (1978) |
| *Puccinellia vaginata* | 38 | | Bell & Bliss (1978) |
| *Luzula nivalis* | 33 | | Bell & Bliss (1978) |
| *Luzula confusa* | 17 | 55 | Bell & Bliss (1978) |
| *Eriophorum vaginatum* | 33 | 33 | Shaver & Cutter (1979) |
| *Eriophorum vaginatum* | 46 | 75 | Chapin *et al.* (1980) |
| *Eriophorum vaginatum* | | 74 | Chapin (1980) |
| *Carex bigelowii* | | 92 | Shaver & Cutter (1979) |
| *C. aquatilis* | | 90 | Chapin (1980) |
| Graminae | 29 | 29 | Miller *et al.* (1982) |
| *Carex* sp. | 29 | 40 | Miller *et al.* (1982) |
| Graminoids | | 98 | Billings *et al.* (1978) |
| Forbs | | | |
| *Petasites* sp. | 39 | 94 | Miller *et al.* (1982) |
| *Rubus chamaemorus* | 30 | 92 | Miller *et al.* (1982) |
| *Rubus chamaemorus* | | 96 | Flower-Ellis (1980) |
| *Rubus chamaemorus* | | 98* | Wallen (1986) |
| *Saxifraga* spp. | 27–80 | 27–80 | Grulke & Bliss (1985) |
| *Cerastium arcticum* | 23 | | Bell & Bliss (1978) |
| *Ranunculus sabinei* | 55 | | Bell & Bliss (1978) |
| *Papaver radicatum* | 47 | | Bell & Bliss (1978) |
| *Cochlearia officinalis* | 20 | | Bell & Bliss (1978) |
| *Cardamine bellidifolia* | 35 | | Bell & Bliss (1978) |
| *Draba corymbosa* | 19 | | Bell & Bliss (1978) |
| *Saxifraga cernua* | 24 | | Bell & Bliss (1978) |
| *Saxifraga flagellaris* | 10 | | Bell & Bliss (1978) |
| *Potentilla hyparctica* | 40 | | Bell & Bliss (1978) |
| *Polygonum viviparum* | | 49–89 | Callaghan (1973) |

affected significantly by the increased levels of N and P supply or temperature (Kummerow & Krause 1982). This suggests that there is little plasticity in the root : shoot quotients of *E. vaginatum*.

Most of the roots of arctic plants are found in the top 10 cm of the soil/humus layer, especially in the moss layer (Miller, Mangan & Kummerow 1982; Dennis 1977; Dennis & Johnson 1970) where temperatures rarely exceed 7 °C (Billings, Peterson & Shaver 1978). Evergreen shrubs are generally very shallow rooted and most of the roots are in the top 10 cm (Miller, Mangan & Kummerow 1982; Wallen 1986), and generally start fine-root growth before deciduous dwarf shrubs (Kummerow *et al.* 1983). *Loiseluria procumbens* has its roots just below the surface enabling the plant to take advantage of the early thaw of the uppermost layers and gain several days more of the growing season before the deeper rooted species can start growing (Bliss 1971). This is probably coupled to the ability of evergreen shrubs to start photosynthesizing earlier in the season than deciduous shrubs as the evergreen shrubs will require roots in thawed soil to allow the uptake of water once the transpiration from leaves has commenced (Karlsson 1985).

The vertical distribution of roots varies between different communities. Miller, Mangan & Kummerow (1982) showed that *Cassiope tetragona*-dominated communities had most roots concentrated in the upper 10 cm whilst fell-field and sedge-moss communities had deeper roots (10–20 cm). Bell & Bliss (1978) showed a similar trend in the High Arctic: in moist habitats, most roots penetrated no deeper than 6 cm whereas in dry habitats, most roots penetrated to 15 cm although some reached 25 cm. On the whole, deciduous shrubs have deep main tap roots and shallower finer roots whilst graminoids have deep straight root systems, often unbranched, that may reach the permafrost.

### *Root longevity*

Problems of infertile soils and the very restricted period for nutrient acquisition within short growing seasons can be overcome to some extent if life spans of roots are long. Long-lived roots increase the efficiency of use of nutrients and carbon invested in root biomass, while overwintering roots are present during early spring thaw to exploit nutrient-rich pockets of unfrozen water. In addition, slow growing long-lived roots are better synchronized with the slow, temperature-limited decomposition rates in tundra soils (Chapin, Barsdate & Barel 1978): the activity of fast growing short-lived roots would be limited by the rate at which nutrients became available.

Only 2% of the tundra flora are therophytes — i.e. annuals — and these are either small plants of wet habitats or hemi-parasites (Callaghan & Emanuelsson 1985). In general, however, the short tundra growing seasons have selected for slow growing long-lived perennial species. Roots of tundra plants would, therefore, also be expected to be long-lived.

The roots of *Eriophorum angustifolium*, *E. vaginatum* and *Arctagrostis latifolia* grow at the surface of the frozen soil as this recedes down the profile towards the permafrost during summer (Bliss 1956; Shaver & Billings 1975; Billings, Shaver & Trent 1976; Billings *et al.* 1977). Roots of *E. angustifolium* die when the soil freezes in winter and are therefore functional for only one growing season (Shaver & Billings 1975). *Phippsia algida* also has a fast turnover of roots with 66% mortality of roots in the first year (Bell & Bliss 1978). In *Dupontia fischeri*, 90% of the lateral roots are produced in the first year of root growth but the root system survives for 5–6 years (Shaver & Billings 1975). *Puccinellia vaginata*, *Alopecurus alpinus* and *Ranunculus sabinei* have roots estimated to survive for 3–4 years (Bell & Bliss 1978) whereas the sedges *Luzula nivalis*, *L. confusa* and *Carex aquatilis* have longer lived roots which can survive for between 7 and 10 years (Bell & Bliss 1978; Shaver & Billings 1975). Similarly, the roots of *Carex bigelowii* are long-lived. Roots over 11 years old receive $^{14}C$ from young photosynthesizing tillers (Fig. 4a;

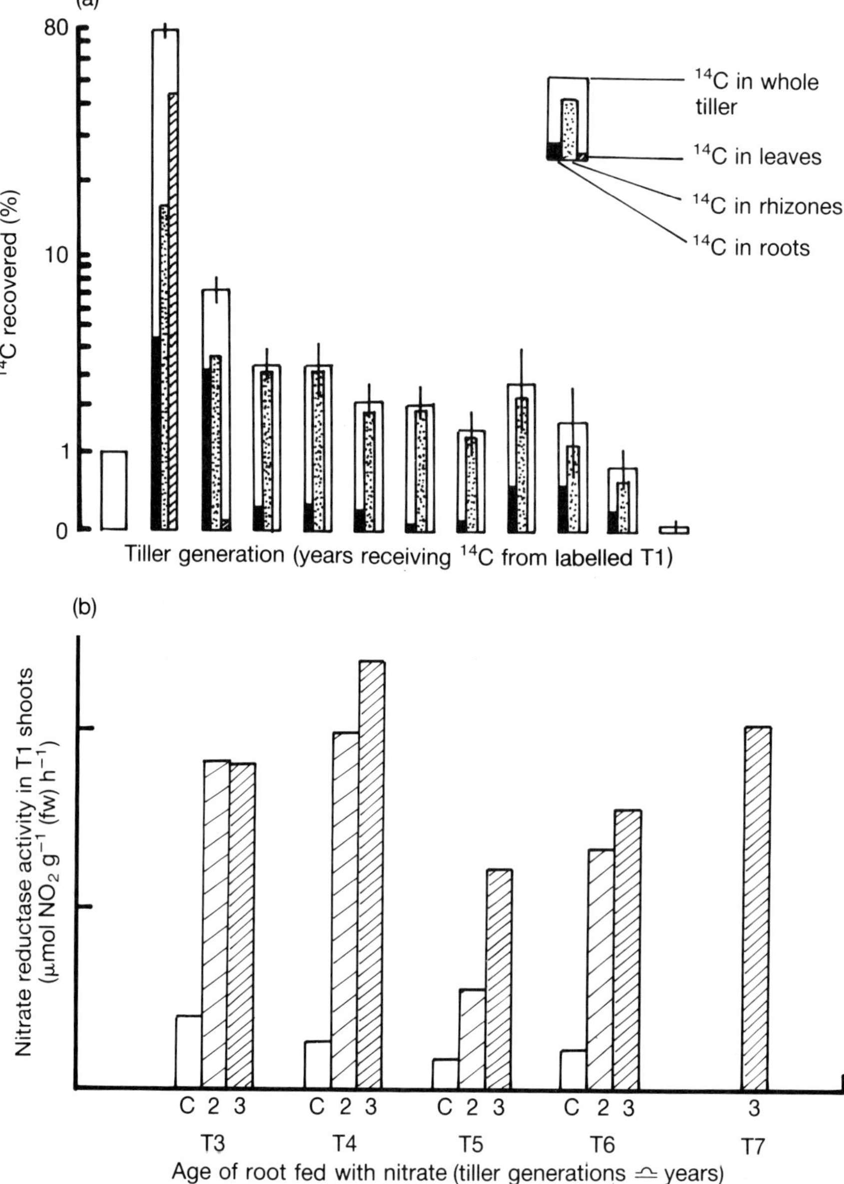


FIG. 4. The prolonged functioning of roots of the sedge *Carex bigelowii*. (a) Import of energy via $^{14}$C by old roots from young tillers which still have a living leaf complement (from Jonsdottir & Callaghan 1988). (b) Uptake and translocation of nitrate, inferred from the induced activity of nitrate reductase in young T1 tillers, related to the age of roots of *Carex bigelowii* (from Jonsdottir & Callaghan, 1990). C = control plants with roots incubated in distilled water; 2 = plants with roots incubated in nitrate for 2 days; 3 = plants with roots incubated in nitrate for 3 days; and T*n* = tiller generation which is approximately equivalent to root age in years. Note that (a) represents the activity of $^{14}$C in each tiller generation when applied to T1, whereas (b) presents the activity of nitrate reductase only in T1 tillers and nitrate was given to the roots of one of the older generations.

Jonsdottir & Callaghan 1988) and, in return, take up and transport at least nitrogen (in both ammonium and nitrate forms) to the young tillers (Fig. 4b; Jonsdottir & Callaghan 1990).

The clonal growth of many tundra graminoids, dwarf shrubs and rhizomatous and stoloniferous herbs is modular, i.e. the clone consists of repeated morphological units (Prevost 1978). Typically, the modular clonal graminoids possess large systems of interconnected tillers representing over about 27 years of growth in the case of *Carex bigelowii*. Few of these are young and able to photosynthesize and most are old and consist of only a below-ground system of rhizome and roots, having lost the leaf complement which usually survives for only 4–5 years. The disparity between the life spans of the shoots and below-ground organs helps to explain the exceptionally high below-ground to above-ground biomass quotients discussed above and also suggests that nutrient and/or water acquisition is more critical than energy capture in the tundra.

The vascular cryptogams *Diphasiastrum complantatum* and *Lycopodium annotinum* also possess long-lived root systems (up to 16 years) which have been shown to take up water and $^{32}P$ up to the age of 12 years (Headley, Callaghan & Lee 1985; 1988a). However, 3–4-year-old roots of *L. annotinum* took up most $^{32}P$ and nitrate and translocated them both apically and distally within the plant (Headley, Callaghan & Lee 1988a; 1989).

Fine roots of shrubs also demonstrate a long period of physiological activity with regard to $P_i$ uptake (Chapin & Tryon 1983). It is however, not surprising that old suberized root systems can take up $P_i$ as this ion is taken up through the symplast and is little affected by the development of the endodermis (Clarkson *et al.* 1968). However, calcium is taken up apparently only through the apoplast (Clarkson *et al.* 1971). This means that some new unsuberised root is necessary for the continued growth of the plant, and experiments preventing new root initiation in grasses have indicated that calcium deficiency can kill these plants (Troughton 1981).

Plants with tap roots probably have the longest lived roots with life spans of about 100 years, e.g. *Silene acaulis* (Callaghan & Emanuelsson 1985; Benedict 1989). Root life span in a tap rooted *Cerastium arcticum* has been estimated at 13 years (Bell & Bliss 1978). Dwarf shrubs also have long life spans but they tend to have tap roots only in dry habitats. In moist habitats, species such as *Empetrum hermaphroditum* and *Vaccinium vitis-idaea* produce short-lived adventitious roots.

### *Plant mobility or foraging*

In some areas, nutrient distribution is very patchy or physical obstacles such as rocks prevent root penetration. Plants which have foraging rhizomes or stolons (Hutchings & Slade 1988; Callaghan *et al.* 1990) may overcome these problems of nutrient and water acquisition by displaying an architecture which:

1 increases the probability of nutrient rich pockets being located;

2 allows extended growing points to receive subsidies of nutrients, water and carbon whilst crossing unsafe microsites where root penetration is impossible; and

3 allows weak competitors to escape intense competition from more aggressive species.

The foraging habit — or guerilla strategy — relies on the production of meristems on rhizomes or stolons which are extended from established and rooted parts of the plant. These growing points are subsidized by the older modules until favourable conditions are encountered and roots are able to develop.

In some species, such as the vascular cryptogams *Lycopodium* and *Diphasiastrum*, growth is deterministic and roots are produced in a regular sequence (one root per 92 cm of horizontal branch in *Lycopodium*) and new roots cannot be formed by old tissues (Callaghan, Svensson & Headley 1986b). Water and nutrients can be moved over a horizontal distance of 64 cm between root and growing point (Headley, Callaghan & Lee 1988a,b). Beyond this distance, the water potential of the growing point decreases until death occurs (Fig. 5). Apical death releases subdominant meristems from inhibition and the direction of growth changes and there is an increased probability of safe microsites being encountered (Svensson & Callaghan 1988). When rooting is possible, the direction of growth is controlled by branching angles (Callaghan *et al.* 1990) and microtopography (Callaghan *et al.* 1986a).

The deterministic growth described above suffers from the disadvantages that the death of young parts of a plant can terminate the growth of a whole clone as no new roots can develop and that nutrient-rich pockets, which exceed the size of

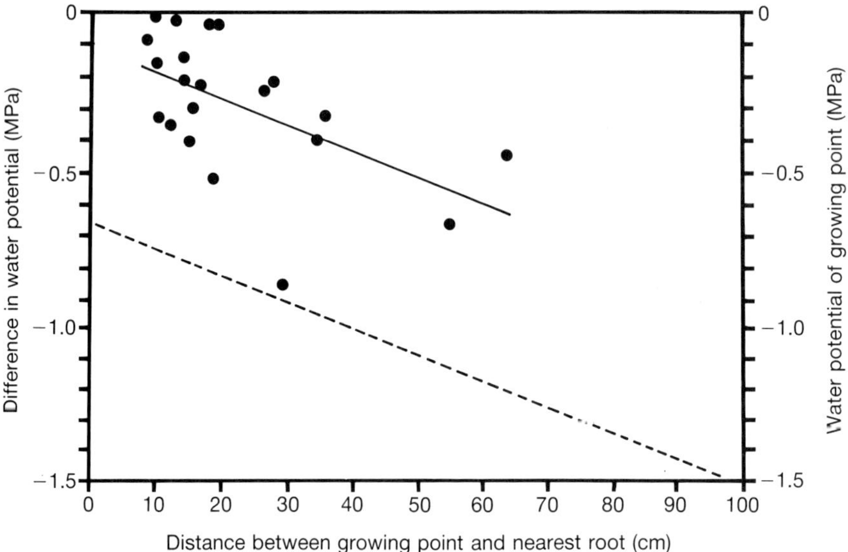


FIG. 5. The difference in water potential between the growing point of a horizontal branch of *Lycopodium annotinum* and the nearest root (solid regression line and circles) and the predicted water potentials (broken line) assuming a mean water potential per root of −0·62 MPa and the regression relationship depicted in the solid line. Assuming that a water potential of −1·5 MPa is lethal to growing points, a maximum potential distance of 97 cm between the growing point and its nearest root is indicated (see Headley, Callaghan & Lee 1988a).

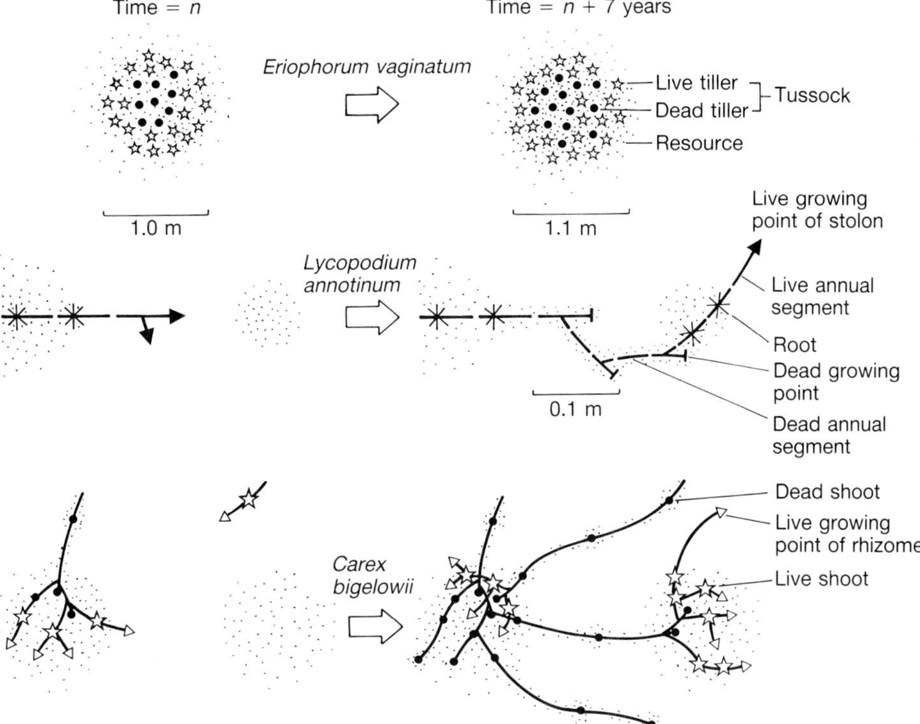


FIG. 6. Diagrammatic representation of the type of mobility of three tundra plant species and their effects on the redistribution of nutrients over a 7-year period. Top: *Eriophorum vaginatum* — non-mobile and little effect on nutrient spatial heterogeneity; middle: *Lycopodium annotinum* — deterministic foraging overcoming obstacles to rooting and decreasing the spatial heterogeneity of nutrients; bottom: *Carex bigelowii* — opportunistic foraging with slight decrease in the spatial heterogeneity of nutrients. From Heal, Callaghan & Chapman, 1989.

individual roots, are not fully exploited once encountered as the foraging plant immediately grows out of them. This trait may, however, reduce competition with more aggressive species better able to exploit the resource than the small roots (5% of total biomass) of the cryptogams.

Other species may be more opportunistic in the way they forage. They may produce adventitious roots and buds from old tissues and survive damage. Also, they may exhibit a 'phalanx' strategy in large nutrient-rich pockets whereby modules with short internodes proliferate and efficiently utilize the resource, e.g. *Carex bigelowii* (Kershaw 1962) and *Glechoma hederacea* (Slade & Hutchings 1987) (Fig. 6).

## *Nutrient recycling*

Once nutrients have been captured by tundra plants, many species conserve them by recycling them from old to young tissues (Chapin & Kedrowski 1983). The

TABLE 3. Estimates of the amount of nutrients recycled within tissues of tundra plants (%). See Jonasson (1989) for a critical evaluation of the weaknesses in making these estimates, particularly in the graminoids

| Species | N | P | K | References |
|---|---|---|---|---|
| *Lycopodium annotinum* | 63 | 64 | 90 | Callaghan (1980) |
| *Ledum palustre* | 56 | 68 | | Stoner *et al.* (1978) |
| *Eriophorum vaginatum* | 52 | 84 | | Stoner *et al.* (1978) |
| *E. angustifolium* | 45 | 61 | 48 | Chapin (1978) |
| *Carex aquatilis* | 53 | 55 | 64 | Chapin (1978) |
| *Dupontia fischeri* | 21 | 44 | 25 | Chapin (1978) |
| *Salix pulchra* | 18 | 24 | | Stoner *et al.* (1978) |

efficiencies of recycling nitrogen and phosphorus vary considerably between species within the same growth form (Table 3). It was originally thought that evergreen plants were particularly efficient in recycling nutrients (Small 1972) but Jonasson (1989) and Jonasson and Chapin (1985) have published evidence to the contrary. There are however, relatively high efficiencies of recycling nitrogen and phosphorus in deciduous and evergreen dwarf shrubs such as *Lycopodium annotinum* (Callaghan 1980; Headley, Callaghan & Lee 1985) and *Ledum palustre* (Stoner *et al.* 1978). Such plants can therefore, survive with little root biomass. Tundra graminoids at Point Barrow begin to return nitrogen and phosphorus to below-ground storage organs before peak biomass is achieved during the growing season (Chapin 1978). The evergreen species *Lycopodium annotinum* recycles nutrients from all of its old tissues with the exception of roots (Headley, Callaghan & Lee 1985).

An important consequence of the foraging strategy and nutrient recycling is the potential for the redistribution of nutrients in the soil and the implications of this for subsequent root development (Fig. 6; Heal, Callaghan & Chapman 1989). Even the slow growing *Lycopodium annotinum* has a mean annual horizontal extension growth of 63 mm and as N, P and K are efficiently recycled from old to young modules, it can be calculated that 10, 16 and 58% of the N, P and K, respectively, taken up in one year by a particular annual growth increment will be moved 31 cm horizontally over a subsequent 5-year period. In inhospitable conditions, some of these nutrients will be deposited on the ground surface as litter and, after decomposition, will be taken up by roots on subsequent foraging modules. This 'over-topping' process is an important mechanism in the primary colonization of hard ground surfaces and rock.

### *Concentrations of nutrients in tissues*

In addition to the recycling of nutrients, another form of nutrient economy adopted by some plants is their ability to function with low concentrations of some nutrients, such as nitrogen, phosphorus and potassium, in their tissues. In general,

TABLE 4. Concentrations of nutrients in roots of tundra plants (% dry weight)

| | N | P | K | References |
|---|---|---|---|---|
| Graminoids | | | | |
| *Eriophorum vaginatum* | 0·27–1·13 | 0·03–0·18 | | Chapin & Tryon (1982) |
| *Eriophorum vaginatum* | 0·7–1·7 | 0·08–0·19 | | Chapin *et al.* (1980) |
| *Carex aquatilis* | 0·97–1·47 | 0·18–0·38 | | Chapin & Tryon (1982) |
| *Deschampsia flexuosa* | 0·8 | 0·07 | 0·2 | Wielgolaski *et al.* (1975) |
| Deciduous shrubs | | | | |
| *Salix pulchra* | 1·48 | 0·14 | | Chapin & Tryon (1982) |
| *Salix pulchra* | 1·0–1·5 | 0·09–0·1 | 0·3–0·4 | Chapin *et al.* (1980) |
| *Salix herbacea* | 1·2 | 0·09 | 0·3 | Wielgolaski *et al.* (1975) |
| *Betula nana* | 1·19 | 0·12 | | Chapin & Tryon (1982) |
| *Betula nana* | 0·9–1·2 | 0·07–0·09 | 0·2–0·3 | Chapin *et al.* (1980) |
| *Vaccinium uliginosum* | 1·25 | 0·12 | | Chapin & Tryon (1982) |
| Evergreen shrubs | | | | |
| *Vaccinium vitis-idaea* | 0·57–1·08 | 0·06–0·11 | | Chapin & Tryon (1982) |
| *Vaccinium vitis-idaea* | 0·8 | 0·08 | 0·2 | Wielgolaski *et al.* (1975) |
| *Ledum palustre* | 0·75–1·21 | 0·06–0·11 | | Chapin & Tryon (1982) |
| *Ledum palustre* | 0·7–0·8 | 0·05–0·08 | 0·15–0·2 | Chapin *et al.* (1980) |
| Forbs | | | | |
| *Rubus chamaemorus* | 1·0–1·4 | 0·07–0·09 | 0·37–0·39 | Chapin *et al.* (1980) |
| *Rubus chamaemorus* | 1·4 | 0·15 | 0·9 | Wielgolaski *et al.* (1975) |
| All roots present | | | | |
| *Empetrum hermaphroditum* heath | 0·67 | 0·09 | 0·02 | Jonasson (1983) |
| *Vaccinium myrtillus* heath | 0·72 | 0·13 | 0·26 | Jonasson (1983) |
| Mixed shrub heath | 1·04 | 0·15 | 0·32 | Jonasson (1983) |

concentrations of nitrogen, phosphorus and potassium are lower in tundra plant species than in temperate plant species (Chapin, Van Cleve & Tieszen 1975; Wielgolaski, Kjelvik & Kallio 1975). Within the life forms of the tundra plants, lichens have the lowest concentrations of nitrogen (0.6%), deciduous dwarf shrubs, monocotyledons and forbs have the highest (1·4–2·5%) and evergreen dwarf shrubs are intermediate (Wielgolaski, Kjelvik & Kallio 1975; Jonasson 1983). However, the concentrations found within the different tissues of *Lycopodium annotinum* span the complete range found in the various tundra plant growth forms (Callaghan 1980). Concentrations of nitrogen, phosphorus and potassium in the roots of the different tundra plant growth forms, follow the trends described above (Table 4).

### *Root morphology and surface to volume relationships*

Lateral roots have much higher uptake rates than primary roots in nearly all of the species investigated (Chapin & Tryon 1982). This may be partly due to their greater surface area to weight quotient and their different age and function. Larger primary roots are probably more important in acquiring water than lateral roots which tend to extend into the moss and upper humus layers where most of the nutrient recycling occurs (Chapin, Van Cleve & Chapin 1979). The roots of arctic species or ecotypes tend to be narrower than equivalent temperate species or ecotypes when grown at the same temperature (Chapin 1974). However, as temperature decreases, the roots become thicker. In their respective environments, therefore, roots of tundra and temperate graminoids probably have similar root surface to volume quotients (Chapin 1978).

Evergreen dwarf shrubs have a much finer root system in the upper layers of the soil but this is unlikely to differ from the situation in temperate regions.

In general, branching in the root systems of tundra plants appears to be absent or simple compared to the situation described in other regions (Fitter pp. 229–243). Thus, the drier polar desert, semi-desert and fell-field habitats are characterized by small fibrous root systems (Bell & Bliss 1978), patterned ground is also characterized by fibrous root systems (Jonasson 1986) and wet habitats are characterized by long, thick unbranched roots of the graminoids. It must also be noted that much of the tundra's vegetation is dominated by cryptogams without roots (e.g. lichens and mosses) and vascular cryptogams (e.g. *Lycopodium sensu lato*) with simple, dichotomously branched root systems are also widespread.

## *Mycorrhizae*

Very little work has been carried out on the mycorrhizae of arctic plants (Miller 1981, 1982) with only a little more carried out on alpine species or communities (Read & Haselwandter 1981). Most mycorrhizae have been found in the deciduous and evergreen dwarf shrubs and some forbs, whilst mycorrhizae tend to be completely absent from the Cyperaceae and Juncaceae (Tester, Smith & Smith 1987; Miller 1982; Harley & Harley 1987). These plant groups are particularly abundant in the Arctic and it may be inferred therefore, that mycorrhizae are unimportant over large areas of arctic vegetation (Table 5), in contrast to vegetation of other regions.

The common and dominant species of sedge over much of the Arctic are *Eriophorum vaginatum*, *E. angustifolium* and *Carex bigelowii*. These species are normally non-mycorrhizal and the few records of mycorrhizae may be doubtful identifications or parasitic infections of necrotic tissue (Tester, Smith & Smith 1987). These sedges tend to have abundant roots hairs, as do *Huperzia selago* and species of *Lycopodium sensu lato* which are also non-mycorrhizal in populations from Torne Lappmark and the English Lake District (Headley 1986). Other non-mycorrhizal species have alternative modes of nutrition such as carnivory and hemi-parasitism (Table 5).

The members of the Ericales are all mycorrhizal and nearly all have ericoid type mycorrhizae including all of the common species of the Arctic (Read 1983; Harley & Harley 1987). It appears that this type of association may lead to the exploitation

TABLE 5. Types of mycorrhizal associations found in tundra plants. The following families are normally non-mycorrhizal: Chenopodiaceae, Portulacaceae, Caryophyllaceae, Cruciferae, Urticaceae, Juncaceae, Cyperaceae, Polygonaceae. T = Trappe (1987); H&H = Harley & Harley (1987); TSS = Tester, Smith & Smith (1987); M = Miller (1982); RH = Read & Haselwandter (1981); ML = Miller & Laursen (1978). VA = vesicular–arbuscular; Ecto = ectomycorrhizal; Eri = ericoid; Ect/End = ecto/endomycorrhizal. * These plants have alternative modes of nutrition

| Non-mycorrhizal plants | Ref. |
|---|---|
| *Woodsia alpina* | H&H |
| *W. ilvensis* | H&H |
| *Arabis alpina* | H&H |
| *Cerastium alpinum* | H&H |
| *C. arcticum* | H&H |
| *Sibbaldia procumbens* | H&H |
| *Saxifraga stellaris* | H&H |
| *Oxyria digyna* | H&H |
| *Melampyrum sylvaticum** | H&H |
| *Veronica alpina* | H&H |
| *Pinguicula alpina** | H&H |
| *Juncus alpinus* | H&H |
| *J. biglumis* | H&H |
| *Carex buxbaumi* | H&H |
| *C. dioica* | H&H |
| *C. bigelowii* | RH, M |
| *Eriophorum angustifolium* | HH |
| *E. vaginatum* | HH, M |
| *Trichophorum alpinum* | HH |
| *Eleocharis palustris* | HH |
| **Weakly mycorrhizal plants** | |
| *Silene acaulis* (VA/Ecto) | RH |
| *Alchemilla alpina* (10% VA) | RH |
| *Saxifraga oppositifolia* (VA) | RH |
| *Luzula spicata* | RH |
| *Carex firma* | RH |
| *C. curvula* | RH |
| **Ericoid-mycorrhizal plants** | |
| *Andromeda polifolia* | HH |
| *Arctostaphylos alpinus* | HH |
| *Calluna vulgaris* | HH |
| *Ledum palustre* | HH |
| *Loiseleuria procumbens* | HH |
| *Phyllodoce caerulea* | HH |
| *Vaccinium myrtillus* | HH |
| *V. uliginosum* | HH |
| *V. vitis-idaea* | HH |
| *Empetrum hermaphroditum* | HH |
| *Diapensia lapponica* | HH |
| *Cassiope tetragona* | R |
| **Ecto/endo-mycorrhizal (Arbutoid)** | |
| *Arctostaphylos uva-ursi* | |

| Ecto-mycorrhizal | Ref. |
|---|---|
| *Polygonum viviparum* | H&H |
| *Betula pubescens* | HH, T |
| *B. nana* | M, HH, T |
| *Salix herbacea* | HH, T |
| *S. lapponum* | HH |
| *S. phyllicifolia* | HH |
| *S. reticulata* | HH, TM |
| *S. arctica* | M, T |
| *Cassiope tetragona* | M |
| *Dryas octopetala* | M |
| *Arctostaphylos rubra* | M |
| *Salix polaris* | T |
| **Pedicularis karei* | ML |
| *Dryas integrifolia* | ML |
| **Vesicular–arbuscular-mycorrhizal plants** | |
| *Dryas octopetala* | HH |
| *Sedum rosea* | HH |
| *Circaea alpina* | HH |
| *Epilobium anagallidifolium* | HH |
| *E. alsinifolium* | HH |
| *Gentiana nivalis* | HH |
| *G. verna* | HH |
| *Myosotis alpestris* | HH |
| *Bartsia alpina** | HH |
| *Veronica fruticans* | HH |
| *Linnaea borealis* | HH |
| *Antennaria dioica* | HH |
| *Cicerbita alpina* | HH |
| *Erigeron borealis* | HH |
| *Gnaphalium norvegicum* | HH |
| *G. supinum* | HH |
| *Hieracium alpinum* | HH |
| *Saussurea alpina* | HH |
| *Tofieldia pusilla* | HH |
| *Juncus trifidus* | HH |
| *Festuca vivipara* | HH |
| *Hierochloe odorata* | HH |
| *Phleum alpinum* | HH |
| *Poa alpina* | HH |
| *Ranunculus nivalis* | ML |
| *Ranunculus palasii* | ML |
| *Ranunculus pygmaeus* | ML |
| *Caltha palustris* | ML |
| *Saxifraga oppositifolia* | ML |
| *Saxifraga punctata* | ML |

of organic nitrogen and phosphorus sources which would normally be unavailable to the vascular plants (Harley & McCready 1981; Read 1983). The vesicular–arbuscula-type mycorrhizal association found in many forb species and some grasses may facilitate the exploitation of a large pool of phosphate beyond the depletion zone of the plant root itself.

The deciduous dwarf shrubs, such as *Salix herbacea*, *S. reticulata* and *Betula nana* are strongly ecto-mycorrhizal and associated with species of Basidiomycete such as *Amanita nivalis*, *Lactarius helvus*, *L. rufus* and *Russula emetica* (Miller 1982; Watling 1988; Trappe 1987).

The mycorrhizal plants do benefit from the association, either through obtaining nutrients from a larger soil volume or from exploiting organic sources which would be otherwise inaccessible. Mycorrhizae may also help in the exclusion of toxic heavy metals (Read 1983). In this regard there is potential for further work on those species which accumulate heavy metals, such as aluminium by *Diappensia lapponica* which is mycorrhizal and *Lycopodium* species which so far have not been shown to be mycorrhizal. Much further research is required to characterize the mycorrhizal associations of arctic plants and to determine the relative importance of the association for the nutrition of the host plant. This is particularly important in the Arctic where a very large proportion of the soil nitrogen and phosphorus is bound up in the organic component, and soil decomposition rates are very slow (Chapin, Barsdate & Barel 1978).

## CONCLUSIONS

The major conclusion of this review must be that little is known about the root systems of tundra plants and a high priority for research must be given to understanding how roots in frost-heave areas survive in these sites of extreme environmental stress and disturbance. The morphology of the roots of tundra plants in terms of their branching patterns, elasticity and tensile strength could be of particular importance, as the typically unbranched and elastic roots of some species could contrast strongly with those of other regions. Similarly, the absence of mycorrhizae from much of the Arctic's vegetation contrasts with the situation in other regions and it is important to investigate how, for example, the *Carex* species compensate for this, and to what extent the production of external phosphatases are important in this group.

Within the review, a series of hypothetical options has been established (Table 1) as an aid to consider the ways in which tundra plants and their roots could have responded to overcome cold, infertile tundra soils. Three predictions were made: (i) that there is no unique combination of options for species in a particular habitat, (ii) that the severity of the stress experienced by a species should be reflected in the number of options exhibited, and (iii) that species exhibiting different options should have an increased chance of coexisting whereas those with similar options should compete.

Table 6 presents the options for a range of tundra plant species. It can be seen

TABLE 6. A between-species comparison of the options to combat infertility in cold tundra soils

| | *Eriophorum vaginatum* | *Carex aquatilis* | *Lycopodium annotinum* | *Ledum palustre* | *Salix pulchra* | *Rubus chamaemorus* |
|---|---|---|---|---|---|---|
| Increased rate of nutrient uptake by roots at low nutrient concentrations and temperatures | Yes | Yes | No | No | Yes | No |
| nmol $P_i$ $h^{-1}$ $g^{-1}$ fresh weight at 5 °C and 5 μmol $P_i$ | 340 | 500 | | | 720 | |
| nmol $P_i$ $h^{-1}$ $g^{-1}$ fresh weight at 10 °C and 0·5 μmol $P_i$ | 489 | 312 | | 25 | 119 | 20 |
| Increased root–shoot biomass | Yes | Yes | No | No | Yes | Yes |
| (root biomass as % total biomass) | 33–46 | | 5 | 4–16 | 22–24 | 30 |
| Increased surface area of roots relative to weight | No | No | No | Yes | Yes | Yes |
| ($cm^2$ $g^{-1}$ fresh weight at 5 °C) | 58 | 72 | | | | |
| Increased root longevity | No | Yes | Yes | Yes | Yes | Yes |
| (years) | 1 | 5–8 | >12 | >1 | >1 | >1 |
| Presence of foraging habit to locate favourable microsites | No | Yes | Yes | No | No | Yes |
| Functioning with low growth rates to reduce nutrient | Yes | Yes | Yes | Yes | No | No |
| requirements (mg $shoot^{-1}$ $year^{-1}$) | 46 | | 31 | 31 | 81 | 150 |
| Increased efficiency of nutrient use | Yes | Yes | Yes | Yes | No | No |
| N (mg $g^{-1}$) | 5–40 | | 4–22 | 10–30 | 10–50 | 10–50 |
| P (mg $g^{-1}$) | 1–6 | | 0.6–3 | 0.5–4 | 1–6 | 0.8–4 |
| Efficient recycling of nutrients | Yes | No | Yes | Yes | No | Yes |
| N in shoot recycled (%) | 52 | 21 | 63 | 56 | 18 | 69 |
| P in shoot recycled (%) | 84 | 44 | 64 | 68 | 24 | 50 |
| Presence of mycorrhizae to forage for N and P beyond zones depleted by roots | No | No | No | Yes | Yes | Yes |
| Ability to function at low temperatures | Yes | Yes | Yes | Yes | Yes | Yes |

that no two species have adopted the same set of options although four of the six species have each adopted seven options out of the possible eleven, and the remaining two species have adopted six and eight options. Only one option has been adopted by all species, i.e. the ability to function at low temperatures, and all options have been adopted by at least one species. Two of the species coexist, i.e. *Eriophorum vaginatum* and *Salix pulchra*, and they share only four options.

Unfortunately, quantitative data are lacking to complete the matrix, but this structure for acquiring and comparing data may help with interpretations of the many and varied responses of tundra plants to their extreme environments.

## ACKNOWLEDGEMENTS

The authors are all grateful to Professor Mats Sonesson, the Director, Abisko Scientific Research Station, Swedish Lapland, for hospitality, facilities and financial assistance to work together on arctic vegetation. Dr Sven Jonasson, University of Copenhagen, kindly gave useful comments on the manuscript.

## REFERENCES

**Barry, R.G., Courtin, G.M. & Labine, C. (1981).** Tundra climates. *Tundra ecosystems: a Comparative Analysis* (Ed. by L.C. Bliss, O.W. Heal & J.S. Moore), pp. 81–114. Cambridge University Press, Cambridge.

**Bell, K.L. & Bliss, L.C. (1978).** Root growth in a polar semidesert environment. *Canadian Journal of Botany*, **56**, 2470–2490.

**Benedict, J.B. (1989).** Use of *Silene acaulis* for dating: the relationship of cushion diameter to age. *Arctic and Alpine Research*, **21**, 91–96.

**Billings, W.D., Shaver, G.R. & Trent, S.W. (1976).** Measurement of root growth in simulated and natural temperature gradients over permafrost. *Arctic and Alpine Research*, **8**, 247–250.

**Billings, W.D., Peterson, K.M., Shaver, G.R. & Trent, A.W. (1977).** Root growth, respiration, and carbon dioxide evolution in an arctic tundra soil. *Arctic and Alpine Research*, **9**, 127–135.

**Billings, W.D., Peterson, K.M. & Shaver, G.R. (1978).** Growth turnover and respiration rates of roots and tillers in tundra graminoids. *Vegetation and Production Ecology of an Alaskan Arctic Tundra* (Ed. by L.L. Tieszen), pp. 415–434. Springer-Verlag, New York.

**Bliss, L.C. (1956).** A comparison of plant development in microenvironments of arctic and alpine tundras. *Ecological Monographs*, **26**, 303–337.

**Bliss, L.C. (1971).** Arctic and alpine plant life cycles. *Annual Review of Ecology and Systematics*, **2**, 405–438.

**Bravo, F.P. & Uribe, E.G. (1981).** Temperature dependence of the concentration kinetics of absorption of phosphate and potassium in corn roots. *Plant Physiology*, **67**, 815–819.

**Brown, R.J.E. (1977).** Permafrost investigations on Truelove Lowland. *Truelove Lowland, Devon Island, Canada: a High Arctic Ecosystem* (Ed. by L.C. Bliss), pp. 13–30. University of Alberta Press, Canada.

**Callaghan, T.V. (1973).** A comparison of the growth of tundra plant species at several widely separated sites. *Merlewood Research and Development Paper No. 53*. Institute of Terrestrial Ecology, Cambridge.

**Callaghan, T.V. (1980).** Age-related patterns of nutrient allocation in *Lycopodium annotinum* from Swedish Lapland. Strategies of growth and population dynamics of tundra plants 5. *Oikos*, **35**, 373–386.

**Callaghan, T.V. (1987).** Plant population processes in arctic and boreal regions. *Ecological Bulletins*, **38**, 58–68.

**Callaghan, T.V. (1988).** Physiological and demographic implications of modular construction in cold environments. *Plant Population Ecology* (Ed. by A.J. Davy, M.J. Hutchings & A.R. Watkinson), pp, 111–135. Blackwell Scientific Publications, Oxford.

**Callaghan, T.V. & Emanuelsson, U. (1985).** Population structure and processes of tundra plants and vegetation. *The Population Structure of Vegetation* (Ed. by J. White), pp. 399–439. Junk, Dordrecht.

**Callaghan, T.V., Headley, A.D., Svensson, B.M., Li Lixian, Lee, J.A. & Lindley, D.K. (1986a).** Modular growth and function in the vascular cryptogam, *Lycopodium annotinum*. *Philosophical Transactions of the Royal Society Series B*, **228**, 195–206.

**Callaghan, T.V., Svensson, B.M. & Headley, A.D. (1986b).** The modular growth of *Lycopodium annotinum*. *Fern Gazette*, **13**, 65–76.

**Callaghan, T.V., Svensson, B.M., Bowman, H., Lindley, D.K. & Carlsson, B.A. (1990).** Models of clonal plant growth based on population dynamics and architecture. *Oikos*, **57**, 257–269.

**Carter, O.G. & Lathwell, D.J. (1967).** Effect of temperature on orthophosphate absorption by excised corn roots. *Plant Physiology*, **42**, 1407–1412.

**Chapin, F.S. III (1974).** Morphological and physiological mechanisms of temperature compensation in phosphate absorption along a latitudinal gradient. *Ecology*, **55**, 1180–1198.

**Chapin, F.S. III (1977).** Nutrient/carbon costs associated with tundra adaptations to a cold nutrient-poor environment. *Proceedings of the Circumpolar Conference on Northern Ecology*, pp. 1183–1194. Natural Research Council of Canada, Ottawa.

**Chapin, F.S. III (1978).** Phosphate uptake and nutrient utilisation by Barrow tundra vegetation. *Vegetation and Production Ecology of an Alaskan Arctic Tundra* (Ed. by L.L. Tieszen), pp. 483–507. Springer-Verlag, New York.

**Chapin, F.S. III (1980).** The mineral nutrition of wild plants. *Annual Review of Ecology & Systematics*, **11**, 233–260.

**Chapin, F.S. III & Bloom, A.J. (1976).** Phosphate absorption: adaptations of tundra graminoids to a low temperature, low phosphorus environment. *Oikos*, **27**, 111–121.

**Chapin, F.S. III & Tryon, P.R. (1982).** Phosphate absorption and root respiration of different plant growth forms from northern Alaska. *Holarctic Ecology*, **5**, 164–171.

**Chapin, F.S. III & Tryon, P.R. (1983).** Habitat and leaf habit as determinants of growth, nutrient absorption and nutrient use by Alaskan taiga forest species. *Canadian Journal of Forest Research*, **13**, 818–826.

**Chapin, F.S. III, & Kedrowski, R.A. (1983).** Seasonal changes in nitrogen and phosphorus fractions and autumn retranslocation in evergreen and deciduous taiga trees. *Ecology*, **64**, 376–391.

**Chapin, F.S. III, Van Cleve, K. & Tieszen, L.L. (1975).** Seasonal nutrient dynamics of tundra vegetation at Barrow, Alaska. *Arctic and Alpine Research*, **7**, 209–226.

**Chapin, F.S. III, Barsdate, R.J. & Barel, D. (1978).** Phosphorus cycling in Alaskan coastal tundra: a hypothesis for the regulation of nutrient cycling. *Oikos*, **31**, 189–199.

**Chapin, F.S. III, Van Cleve, K. & Chapin, M.C. (1979).** Soil temperature and nutrient cycling in the tussock growth of *Eriophorum vaginatum*. *Journal of Ecology*, **67**, 169–189.

**Chapin, F.S. III, Tieszen, L.L., Lewis, M.C., Miller, P.C. & McCown, B.H. (1980).** Control of tundra plant allocation patterns and growth. *An Arctic Ecosystem. The Coastal Tundra at Barrow, Alaska* (Ed. by J. Brown, P.C. Brown, P.C. Miller, L.L. Tieszen & F.L. Bunnell), pp. 140–185. Dowden, Hutchinson & Ross Inc, Stoudsberg.

**Clarkson, D.T., Sanderson, J. & Russel, R.S. (1968).** Ion uptake and root age. *Nature*, **220**, 805–806.

**Clarkson, D.T., Robards, A.W. & Sanderson, N.J. (1971).** The tertiary endodermis in barley roots; fine structure in relation to radial transport of ions and water. *Planta*, **96**, 292–305.

**Clarkson, D.T., Hall, L.C. & Roberts, J.K.M. (1980).** Phospholipid composition and fatty acid desaturation in the roots of rye during acclimation to low temperature. Positional analysis of fatty acids. *Planta*, **149**, 464–471.

**Dennis, J.G. (1977).** Distribution patterns of belowground standing crop in arctic tundra at Barrow, Alaska. *Arctic and Alpine Research*, **9**, 113–127.

**Dennis, J.G. & Johnson, P.L. (1970).** Shoot and rhizome-root standing crops of tundra vegetation at Barrow, Alaska. *Arctic and Alpine Research*, **2**, 253–266.

**Drew, M.C. (1987).** Function of root tissues in nutrient and water movement. In *Root Development and*

*Function* (Ed. by P.J. Gregory, J.V. Lake & D.A. Rose), pp. 71–102. Cambridge University Press, Cambridge.

**Earnshaw, M.J. (1981).** Arrhenius plots of root respiration in some arctic plants. *Arctic and Alpine Research*, **13**, 425–430.

**Everett, K.R., Vassiljevskaya, V.D., Brown, J. & Walker, B.D. (1981).** Tundra and analogous soils. *Tundra Ecosystems: a Comparative Analysis* (Ed. by L.C. Bliss, O.W. Heal & J.J. Moore), pp. 139–179. Cambridge University Press, Cambridge.

**Fahey, B.D. (1979).** Frost heaving of soils at two locations in Southern Ontario, Canada. *Geoderma*, **22**, 119–126.

**Flower-Ellis, J.G.K. (1980).** Diurnal dry weight variation and dry matter allocation of some tundra plants, II. *Rubus chamaemorus* L. In *Ecology of a Subarctic Mire* (Ed. by M. Sonesson). *Ecological Bulletins*, **30**, 163–179.

**French, H.M. (1987).** Preiglacial geomorphology in North America: current research and future trends. *Ecological Bulletins*, **38**, 5–16.

**Grulke, N.E. & Bliss, L.C. (1985).** Growth forms, carbon allocation and reproductive patterns of high arctic Saxifrages. *Arctic and Alpine Research*, **17**, 241–250.

**Harley, J.L. & McCready, C.C. (1981).** Phosphate accumulation in Fagus mycorrhizas. *New Phytologist*, **89**, 75–80.

**Harley, J.L. & Harley, E.L. (1987).** A check-list of mycorrhiza in the British Flora. *New Phytologist*, **105**, 1–102.

**Headley, A.D. (1986).** *The comparative autecology of some European species of Lycopodium sensu lato.* Ph.D. thesis, University of Manchester.

**Headley, A.D., Callaghan, T.V. & Lee, J.A. (1985).** The phosphorus economy of the evergreen tundra plant, *Lycopodium annotinum*. *Oikos*, **45**, 235–245.

**Headley, A.D., Callaghan, T.V. & Lee, J.A. (1988a).** Water uptake and movements in the evergreen clonal plants *Lycopodium annotinum* L. and *Diphasiastrum complanatum* (L.) Holub. *New Phytologist*, **110**, 487–495.

**Headley, A.D., Callaghan, T.V. & Lee, J.A. (1988b).** Phosphate and nitrate movement in the clonal plant *Lycopodium annotinum* L. and *Diphasiastrum complantum* (L.) Holub. *New Phytologist*, **110**, 497–502.

**Headley, A.D., Callaghan, T.V. & Lee, J.A. (1989).** A field-based demographic approach to nutrient uptake and movement in a clonal perennial plant. *Nutrient Cycling in Terrestrial Ecosystems: Field Methods, Applications and Interpretation* (Ed. by A.F. Harrison, P. Ineson & O.W. Heal), pp. 421–429. Elsevier Applied Science, London.

**Heal, O.W., Callaghan, T.V. & Chapman, K. (1989).** Can population and process ecology be combined to understand nutrient cycling? *Ecology of Arable Land. Developments in Plant and Soil Science* (Ed. by M. Clarholm & L. Bergstrom), pp. 205–216. Kluwer, Dordrecht.

**Hutchings, M.J. & Slade, A.J. (1988).** Morphological plasticity, foraging and integration in clonal perennial herbs. *Plant Population Ecology* (Ed. by A.J. Davy, M.J. Hutchings & A.R. Watkinson), pp. 83–109. Blackwell Scientific Publications, Oxford.

**James, P.A. (1971).** The measurement of frost-heave in the field. *British Geomorphological Research Group Technical Bulletin 8*. GeoAbstracts, Norwich, 43pp.

**Jonasson, S. (1983).** Nutritional content and dynamics in north Swedish shrub tundra areas. *Holarctic Ecology*, **6**, 295–304.

**Jonasson, S. (1986).** Influence of frost heaving on soil chemistry and on the distribution of plant growth forms. *Geografiska Annaler*, **68**, 185–195.

**Jonasson, S. (1989).** Implications of leaf longevity, leaf nutrient reabsorption and retranslocation for resource economy of five evergreen plant species. *Oikos*, **56**, 121–131.

**Jonasson, S. & Chapin, F.S. III. (1985).** Significance of sequential leaf development for nutrient balance in the cottonsedge, *Eriophorum vaginatum* L. *Oecologia (Berlin)* **67**, 511–518.

**Jonsdottir, I.S. & Callaghan, T.V. (1988).** The interrelationships between different generations of interconnected tillers of *Carex bigelowii*. *Oikos*, **52**, 120–128.

**Jonsdottir, I.S. & Callaghan, T.V. (1990).** Intraclonal translocation of ammonium and nitrate nitrogen in *Carex bigelowii* using $^{15}N$ and nitrate reductase assays. *New Phytologist*, **114**, 419–428.

**Karlsson, P.S. (1985).** Photosynthetic characteristics and leaf carbon economy of a deciduous and an evergreen dwarf shrub: *Vaccinium uliginosum* L. and *V. vitis-idaea* L. *Holarctic Ecology*, **8**, 9–17.

**Kedrowski, R.A. & Chapin, F.S. III (1978).** Lipid properties of *Carex aquatilis* from hot spring and permafrost-dominated sites in Alaska: Implications for nutrient requirements. *Physiologia Plantarum*, **44**, 231–237.

**Kershaw, K.A. (1962).** Quantitative ecological studies from Landmannahellir, Iceland. II. The rhizome behaviour of *Carex bigelowii* and *Calamagrostis neglecta*. *Journal of Ecology*, **50**, 171–179.

**Kjelvik, S. & Karenlampi, L. (1975).** Plant biomass and primary production of Fennoscandian subarctic and subalpine forests and of alpine willow and heath ecosystems. *Fennoscandian Tundra Ecosystems. Part 1. Plants and Microorganisms* (Ed. by F.E. Wielgolaski), pp. 111–120. Springer-Verlag, Berlin.

**Klepper, B. (1987).** Origin, branching and distribution of root systems. In *Root Development and Function* (Ed. by P.J. Gregory, J.V. Lake & D.A. Rose), pp. 71–102. Cambridge University Press, Cambridge.

**Kroehler, C.J. & Linkins, A.E. (1988).** The root surface phosphatases of *Eriophorum vaginatum*. Effects of temperature, pH, substrate concentration and inorganic phosphorus. *Plant and Soil*, **105**, 3–10.

**Kummerow, J. & Krause, D.A. (1982).** The effects of variable nitrogen and phosphorus concentrations on *Eriophorum vaginatum* tillers grown in nutrient solutions. *Holarctic Ecology*, **5**, 187–193.

**Kummerow, J., Ellis, B.A., Kummerow, S. & Chapin, F.S. III (1983).** Spring growth of shoots and roots in shrubs of an Alaskan Muskeg. *American Journal of Botany*, **70**, 1509–1515.

**Larcher, W. (1981).** Aims, methods and results of phytoecological studies in ecosystems of the Tyrolean Alps. *Botanische Zhurnal*, **66**, 1114–1134.

**Larcher, W. (1985).** Winter stress in high mountains. In *Establishment and Tending of Subalpine Forest; Research and Management* (Ed. by H. Turner & W. Tranquillini), pp. 11–19. Proceedings of 3rd IUFRO Workshop 1984. Eig. Anst. Forstl. Versuchsives, Berlin.

**Mackay, J.R. (1981).** Active layer slope movement in continuous permafrost environment, Garry Island, Northwest Territories, Canada. *Canadian Journal of Earth Science*, **18**, 1666–1680.

**Miller, Jr., O.K. (1981).** Mycorrhizae, mycorrhizal fungi and fungal biomass in subalpine tundra at Eagle Summit, Alaska. *Holarctic Ecology*, **5**, 125–134.

**Miller, Jr., O.K. & Laursen, G.A. (1978).** Ecto- and endomycorrhizae of arctic plants at Barrow, Alaska. In *Vegetation and Production Ecology of an Alaskan Arctic Tundra* (Ed. by L.L. Tieszen), pp. 229–238. Springer Verlag, New York.

**Miller, P.C. (1982).** The availability and utilization of resources. Tundra ecosystems (Ed. by P.C. Miller). *Holarctic Ecology*, **5**, 83.

**Miller, P.C., Stoner, W.A. & Ehleringer, J.R. (1978).** Some aspects of water relations of arctic and alpine regions. In *Vegetation and Production Ecology of an Alaskan Arctic Tundra* (Ed. by L.L. Tieszen), pp. 343–358. Springer Verlag, New York.

**Miller, P.C., Mangan, R. & Kummerow, J. (1982).** Vertical distribution or organic matter in eight vegetation types near Eagle Summit, Alaska. *Holarctic Ecology*, **5**, 117–124.

**Oberauer, S. & Miller, P.C. (1982).** Growth of Alaskan tundra plants in relation to water potential. *Holarctic Ecology*, **5**, 194–199.

**Osmond, D.L., Wilson, R.F. & Raper, C.D. (1982).** Fatty acid composition and nitrate uptake of soybean roots during acclimation to low temperature. *Plant Physiology*, **70**, 1689–1693.

**Perfect, E., Miller, R.D. & Burton, B. (1988).** Frost upheaval of overwintering plants: A quantitative study of displacement process. *Arctic and Alpine Research*, **20**, 70–75.

**Prevost, M.F. (1978).** Modular construction and its distribution in tropical woody plants. *Tropical Trees as Living Systems* (Ed. by P.B. Tomlinson & H. Zimmerman), pp. 223–231. Cambridge University Press, Cambridge.

**Read, D.J. (1983).** The biology of mycorrhiza in the *Ericales*. *Canadian Journal of Botany*, **61**, 985–1004.

**Read, D.J. & Haselwandtler, K. (1981).** Observations on the mycorrhizal status of some alpine plant communities. *New Phytologist*, **88**, 341–352.

**Robards, A.W. & Clarkson, D.T. (1984).** Effects of chilling temperatures on root cell membranes as viewed by freeze-fracture electron microscopy. *Protoplasma*, **122**, 75–85.

**Rosswall, T. & Heal, O.W. (Eds) (1975).** Structure and function of tundra ecosystems. *Ecological Bulletins*, **20**, 450 pp.

**Savile, D.B.O. (1972).** Arctic adaptations in plants. *Canadian Department of Agriculture Monographs*, **6**, 81 pp.

**Shaver, G.R. & Billings, W.D. (1975).** Root production and root turnover in a wet tundra ecosystem, Barrow, Alaska. *Ecology*, **56**, 401–409.

**Shaver, G.R. & Billings, W.D. (1976).** Carbohydrate accumulation in tundra graminoid plants as a function of season and tissue age. *Flora*, **165**, 247–267.

**Shaver, G.R. & Cutter, J.C. (1979).** The vertical distribution of live vascular phytomass in cotton grass tussock tundra. *Arctic Alpine Research*, **11**, 335–342.

**Slade, A.J. & Hutchings, M.J. (1987).** The effects of nutrient availability on foraging in the clonal herb *Glechoma hederacea. Journal of Ecology*, **75**, 639–650.

**Small, E. (1972).** Photosynthetic rates in relation to nitrogen recycling as an adaptation to nutrient deficiency in peat bog plants. *Canadian Journal of Botany*, **50**, 2227–2233.

**Stoner, W.A., Miller, P.C., Richards, S.P. & Barkley, S.A. (1978).** Internal nutrient cycling as related to plant life-form — a simulation approach. In *Environmental Chemistry and Cycling Processes* (Ed. by D.C. Adriana & I.L. Brisbin), pp. 165–181. D.O.E. Symposium Series Conf. — 760429. Technical Information Center, U.S. Department of Energy, Washington, D.C.

**Stoner, W.A., Miller, P. & Miller, P.C. (1982).** Seasonal dynamics and standing crops of biomass and nutrients in a subarctic tundra vegetation. *Holarctic Ecology*, **5**, 172–179.

**Svensson, B.M. & Callaghan, T.V. (1988).** Apical dominance and the simulation of metapopulation dynamics in *Lycopodium annotinum. Oikos*, **51**, 331–342.

**Teeri, J.A. (1973).** Desert adaptations of a high arctic plant species. *Science*, **179**, 496–497.

**Tester, M., Smith, S.E. & Smith, F.A. (1987).** The phenomenon of 'nonmycorrhizal' plants. *Canadian Journal of Botany*, **65**, 419–431.

**Thomas, W.A. & Grigal, D.F. (1976).** Phosphorus conservation by evergreens of mountain laurel. *Oikos*, **27**, 19–26.

**Trappe, J.M. (1987).** Phylogenetic and ecological aspects of mycotrophy in the angiosperms from an evolutionary standpoint. In *Ecophysiology of VA mycorrhizal plants* (Ed. by G. Safir), pp. 5–25. CRC Press, Boca Raton, Florida.

**Troughton, A. (1981).** Length of life of grass roots. *Grass and Forage Science*, **36**, 117–120.

**Wager, H.G. (1938).** Growth and survival of plants in the Arctic. *Journal of Ecology*, **26**, 390–410.

**Wallen, B. (1986).** Above- and belowground drymass of the three main vascular plants on hummocks on a subarctic peat bog. *Oikos*, **46**, 51–56.

**Watling, R. (1988).** Presidential address: a mycological kaleidoscope. *Transactions of the British Mycological Society*, **90**, 1–28.

**Webber, P.J. (1977).** Belowground tundra research: A commentary. *Arctic and Alpine Research*, **9**, 105–111.

**Wielgolaski, F.E. (1972).** Vegetation types and primary production in tundra. *Proceedings IVth International Meeting on the Biological Productivity of Tundra* (Ed. by F.E. Wielgolaski & T. Rosswall), pp. 9–34. Tundra Biome Steering Committee, Stockholm.

**Wielgolaski, F.E., Kjelvik, S. & Kallio, P. (1975).** Mineral content of tundra and forest plants. Fennoscandian Tundra Ecosystems. Part 1. *Plants and Microorganisms* (Ed. by F.E. Wielgolaski), pp. 316–332. Springer-Verlag, Berlin.

**Wielgolaski, F.E., Bliss, L.C., Svoboda, J. & Doyle, G. (1981).** Primary production of tundra. *Tundra Ecosystems: a Comparative Analysis* (Ed. by L.C. Bliss, O.W. Heal & J.J Moore), pp. 187–225. Cambridge University Press, Cambridge.

**Wein, R.W. & Bliss, L.C. (1974).** Primary production in arctic cottongrass tussock tundra communities. *Arctic Alpine Research*, **6**, 261–274.

**Williams, P.J. (1976).** Volume change in frozen soils. *Laurits Bjerrum Memorial Volume*, pp. 233–246. Norwegian Geotechnical Institute, Oslo.

**Williams, P.J. (1977).** General properties of freezing soils. *Soil Freezing and Highway Construction* (Ed. by P.J. Williams & M. Fremond), pp. 7–12. Paterson Centre, Carleton University, Ottawa.

# Underground biomass and its influence on soil shear strength in a grazed and an ungrazed German coastal marsh

M. SCHOLAND, F.A. AUSTENFELD AND D.J. VON WILLERT
*Institut für Angewandte Botanik, Hindenburgplatz 55, D-4400 Münster, Germany*

## SUMMARY

1 The underground plant biomass of a grazed and an ungrazed coastal salt marsh area in Niedersachsen was compared during the main part of the growing season. The rate of biomass declined with depth and differed significantly between grazed and ungrazed plots.

2 Shear strength, a major soil physical parameter influencing resistance to erosion, was also measured at both sites. It increased with depth and was generally higher in the surface layer of grazed compared with ungrazed areas.

3 Under the prevailing conditions of the wet summer of 1987 there was no clear correlation between shear strength and underground biomass.

## INTRODUCTION

In Germany coastal marshes are of great importance for the security of the dikes. Traditionally the marshes are managed by grazing and cutting for hay or silage (Kempf, Lamp & Prokosch 1987). Rahmann *et al.* (1987) showed that during agricultural uses the flora and fauna of salt marshes changed and that underground biomass decreased substantially. Since the amount of the organic matter, such as roots and litter fall, considerably affects soil texture (Jamison 1954; Schiechtl 1983) the management of the coastal salt marshes, e.g. grazing, hay-making, cutting grass for silage or other traditional agricultural uses, should be optimized to aid dike protection and conservation.

Shear strength is a physical parameter which determines the resistance of the soil to erosion caused by waves and streaming water. It is an often used and easy to measure quantity (Zanke 1982). Since there is no doubt that plants by virtue of their roots diminish erosion we aimed to quantify this protective ability by comparing soil shear strength with the corresponding underground plant biomass. Both parameters were therefore determined in soil profiles during the growing season when underground plant biomass was expected to vary most (Ranwell 1961; Tyler 1971; Jefferies & Perkins 1977; Jensen 1980; Hussey & Long 1982; Groenendijk & Vink-Lievaart 1987). Since grazing has large effects on the vegetation (Beeftink 1979; Irmler & Heydemann 1983; Jensen 1985) it should also alter the amount of roots in the soil. Consequently we compared a grazed with an ungrazed plot in the salt marshes.

## MATERIALS AND METHODS

### *Research area*

The Leybucht is situated in the north-westerly part of Niedersachsen, 30 km north of Emden. This little seabay is attached to the Ems–Dollart Estuary.

After building the 'Störtebeker dike' in 1950 a salt marsh developed. Today, the breadth of the salt marsh varies between 800 and 1200 m. Typical marsh vegetation communities are found there (Gross, unpublished).

The research areas belong to the Puccinellietum maritimae and are situated about 50–70 cm above middle tidemark level. The marsh is randomly inundated by tides (salinity 2·8–3·2%), mainly during autumn and winter. The area is drained artificially by ditches, which are about 40–50 cm deep and 110 cm wide. The distance between two ditches is about 10 m.

The research areas had been established in 1980 in order to study the influence of different grazing activities on the fauna and flora of the communities. Each research area encloses 10 ha. Since 1980 these plots have remained either ungrazed or grazed by one cow per ha between May and October.

### *Underground biomass*

Between March and July 1987 soil samples were taken monthly. Samples were taken 1–2 m from a ditch. Preliminary studies in autumn 1986 showed very low amounts of plant biomass in layers below a depth of 30 cm. Consequently cores were collected, with a soil auger, down to a depth of 30 cm and divided into six sections (0–5 cm, 5–10 cm, 10–15 cm, 15–20 cm, 20–25 cm and 25–30 cm). Each piece had a volume of 125 $cm^3$ and was stored in a refrigerator until isolation of roots started (Schuurman & Goedewaagen 1971). Four parallel samples were taken on each occasion and at each site.

The underground root biomass (living plus dead material) was separated from soil particles by washing (Böhm 1979; Heringa, Groenwold & Schoonderbeek 1980). After shaking the soil cores for several hours in 1 M oxalic acid — the gas bubbles of carbon dioxide formed by the reaction of carbonate with oxalic acid helped to disperse the soil aggregates — roots and soil were separated by a jet of water. A sieve with a mesh size of 0·5 mm was used to retain the root material. The isolated roots were thoroughly cleaned, dried at 95 °C, and weighed.

### *Soil shear strength*

The shear strength was measured with a vane tester of the Fa. Geonor at the same time and the same places where the soil cores were taken. Four parallel measurements were made in depths of 10, 20, 30, 30 and 50 cm.

## RESULTS

### *Underground biomass*

In the ungrazed area (Fig. 1) the underground biomass concentration was highest in the top layer (0–5 cm) and decreased rapidly with depth. The initial steep decline stopped at 10–15 cm and was followed by a rather slow further decrease of biomass. This distribution pattern was found at very different sampling places and at all sampling times. Hence Fig. 1 represents the typical distribution of underground biomass of ungrazed salt marshes. The highest underground biomass concentration was found in April in a depth of 0–5 cm — 28·9 kg dry weight $m^{-3}$.

Figure 2 shows that in the grazed area the underground biomass distribution differed significantly from that of the ungrazed area. Again the highest biomass concentration was found in the top layer and the biomass concentration decreased with increasing depth, but here the decrease was steady with a fairly constant slope. This distribution pattern was typical and was obtained at all sites of the grazed plot and at all sampling times.

### *Soil shear strength*

Figure 3 shows that in the ungrazed area the lowest shear strength occurred in the top layer (0–5 cm). With increasing depth shear strength increased; the greatest

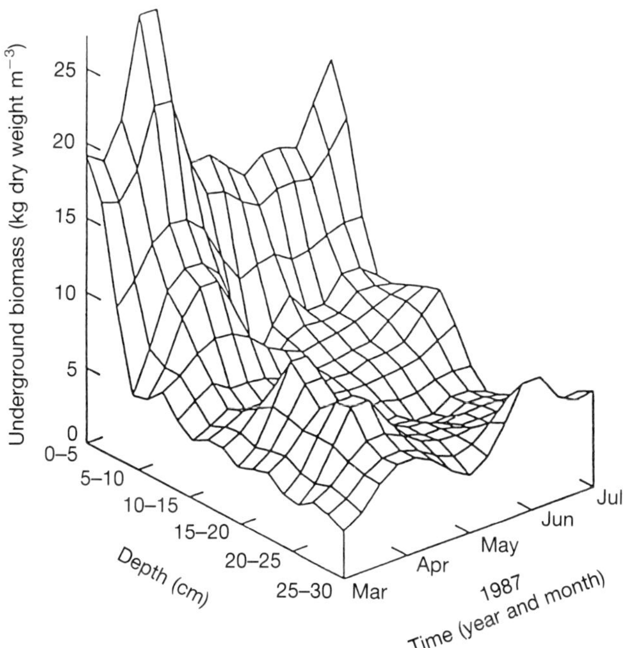


Fig. 1. Vertical distribution of underground biomass concentration in an ungrazed salt marsh.

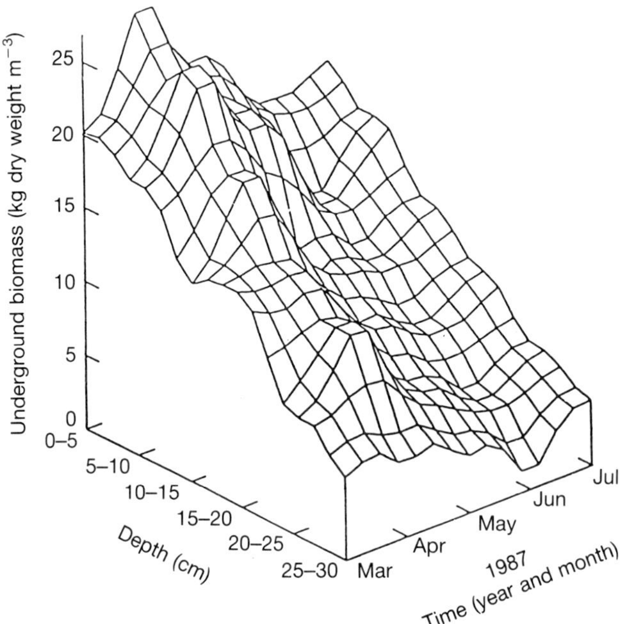


FIG. 2. Vertical distribution of underground biomass concentration in a salt marsh plot grazed by one cow per hectare.

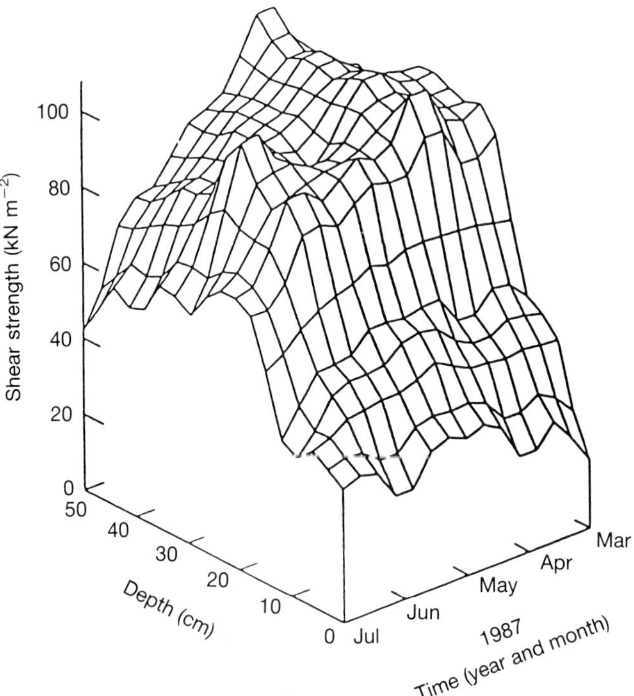


FIG. 3. Vertical distribution of shear strength in an ungrazed salt marsh.

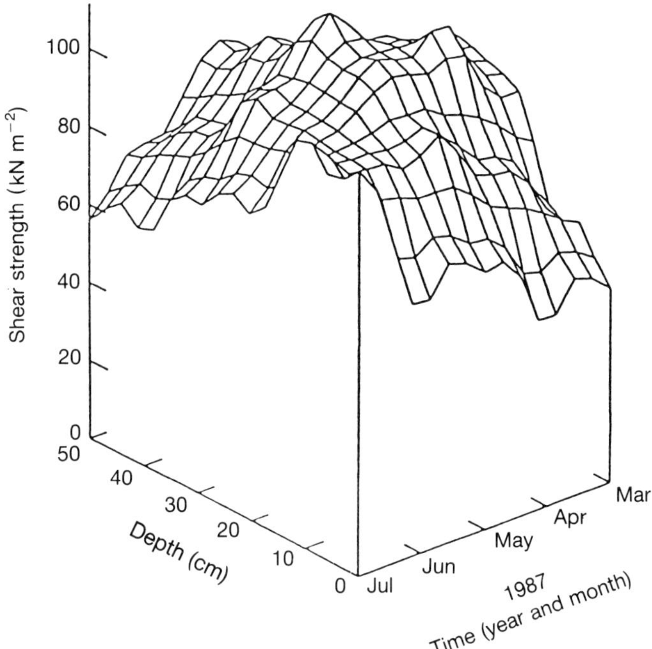


Fig. 4. Vertical distribution of shear strength in a salt marsh plot grazed by one cow per hectare.

increment was found between 20 and 30 cm. Shear strength did not vary in the upper layers (0–20 cm) during the measuring period. In layers deeper than 20 cm shear strength decreased from April to July.

Generally the shear strength of the upper layers was higher in the grazed than in the ungrazed area (Fig. 4). Furthermore the upper soil of the grazed area was characterized by an increasing shear strength during the measuring period.

Figure 5 illustrates that there was no clear correlation between soil shear strength and biomass content. In the upper soil layers the underground biomass content varied greatly in both grazed and ungrazed areas while the shear strength remained constant. For this layer the shear strength of the grazed salt marsh was higher than that of the ungrazed area. At 10 cm depth the biomass content ranged between 3·8 and 7·8 kg dry weight $m^{-3}$ in the ungrazed area and between 10·2 and 19·5 kg dry weight $m^{-3}$ in the grazed area. The corresponding shear strengths for the ungrazed plot were lower. At 20 cm depth there was no difference in biomass content and shear strength between grazed and ungrazed areas. For both biomass content varied between 3·0 and 13·1 kg dry weight $m^{-3}$ and shear strength between 71 and 110 kN $m^{-2}$.

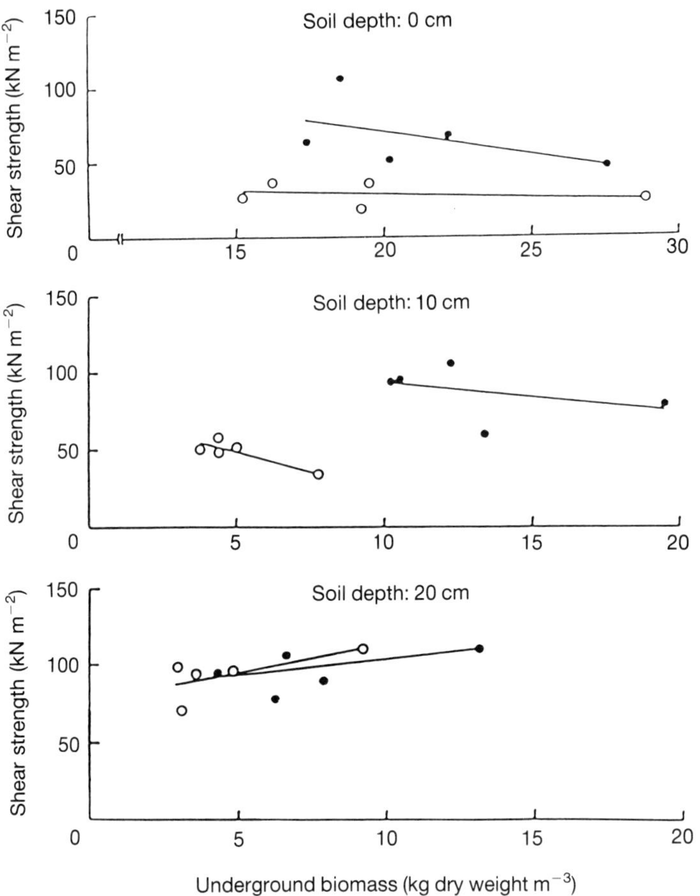


FIG. 5. Correlation between shear strength and underground biomass concentration for three soil depths in an ungrazed (○) and a grazed plot (●) of a salt marsh.

## DISCUSSION

The underground biomass concentration in the Leybucht salt marsh soils was of the same magnitude as that observed for salt marsh soils of North Carolina (Stroud 1976) and Mississippi (De la Cruz & Hackney 1977).

The vertical distribution of the underground biomass in the grazed and ungrazed areas differed significantly: whereas in the grazed area biomass decreased at a constant rate with increasing depth it decreased in a biphasic pattern in the ungrazed area. Similar distribution patterns of underground biomass have been reported for salt marsh soils of Georgia, Delaware and Maine (Gallagher & Plumley 1979). These authors supposed that microbiological activity was responsible for the different underground biomass profiles. Ketner (1972), who studied underground biomass concentration in salt marsh soils on the island of Terschelling

(Netherlands) pointed out that roots in ungrazed salt marsh areas grew deeper than in grazed areas. This can well result in different biomass distribution patterns in grazed and ungrazed salt marsh soils.

The higher shear strength of grazed areas may be due to the solidification by cattle. Jensen (1985) estimated that a standing cow exerts a pressure of approximately 1·56 kg $cm^{-2}$ on the soil. Assuming that the cow walks 4 km per day he calculated that each point of a pasture stocked with one cow per hectare can be trodden once within 80 days. With increasing depth shear strength increased. Often there were great oscillations in the amounts, perhaps being due to variations in climate conditions as suggested by Waldron (1977).

There was no significant correlation between underground biomass and shear strength. It is possible that the influence of underground biomass on shear strength is so small that under the conditions of the experiment (climate, prevailing field conditions) no effect could be resolved; the influence of depth on shear strength is clearly greater than that of biomass. It seems to be obvious that grazing cattle can protect the salt marsh due to solidification and thus increase the resistance of the soil against erosion. Other authors however have emphasized the negative influence of agricultural management on salt marshes (Beeftink 1979; Irmler & Heydemann 1983; Jensen 1985).

## ACKNOWLEDGEMENTS

This investigation was initiated and supported by the Bauamt für Küstenschutz in Norden.

## REFERENCES

**Beeftink, W.G. (1979).** The structure of salt marsh communities in relation to environmental disturbances. *Ecological Process in Coastal Environments* (Ed. by R.L. Jefferies & A.J. Davy), pp. 77–93. European Ecological Symposium 1977, Norwich.

**Böhm, W. (1979).** Methods of studying root systems. *Ecological Studies*, Vol. 33 Springer Verlag, Berlin.

**De la Cruz, A.A. & Hackney, C.T. (1977).** Energy value, elemental composition and productivity of belowground biomass of a Juncus tidal marsh. *Ecology*, **58**, 1165–1170.

**Gallagher, J.L. & Plumley, F.G. (1979).** Underground biomass profiles and productivity in Atlantic coastal marshes. *American Journal of Botany*, **66**, 156–161.

**Groenendijk, A.M. & Vink-Lievaart, M.A. (1987).** Primary production and biomass on a Dutch salt marsh: emphasis on the below-ground component. *Vegetatio*, **70**, 21–27.

**Heringa, J.W., Groenwold, J. & Schoonderbeek, D. (1980).** An improved method for the isolation and the quantitative measurement of crop roots. *Netherland Journal of Agricultural Science*, **28**, 127–134.

**Hussey, A. & Long, S.P. (1982).** Seasonal changes in weight of above- and below-ground vegetation and dead plant material in a salt marsh at Colne Point, Essex. *Journal of Ecology*, **70**, 757–771.

**Irmler, U. & Heydemann, B. (1983).** Die ökologische Problematik der Beweidung von Salzwiesen an der Niedersächsischen Küste — am Beispiel der Leybucht. Bericht des Landesverwaltungsamtes Hannover, Abt. Naturschutz.

**Jamison, V.C. (1954).** Effect of some soil conditions on friability and compactibility of soils. *Proceedings of the Soil Science Society of America*, **18**, 391–394.

**Jefferies, R.L. & Perkins, N. (1977).** The effect on the vegetation of the additions of inorganic nutrients to salt marsh soils at Stiffkey, Norfolk. *Journal of Ecology*, **65**, 865–882.

**Jensen, A. (1980).** Seasonal changes in near infrared reflectance ratio and standing crop biomass in a salt marsh community dominated by *Haliminione portulacoides* (L.) Aellen. *New Phytologist*, **86**, 57–67.

**Jensen, A. (1985).** The effect of cattle and sheep grazing on salt marsh vegetation at Skallingen, Denmark. *Vegetatio*, **60**, 37–48.

**Kempf, N., Lamp, J. & Prokosch, P. (Eds) (1987).** Salzwiesen: Geformt von Küstenschutz, Landwirtschaft oder Natur? *Tagungsbericht 1 der Umweltstiftung WWF-Deutschland.*

**Ketner, P. (1972).** *Primary production of salt marsh communities on the island of Terschelling in the Netherlands*. Ph.D. thesis, University of Nijmwegen.

**Rahmann, M., Rahmann, H., Kempf, N., Hoffmann, B. & Gloger, H. (1987).** Auswirkungen unterschiedlicher landwirtschaftlicher Nutzung auf die Flora und Fauna der Salzwiesen an der ostfriesischen Wattenmeerküste. *Senckenbergiana Maritima*, **19**, 163–193.

**Ranwell, D.S. (1961).** Spartina salt marshes in Southern England. I. The effects of sheep grazing at the upper limits of Spartina marsh in Bridgewater Bay. *Journal of Ecology*, **49**, 325–340.

**Schiechtl, H. (1983).** Pflanzen als Mittel zur Bodenstabilisierung. *Root Ecology and its Practical Application* (Ed. by L. Kutschera), pp. 703–708. International Symposium Gumpenstein, 1982, Austria.

**Schuurman, J.J. & Goedewaagen, M.A.J. (1971).** *Methods for the Examination of Root Systems and Roots* 2nd edn. Wageningen, Pudoc.

**Stroud, L.M. (1976).** *Net primary production of belowground material and carbohydrate patterns of two height forms of Spartina alterniflora in two North Carolina marshes*. Ph.D. thesis, University of Raleigh.

**Tyler, G. (1971).** Distribution and turnover of organic matter and minerals in a shore meadow ecosystem. *Oikos*, **22**, 265–291.

**Waldron, L.J. (1977).** The shear resistance of root-permeated homogeneous and stratified soil. *Journal of the Soil Science Society of America*, **41**, 843–849.

**Zanke, U. (1982).** *Grundlagen der Sedimentbewegung*. Springer Verlag, Berlin.

# Structure and function in desert root systems

P.W. RUNDEL AND P.S. NOBEL

*Department of Biology and Laboratory of Biomedical and Environmental Sciences, University of California, Los Angeles, California 90024, USA*

## SUMMARY

1 Low and unpredictable rainfall, highly variable infiltration capacities and divergent patterns of hydrologic flow all combine to produce strong spatial and temporal heterogeneity in soil moisture availability for desert ecosystems. Divergent patterns of rooting architecture in desert plants results from this inhomogeneity.

2 Competitive interactions of desert plants at the community level are strongly influenced by rooting architecture and phenology of growth. Symbiotic nitrogen fixation and mycorrhizal fungi provide important attributes of plant nutrition for individual species and impact patterns of nutrient flow for biological communities.

3 Root systems of succulent plants are shallow, averaging only about 10 cm in depth for various species of cacti. Simulation modelling suggests that such shallow rooting leads to enhanced water uptake under the sporadic and light rainfall conditions characteristic of deserts. The majority of water uptake over the first few days following rain is from established roots which rehydrate readily.

4 Roots of *Agave deserti* and *Ferocactus acanthodes* exhibit rectifier-like behaviour, facilitating water uptake from a wet soil but greatly limiting water loss from the succulent shoots to a drying soil.

5 Desert phreatophytes with root systems extending to groundwater pools are able to establish high levels of net primary production by largely decoupling themselves from the normal limiting factor of water availability. Many phreatophytic legumes also utilize symbiotic nitrogen fixation to further reduce limitations for growth with good water availability.

## INTRODUCTION

Desert and semi-desert regions, comprising nearly 30% of the earth's surface, are characterized by low and variable precipitation, low relative humidity, high solar irradiance, and high soil surface temperatures. All of these factors combine their influences to limit the availability of soil moisture. Such arid-region soils provide a particularly unfavourable environment for root growth. The movement of water and nutrients from soil to root surfaces is often slow or negligible, dry and cemented soils offer a difficult medium for root penetration, and low soil humidities provide difficult conditions for the maintenance of root turgidity and growth.

Despite these difficult environmental conditions, desert and semi-desert regions support a surprising biomass and diversity of plant species. Even more remarkable are the variety of plant growth-forms present in many desert communities. Stressful environments are commonly dominated by a single plant growth-form that provides the most evolutionarily successful solution to adaptation to specific suites of environmental stresses. Thus, deciduous forests with a predictable winter stress are dominated by winter-deciduous, broad-leaved trees. Temperate forest climates with low nutrient availability have a dominance of needle-leaf, evergreen conifers. Evergreen, sclerophyllous shrubs are characteristic of all five Mediterranean-type ecosystems of the world.

Deserts, with extreme but unpredictable drought stress, provide a very different situation. Evergreen shrubs, drought-deciduous shrubs, winter-deciduous low trees or shrubs, CAM succulents, and ephemerals may each be dominant in local areas, or may be codominant in a single stand. We hypothesize that it is the inhomogeneity of limited water resources in desert soils that allows such a diversity of growth-forms to coexist.

Although light desert rains penetrate only to relatively shallow depths, there are many geomorphological and hydrological factors in desert ecosystems that provide conditions promoting the inhomogeneity of soil water resources at greater depths. The porous nature of stony pediments and sand dunes in desert regions produces rapid rates of rainfall infiltration to relatively great depths. Less porous rocky hillsides have little infiltration, except in local pockets of water accumulation, and promote rapid run-off onto upper bajada slopes (Fig. 1). Such run-off collects in ephemeral drainages and arroyos, becoming rapidly concentrated into major drainage channels or into sheet flow over lower bajada slopes and flats. This water may percolate locally into ground water pools, flow into playas, or drain externally to the coast. All of these conditions lead to a complex array of soil moisture pools, varying quantitatively, spatially, and temporally.

In this chapter we first discuss the architecture of desert root systems, the phenological and allocation strategies of such root systems, and the biological

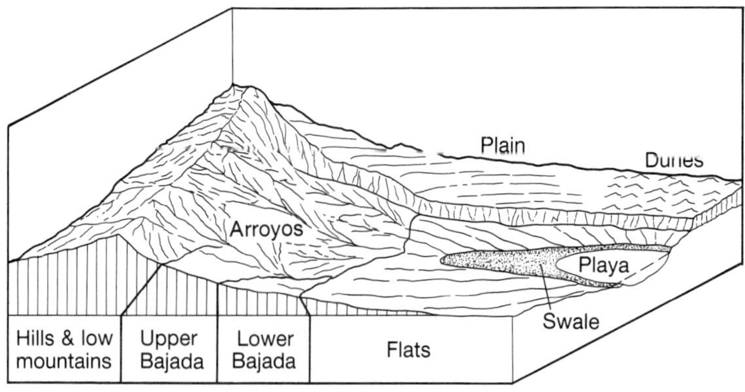


FIG. 1. Geomorphic habitat distribution in desert regions (from Ludwig 1977).

environment of desert soils. We will then discuss the ecological relationships in more detail for two special groups of desert plants — shallow-rooted CAM succulents and deep-rooted phreatophytes.

## STRUCTURE OF DESERT ROOT SYSTEMS

The earliest detailed studies of root systems in desert plants (Cannon 1911, 1913, 1921, 1924; Markle 1917; Waterman 1923) focused on general canopy architecture of these root systems and on the possible adaptive role of this architecture. Before the pioneering early study of Cannon (1911), the prevailing idea was that root systems of desert and semi-desert plants were characteristically of great length and penetrated to great depths in the soil. This form of root structure was considered essential for a desert plant to take up sufficient moisture to survive long dry seasons (Volkens 1887; Schimper 1903). The only major exceptions to this form of expansive root system noted by early naturalists were for the shallow, fibrous roots of desert succulents (Volkens 1887).

Cannon (1911), in his studies of the root systems of Sonoran Desert plants in Arizona, grouped perennial species into three broad types. Generalized root systems were those with both well-developed tap roots and lateral root systems. In this common type, which is widespread in all types of desert habitats, he included such genera as *Encelia, Acacia, Lycium, Prosopis* and *Larrea*. A second category included those species whose rooting architecture is characterized by a prominent tap root, with only minor development of laterals. Examples of this can be seen in *Ephedra, Condalia* and *Koeberlina*. These species are most characteristic of wash habitats where water is available at depth. The final group included species where lateral roots are predominant with little or no development of a tap root. *Yucca*, most Cactaceae, *Krameria* and *Jatropha* form this group, which is common on bajadas. Cannon (1911) also studied the root systems of both winter and summer annuals. While the depth of these annual root systems seldom exceeds 15 cm, the characteristic root architecture fits that of the generalized perennials.

There have been attempts to classify root systems in more detail (see, for example, Cannon 1949), but none of these has found wide adoption. Zohary (1961), stressing the enormous plasticity and adaptability of desert root systems, distinguished ten main types of root systems in the desert regions of the Near East. These types differed in their depth, production of laterals, and specialized modifications.

### *Ecological approaches to root classification*

A more ecological approach to the classification of desert roots is to group species into growth-form categories that reflect the general structure of their below-ground architecture. These groups include shallow-rooted ephemerals, shallow-rooted perennials, deep-rooted perennials with superficial laterals, and deep-rooted phreatophytes.

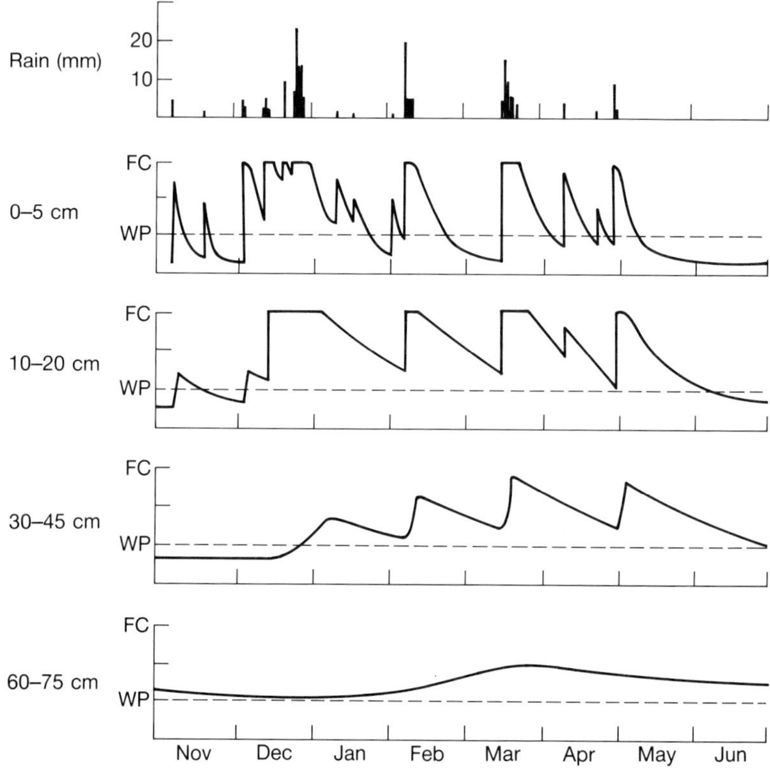


FIG. 2. Dynamics of soil water movement with depth in a desert shrubland. FC = field capacity (−0.01 MPa), WP = permanent wilting point (−1·5 MPa). Adapted from Noy-Meir (1973).

### *Shallow-rooted ephemerals*

The shallow-rooted ephemerals are characterized by shallow and poorly-branched root systems, growing just deep enough to utilize the moisture from light rains that penetrate only the upper 10–30 cm of soil and are subject to evaporative loss (Fig. 2). Root depths no greater than 15 cm are characteristic of these ephemerals, although some species may reach to a depth of 50 cm (Cannon 1911; Evenari, Shanan & Tadmore 1971; Forseth *et al.* 1984). Shallow-rooted ephemerals, whether winter or summer ephemerals, limit their active growth to relatively few weeks of the year when soil moisture conditions are most favourable (Mulroy & Rundel 1977). These so-called *drought evaders* do not require deeper soil moisture pools to complete their life cycle.

The root systems of winter annuals can generally be easily distinguished from those of summer annuals (Cannon 1911). Although the depths of the root systems are about the same, summer annuals commonly have a generalized root form with relatively well-developed and coarser laterals and with a main root that is frequently forked. Winter annuals are more specialized, with a prominently developed tap root and meagre development of thin laterals. These morphologies

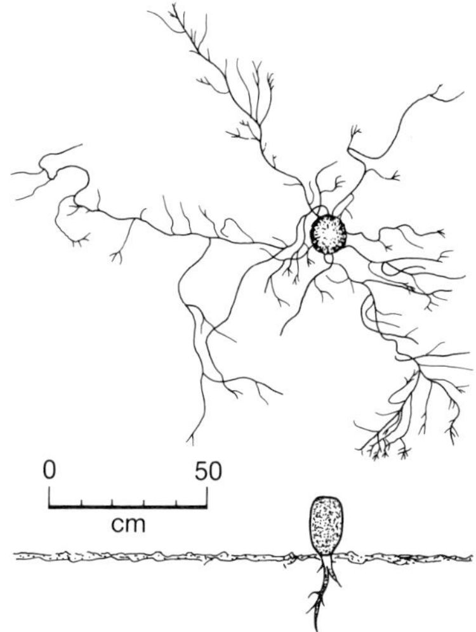


FIG. 3. Horizontal and vertical distribution of roots in the cactus *Copiapoa cinerea* var. *columna-alba* from the Atacama Desert of northern Chile (from Gulmon *et al.* 1979).

suggest that the absorbing root surface of summer annuals should be greater than that of winter annuals. Cannon (1911) noted that three winter annuals that tended to have a generalized root system more similar to those of summer annuals (*Bowlesia, Parietaria* and *Malva)* were limited in his area near Tucson, Arizona, to relatively mesic sites. Winter annuals with the more characteristic root morphology had wider patterns of distribution.

### *Shallow-rooted perennials*

The second ecological grouping of root development in desert plants is the shallow-rooted perennials. The best examples of this group are in the Cactaceae and other families of CAM succulents, where extensive, finely-branched root systems may be restricted to the upper 20 cm of soil (Fig. 3). These shallow lateral roots are potentially subject to extremes of temperature and desiccation because of their position in the soil, which is the subject of special focus later in this paper. While succulents are the most obvious examples of this group, shallow root systems are also common in many low desert shrubs. Good examples can be seen in *Zygophyllum dumosum* and *Retama raetum* from the Negev Desert (Fig. 4).

### *Deep-rooted perennials with superficial laterals*

During periods of prolonged drought stress, which may last for a year or more in many desert areas, woody shrubs require deeper stores of soil moisture. This moisture is harvested by well-developed tap roots or by a number of vertically descending lateral roots. Deeply rooted is, of course, a relative expression.

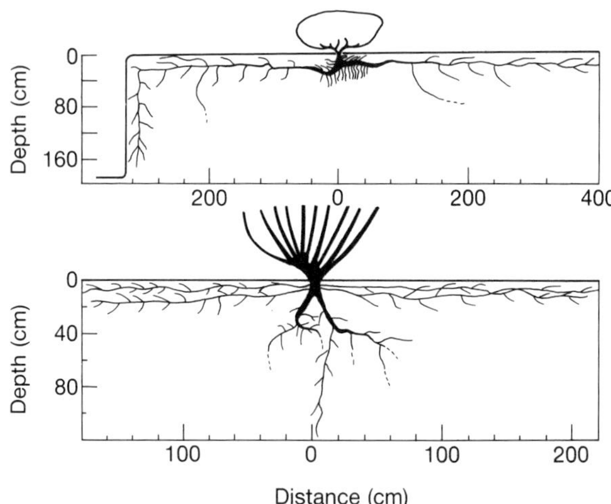


FIG. 4. Rooting profiles of *Zygophylum dumosum* (top) and *Retama raetum* (bottom) in the Negev Desert of Israel. Adapted from Evenari, Shanan & Tadmore (1971).

Depending on the depth to which water from heavy rains or sheet flow penetrates, deep roots of species in this group may extend to depths from 50 cm to several metres. In all cases, these deep soil water reserves reached by roots are ephemeral in nature and are replenished only irregularly. Such perennials, in addition to their deep root system, have a well-developed lateral root system with adventitious roots that can utilize soil moisture from light rains in shallow layers. Examples of this type of root system can be seen for certain shrubs in the Mojave Desert of North

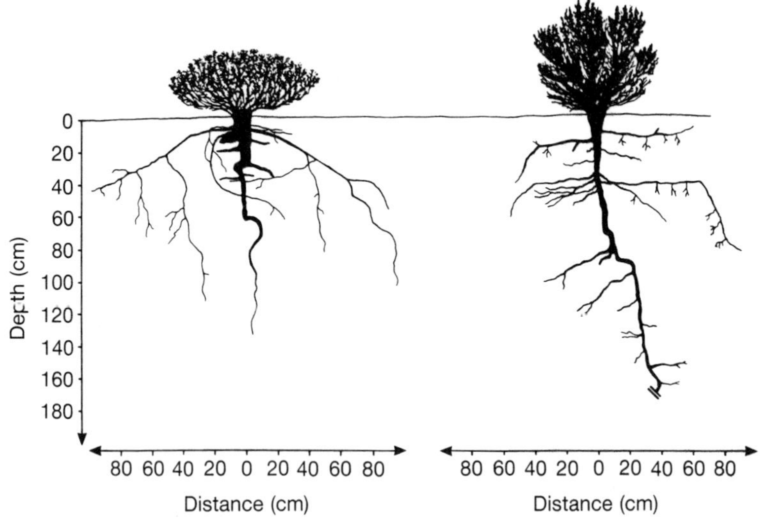


FIG. 5. Rooting profiles in *Haplopappus cooperi* and *Chrysothamus teretifolius* in the Mojave Desert of California (from Manning & Barbour 1988).

America (Fig. 5) and most *Artemisia* species (Orshan 1972; Tabler 1964; Sturges 1977). Such a root system can also develop in perennial desert grasses that have an extensive network of fibrous surface roots extending many metres laterally, whereas vertical roots penetrate to several metres in depth (Price 1911; Zohary 1961). Often the lateral rooting area of perennial shrubs extends over an area that is ten–fifteen times the area of the canopy (Evanari, Shanan & Tadmore 1971).

### *Deeply rooted phreatophytes*

Deeply rooted perennials are often able to reach semi-permanent ground water resources 3–10 m or more below the soil surface, thereby rendering themselves relatively independent of shallow moisture resources. Deep tap roots or laterals in these species may branch copiously in the capillary fringe above the water table. Because roots cannot grow through dry soil, phreatophytes can only develop their deep root systems under conditions where moisture infiltrates through the soil column to the capillary fringe for at least brief periods. Complete saturation of the soil column is not necessary if channels or fracture lines are present. Because of this requirement, phreatophytes are usually restricted to dune areas, porous alluvial soils that receive irregular sheet flow run-off, desert washes, and playas. At the higher rainfall margins of desert regions, many phreatophytic species will occur without the presence of a ground-water table, exhibiting root systems like those in the previous category.

Remarkable records of maximum rooting depths have been reported for desert phreatophytes. Phillips (1963) documented roots of *Prosopis* at a depth of 53 m in a mine excavation in Arizona, and *Zizyphus lotus* in Morocco has been reported to have roots reaching 60 m in depth (Zohary 1961). Such records may be extreme, but *Atriplex halimus, Retama raetum* (Fig. 4) and *Tamarix aphylla* reach to 8, 10 and 20 m depth, respectively, in the Near East (Zohary 1961). *Prosopis tamarugo* in the Pampa del Tamarugal of northern Chile has roots tapping ground water at 12 m (Mooney *et al.* 1980).

## *Community rooting architecture*

While there have been numerous studies of the rooting architecture of individual desert shrubs, there has been remarkably little research focused on rooting architecture at a community level. One excellent example of this type of approach comes from the work of Cody (1986, 1987), who looked at spatial relationships among species of shrubs and cacti in the Mojave Desert of California. He observed that there are distinct neighbour preferences such that some species occur in proximity much more often than would be expected by chance alone, while others are significantly less likely to occur in close proximity. Among moderate-sized shrubs, for example, *Cassia armata* and *Hymenoclea salsola* tend to be separated from conspecifics but demonstrate a significant preference for each other. Three smaller shrubs, *Acamptopappus sphaerocephalus, Haplopappus cooperi* and *Salvia*

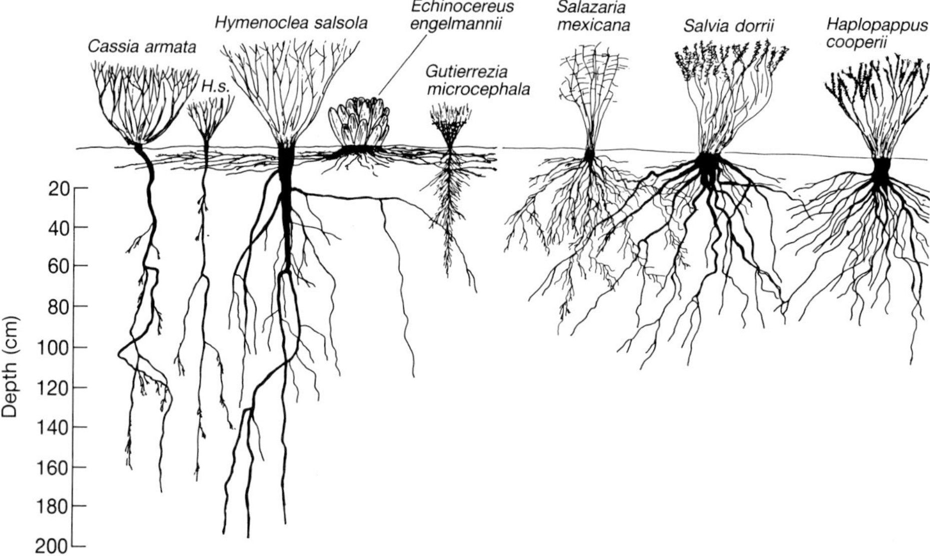


FIG. 6. Community rooting profiles for seven shrub and semi-shrub species in the Mid Hills, eastern Mojave Desert (from Cody 1986).

*dorrii* are clumped in distribution in openings among the larger shrubs, but each significantly avoids the other two species.

Excavations have revealed a diversity of rooting architecture in these shrubs that is far greater than the architectural variation of the above-ground canopies. Species with prominent tap roots such as *Cassia* and *Hymenoclea* compete less in shallow soil horizons, but are relatively complementary in deeper horizons because of the interspecific differences in the depth of lateral root spread (Fig. 6). All three of the smaller shrubs have similar fibrous patterns of root distribution and thus compete directly for water.The observed clumping in conspecifics is likely the result of historical patterns of seedling establishment.

Cody (1986, 1987) has also documented a similar relationship between spacing patterns and rooting architecture in four species of Mojave Desert succulents that are community dominants in the Granite Mountains of California. Two large species of *Opuntia, O. ramosissima* and *O. acanthocarpa*, tend to avoid their own and each other's company. A third species, the smaller *O. echinocarpa*, interacts negatively with *O. acanthocarpa,* whereas *Yucca schidigera* is a preferred neighbour for all three species. All of the *Opuntia* species have similar spreading root systems of moderate depth, while *Y. schidigera* has a shallow and fibrous root system of very different form (Fig. 7). Again, species with complementary and dissimilar rooting architectures apparently compete less and thus become preferred neighbours.

These studies in the Mojave Desert suggest that interspecific coexistence, and thus diversity, in desert ecosystems may be due in part to differences in water

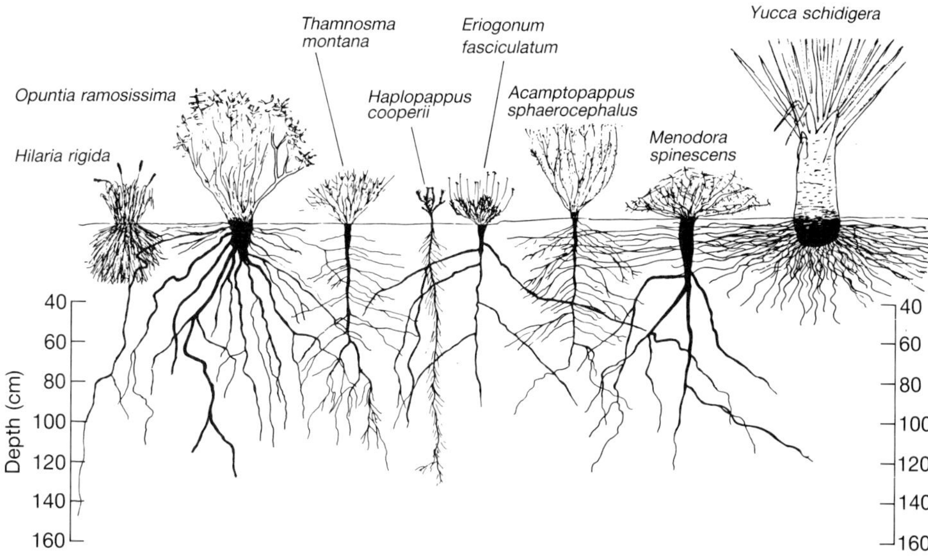


FIG. 7. Community rooting profiles for eight shrub species in the Granite mountains of the eastern Mojave Desert (from Cody 1986).

uptake strategies expressed as variations in rooting architecture. Cody (1986) points out that the highest levels of alpha shrub diversity in the Mojave Desert occur on coarse alluvial fans where water infiltrates rapidly and is available deep into the soil profile

The open spacing patterns of desert shrubs and succulents have made these communities popular sites for studies of interspecific and intraspecific competition. The mechanism for such interactions is generally assumed to be interference competition for soil moisture, and there are a number of manipulation studies that have demonstrated such an effect (Fonteyn & Mahall 1981; Robberecht, Mahall & Nobel 1983; Ehleringer 1984). The desert competition literature has been reviewed in detail (Barbour 1979; Caldwell 1987). Future advances in our understanding of below-ground mechanisms of competition will come from both theoretical models of competition for limited resources, experimental studies, and improved knowledge of rooting architecture (Tilman 1988; Caldwell & Richards 1986; Cody 1986).

### *Root : shoot ratios*

An old paradigm about root systems in desert perennials is the expectation of high ratios of roots to shoots. This is based on the assumption that the limiting nature of water availability in desert soils must require a large amount of absorbing root biomass per unit of physiologically active above-ground tissue. High root : shoot ratios have been reported for shrubs in cold desert ecosystems. Measurements of cold desert shrublands by Rodin & Bazilevich (1967) found root : shoot ratios of

TABLE 1. Root to shoot ratios (dry weight basis) for mature cold-desert shrublands

| Community | Location | Root: shoot | Reference |
|---|---|---|---|
| *Artemisia/Anabasis* | USSR | 6·70 | Rodin & Bazilevich (1967) |
| *Artemisia* | USSR | 3·55 | Rodin & Bazilevich (1967) |
| *Artemisia/Poa* | USSR | 5·11 | Rodin & Bazilevich (1967) |
| *Anabasis* | USSR | 5·11 | Rodin & Bazilevich (1967) |
| *Halaxylon/Salsola* | USSR | 7·33 | Rodin & Bazilevich (1967) |
| *Atriplex confertifolia* | Utah | 3·15 | Osmond, Björkman & Anderson (1980) |
| *Ceratoides lanata* | Utah | 6·67 | Osmond, Björkman & Anderson (1980) |
| *Artemisia tridentata* | Idaho | 0·40 | Pearson (1966) |
| *Artemisia tridentata* | Idaho | 1·80 | Pearson (1966) |

3·6–7·3 (Table 1). Studies in the Great Basin of Utah in the United States have found root : shoot ratios of 3·2 for an *Atriplex confertifolia* community and 6·7 for a *Ceratoides lanata* community (Osmond, Björkman & Anderson 1980). Lower ratios, however, have been reported for cold *Artemisia* shrublands in Idaho (Pearson 1966).

For warm desert ecosystems, shrubland root : shoot ratios are surprisingly low. Studies from a number of desert areas in North America, the Middle East, and Australia have all found individual species values that most commonly range from 0·5 to 1·0 (Table 2). These values are only slightly higher than the 0·26 typical of tropical and temperate forest trees (Rodin & Bazilevich 1967), where trunks form large amounts of woody tissue above ground. Warm desert root : shoot ratios are quite comparable to such ratios for temperate forest and chaparral shrubs (Whittaker 1962; Whittaker & Woodwell 1969; Kummerow 1981).

The root : shoot ratios of desert annuals have been found to range from 0·14 to 0·28, although these values decline late in flowering (Bell, Hiatt & Niles 1979). The small root : shoot ratio of desert annuals is not merely a function of their small size. Shrub seedlings of similar size produce very extensive root systems with deep tap roots during their first few months of growth. Carbon isotope studies have provided data indicating that the water-use efficiency of seedlings that survive their first year are lower than efficiencies in mature shrubs (P.W. Rundel, unpublished data).

While this discussion of root : shoot ratios has stressed mean species values, such ratios for any species can vary considerably. Rooting architecture varies with substrate and resource availability in particular. Shrubs on hillsides, ridgetops, and fine-textured alluvial soils have shallower root systems than conspecifics in other habitats with deeper or less-impervious soil profiles. Rooting depth of laterals is generally a function of the depth of penetration of rainfall and/or run-off. Variation in root : shoot ratios of *Larrea tridentata* has been discussed by Barbour (1973) and of *Artemisia tridentata* by Tabler (1964).

## PHYSIOLOGY OF ROOT GROWTH

There has been very little study of the physiology of root growth in desert plants. Roots show preferential growth in wetter zones of the soil profile (Hellmers *et al.*

TABLE 2. Root to shoot ratios in communities of mature shrub species in warm-desert ecosystems

| Community/species | Location | Root: shoot | Reference |
|---|---|---|---|
| Mojave Desert wash | California | | Garcia-Moya & McKell (1970) |
| *Ambrosia dumosa* | | 0·60 | |
| *Brickellia incana* | | 0·50 | |
| *Cassia armata* | | 0·70 | |
| *Ephedra nevadensis* | | 1·20 | |
| *Eriogonum fasciculatum* | | 0·40 | |
| *Hymenoclea salsola* | | 0·70 | |
| *Krameria grayi* | | 0·40 | |
| *Larrea tridentata* | | 0·30 | |
| Mojave Desert | Nevada | | Wallace, Bamberg & Chew (1974) |
| *Acamptopappus shockleyi* | | 0·50 | |
| *Ambrosia dumosa* | | 1·01 | |
| *Atriplex confertifolia* | | 0·37 | |
| *Atriplex canescens* | | 0·58 | |
| *Ephedra nevadensis* | | 0·72 | |
| *Grayia spinosa* | | 0·61 | |
| *Krameria parvifolia* | | 0·69 | |
| *Larrea tridentata* | | 1·08 | |
| *Lycium andersonii* | | 0·73 | |
| *Lycium pallidum* | | 1·42 | |
| Sonoran Desert | Arizona | | Chew & Chew (1965) |
| *Larrea tridentata* | | 0·23 | |
| Chihuahuan Desert | New Mexico | | Ludwig (1977) |
| *Chilopsis linearis* | | 1·40 | |
| *Ephedra trifurcata* | | 0·70 | |
| *Fallugia paradoxa* | | 0·60 | |
| *Flourensia cernua* | | 1·10 | |
| *Larrea tridentata* | | 0·90 | |
| *Parthenium incanum* | | 0·40 | |
| *Prosopis glandulosa* | | 1·30 | |
| *Yucca elata* | | 1·00 | |
| Negev Desert | Israel | | Evenari *et al.* (1975) |
| *Artemisia herba-alba* | | 0·91 | |
| *Hammada scoparia* | | 1·06 | |
| *Zygophyllum dumosum* | | 0·84 | |
| Australian Desert | New South Wales | | Charley & Cowling (1968) |
| *Atriplex vesicaria* | | 0·41 | |
| Australian Desert | Queensland | | Burrows (1972) |
| *Eremophila gilesii* | | 0·33 | |

1955; Evenari, Shanan & Tadmore 1971; Batanouny & Abdel Wahab 1973), but there is no evidence of any ability of roots to grow through dry soils to such zones. Most of the observations of highly branched roots in moist pockets of soil can be explained by roots extending preferentially into soils of higher water content because of reduced mechanical impedance (Drew 1979). However, there may be a more direct physiological limitation of root expansion into dry soils.

Shallow-rooted desert plants appear to have higher optimum temperatures for root growth than do more mesic plants. Roots of species of *Opuntia* barely begin extension at 16 °C and reach an optimum at 30 °C, whereas the optimum temperature for extension in temperate grasses is around 10 °C (Cannon 1916). High temperatures are known to have a strong effect on root distribution by altering geotropic responses. Roots of many crop plants grow horizontally at moderate temperatures but turn vertically downward above and below this temperature.

Several studies have suggested that soil aeration may be a more important factor limiting root growth than either moisture or problems of mechanical impedance (Cannon 1925; Drew 1979). Aeration appears to correlate well with root distribution in a number of shrubs in central and northern Iraq (Harris 1960). Lunt, Letey & Clark (1973) studied the oxygen requirements for root growth in three species of Mojave Desert shrubs and found that *Artemisia tridentata* and *Larrea tridentata* have unusually high oxygen requirements for root growth. They suggested that the general exclusion of these two species from fine-textured and poorly-drained soils was a reflection of this high oxygen requirement. Roots of *Ambrosia dumosa* require lesser amounts of oxygen and do not show these soil restrictions. The widespread occurrence of deep root systems in desert phreatophytes, as previously discussed, suggests that soil depth by itself may not be a limiting factor for aeration.

## *Root phenology*

Detailed studies of phenology in desert root systems are largely lacking. Careful quantitative studies of the growth and phenology of desert root systems can best be seen in studies with three species of Great Basin shrubs — *Atriplex confertifolia, Ceratoides lanata* and *Artemisia tridentata* (Fernandez & Caldwell 1975). Although the root morphologies (colour, degree of branching, degree of elongation and diameter) are distinctly different in each species, the general patterns of root growth and phenology are common to all three. The initiation of significant growth activity in the spring begins a few days earlier with roots than with above-ground shoots, and active root growth extends several weeks after shoot growth ceased, particularly in *Atriplex*. Small amounts of intermittent root growth extend into the early winter when upper soil layers are frozen (Fig. 8).

The general pattern of activity of the root system varies through the year with soil depth. The pattern primarily corresponds to the seasonal depletion of soil moisture at various depths through the growth cycle. High temperatures in the upper soil horizons during the summer months is also a factor limiting root growth (Moore, White & Caldwell 1972; Fernandez & Caldwell 1975; Jordan & Nobel 1984).

Although the root system as a whole maintains a pattern of growth through much of the year, apical meristems of individual roots (except for primary extension roots) are seldom active for more than 2 weeks for the three Great Basin shrubs (Fernandez & Caldwell 1975). A dense zone of root hairs develops behind the active meristem tips, and these root hairs undergo apparent suberization within 1–2 weeks. The root hairs are persistent in all three shrubs and may last for more than 1 year in *Atriplex* (Fernandez & Caldwell 1975). There are few data to suggest how effective older and apparently suberized root tips may be in the absorption of water and nutrients for desert plants. Tracer studies on roots of *Atriplex vesicaria* in Australia have indicated that nutrient uptake largely occurs in new root tissues (Cowling 1969).

During the summer months, root extension may continue to occur in *Atriplex* communities where soil water potentials have dropped to less than −7 MPa at a depth of 40 cm (Fig. 8). Root activity at low water potentials can be an important adaptive mechanism in these cold desert shrubs. The combination of greatly

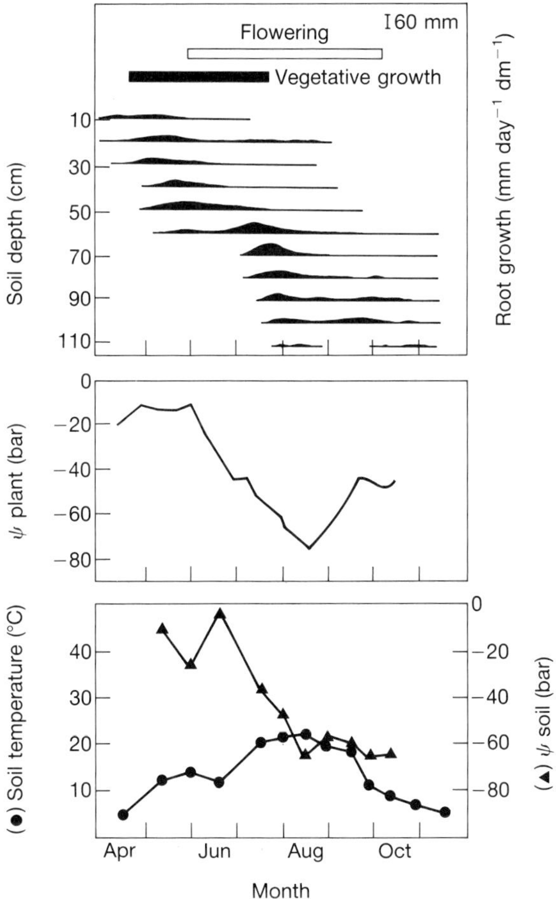


FIG. 8. Seasonal rate of root growth beneath the canopy of *Artiplex confertifolia* in relation to soil depth and seasonal changes in soil temperature and water relations in the Great Basin Desert, Utah. Redrawn from Fernandez & Caldwell (1975).

reduced hydraulic conductivity of desert soils at very low water potentials and rapid apparent suberization of fine-root elements after formation are compelling reasons for an adaptive strategy of continual re-exploration of the same soil volume through root growth extension and the formation of new root hairs (Caldwell & Camp 1974; Caldwell & Fernandez 1975). The maintenance of root growth activity during summer drought may be a decisive factor in allowing photosynthetic production to continue at low water potentials.

If large root : shoot ratios are of such critical importance to cold desert shrubs, why should warm desert shrubs be characterized by ratios less than one? Seasonal water stress is generally higher in warm desert systems, suggesting a need for more, not less, root tissue. Caldwell & Fernandez (1975) speculate that low-temperature limitations to productivity in cold deserts may be the key factor. Many warm-desert shrubs shed their leaves during periods when water stress is most limiting and become physiologically inactive. The maintenance of seasonal productivity rates in cold-desert shrubs may require the maximization of the length of growing season by continuing physiological activity as long into the summer drought period as possible. An extensive and diffuse root system may be required to maintain water uptake into late summer. Comparative data on the structure and physiological limitations for water uptake of cold- and warm-desert root systems are crucially needed. Also, almost nothing is known about the productivity and turnover of root systems in warm-desert shrubs.

### *Symbiotic nitrogen fixation*

Symbiotic nitrogen fixation associated with root nodules may provide a significant input of this nutrient to desert ecosystems where woody legumes are a major component of the vegetation cover. Indeed, virtually all of the major semi-arid and arid regions of the world have large areas of land dominated by woody legumes. Species of *Prosopis* and *Acacia* are particularly important in this regard (Rundel 1989). Thorn forest vegetation at tropical margins of deserts are often overwhelmingly dominated by woody legumes. In the southwestern United States alone more than 30 million hectares are dominated by *Prosopis* (Parker & Martin 1952).

Not all legumes can be assumed to fix nitrogen. A large percentage of the subfamily Caesalpinoideae do not form nodules or fix nitrogen, and nodulation appears to be facultative in a number of members of the Papilionoideae and Mimosoideae (Allen & Allen 1981).

A recent literature review of symbiotic fixation in desert ecosystems (Farnsworth, Romney & Wallace 1978) suggests that fixation in woody legumes is rare or absent in such habitats. More recent intensive studies of nitrogen dynamics in woody legumes in the Sonoran Desert of southern California have demonstrated, however, that nodulation is widespread and that nitrogen fixation by these nodules provides a large annual input of nitrogen to these ecosystems (Shearer *et al.* 1983). Symbiotic fixation by herbaceous legumes may also represent a significant input of nitrogen in desert habitats where such annuals or herbaceous perennials provide a large biomass.

Although the major source of symbiotic fixation in arid regions is clearly *Rhizobium* nodules in legumes, a number of non-legumes also have nitrogen-fixing root nodules. The micro-organisms responsible for fixation in these other nodules are actinomycetes. Few of these non-legume nitrogen fixers are true desert plants, but a number of semi-arid zone shrubs are in this group. Species of *Casuarina* (Casuarinaceae) are important nitrogen fixers in many arid and semi-arid ecosystems in Australia. Two species of *Purshia* (Rosaceae) in the western United States are widespread nitrogen fixers in less xeric areas of the Mojave and Sonoran deserts. Species of *Cercocarpus* (Rosaceae) with active nodules may also be present in similar habitats. Generally, however, non-legume nitrogen fixers are rare in the drier desert areas of the world.

A number of genera that reportedly fix nitrogen may be incorrect. These include reports of nodulation and fixation in a variety of desert genera including *Artemisia* (Compositae), *Chrysothamnus* (Compositae), *Opuntia* (Cactaceae), *Mertensia* (Boraginaceae), *Krameria* (Krameriaceae), *Lomatium* (Umbelliferae), *Viola* (Violoceae) and *Castelleja* (Saxifragaceae) (Farnsworth & Hammond 1968; Krebill & Muir 1974; Farnsworth, Romney & Wallace 1976). Careful studies of *Artemisia* in particular have been unable to confirm nodulation or nitrogen fixation (Wullstein & Harker 1982). The reports of nodulation are likely due to observations of insect galls or haustoria. Similarly, reports of nodules in a number of genera of Zygophyllaceae, *Fagonia, Tribulus* and *Zygophyllum* need to be confirmed before they can be accepted.

### *Mycorrhizae*

The importance of mycorrhizae to vascular plant nutrition has been well documented, and arid lands are no exception. Although there has been relatively little physiological and ecological research on mycorrhizae in desert and semidesert ecosystems (Trappe 1981; Miller 1987), the widespread distribution of mycorrhizal symbioses is evident (Khan 1974; Miller 1979; Bethlenfalvay, Dakessian & Pakovsky 1984; Bloss 1986). Mycorrhizal spores apparently survive in the environmental extremes of drought and temperature that characterize desert soils. Also remarkable is the depth to which roots can maintain mycorrhizal symbioses. *Prosopis* roots recovered from a depth of 12 m in New Mexico were mycorrhizal (Virginia, Jenkins & Jarrell 1986).

The primary ecological role of mycorrhizae is generally considered to be in allowing roots access to nutrients at greater rates or from sources unavailable to the roots alone. Phosphorus nutrition is particularly important in this respect. Less clear, however, is the significance of mycorrhizae in plant–water relations. Studies of semi-arid grasses have shown that mycorrhizal infection can halve whole-plant resistance to water transport, thereby enhancing leaf conductance and photosynthetic capacity (Allen *et al.* 1981).

Several families of prominent desert plants are notable for the common absence of mycorrhizae. These include the Cruciferae, Polygonaceae, Chenopodiaceae, Zygophyllaceae and Cactaceae. The extent of the non-mycorrhizal condition in these families is still controversial and requires more study to resolve. There is evidence that non-mycorrhizal species predominate in the early stages of succession in disturbed arid-land soils (Miller 1979, 1987).

## ROOT DYNAMICS IN DESERT SUCCULENTS

In this section we will focus on agaves and cacti and will consider water uptake and its associated costs from a quantitative point of view. Teleologically speaking, roots should maximize water uptake, minimize costs in terms of carbon, avoid possible water loss from the plant during drought, and avoid high-temperature damage. As we shall see, roots of agaves and cacti fit such criteria.

To understand adaptations of agaves and cacti to desert conditions that are universally characterized by low annual precipitation, we need to know at the very least three aspects of roots: (i) their deployment with depth, usually expressed as root length or root dry weight in various soil layers; (ii) the relation between length or dry weight of such roots and their area across which water and mineral uptake occurs; and (iii) the hydraulic conductivity of the roots, which relates water uptake to differences in water potential. The root hydraulic conductivity $L_P$ ($\mathrm{m\,s^{-1}\,MPa^{-1}}$) can be defined as follows:

$$J_V = L_P(\Psi_{rs}^{soil} - \Psi_x^{root}), \tag{1}$$

where $J_V$ is the volume flux density of water at the root surface ($\mathrm{m^3}$ of water $\mathrm{m^{-2}}$ of root surface area $\mathrm{s^{-1}}$), $\Psi_{rs}^{soil}$ (MPa) is the soil water potential at the root surface, and $\Psi_x^{root}$ is the water potential in the root xylem. The drop in water potential from the root xylem to the shoot is generally small for agaves and cacti of relatively short stature. In such cases, $\Psi_x^{root}$ is similar to $\Psi_x^{shoot}$, which is usually above $-0{\cdot}5$ MPa for agaves and cacti with their succulent, water-storage tissues (Nobel 1988).

Water uptake occurs both by mature, established roots and by rain roots that are induced upon wetting of the soil. Established roots are often brown and can survive drought even though they often have a shrivelled appearance then, whereas the whitish, fragile, rain roots, which generally occur as branches on the established roots, are generally shed during drought (Nobel 1988). However, some roots induced by rain occur as extensions of the main root system or develop directly from the stem. Some of these roots eventually become established roots, although the details of the transition from rain to established roots require more research attention. In any case, we will next modify eqn (1) to indicate that the total water uptake by an entire root system of surface area $A^{total}$ is contributed by both established roots (area of $A^{est}$) and rain roots (area of $A^{rain}$):

$$J_V A^{total} = (L^{est}A^{est} + L^{rain}A^{rain})\,\Delta\Psi \tag{2}$$

where $\Delta\Psi$ is the drop in water potential from the root surface to the root xylem.

*Root deployment*

Roots of agaves and cacti are quite shallow, the mean depth averaging about 8 cm for *Echinocereus engelmannii* (Cody 1986), 10 cm for *Ferocactus wislizenii* (Cannon 1911), 12 cm for *Opuntia discata* (Cannon 1911), and 20–30 cm for *Carnegiea gigantea* (Cannon 1911). The two species in this category whose roots have received the most research attention, *Agave deserti* and *Ferocactus acanthodes*, have mean root depths in the north-western Sonoran Desert averaging 9 cm (Nobel 1976, 1977, 1984; Jordan & Nobel 1984; Hunt & Nobel 1987a). Although the ratio of rain roots to established roots varies with the soil water status, rain roots have essentially the same distribution with depth as do established roots for both species (Fig. 9). The roots tend to be absent from the upper 3 cm of the soil, except directly under the part of the stem from which they emanate. The maximum depth of roots of *A. deserti* and *F. acanthodes* is less than 25 cm (Fig. 9), similar to observations on most other cacti (Cannon 1911).

The absence of roots from the upper part of the soil apparently relates to temperature extremes, especially high temperatures. In particular, the surface temperatures of unshaded desert soil during dry periods can exceed 70 °C (Buxton 1925; Hadley 1970; Oke 1978; Korner & Cochrane 1983). How rapidly such surface temperatures are attenuated down through the soil is characterized by the 'damping depth', which equals the depth in the soil where the amplitude of the daily oscillation in temperature has decreased to $1/e$ or 0·37 of the value at the soil surface. For both wet and dry soils from the north-western Sonoran Desert, the damping depth is 10 cm (Nobel & Geller 1987). Assuming that the daily oscillation in temperature decreases exponentially with depth (Campbell 1977; Nobel 1983),

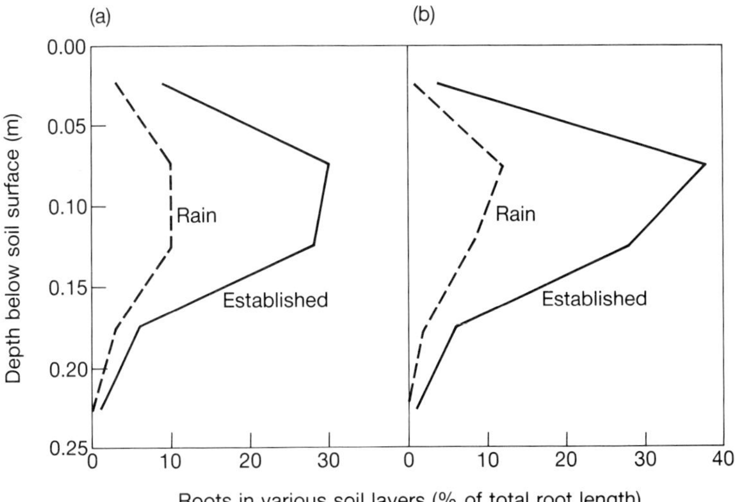


FIG. 9. Distribution with soil depth of established and rain roots in *Agave deserti* and *Ferocactus acanthodes* (from Jordan & Nobel 1984; Hunt & Nobel 1987a).

the maximum soil temperature at any depth can be predicted once the maximum and minimum temperatures at the soil surface and the damping depth are known. Using such an approach and incorporating the measured tolerances of roots of agaves and cacti to high temperature, the absence of roots in the upper 3 cm or so of unshaded soil can be explained because of the high temperatures there (Jordan & Nobel 1984; Nobel *et al.* 1986; Nobel 1988, 1991).

### *Rectification properties*

Equation (1) epitomizes one of the dilemmas faced by all plants but which becomes particularly acute for agaves and cacti with their succulent, water-storage tissues. Specifically, when the soil dries and hence $\Psi_{rs}^{\text{soil}}$ becomes much less than $\Psi_{x}^{\text{root}}$, substantial water movement could occur from the plant back to the soil. The potentially lethal loss of water is avoided by the substantial changes that occur in the root hydraulic conductivity.

Under wet conditions, $L_P$ for the terminal portion of a primary root is about $3 \times 10^{-7}\,\text{m}\,\text{s}^{-1}\,\text{MPa}^{-1}$ for *A. deserti* and *F. acanthodes* (Nobel 1989). Upon exposing such roots to air, $L_P$ decreases over sixty-fold in 16–24 h (Nobel & Sanderson 1984). Instead of a substantial water loss from these succulent plants to a drying soil, the large decrease in $L_P$ eliminates this potentially-lethal plant dehydration. When the root is rewet following dehydration, $L_P$ increases in about 16 h to the value it originally had under wet conditions for both species, facilitating water uptake under wet conditions. Roots of *A. deserti* and *F. acanthodes* therefore tend to act as rectifiers, having a high hydraulic conductivity favouring water uptake when the soil is wet but having a much lower $L_P$ when the soil is dry, thereby minimizing water loss under desiccating conditions.

Little is known about the anatomical details underlying this rectifier-like action. Although cells within the stele are little changed from their appearance under wet conditions, cells of the epidermis, exodermis, and cortex of mature roots of *A. deserti* and *F. acanthodes* become dehydrated and can appear collapsed during drought. The cells of the cortex can become rehydrated following prolonged drought, but they do not appear normal under a light microscope and no cytoplasmic streaming is observed (P.S. Nobel, unpublished observations). Mature roots of *A. deserti* can have a few layers of living cells surrounding the stele that are apparently sclerenchyma, and analogous layers of tangentially flattened cells surround the endodermis (or perhaps form from it) for *F. acanthodes* and in this case appear to be periderm (Nobel 1991). When desiccated, such layers may conduct water less readily than when hydrated, just as dehydrated cells of the epidermis, exodermis, and cortex may also not readily conduct water whereas the hydrated cells presumably do.

### *Water uptake following rainfall*

Agaves and cacti respond rapidly to rainfall because of their shallow root systems and also changes that occur to the roots (Cannon 1911; Drew 1979). A rain root

6 mm long has been detected only 5 h after watering *A. deserti* (Nobel 1988). New roots are visible 8 h after watering for *Opuntia puberula* (Kausch 1965) and within 24 h for other cacti (Walter & Stadelmann 1974). Water uptake and stomatal opening can be induced in 12 h after drought is broken by rainfall for *A. deserti* (Nobel 1976), 24–36 h after a rainfall for *Opuntia basilaris* (Szarek, Johnson & Ting 1973; Szarek & Ting 1975), and within 36 h for *F. acanthodes* (Nobel 1977). It is logical to suppose that rain roots account for the substantial water uptake by desert succulents soon after rainfall interrupts a drought.

Rain roots of *A. deserti* and *F. acanthodes* have an approximately five-fold higher hydraulic conductivity than do 4-month-old established roots and over fifteen-fold higher than do 2-year-old established roots (Nobel 1991), which favours water upake by the rain roots. Yet based on eqn (2) we must also consider the relative area of rain and established roots to predict water uptake; such area begins at zero for the rain roots. Over 90% of the water taken up during the first 24 h after rewetting droughted plants occurs through existing established roots that have become rehydrated (Fig. 10). By 4 days after the soil has become rewet, water uptake by rain roots with their relatively high $L_P$ has the same rate as that of the rehydrated established roots. Therefore, over 75% of the water taken up by *A. deserti* for the first 4 days after rainfall interrupts a drought is by the established roots, whose $L_P$ is greatly increased by the rehydration, which is an example of the rectifier-like action discussed above.

### *Simulation studies on root deployment*

To predict water uptake using eqn (2), the soil in the vicinity of a plant can be divided into a series of subvolumes (Hunt & Nobel 1987a). The area of both established roots and rain roots for each subvolume is based on field excavations.

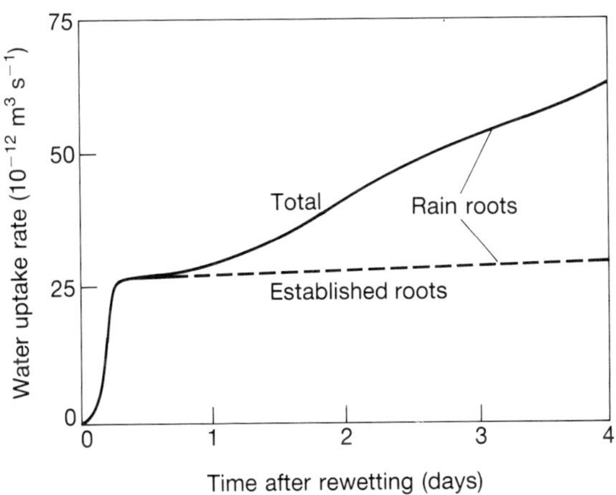


FIG. 10. Water uptake rate by established and rain roots of *Agave deserti* and *Ferocactus acanthodes* in the first four days after rainfall on dry soil (from Nobel 1988).

Using $L_P$ measurements obtained in the laboratory, water movement up to the root surfaces in each subvolume is then calculated using the appropriate form of Darcy's law, which describes water flow in the soils and certain assumptions about the $\Psi$ values (Nobel 1983; Hunt & Nobel 1987a).

For a dry year, an average year, and a wet year, over two-thirds of the water uptake by *Ferocactus acanthodes* occurs from roots in soil layers from 5–15 cm deep (Fig. 11a). The uptake pattern is shifted to the upper layers for dry years and to lower layers for wet years. The natural root distribution leads to an uptake of 28% of the annual precipitation for the dry year, 25% for an average year and 21% for the wet year (Hunt & Nobel 1987a).

For simulations with all the roots in a single, 5-cm-thick layer, the percentage of annual precipitation taken up is lower (Fig. 11b) than for the natural root distribution; the maximum percentages for these single layer distributions are only 14% for the dry year, 16% for the average year, and 17% for the wet year. For the wet year, the various single layers in which the roots can occur are approximately equivalent with respect to water uptake, whereas for the dry year the upper layers lead to maximum water uptake (Fig. 11b). For a year with average rainfall, simulations having all the roots in a single layer show that water uptake decreases when the single layer is deeper than 15 cm. Thus, the observed shallow root systems for *F. acanthodes* (Fig. 9b) leads to the uptake of a substantial fraction of the incident rainfall in dry and average years (<200 mm annual rainfall). Use of such simulation models helps quantitatively show that shallow roots allow desert succulents to take advantage of light rains. Indeed, the root deployment with depth of these species is essentially optimal for water uptake.

## *Cost/benefit analyses*

One of the most important costs for roots is carbon loss by respiration, both for growth and for maintenance (Caldwell & Fernandez 1975). Carbon absorbed by the shoot as $CO_2$ is also structurally incorporated into roots, including the periodically-shed rain roots. About 11% of annual shoot $CO_2$ uptake for *A. deserti* and 19% for *F. acanthodes* is incorporated into established and rain roots (Jordan & Nobel 1984), which may be lower percentages than for most desert perennials, as root turnover of a cold-desert community can be three times greater in terms of carbon than the above-ground growth (Caldwell & Camp 1974). Based on measured root respiration rates, adjusted for temperatures and water conditions that occur in the field for representative 1 year periods, approximately 21% of annual shoot $CO_2$ uptake by *A. deserti* is used for maintenance and growth respiration by the roots and 27% is used by *F. acanthodes* (Nobel 1991). Therefore, 32–46% of the annual net $CO_2$ uptake by the shoots of these two species is diverted to the roots.

We now turn to the carbon costs of the roots with respect to water uptake, focusing on the differences between rain roots and established roots of *A. deserti*. An increase in water uptake is the benefit from root production, and total root respiration plus $CO_2$ incorporated into the root is the cost in terms of $CO_2$. We can convert $CO_2$ costs into water costs by dividing the former by the shoot water-use efficiency (net $CO_2$ uptake divided by transpirational water loss). Because rain roots have a higher respiration rate per unit dry weight than do established roots (about forty-fold higher for maintenance under some conditions; Hunt, Zakir & Nobel 1987) and because rain roots are periodically shed, their daily carbon costs are higher than for established roots per unit root dry weight. On the other hand, rain roots have a higher hydraulic conductivity and can lead to more water uptake per unit dry weight than do established roots. We will specifically consider an *A. deserti* under field conditions where rain roots are shed three times a year in response to drying conditions. Models are used to predict water uptake (Hunt & Nobel 1987a,b), respiration is directly measured (Hunt, Zakir & Nobel 1987), and the water-use efficiency is assumed to average 0·0164 mol $CO_2$ (mol water)$^{-1}$ for the 1 year period for *A. deserti* (Nobel 1976).

As the mass of established roots or rain roots is increased, water uptake by *A. deserti* is increased. Water uptake reaches 90% of maximum at about 90 g of established roots and at 15 g of rain roots. At such root masses, water uptake is

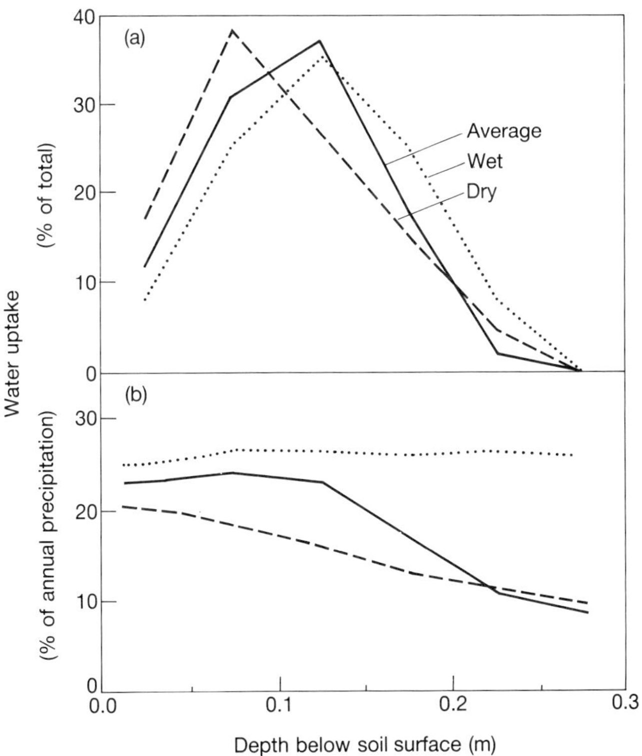


FIG 11. Water uptake in relation to soil depth for *Ferocactus acanthodes* for dry, average and wet years. (a) Per cent of total uptake. (b) Per cent of annual precipitation. (From Hunt & Nobel 1987a.)

60% higher for the rain roots even though they have six-fold less mass than do the established roots because of the much higher hydraulic conductivity of the rain roots. The marginal water revenue (water gain for an increment of root mass, which is the slope of the revenue curve; Bloom, Chapin & Mooney 1985) equals the marginal cost at an established root mass of 130 g, which is close to the actual mass of established roots measured in the field of 112 g. The mass of established roots for *A. deserti* therefore is close to optimal based on a cost/benefit model using water as the currency. The marginal water revenue for a rain root equals the marginal water cost at a mass of 8 g, which is about double the observed mass of the rain roots for a year with 105 wet days (Hunt, Zakir & Nobel 1987) but similar to the 7 g predicted for a plant of this size for other years (Nobel 1991). Again the cost/benefit model (Bloom, Chapin & Mooney 1985) helps rationalize the observed rain root mass for *A. deserti*, even though somewhat fewer rain roots may be produced than lead to optimal water uptake.

In summary, the marginal cost of root production using water as the currency, as seems appropriate for desert plants but which may have much wider applicability, equals the marginal revenue of water uptake at the observed mass for established roots and close to that for rain roots of *A. deserti*. Such cost/benefit models may also prove quite useful for interpreting root deployment strategies of other plant groups.

## ROOT FUNCTION IN DESERT PHREATOPHYTES

Deep-rooted desert phreatophytes, particularly woody legumes, have been studied in considerable detail as a model system for investigating the interplay of tissue water relations and nitrogen availability in limiting net primary production in arid and semi-arid ecosystems. Woody legumes often provide an exception to the general ecological paradigm that primary production is very low in desert ecosystems. The genus *Prosopis*, for example, dominates millions of square kilometres of arid woodlands in North America and South America. Biomass and productivity in these stands of *Prosopis* far exceed values that might be expected from regional levels of precipitation (Rundel *et al.* 1982). The key element in the ecological success of *Prosopis*, and hypothetically in many species of *Acacia* and other legumes in Old World deserts, is the ability to decouple itself from the normal limiting factors of water and nitrogen availability.

### Prosopis *root architecture*

*Prosopis glandulosa*, which has been studied in detail in the Sonoran and Chihuahuan deserts of the United States, grows as a facultative phreatophyte. In semi-arid rangelands where annual precipitation exceeds 250 mm, and where ground water pools are absent, *Prosopis* grows successfully with multiple lateral roots little more than a metre in depth (R.A. Virginia, unpublished data). Where ground water pools are present, however, or around playa lake basins where soil moisture

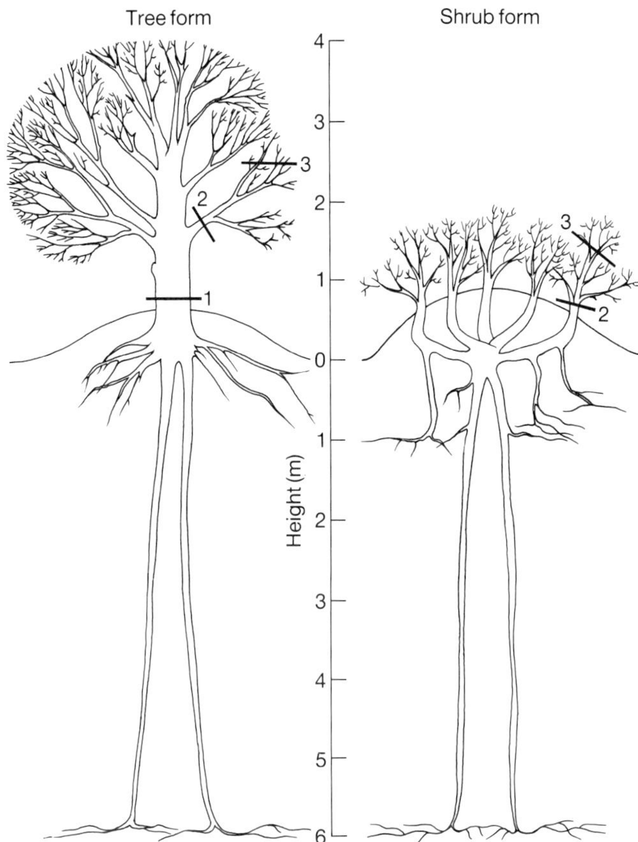


FIG. 12. Below-ground architecture in tree and shrub forms of *Prosopsis glandulosa* in the Sonoran Desert at Harper's Well, California.

collects, *Prosopis* forms multiple tap roots that reach these water pools. Rooting depths of 6–12 m are common in species of *Prosopis* (Fig. 12), and we have already mentioned a record depth of 53 m for *P. glandulosa* (Phillips 1963). The underground distribution of root biomass with depth in this species varies greatly with the nature of the habitat (Fig. 13).

The mechanisms that permit root growth of *Prosopis* (and other phreatophytes) to great depth have not been studied. From field studies and greenhouse experiments, we infer that successful establishment of phreatophytic populations of *Prosopis* is a very rare event. Because roots will not penetrate significant distances through dry soil, establishment of roots down to the capillary fringe of ground water pools requires a wetted conduit of soil. Such conditions can occur in wet years or in years with surface flooding from higher elevations when fracture lines in the soil profile offer a preferential channel for water movement. *Prosopis* tap roots in the field often follow a tortuous pathway through the soil profile rather than following a simple gravitational cue to grow straight down. Where wetted profiles

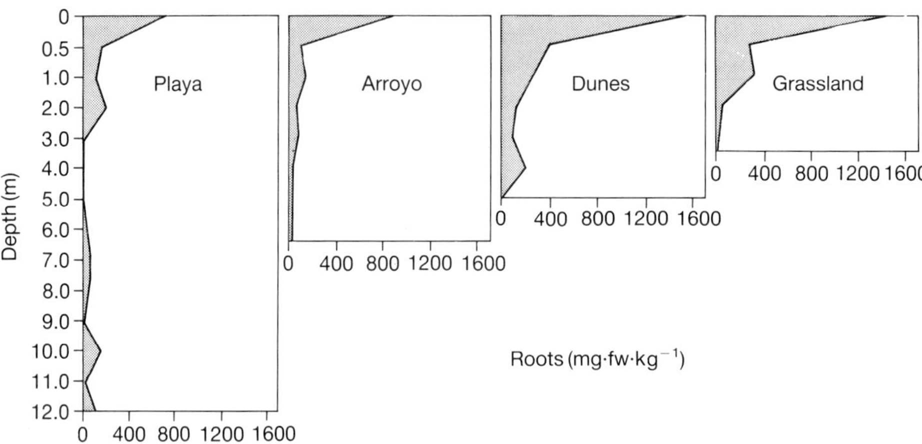


FIG. 13. Profiles of root biomass with depth in soil in four habitats at the Jornada Range, New Mexico (R.A. Virginia, unpublished observation).

exist, root growth of seedlings can be extremely rapid. For example, freshly germinated seedlings of *Prosopis* can grow roots to a depth of 2 m in eight weeks or less under greenhouse conditions (R.A. Virginia & W.M. Jarrell, unpublished data).

## Prosopis *water relations*

Despite the fact that *Prosopis glandulosa* commonly taps a permanent pool of ground water, and thus avoids variations in seasonal water availability, its midday leaf water potentials in summer commonly reach surprisingly low levels of −4·5 MPa (Nilsen, Rundel & Sharifi 1981; Nilsen, Rundel & Sharifi 1982). This low $\Psi^{leaf}$ results from the fact that the maximum leaf area occurs at this time of the year, producing very high levels of transpirational water loss. Two important mechanisms of physiological drought tolerance permit *Prosopis* to maintain positive turgor potentials during the summer months. One is seasonal variation in the stomatal control of water loss. Leaf conductance was found to be correlated linearly with vapour pressure deficit (VPD) at water potentials above −4 MPa, but the slope of this relationship changed through the year (Nilsen *et al.* 1983). A second mechanism of adaptation to drought in *P. glandulosa* is a seasonal osmotic adjustment to allow for the maintenance of favourable water potential gradients in summer. Once leaves matured, however, there was no evidence of seasonal change in cellular elasticity to adapt to increased drought stress (Nilsen *et al.* 1983).

While *Prosopis* is an interesting model system, the water relations of other phreatophytic legumes in the Sonoran Desert are different (Nilsen *et al.* 1984). Some species, as for example *Psorothamnus (Dalea) spinosus*, maintain relatively high leaf water potentials of −2·5 MPa or more throughout the summer months, while other species of phreatophytic legumes at the same site achieve lower potentials (F.C. Meinzer, unpublished data).

### Prosopis *nutrient relations*

Because most desert soils have limited nitrogen pools, primary production may be limited by the availability of this element when water is non-limiting. *Prosopis* and many other woody legumes in arid lands, as previously discussed, use *Rhizobium* nodules to fix nitrogen symbiotically. Stands of *P. glandulosa* in the Sonoran Desert of California fix up to 65% of their tissue nitrogen in this manner (Shearer *et al.* 1983). On a stand basis, this may account for 45 kg N $ha^{-1}$ or more of fixed nitrogen.

Because nitrate-nitrogen is stable in desert ecosystems where insufficient water is available for run-off, leaching or denitrification losses, high levels of soil nitrate can accumulate. Large amounts of nitrogen (1·7 g $kg^{-1}$ soil, 1020 g $m^{-2}$) can occur in the upper 30 cm of the soil profile beneath canopies of *P. glandulosa* (Virginia & Jarrell 1983). Nitrate-nitrogen accounts for 25% of this total. In addition to nitrogen, significant quantities of organic carbon, calcium, magnesium, phosphorus and potassium also accumulate in surface soils beneath *Prosopis* canopies (Barth & Klemmedson 1978; Tiedemann & Klemmedson 1973a,b; Virginia & Jarrell 1983). Accumulations of cations and phosphorus result from the concentration of elements that were absorbed in litter-fall over the entire root zone and from cation exclusion during selective uptake of salts (Virginia 1986). All of these effects have a profound effect on soil microbial processes and on the availability of nutrients to other desert species.

## CONCLUSIONS

In the extreme soil environment of desert ecosystems, which often reach very high temperatures near the surface but experience only irregular pulses of soil moisture and nutrient availability, the physiological and ecological responses of root structure and function are of critical importance. Despite this importance, we know very little about the biology of these below-ground systems beyond a superficial understanding of root architecture and a few measurements of root biomass and turnover. In the same way that leaves of desert plants exhibit many unusual structural and functional characteristics not common in species of more mesic habitats, we would expect equally interesting traits of root systems to occur. With an increasing focus on experimental and physiological modelling, we are beginning to understand the functional attributes of root systems in desert succulents, deep-rooted phreatophytes, and cold-desert shrubs. Complementary studies are greatly needed with other groups of desert plants.

Many important questions on the functional relationships of desert roots remain to be answered. With respect to water relations, we need to know much more about the hydraulic architecture of root systems, about the significance of mycorrhizae in soil–root fluxes of water, and about the functional significance of patterns of rooting architecture. Metabolic studies are needed to understand better the maintenance respiration of desert roots, the longevity of individual root elements, the environmental, physiological and hormonal controls on root/shoot carbon allocation, the direction and branching of rooting, and the suberization of roots. Finally, there are a wide range of nutritional questions that also require new investigations. These include the absorption kinetics of individual nutrients from oligotrophic soils with irregular moisture availability, the significance of individual nutrients as limiting factors, and the broader ecosystem question of nutrient concentration and/or depletion around long-lived perennials. This last topic is particularly critical for studies of woody legumes that symbiotically fix nitrogen.

Addressing these questions will require a vertically integrated program of research including studies at the biochemical, physiological, ecological, community, ecosystem, and even landscape level of study. Detailed physical and physiological models of moisture and nutrient uptake in agricultural systems should be extended to studies of desert root systems, and these models coupled with an ecosystem perspective on the role of desert vegetation in the biogeochemical cycles of carbon, nitrogen, and other elements. Such an approach will be particularly important with increasing international interest in problems of global climate change and associated problems of desertification.

## REFERENCES


**Allen, O.N. & Allen, K.K. (1981).** *The Leguminosae: A Source Book of Characteristics, Uses and Nodulation*. University of Wisconsin Press, Madison.

**Allen, M.F., Smith, W.K., Moore, T.S. & Christensen, M. (1981).** Comparative water relations and photosynthesis of mycorrhizal and non-mycorrhizal *Bouteloua gracilis* H.B.K. Lag ex steud. *New Phytologist*, **88**, 683–693.

**Barbour, M.G. (1973).** Desert dogma reexamined: root/shoot productivity and plant spacing. *American Midland Naturalist*, **89**, 41–57.

**Barbour, M.G. (1979).** Plant–plant interactions. *Arid Land Ecosystems: Structure, Functioning and Management* (Ed. by D.W. Goodall & R.A. Perry), Vol. 2, pp. 33–49. Cambridge University Press, Cambridge.

**Barth, R.C. & Klemmedson, J.O. (1978).** Shrub-induced spatial patterns of dry matter, nitrogen, and organic carbon. *Soil Science Society of America Journal*, **42**, 804–809.

**Batanouny, K.H. & Abdel Wahab, A.M. (1973).** Ecophysiological studies of desert plants. VII. Root penetration of *Leptadenia pyrotechnica* (Forsk.) Desne. in relation to its water balance. *Oecologia*, **11**, 151–161.

**Bell, K.L., Hiatt, H.D. & Niles, W.E. (1979).** Seasonal changes in biomass allocation in eight winter annuals of the Mojave Desert. *Journal of Ecology*, **67**, 781–787.

**Bethlenfalvay, G.J., Dakessian, S. & Pakovsky, R.S. (1984).** Mycorrhizae in a southern Californian desert — ecological implications. *Canadian Journal of Botany*, **62**, 519–524.

**Bloom, A.J., Chapin III, F.S. & Mooney, H.A. (1985).** Resource limitation in plants — an economic analogy. *Annual Review of Ecology and Systematics*, **16**, 363–392.

**Burrows, W.H. (1972).** Productivity of an arid zone shrub (*Eremophila gilesii*) community in south-western Queensland. *Australian Journal of Botany*, **20**, 317–329.

**Buxton, P.A. (1925).** The temperature of the surface of deserts. *Journal of Ecology*, **12**, 127–134.

**Caldwell, M.M. (1987).** Competition between root systems in plant communities. *Root Development and Function* (Ed. by P.J. Gregory, J.V. Lake & D.A. Rose), pp. 167–186. Cambridge

University Press, Cambridge.

**Caldwell, M.M. & Camp, L.B. (1974).** Belowground productivity of two cool desert communities. *Oecologia*, **17**, 123–130.

**Caldwell, M.M. & Fernandez, O.A. (1975).** Dynamics of Great Basin shrub root systems. *Environmental Biology of Desert Organisms* (Ed. by N.F. Hadley), pp. 38–51. Dowden, Hutchinson & Ross, Stroudsburg.

**Caldwell, M.M. & Richards, J.H. (1986).** Competing root systems: morphology and models of absorption. *On the Economy of Plant Form and Function* (Ed. by T.J. Givnish), pp. 251–273. Cambridge University Press, Cambridge.

**Campbell, G.S. (1977).** *An Introduction to Environmental Biophysics*. Springer-Verlag, New York, 159 pp.

**Cannon, W.A. (1911).** The root habits of desert plants. *Carnegie Institution of Washington, Publication No. 131*, 96 pp.

**Cannon, W.A. (1913).** Botanical features of the Algerian Sahara. *Carnegie Institution of Washington, Publication No. 178*, 1–81.

**Cannon, W.A. (1921).** Plant habits and habitats in the arid portions of South Australia. *Carnegie Institution of Washington, Publication No. 308*, 137 pp.

**Cannon, W.A. (1924).** General and physiological features of the vegetation of the more arid portions of southern Africa, with notes on the climatic environment. *Carnegie Institution of Washington, Publication No. 354*, 159 pp.

**Cannon, W.A. (1925).** Physiological features of roots, with especial reference to the relation of roots to aeration of the soil. *Carnegie Institution of Washington, Publication No. 368*, 168 pp.

**Cannon, W.A. (1949).** A tentative classification of root systems. *Ecology*, **30**, 542–548.

**Charley, J.L. & Cowling, S.W. (1968).** Changes in soil nutrient status resulting from overgrazing and their consequences in plant communities of semi-arid areas. *Proceedings of the Ecological Society of Australia*, **3**, 28–38.

**Chew, R.M. & Chew, A.E. (1965).** The primary productivity of a desert shrub (*Larrea tridentata*) community. *Ecological Monographs*, **35**, 355–375.

**Cody, M.L. (1986).** Structural niches in plant communities. *Community Ecology* (Ed. by J. Diamond & T.J. Case), pp. 381–405. Harper & Row, New York.

**Cody, M.L. (1987).** Spacing patterns in Mojave Desert plant communities: near-neighbor analyses. *Journal of Arid Environments*, **11**, 199–217.

**Cowling, S.W. (1969).** *A study of the vegetation activity patterns in a semi arid environment*. Ph.D. thesis. University of New England, Armidale.

**Drew, M.C. (1979).** Root development and activities. *Arid-land Ecosystems: Structure, Functioning, and Management* (Ed. by R.A. Perry & D.W. Goodall), Vol. I, pp. 573–606. Cambridge University Press, Cambridge.

**Ehleringer, J.R. (1984).** Intraspecific competitive effects on water relations, growth and reproduction in *Encelia farinosa*. *Oecologia*, **63**, 153–158.

**Evenari, M., Shanan, L. & Tadmore, N. (1971).** *The Negev: Challenge of a Desert*. Harvard University Press, Cambridge.

**Evenari, M., Kappen, L., Buschbom, U. & Lange, O.L. (1975).** Adaptive mechanisms in desert plants. *Physiological Adaptation to the Environment* (Ed. by F.J. Vernberg), pp. 111–121. Intext Educational Publications, New York.

**Farnsworth, R.B. & Hammond, M.W. (1976).** Root nodules and isolation of endophyte on *Artemesia ludoviciana*. *Proceedings of the Utah Academy of Sciences*, **45**, 182–188.

**Farnsworth, R.B., Romney, E.M. & Wallace, A. (1976).** Implications of symbiotic nitrogen fixation by desert plants. *Great Basin Naturalist*, **36**, 65–80.

**Farnsworth, R.B., Romney, E.M. & Wallace, A. (1978).** Nitrogen fixation by microfloral–higher plant associations in arid to semiarid environments. *Nitrogen in Desert Ecosystems* (Ed. by N.E. West & J.J. Skujins), pp. 17–19. Dowden, Hutchinson & Ross, Stroudsburg.

**Fernandez, O.A. & Caldwell, M.M. (1975).** Phenology and dynamics of root growth of three cool semi-desert shrubs. *Journal of Ecology*, **63**, 703–714.

**Fonteyn, P.J. & Mahall, B.E. (1981).** An experimental analysis of structure in a desert plant community. *Journal of Ecology*, **69**, 883–896.

**Forseth, F.N., Ehleringer, J., Werk, K.S. & Cook, C.S. (1984).** Field water relations of Sonoran Desert annuals. *Ecology*, **65**, 1436–1444.

**Garcia-Moya E. & McKell, C.M. (1970).** Contribution of shrubs to the nitrogen economy of a desert-wash plant community. *Ecology*, **51**, 81–88.

**Gulmon, S.L., Rundel, P.W., Ehleringer, J.R. & Mooney, H.A. (1979).** Spatial relations and competition in a chilean desert cactus. *Oecologia*, **44**, 40–43.

**Hadley, N.F. (1970).** Micrometeorology and energy exchange in two desert arthropods. *Ecology*, **51**, 434–444.

**Harris, S.A. (1960).** The distribution of certain plant species in similar desert and steppe soils in central and northern Iraq. *Journal of Ecology*, **48**, 97–105.

**Hellmers, H., Horton, J.S., Juhren, G. & O'Keefe, J. (1955).** Root systems of some plants in Southern California. *Ecology*, **36**, 667–678.

**Hunt, E.R., Jr & Nobel, P.S. (1987a).** A two-dimensional model for water uptake by desert succulents: implications of root distribution. *Annals of Botany*, **59**, 559–569.

**Hunt, E.R., Jr & Nobel, P.S. (1987b).** Allometric root/shoot relationships and predicted water uptake for desert succulents. *Annals of Botany*, **59**, 571–577.

**Hunt, E.R., Jr., Zakir, N.J.D. & Nobel, P.S. (1987).** Water costs and water revenues for established and rain-induced roots of *Agave deserti*. *Functional Ecology*, **1**, 125–129.

**Jones, R. & Hodgkinson, K.C. (1970).** Root growth of rangeland chenopods: morphology and production of *Atriplex nummularia* and *Atriplex vesicaria*. *The Biology of* Atriplex (Ed. by R. Jones), pp. 77–85. CSIRO Division of Plant Industry, Canberra.

**Jordan, P.W. & Nobel, P.S. (1984).** Thermal and water relations of roots of desert succulents. *Annals of Botany*, **54**, 705–717.

**Kausch, W. (1965).** Beziehungen zwischen Wurzelwachstum, Transpiration und $CO_2$-Gaswechsel bei einigen Kakteen. *Planta*, **66**, 229–238.

**Khan, A.G. (1974).** The occurrence of mycorrhizas in halophytes, hydrophytes and xerophytes, and of *Endogyne* spores in adjacent soils. *Journal of General Microbiology*, **81**, 7–14.

**Korner, Ch. & Cochrane, P. (1983).** Influence of plant physiognomy on leaf temperature on clear midsummer days in the Snowy Mountains, south-eastern Australia. *Acta Oecologica/Oecologia Plantarum*, **4**, 117–124.

**Krebill, R.G. & Muir, J.M. (1974).** Morphological characterization of *Frankia purshiae*, the endophyte in root nodules of bitterbrush, *Purshia tridentata*. *Northwest Science*, **48**, 266–268.

**Kummerow, J. (1981).** Structure of roots and root systems. In *Mediterranean-Type Shrublands* (Ed. by F. diCastri, D.W. Goodale & R.L. Specht), pp. 269–288. Elsevier, Amsterdam.

**LeHouerou, H.N. (1972).** Africa — the Mediterranean region. In *Wildland Shrubs — Their Biology and Utilization* (Ed. by C.M. McKell, J.P. Blaisdell & J.R. Goodin), pp. 26–36. USDA Forest Service, Ogden, Utah.

**Ludwig, J.A. (1977).** Distributional adaptations of root systems in desert environments. *The Below-ground Ecosystem: A Synthesis of Plant-Associated Processes* (Ed. by J.K. Marshall), pp. 85–91. Colorado State University Range Science Series No. 26, Fort Collins.

**Lunt, O.R., Letey, J. & Clark, S.B. (1973).** Oxygen requirements for root growth in three species of desert shrubs. *Ecology*, **54**, 1356–1362.

**Markle, M.S. (1917).** Root systems of certain desert plants. *Botanical Gazette*, **64**, 177–205.

**Miller, R.M. (1979).** Some occurrences of vesicular–arbuscular mycorrhiza in natural and disturbed ecosystems of the Red Desert. *Canadian Journal of Botany*, **57**, 619–623.

**Miller, R.M. (1987).** The ecology of vesicular–arbuscular mycorrhizae in grass and shrublands (Ed. by G.R. Safir), pp. 135–170. *Ecophysiology of VA-mycorrhizal plants*. CRC Press, Boca Raton.

**Mooney, H.A., Gulmon, S.L., Rundel, P.W. & Ehleringer, J.R. (1980).** Further observations on the water relations of *Prosopis tamarugo* of the northern Atacama Desert. *Oecologia*, **44**, 177–180.

**Moore, R.T., White, R.S. & Caldwell, M.M. (1972).** Transpiration of *Atriplex confertifolia* and *Eurotia lanata* in relation to soil, plant, and atmospheric moisture stresses. *Canadian Journal of Botany*, **50**, 2411–2418.

**Mulroy, T.W. & Rundel, P.W. (1977).** Annual plants: adaptations to desert environments. *Bioscience*, **27**, 109–114.

**Nilsen, E.T., Rundel, P.W. & Sharifi, M.R. (1981).** Summer water relations of the desert phreatophyte *Prosopis glandulosa* in the Sonoran desert of California. *Oecologia (Berlin)*, **50**, 271–476.

**Nilsen, E.T., Rundel, P.W. & Sharifi, M.R. (1982).** Productivity in native stands of *Prosopis glandulosa*, mesquite in the Sonoran desert of southern California. In *California Riperian Ecosystems* (Ed. by R.E. Werner), pp. 722–727. University of California Press, Berkeley.

**Nilsen, E.T., Sharifi, M.R., Rundel, P.W., Jarrell, W.M. & Virginia, R.A. (1983).** Diurnal and seasonal water relations of the desert phreatophyte *Prosopis glandulosa* (Honey mesquite) in the Sonoran desert of California. *Ecology*, **64**, 1381–1393.

**Nobel, P.S. (1976).** Water relations and photosynthesis of a desert CAM plant, *Agave deserti. Plant Physiology*, **58**, 576–582.

**Nobel, P.S. (1977).** Water relations and photosynthesis of a barrel cactus, *Ferocactus acanthodes*, in the Colorado Desert. *Oecologia*, **27**, 117–133.

**Nobel, P.S. (1983).** *Biophysical Plant Physiology and Ecology*. W.H. Freeman, New York, 608 pp.

**Nobel, P.S. (1984).** Productivity of *Agave deserti*: measurement by dry weight and monthly prediction using physiological responses to environmental parameters. *Oecologia*, **64**, 1–7.

**Nobel, P.S. (1988).** *Environmental Biology of Agaves and Cacti*. Cambridge University Press, New York, 270 pp.

**Nobel, P.S. (1991).** Ecophysiology of roots of desert plants, with special emphasis on agaves and cacti. *Roots: The Hidden Half* (Ed. by Y. Waizel, U. Kafkaki & A. Eshel). Marcel Dekker, New York.

**Nobel, P.S. & Sanderson, J. (1984).** Rectifier-like activities of roots of two desert succulents. *Journal of Experimental Botany*, **35**, 727–737.

**Nobel, P.S. & Geller, G.N. (1987).** Temperature modelling of wet and dry desert soils. *Journal of Ecology*, **75**, 247–258.

**Nobel, P.S., Geller, G.N., Kee, S.C. & Zimmerman, A.D. (1986).** Temperatures and thermal tolerances for cacti exposed to high temperatures near the soil surface. *Plant, Cell and Environment*, **9**, 279–287.

**Oke, T.R. (1978).** *Boundary Layer Climates*. Methuen, London, 372 pp.

**Orshan, G. (1972).** Morphological and physiological plasticity in relation to drought. *Wildland Shrubs — Their Biology and Utilization* (Ed. by C.M. McKell, J.P. Blaisdell & J.R. Goodin), pp. 245–254. USDA Forest Service, Ogden, Utah.

**Osmond, C.B., Björkman, O. & Anderson, D.J. (1980).** *Physiological Processes in Plant Ecology: Toward a Synthesis with* Atriplex. Springer Verlag, Berlin.

**Parker, K.W. & Martin, S.G. (1952).** The mesquite problem on southern Arizona range. USDA Circular 968, 90 pp. Government Printing Office, Washington D.C.

**Pearson, L.C. (1966).** Primary productivity in a northern desert area. *Oikos*, **15**, 221–228.

**Phillips, W.S. (1963).** Depth of roots in soil. *Ecology*, **44**, 424.

**Price, J.R. (1911).** The roots of some North African desert grasses. *New Phytologist*, **10**, 328–340.

**Robberecht, R., Mahall, B.E. & Nobel, P.S. (1983).** Experimental removal of intraspecific competitors — effects on water relations and productivity of a desert bunchgrass, *Hilaria rigida*. *Oecologia*, **60**, 21–24.

**Rodin, L.E. & Bazilevich, N.I. (1967).** *Production and Mineral Cycling in Terrestrial Vegetation*. Oliver & Boyd, London.

**Rundel, P.W. (1989).** Ecological success in relation to plant form and function in the woody legumes. *Advances in Legume Biology* (Ed. by C.H. Stirton & J.L. Zarucchi), pp. 377–398. Missouri Botanical Garden St. Louis.

**Rundel, P.W., Nilsen, E.T., Sharifi, M.R., Virginia, R.A., Jarrell, W.M., Kohl, D.H. & Shearer, G.B. (1982).** Seasonal dynamics of nitrogen cycling for a *Prosopis* woodland in the Sonoran desert. *Plant and Soil*, **67**, 343–353.

**Schimper, A.F.W. (1903).** *Plant Geography Upon a Physiological Basis*. Clarendon Press, Oxford.

**Shearer, G., Kohl, D.H., Virginia, R.A., Bryan, B.A., Skeeters, J.L., Nilsen, E.T., Sharifi, M.R. & Rundel, P.W. (1983).** Estimates of $N_2$-fixation from variation in the natural abundance of $^{15}N$ in Sonoran Desert ecosystems. *Oecologia (Berlin)*, **56**, 365–373.

**Sturges, D.L. (1977).** Soil water withdrawal and root characteristics of big sagebrush. *American Midland Naturalist*, **98**, 257–274.

**Szarek, S.R. & Ting, I.P. (1975).** Photosynthetic efficiency of CAM plants in relation to $C_3$ and $C_4$ plants. *Environmental and Biological Control of Photosynthesis* (Ed. by R. Marcelle), pp. 289–297. W. Junk, The Hague.

**Szarek, S.R., Johnson, H. & Ting, I.P. (1973).** Drought adaptation in *Opuntia basilaris* — significance of recycling carbon through Crassulacean acid metabolism. *Plant Physiology*, **52**, 539–541.

**Tabler, R.D. (1964).** The root system of *Artemisia tridentata* at 9500 feet in Wyoming. *Ecology*, **24**, 125–126.

**Tiedemann, A.R. & Klemmedson, J.O. (1973a).** Effect of mesquite on physical and chemical properties of soil. *Journal of Range Management*, **26**, 27–29.

**Tiedemann, A.R. & Klemmedson, J.O. (1973b).** Nutrient availability in desert grassland soils under mesquite (*Prosopis glandulosa*) and adjacent open areas. *Soil Science Society of America Journal*, **37**, 107–111.

**Tilman, D. (1988).** *Plant Strategies and the Dynamics and Structure of Plant Communities*. Princeton University Press, Princeton.

**Trappe, J.M. (1981).** Mycorrhizae and productivity of mid- and semi-arid rangelands. *Advances in Food Producing Systems for Arid and Semi-arid Lands*, pp. 581–599. Academic Press, New York.

**Virginia, R.A. (1986).** Soil development under legume tree canopies. *Forest Ecology and Management*, **16**, 69–79.

**Virginia, R.A. & Jarrell, W.M. (1983).** Soil properties in a mesquite-dominated Sonoran desert ecosystem. *Soil Science Society of America Journal*, **47**, 138–144.

**Virginia, R.A., Jenkins, M.B. & Jarrell, W.M. (1986).** Depth of root symbiont occurrence in soil. *Biology Fertility and Soils*, **2**, 127–130.

**Volkens, G. (1887).** *Die Flora der Ägyptisch–arabischen Waste, auf Grund Anatomische-Physiologischer Forschungen Dargestellt.* Borntrager, Berlin, 156 pp.

**Wallace, A., Bamberg, S.A. & Chew, J.W. (1974).** Quantitative studies of roots of perennial plants in the Mojave Desert. *Ecology*, **55**, 1160–1162.

**Walter, H. & Stadelmann, E. (1974).** A new approach to the water relations of desert plants. *Desert Biology* (Ed. by G.W. Brown), Vol. 2, pp. 214–302. Academic Press, New York.

**Waterman, W.C. (1923).** Development of root systems under dune conditions. *Botanical Gazette*, **68**, 22–58.

**Whittaker, R.H. (1962).** Net production relations of shrubs in the Great Smokey Mountains. *Ecology*, **43**, 357–376.

**Whittaker, R.H. & Woodwell, G.M. (1969).** Structure, production, and diversity of the oak–pine forest at Brookhaven, New York. *Journal of Ecology,* **57**, 215–219.

**Wullstein, L.H. & Harker, A. (1982).** Nonconfirmation of nodulation in *Artemisia ludoviciana. American Journal of Botany*, **69**, 160–162.

**Zohary, M. (1961).** On the hydro-ecological relations of the near east desert vegetation. *UNESCO Arid Zone Research*, **16**, 198–212.

# Root Systems and Plant Interspecific Effects

# Root plasticity, nitrogen capture and competitive ability

J.P. GRIME, B.D. CAMPBELL, J.M.L. MACKEY
AND J.C. CRICK
*Unit of Comparative Plant Ecology, Department of Plant Science, The University, Sheffield S10 2TN, UK*

## SUMMARY

1 Plant strategy theory predicts that mineral nutrition on fertile soils will involve 'active foraging' which consists of patch exploitation by morphologically dynamic root systems, in which the life span of individual fine roots is short. On infertile soils it is predicted that root systems will be less dynamic but by remaining functional throughout the year will be capable of intercepting brief mineralization pulses even when these coincide with harsh environmental conditions. It is further predicted that within plant communities on fertile soils, dominant and subordinate species will differ in the scale at which patchiness in mineral nutrients is exploited.

2 Experimental tests of these predictions are described. These involve partitioned containers allowing presentation of mineral nutrients in various spatial and temporal patterns and a new technique which maintains patchiness without partitions.

3 The results confirm the superior ability of species associated with fertile soils to capture nitrogen from the undepleted sectors of a nutritionally-heterogeneous rooting volume. This capacity is due to high rates of dry matter production and high specific N-absorption rates and is not the result of greater plasticity in the partitioning of dry matter between parts of the root system located in rich and poor sectors.

4 Evidence has been obtained of the differential ability of the roots of slow-growing plants from infertile soils to exploit brief episodes of mineral nutrient enrichment; this appears to be associated with low rates of tissue turnover and the capacity of the roots to remain viable under chronic mineral nutrient stress.

5 An inverse relationship was detected between the proportion of the root dry matter allocated to the undepleted sectors of a patchy rooting volume and the status of the species in an experimental plant community synthesized in productive conditions. It is concluded that this relationship is the result of a difference in the scale and precision with which dominant and subordinate plants exploit the soil mosaic. 'Coarse-grained' foraging in potentially dominant species such as *Arrhenatherum elatius* involves modification of the partitioning of assimilate between root and shoot and the production of a robust and extensive root system drawing upon a large volume of soil. In contrast, the smaller and predominantly fine-rooted systems of subordinates such as *Campanula rotundifolia* exhibit 'fine-

grained' adjustment of root development to local patchiness in mineral nutrient supply.

## INTRODUCTION

During their development, the majority of vascular plants exhibit some capacity to modify the size and morphology of the root system in response to environmental cues. This plasticity may be manifested as changes in the allocation of dry matter between root and shoot (Aung 1974) or between different sectors of the root system (Viets 1965; Drew, Saker & Ashley 1973) or may take the form of changes in the architecture of fine roots (Fitter 1987, pp. 239–241). If we are to understand the functional significance of these various types of morphological plasticity it will be necessary to devise and test theories which relate the nature and extent of root plasticity to natural selection in particular plant habitats.

With respect to both roots and shoots, plant strategy theory (Grime 1974, 1979) provides a coherent basis for prediction of the form and magnitude of the plastic responses which may be expected to enjoy a selective advantage in specific ecological circumstances; some general predictions and the results of experimental tests have been reviewed by Grime, Crick & Rincon (1986).

In this paper we narrow the perspective to focus upon root plasticity and its role in competition for resources and as an influence upon the relative abundance of plant populations within communities. The first section uses plant strategy theory to predict two major axes of variation in root plasticity. The second describes experiments which measure root plasticity and nitrogen capture under controlled 'patch' and 'pulse' conditions of mineral nutrient supply. The final section examines the relationship between patch exploitation by roots and the dominance hierarchy developed within an experimental plant community.

## THEORY

### *Habitat, strategy and plasticity*

In the triangular model of primary plant strategies (Grime 1974) it is proposed that there are three extremes of evolutionary specialization (ruderals, competitors and stress-tolerators) each characterized by a set of traits within which distinctive forms of plasticity are of major importance.

The *competitors* (herbaceous plants, shrubs and trees of stable productive habitats) are associated with a rapidly expanding plant biomass; here success depends upon the ability to sustain high rates of resource capture above and below ground. Morphological plasticity in the development of shoots and roots, together with the continuous replacement of leaves and roots, brings about a continuous adjustment in the spatial distribution of the absorptive surfaces above and below ground. Plasticity in the competitor is thus part of an 'active foraging' mechanism (Grime 1979) whereby high rates of resource capture are achieved through the ability to locate functional leaves and roots in the resource-rich zones of an environment which is progressively depleted by the activity of the plant and its competing neighbours.

The dynamic root systems and leaf canopies of the competitor involve high rates of reinvestment of captured resources in the construction of new leaves and roots, and further costs arise from the high rates of herbivory often experienced by the weakly-defended ephemeral tissues of many competitors (Grime 1979; Coley 1983; Coley, Bryant & Chapin 1985). We suspect therefore that pronounced morphological plasticity and active foraging will be of selective advantage only where they allow access to large reserves of light energy, water and mineral nutrients. Morphological plasticity would be expected to occur on a drastically reduced scale in habitats or niches where productivity is chronically low and the supply of the limiting resource (usually mineral nutrients, more rarely light or water) is brief and unpredictable. In *stress tolerators* (slow-growing lichens, bryophytes, herbs and small shrubs and trees of continuously harsh environments) it is predicted that survival will depend primarily upon the capacity to capture *and retain* scarce resources. It is further predicted that the leaves and roots of stress tolerators will be comparatively long-lived structures in which plasticity is expressed mainly through reversible physiological changes (acclimation) which maintain functional integrity over the long life spans of individual organs and facilitate exploitation of resource pulses (e.g. nutrient mineralization from decomposition events).

In *ruderals* (ephemeral herbs and bryophytes of productive but frequently and severely disturbed habitats) the most common expression of plasticity is predicted to be in the form of the premature development of reproductive structures in response to environmental stresses, a mechanism which in conjunction with other features of the ruderal (rapid potential growth rate, early flowering, rapid seed maturation) appears to increase the probability that *some* offspring will be produced in circumstances where the life of the parent is short. Even in ruderals of very short life-span, however, reproduction is preceded by a vegetative phase during which 'active foraging' in response to localized resource depletion is predicted to be an important feature.

### *Community structure and plasticity*

In the preceding section it has been argued that habitat productivity is a major selective force upon the form of plasticity exhibited by plants and it is predicted that along the gradient of increasing habitat productivity there will be a progressive transition from a strategy involving physiological acclimation and pulse capture to those characterized by high morphological plasticity, 'active foraging' and patch exploitation. However, at each point along the productivity gradient another dimension may be expected to occur in the forces of natural selection influencing plasticity. This relates to the structure of vegetation itself and arises from the presence of 'dominants' and 'subordinates' in most plant communities; following Grime (1987), these may be defined as follows:

1 *Dominants*: constituent species which not only make a major contribution to the total biomass of the community but also tend (as an individual or a population) to ramify throughout the edaphic and aerial environments, monopolize resource capture and influence the identity, quantity and local distributions of the associated flora and fauna.

2 *Subordinates*: plants with specializations of various kinds which limit their ability to monopolize the edaphic and aerial environments. These specializations may be in morphology, phenology or life history and, at their most conspicuous, restrict activity to part of the environment or part of the growing season.

The distinction between dominants and subordinates has several important implications for theories of evolution and vegetation dynamics (Grime 1987) and some of these are re-examined at the end of this paper. Here, however, it is necessary only to predict the way in which plasticity will vary according to whether the species functions as a dominant or subordinate. In potential dominants it is predicted that morphological responses to resource depletion (foraging) will be comparatively 'coarse grained', will often occur through changes in the allocation of dry matter between root and shoot, and will have the effect of maximizing resource capture from the environment as a *whole*. By contrast, foraging in subordinates is expected to be 'fine grained', to involve a high degree of 'within-shoot' and 'within-root' plasticity and to result in maximizing resource capture from *parts* of the environmental mosaic.

## MEASUREMENTS OF ROOT PLASTICITY AND NITROGEN CAPTURE UNDER CONTROLLED CONDITIONS

### *Objectives*

The majority of experimental investigations of plasticity, including those specifically concerned with roots, have involved coarse manipulation of the environment of the developing plant and there has been strong emphasis upon measurements of response to rates of resource supply maintained constant through space and time. In a physiological context this research strategy is unobjectionable since it provides good prospects for a secure deductive step from developmental response to environmental cause. However, for the purposes of comparative plant ecology, measurements of plant development under various uniform conditions of resource supply will often fail to reveal the critical differences in plasticity and the attendant costs or benefits which are manifested only in 'patch' or 'pulse' conditions.

With respect to both mineral nutrients (Wiersum 1958; Drew, Saker & Ashley 1973; Drew 1975) and light (Grime & Jeffrey 1965), there have been attempts to compare developmental responses to resource patchiness, and in a small number of studies (e.g. Woods & Turner 1971; Rorison & Gupta 1974) the capacity to exploit a resource pulse has been examined. Building on these pioneer investigations, there is now a clear need for experiments of sufficient duration and sophistication to expose the costs and benefits (in terms of resource capture, resource retention, survivorship or reproductive fitness) of responses to particular spatial and temporal patterns of resource supply. As a contribution towards this objective, we now briefly describe three recent comparative experiments in which root development and nitrogen capture have been examined in plants subjected to patch and/or pulse conditions of mineral nutrient supply.

### *A 'patch' and 'pulse' experiment using partitioned containers*

In an investigation described in detail by Crick & Grime (1987), morphological plasticity and nitrogen capture were compared in two ecologically-contrasted wetland species; these were *Agrostis stolonifera*, a species of productive habitats, and *Scirpus sylvaticus*, which is restricted to conditions of low soil fertility. At the beginning of the 27-day experiment each plant was placed in an individual container of nutrient solution in which the root system was equally distributed between nine watertight radial sectors. Over the course of the experiment five mineral nutrient treatments were applied (Fig. 1). Two treatments (1 and 2) represented the maximum and minimum rates of supply and provided the same level [full or N/100 Rorison solution (Hewitt 1966), respectively] to all nine compartments. The three remaining treatments (3, 4, 5) supplied the same total amount of mineral nutrients to each plant but these were presented in different spatial and temporal

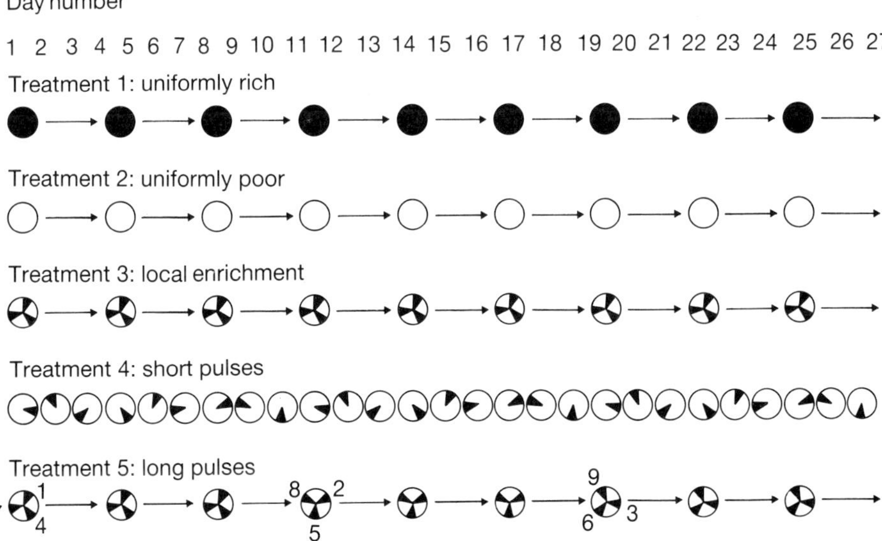


FIG. 1. Summary of treatments presenting five patterns of mineral nutrient supply over 27 days. Circle represents plan view of a root growth arena with nine compartments. Solid area = high nutrient solution (full Rorison solution); open area = low nutrient solution (N/100 Rorison solution).

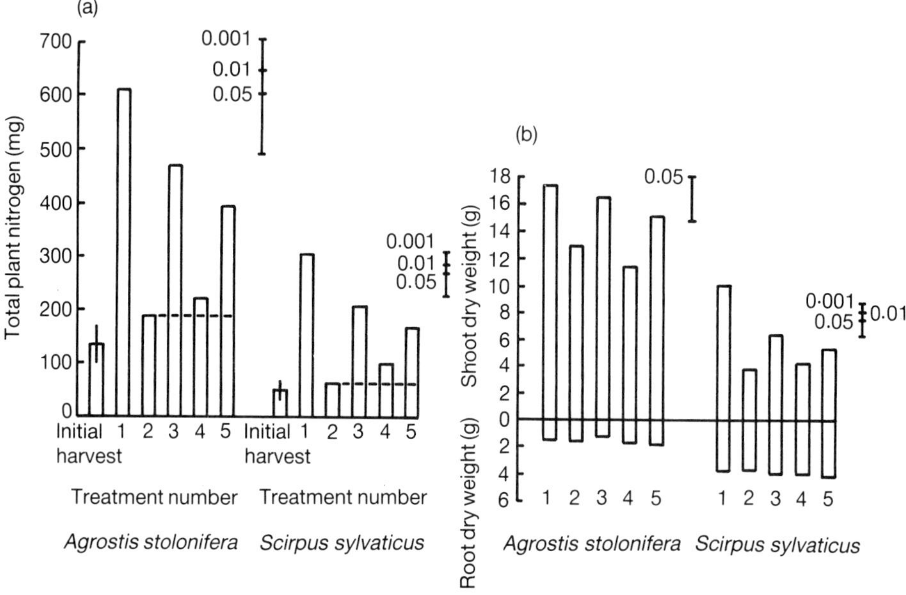


FIG. 2. (a) Total nitrogen content and (b) shoot and root dry weights of *Agrostis stolonifera* and *Scirpus sylvaticus* subjected to the five patterns of mineral nutrient supply described in Fig. 1. Vertical bars represent the LSD at $P<0{\cdot}05$, $P<0{\cdot}01$ and $P<0{\cdot}001$ between treatment means. In (a) 95% confidence limits are indicated as vertical lines on the columns representing mean nitrogen content of plants harvested at the commencement of the experiment.

patterns. Nutrient-rich patches were provided in the form of constantly high-nutrient compartments with low-nutrient compartments interspersed (treatment 3). In treatment 4 high-nutrient pulses, each of 24 h duration and restricted to one compartment at a time, were inserted in a random sequence. Longer pulses were provided in treatment 5 in a regime involving simultaneous enrichment of three compartments over periods of 9 days.

Figure 2 describes the total nitrogen contents and dry weights of the two species at the beginning of the experiment and in plants exposed to each treatment. It is clear that at both high and low external concentrations nitrogen uptake in *A. stolonifera* was consistently greater than that achieved by *S. sylvaticus* although in both species the treatments could be arranged in the series $1>3>5>4>2$. Reference to Fig. 2b reveals that the superior nitrogen uptake of *A. stolonifera* was achieved despite a very much lower allocation of dry matter to root development. It is also apparent that *A. stolonifera* was better able than *S. sylvaticus* to sustain nitrogen uptake and dry matter production where nutrient enrichment was in the form of spatially predictable patches (treatment 3). By calculating the increments of dry weight allocated to neighbouring high- and low-nutrient compartments of the treatment 3 containers over the course of the experiment it was possible to derive a 'root concentration index' (RCI), reflecting the abilities of the

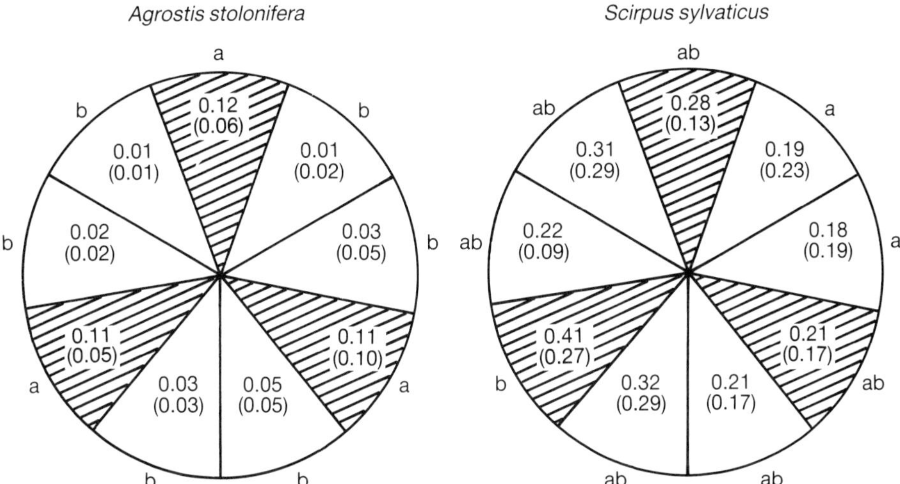


FIG. 3. Comparison of the ability of two wetland species of contrasted ecology and potential growth rates (*Agrostis stolonifera* — rapid; *Scirpus sylvaticus* — slow) to modify allocation between different sectors of the root system when exposed to a nutritionally-patchy rooting medium. Each plant was grown in uniform nutrient-sufficient solution culture for 10 weeks, then exposed to patchy conditions by distributing the root system equally between nine radial compartments, three of which (shaded) contained full strength Rorison nutrient solution. The remaining compartments (open) were provided with 1/100 Rorison solution. The value in each compartment is the mean increment of root dry matter to the compartment over 27 days (95% confidence value in brackets). In each species compartments differing significantly ($P < 0.05$) in root increment have no letters in common. Root concentration index (RCI) is calculated by dividing the mean root increment in compartments receiving full strength solution by the mean increment in compartments containing 1/100 solution (Crick & Grime 1987).

two species rapidly to modify the spatial distribution of the root system (Fig. 3). The mean root concentration index for *A. stolonifera* was 5·72 and for *S. sylvaticus* was 1·42. Since *S. sylvaticus* produced significantly more root dry weight in terms of absolute and percentage increment in treatment 3, than did *A. stolonifera*, the high root concentration index in the latter could not be related to a larger amount of root production during the experiment. We may conclude therefore that the more effective capture of the spatially predictable nitrogen in treatment 3 by *A. stolonifera* was the result of higher specific absorption rates and greater morphological plasticity in root development.

Despite the generally higher rates of nitrogen uptake exhibited by *S. stolonifera*, closer examination of the data reveals a circumstance in which greater uptake was achieved by *S. sylvaticus*. In Fig. 2a a dashed line has been inserted so that nitrogen capture in treatments 3, 4 and 5 can be examined as the excess over that observed in treatment 2. This allows comparison of the extent to which the two species were able to capture additional nitrogen provided in the same quantity but presented in three patterns, i.e. predictable patches maintained over the duration of the experiment (treatment 3), short localized and spatially and temporally unpredictable pulses (treatment 4) or longer and more predictable pulses (treatment 5). The results show that although *A. stolonifera* obtained considerably more of the 'relatively predictable' additional nitrogen provided in treatments 3 and 5, the species was marginally inferior to *S. sylvaticus* in its ability to capture the 'unpredictable' additional nitrogen of treatment 4.

### *An experiment varying the duration of the pulse*

As explained in the theoretical section (pp. 382–384), it is predicted from strategy theory that plants which are successful in chronically unproductive habitats will display an unusual ability to exploit brief pulses of mineral nutrient availability. Some evidence in support of this hypothesis has been presented already in Fig. 2b where it is shown that the massive but relatively undynamic and (in terms of specific absorption rate) relatively ineffective root system of *Scirpus sylvaticus* is marginally superior to that of *Agrostis stolonifera* in its ability to capture nitrogen from localized 24 h pulses of nutrient enrichment.

In a longer and more detailed study of pulse exploitation (Campbell & Grime 1989) nitrogen capture and dry matter production were compared in two grasses (*Arrhenatherum elatius* spp. *bulbosus, Festuca ovina*) of contrasted ecology. This solution-culture experiment extended over 42 days and involved ten treatments, in which episodes of mineral nutrient enrichment, applied to the whole root system within successive 6-day cycles throughout the experiment, varied in duration from 60 s to 6 days.

The results, expressed in Fig. 4 as relative growth rates and specific N-absorption

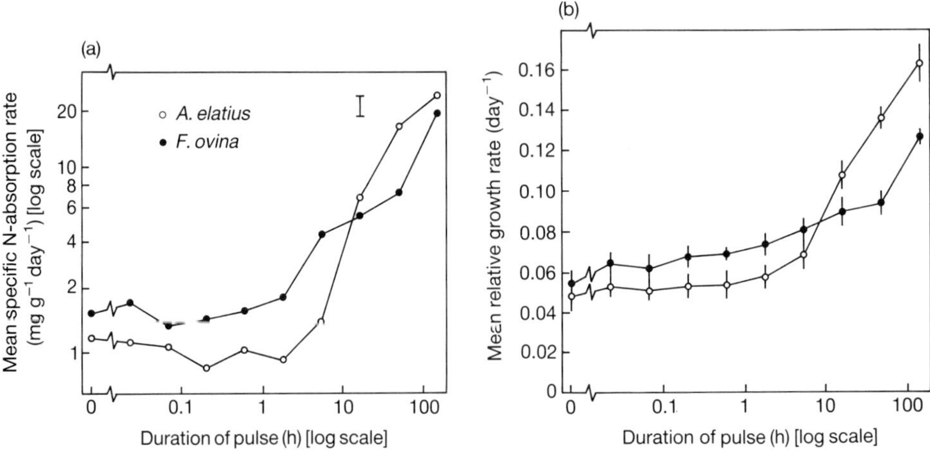


FIG. 4. (a) Mean specific N-absorption rate of *Arrhenatherum elatius* and *Festuca ovina* plants exposed once every 6 days to nutrient pulse treatments of differing duration. Vertical bar is LSD ($P > 0{\cdot}05$) for comparing means on logarithmic scale. (b) Mean relative growth rate of *Arrhenatherum elatius* and *Festuca ovina* plants exposed once every 6 days to nutrient pulse treatments of differing duration. Means are shown ±95% confidence limits.

rates, reveal that in accordance with its confinement in the field to infertile soils, the performance of *F. ovina* was superior to that of the potentially fast-growing species *A. elatius* in treatments providing brief pulses (0·1–10 h duration). Above this range the effect of increasing the length of the enrichment episode was to switch the advantage sharply in favour of *A. elatius*.

Visual inspection of the plants of *A. elatius* exposed to brief nutrient pulses revealed symptoms of nitrogen deficiency in the shoots and prematurely senescent foliage. Shoot tissue turnover rates under nutritional stress were clearly higher in *A. elatius* than in *F. ovina* and it seems reasonable to assume that there was also faster deterioration of *A. elatius* roots resulting in the lower specific N-absorption rates. This interpretation is consistent with the theoretical prediction (p. 383) that a capacity to maintain viable tissues during periods of chronically low nutrient availability is an important component of the survival mechanism of plant species attuned to nutrient-poor soils.

### *An experiment providing patches without partitions*

Experiments such as that described earlier in this paper which employ solid partitions to maintain nutritionally-distinct sectors within the same root system allow measurement of the extent to which roots in nutrient-rich compartments proliferate and compensate for deprivation of other parts of the root system by locally-enhanced uptake. An important limitation may be identified, however, in that the partitions prevent interpenetration of roots between neighbouring rich and poor compartments. In an important respect, therefore, such techniques fail to reproduce the conditions prevailing in natural soils and ignore an important component of root response.

Recently a new approach has been developed to maintain sharply-defined patchiness in mineral nutrient status without recourse to partitions. A description of this technique, its effectiveness and its potential applications has been published (Campbell & Grime 1988); here only the main principles will be described.

In the new technique, solutions of known nutrient concentration are supplied continuously and at matched rates of flow to precise locations on the surface of a cylindrical mass of free-draining sand. By releasing the solutions at four points equidistant from the centre and on opposite radii, symmetrical wetting patterns are created such that the fluid within each quadrant is maintained at a known nutrient concentration from the sand surface down to the base of the container. This steady flow pattern in the saturated sand results from the symmetry of the system preventing convective flow between quadrants. A relatively fast turnover time, coupled with the relatively low diffusion rates of ions in solution, ensures little mixing between quadrants due to diffusive flow. By this means, the nutrient concentrations in four equal and sharply-defined sectors of the sand mass can be manipulated independently. Measurements of root development and mineral nutrient capture by a plant situated at the centre of the sand surface can then be used to analyse

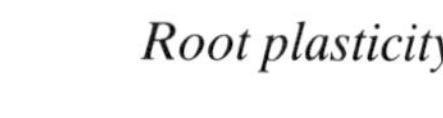

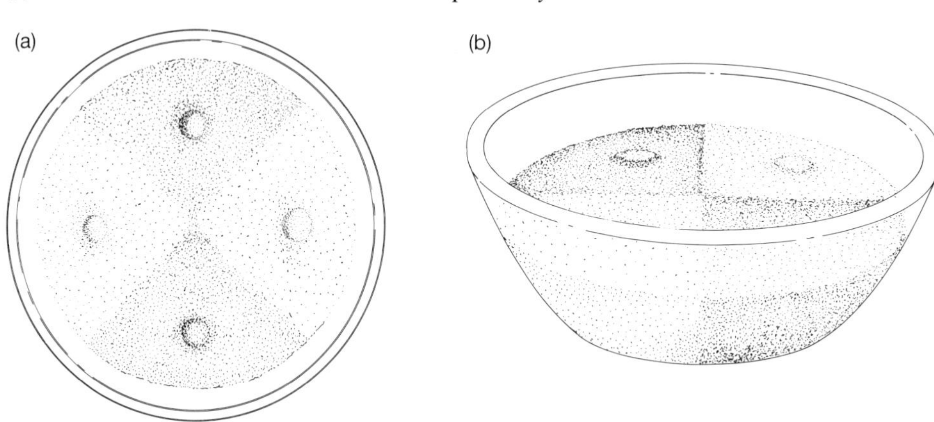


FIG. 5. Distinct nutrient distribution pattern produced by the apparatus. The solution supplied to one pair of alternate nipples contained a dye. (a) Vertical view; (b) lateral view.

responses to either spatial or temporal variations in the mineral nutrient concentrations provided in local sectors of the rhizosphere. The remarkable constancy and sharp definition of the patchiness obtainable by this technique has been confirmed by chemical analysis (Campbell & Grime 1988) and by the use of dyes (Fig. 5).

A screening programme which is now in progress uses the new technique to compare the capacity of various plant species to adjust root development and maintain mineral nutrient capture when depletion zones of standardized dimensions and nutrient concentration are introduced to the rhizosphere. The procedure involves an initial period of 21 days during which all parts of the seedling root system are provided with high levels of mineral nutrients. This is followed by two weeks in which one set of plants continue to receive uniformly high concentrations, whilst another set is subjected to severe localized depletion in two opposite quadrants which together occupy 50% of the total rooting volume available to the seedling. Harvests conducted at the beginning and end of the second phase of the experiment permit measurements of the increment to root dry matter in each quadrant and of the increments to total root and shoot weight and to total nitrogen content in the control plants and in the plants subjected to depletion.

In Table 1 data are presented for eleven common herbaceous species of contrasted ecology. The list includes plants restricted either to fertile or infertile soils and both dominant and subordinate components of herbaceous plant communities are represented. The results show a marked capacity in all the species for swift adjustment of dry matter allocation when the root systems were exposed to zones of mineral nutrient depletion. The proportion of the root increment allocated to the undepleted sectors varied over the range 0·63–0·93, with no consistent difference between species of fertile and infertile soils. However, when the species are compared with respect to the absolute increment of dry matter to the undepleted sectors of the rooting volume it is confirmed, as we might expect, that the highest values occurred in potentially fast-growing species of fertile habitats (*Urtica dioica*

TABLE 1. Comparison of the abilities of eleven common herbaceous species to modify the distribution of roots and maintain nitrogen capture when standardized depletion zones are introduced to the rhizosphere. Each seedling was grown for 3 weeks at uniformly high nutrient concentration (N/10 Rorison solution), followed by a further 2 weeks, during which nutrient concentration was reduced (N/1000 Rorison solution) in two quadrants occupying 50% of the total rooting volume. Root development in the undepleted quadrants during the depletion phase is expressed as the total increment to root dry weight in the undepleted quadrants and as a proportion of the total increment to the root system during the last 2 weeks of the experiment. Allocation to the whole root system in the last 2 weeks of the experiment is calculated for depleted and control plants as a proportion of the total increment to plant dry weight during this period. Nitrogen capture is expressed as the total increment during the depletion phase and as a proportion of that absorbed during the same period by control plants which continued to receive a uniformly high nutrient supply. Estimates of seedling relative growth rate (RGR) under productive conditions are derived from Grime & Hunt (1975) and of seed weight from Grime *et al.* (1981). Confidence limits of 95% are included in brackets

| Species | RGR | Seed weight (mg) | Soil fertility | Proportion of dry weight increment allocated to root | | Root increment in undepleted quadrants | | Nitrogen increment | |
|---|---|---|---|---|---|---|---|---|---|
| | | | | Control | Depleted | Dry weight (mg) | Proportion of total | Total (mg) | % Control |
| *Arrhenatherum elatius* | 1·30 | 2·39 | Moderate–high | 13·96 (2·93) | 17·08 (3·15) | 42·18 (15·28) | 0·63 (0·12) | 13·61 (4·75) | 148 |
| *Bromus erectus* | 1·04 | 4·23 | Low | 18·94 (3·23) | 19·91 (3·04) | 19·89 (6·66) | 0·76 (0·14) | 5·13 (2·61) | 98 |
| *Campanula rotundifolia* | 0·81 | 0·07 | Low | 16·83 (5·10) | 16·14 (6·30) | 2·95 (1·44) | 0·93 (0·13) | 0·88 (0·36) | 70 |
| *Cerastium fontanum* | 1·46 | 0·16 | Moderate–high | 6·09 (1·27) | 6·90 (1·67) | 3·58 (2·08) | 0·80 (0·15) | 2·23 (0·89) | 66 |
| *Chenopodium album* | 2·12 | 0·77 | High | 7·97 (1·75) | 8·47 (1·22) | 15·55 (4·26) | 0·79 (0·08) | 8·70 (4·79) | 70 |
| *Festuca ovina* | 1·00 | 0·38 | Low | 10·94 (2·08) | 12·48 (2·60) | 2·16 (0·92) | 0·79 (0·05) | 1·03 (0·33) | 112 |
| *Koeleria macrantha* | 0·94 | 0·30 | Low | 18·40 (2·70) | 19·70 (2·00) | 6·70 (1·30) | 0·80 (0·12) | 1·88 (0·64) | 71 |
| *Lolium perenne* | 1·30 | 1·79 | High | 13·74 (3·37) | 16·05 (4·78) | 28·57 (4·89) | 0·68 (0·05) | 7·25 (4·59) | 60 |
| *Poa annua* | 2·70 | 0·26 | High | 13·32 (1·50) | 14·55 (1·17) | 17·77 (5·32) | 0·72 (0·05) | 7·66 (2·15) | 73 |
| *Poa trivialis* | 1·40 | 0·09 | High | 16·69 (3·23) | 16·95 (1·58) | 19·51 (10·00) | 0·82 (0·06) | 5·49 (3·61) | 79 |
| *Urtica dioica* | 2·35 | 0·19 | High | 13·65 (2·79) | 15·17 (1·73) | 54·75 (24·65) | 0·78 (0·05) | 13·19 (4·60) | 53 |

and *Chenopodium album*) and in plants which receive a strong impetus to dry matter production from large seed reserves (*Arrhenatherum elatius, Bromus erectus* and *Lolium perenne*).

Table 1 also contains estimates of nitrogen uptake during the last 2 weeks of the experiment by the plants subjected to localized nutrient depletion. The results confirm that nitrogen capture was directly related to the total dry matter increment to the root component in the undepleted quadrants. When the nitrogen increment during the depletion phase is compared to that achieved by the control plants over the same period, a marked reduction is apparent in most of the species and is most pronounced in the shallow-rooted species *Urtica dioica*. However, in two of the large, deep-rooted and potentially dominant grasses (*Bromus erectus* and *Arrhenatherum elatius*), no reduction occurred and, perhaps surprisingly, the latter species obtained substantially more nitrogen from the treatment incorporating depletion zones. This effect coincided with a comparatively large increase in the proportion of dry matter allocated to the root (Fig. 6d). We postulate that the most likely explanation for the relative insensitivity of *B. erectus* and *A. elatius* to the depletion zones is related to the potential of their extensive root systems to spread throughout the rooting volume and to exploit a high proportion of the undepleted quadrants. It is notable that the enhanced nitrogen capture by *A. elatius* in the presence of depletion zones was achieved with only a modest proportional shift in allocation of dry matter to the undepleted quadrants. This suggests that in comparison with species (e.g. *Campanula rotundifolia, Cerastium fontanum* and *Poa trivialis*) which exhibited a higher root concentration index mainly through local proliferation of fine roots, the two main responses of *A. elatius* to depletion were to stimulate increased allocation to the root at the expense of the shoot and to extend trunk roots into the more remote parts of the undepleted quadrants.

The significance of these species differences will now be considered in relation to interspecific competition and plant community structure.

## ROOT PLASTICITY, COMPETITION AND COMMUNITY STRUCTURE

In the theoretical section of this paper it is predicted that competition for mineral nutrients (and other resources) attains maximum importance as a determinant of vegetation structure under conditions of high soil fertility and plant productivity and it is further predicted that in these circumstances, competitive ability will be strongly related to patch exploitation and active foraging (i.e. high morphological plasticity in the deployment of leaves and roots). Strategy theory also suggests that there will be differences in the spatial scale at which the foraging behaviour of dominant and subordinate members of plant communities will be expressed; it is predicted that dominants will tend to maximize resource capture from the habitat at large, whereas subordinates will tend to exploit parts only of the habital resource mosaic.

A preliminary test of some of these predictions can be attempted by reference

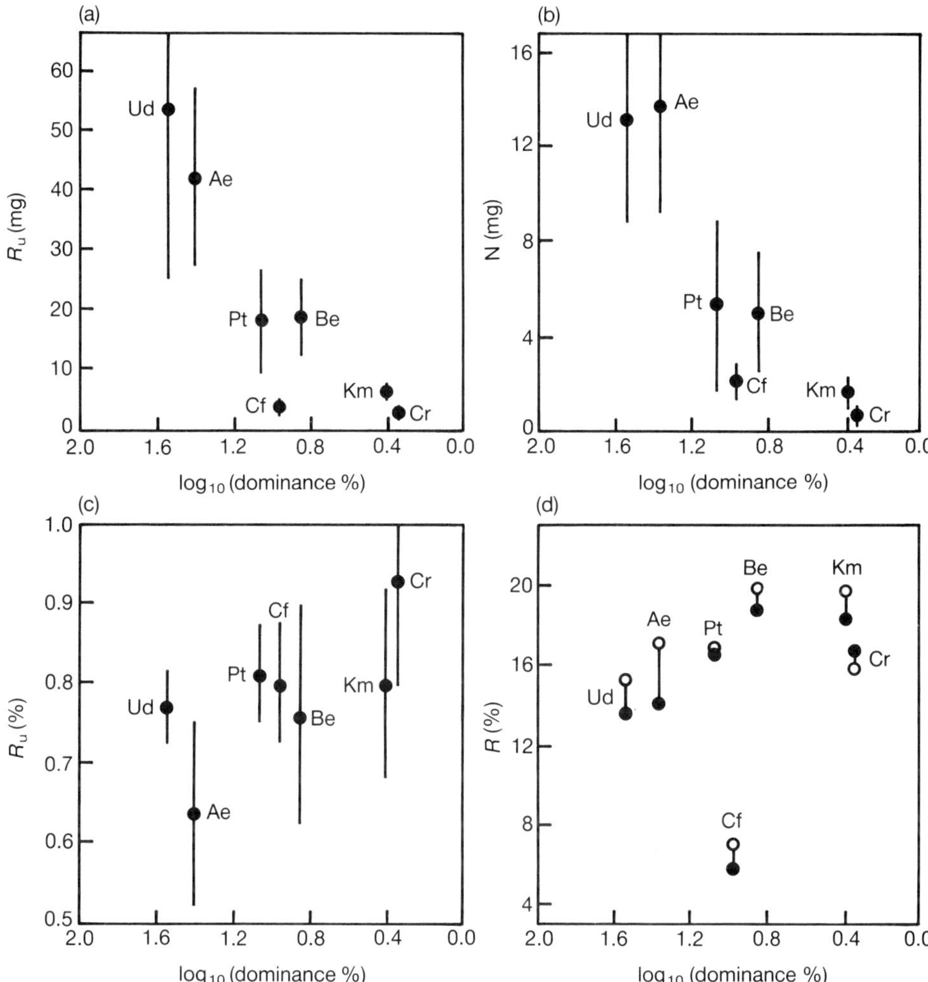


FIG. 6. (a)–(d) The relationship between species ranking in an experimental community and the ability to modify root distribution and maintain nitrogen capture in a nutritionally-heterogeneous rooting volume. In each figure species are ranked according to the shoot biomass attained in the experimental community after 16 weeks growth in a productive greenhouse environment. The remaining data are derived from Table 1. $R_u$ = increment of root dry weight to undepleted quadrants. N = total increment of nitrogen during the depletion phase. $R$ (%) = proportion of dry weight increment allocated to roots in depleted (○) and control (●) treatments. Ae = *Arrhenatherum elatius*; Be = *Bromus erectus*; Cr = *Campanula rotundifolia*; Cf = *Cerastium fontanum*; Km = *Koeleria macrantha*; Pt = *Poa trivialis*; Ud = *Urtica dioica*. Confidence limits of 95% are indicated by the vertical lines.

to the results of an experiment (Campbell 1988) in which a plant 'community' was synthesized in sand culture from a mixture of seedlings of common species, each planted at the same density. The community was allowed to develop under greenhouse conditions of high potential productivity for a period of 16 weeks, before harvesting the shoots and measuring the contribution of each species to the total above-ground biomass. Seven of the species in the mixture were included in the root plasticity screening programme for which data are presented in Table 1 and the same seed sources were used in the two operations. It is therefore possible to examine the relationship between root plasticity in the screening test and the status achieved by the species in the experimental community.

In Fig. 6, the ranking of the six species in the community is compared with their rooting behaviour and nitrogen capture in the experimental treatment involving the depletion zones. It is immediately apparent (Fig. 6a) that there is a strong positive relationship between the absolute increment of dry matter to the undepleted sectors of the rooting volume in the plasticity test and the performance of each species in the experimental mixture. Relative abundance in the mixture is also directly correlated with the capacity for nitrogen uptake in the plasticity test (Fig. 6b). Inspection of Fig. 6c reveals an inverse relationship between the proportion of the root dry weight increment allocated to the undepleted quadrants and ranking in the experimental community. This strongly suggests that 'within root' plasticity *per se* is not responsible for dominance of productive habitats. In Fig. 6d, there is evidence of a general tendency for increased allocation to the root system at the expense of the shoot in response to mineral nutrient depletion; this effect is most pronounced in *Arrhenatherum elatius*. We conclude that high rates of nitrogen capture and the tendency for monopoly of the community in species such as *Arrhenatherum elatius* are associated with high rates of carbon assimilation, modifications in root : shoot partitioning, and the construction of an extensive root system which permeates a large rooting volume but is not as finely adjusted to local patchiness in mineral nutrient status as the more plastic root system of subordinate species such as *Campanula rotundifolia*.

## DISCUSSION AND CONCLUSIONS

In an earlier review of the ecological significance of the potentially rapid rates of dry matter production in plants of fertile habitats (Grime 1979) it was suggested that this attribute would allow 'rapid adjustments in morphology in response to local depletions in resources, arising during competition'. It was pointed out that rates of response were greater in potentially fast-growing species but a source of uncertainty was identified in the statement that 'it is not yet clear to what extent such differences are the result of different genetic limits to flexibility in individual leaves and roots or arise from the fact that, in many slow-growing plants, growth is intermittent and organs at a morphogenetically responsive stage of development form a small proportion of the biomass'.

In the absence of plasticity studies on lichens, bryophytes and vascular plants from the harshest environments, this problem remains unresolved. It is interesting to note, however, that Table 1 of this paper contains no evidence of a difference between fast- and slow-growing species in flexibility of dry matter allocation within root systems exposed to patchiness in mineral nutrient supply. This suggests that the rapid concentration of roots in undepleted zones and the superior competitive ability for mineral nutrients exhibited by fast-growing species on fertile soils depends critically upon high rates of root dry matter production and high specific absorption rates and is not the result of greater flexibility in dry matter allocation between parts of the root system located in rich and poor sectors. The data in Table 1 conflict with the evidence obtained in the first experiment described in this paper and suggest the need for caution in interpretation of the results of experiment employing solid partitions.

There is evidence in Fig. 6c that the extent of the diversion of dry matter into the undepleted mineral nutrient patches, when expressed on a proportional basis, is negatively related to the status attained by the species in a plant community synthesized on fertile soil. This suggests that dominant and subordinate components of a productive community are distinguished primarily by the scale and precision with which the dynamic resource mosaic within the habitat is exploited. At one extreme, *Arrhenatherum elatius* appears to monopolize resource capture within a large soil volume by the development of an extensive root system which includes large arterial roots. Such 'coarse-grained foraging' is clearly incompatible with the precise location of root dry matter in undepleted zones; by comparison, the smaller and predominantly fine-rooted systems of subordinates, such as *Campanula rotundifolia*, exhibit 'fine-grained foraging' in which there is a more intensive and precise exploitation of a localized part of the resource mosaic.

This interpretation is clearly compatible with those theories of vegetation dynamics (Grime 1973, 1987; Grubb, Kelly & Mitchley 1982) which recognize the mechanism of species coexistence in perennial herbaceous communities as an oscillating equilibrium between (i) the tendency, in the absence of vegetation disturbance, for potential dominants to monopolize resource capture and drive the system towards monoculture, and (ii) the tendency, following disturbance and debilitation of potential dominants, for subordinates to generate diversity by exploiting individual parts of the resource mosaic.

Two of the experiments described in this paper provide evidence that the root systems of slow-growing plants of infertile soils have an unusual ability to exploit brief pulses of mineral nutrient enrichment and in the experiment comparing *Arrhenatherum elatius* and *Festuca ovina* it is concluded that the lower rate of tissue turnover in *F. ovina* and the capacity of its roots to remain viable under chronic mineral nutrient stress is of critical importance in this respect. Here it may be important to bear in mind that many of the mineral nutrient flushes which occur on infertile soils are preceded by extremes of temperature and moisture supply and coincide with conditions which are not conducive to rapid root growth (Davison 1964; Gupta & Rorison 1975; Taylor, De-Felice & Havill 1982). Accordingly, there is now an urgent need for long-term studies which examine mineral nutrient pulse exploitation by fast- and slow-growing plants under experimental conditions which manipulate factorally nutritional, climatic and biotic impacts upon root survival and functional efficiency. Also, since hierarchies of plant size and abundance are frequently observed in plant communities on infertile soils there is a requirement for studies which compare the spatial scale at which mineral nutrient pulses are intercepted by dominant and subordinate species.

## ACKNOWLEDGEMENTS

The authors are particularly grateful to Mr S.R. Band and Mr R. Taylor for technical assistance in the experimental work and to Dr R. Hunt and Dr S.H. Hillier for their involvement in the data analysis. The studies described in this paper are part of the Integrated Screening Programme which is funded by the Natural Environment Research Council.

## REFERENCES

**Aung, L.G. (1974).** Root–shoot relationships. In *The Plant Root and its Environment* (Ed. by E.W. Carson), pp. 29–61. University Press, Virginia.

**Campbell, B.D. (1988).** *Experimental tests of C–S–R strategy theory*. Ph.D. thesis, University of Sheffield.

**Campbell, B.D. & Grime, J.P. (1989).** A new method of exposing developing root systems to controlled patchiness in mineral nutrient supply. *Annals of Botany*, **63**, 395–400.

**Coley, P.D. (1983).** Herbivory and defensive characteristics of tree species in a lowland tropical forest. *Ecological Monographs*, **53**, 209–233.

**Coley, P.D., Bryant, J.P. & Chapin, F.S. (1985).** Resource availability and plant antiherbivore defence. *Science*, **230**, 895–899.

**Crick, J.C. & Grime, J.P. (1987).** Morphological plasticity and mineral nutrient capture in two herbaceous species of contrasted ecology. *New Phytologist*, **107**, 403–414.

**Davison, A.W. (1964).** *Some factors affecting seedling establishment in calcareous soils*. Ph.D. thesis, University of Sheffield.

**Drew, M.C. (1975).** Comparison of the effects of a localized supply of phosphate, nitrate, ammonium and potassium on the growth of the seminal root system, and the shoot, in barley. *New Phytologist*, **75**, 479–490.

**Drew, M.C., Saker, L.R. & Ashley, T.W. (1973).** Nutrient supply and the growth of the seminal root system in barley. I. The effect of nitrate concentration on the growth of axes and laterals. *Journal of Experimental Botany*, **24**, 1189–1202.

**Fitter, A.H. (1987).** An architectural approach to the comparative ecology of plant root systems. *New Phytologist*, **106**, 61–77.

**Grime, J.P. (1973).** Competitive exclusion in herbaceous vegetation. *Nature*, **242**, 344–347.

**Grime, J.P. (1974).** Vegetation classification by reference to strategies. *Nature*, **250**, 25–31.

**Grime, J.P. (1979).** *Plant Strategies and Vegetation Processes*. John Wiley, Chichester.

**Grime, J.P. (1987).** Dominant and subordinate components of plant communities: implications for succession, stability and diversity. In *Colonization, Succession and Stability* (Ed. by A.J. Gray, M.J. Crawley & P.J. Edwards), pp. 413–428. Blackwell Scientific Publications, Oxford.

**Grime, J.P. & Jeffrey, D.W. (1965).** Seedling establishment in vertical gradients of sunlight. *Journal of Ecology*, **53**, 621–642.

**Grime, J.P. & Hunt, R. (1975).** Relative growth rate: its range and adaptive significance in a local flora. *Journal of Ecology*, **63**, 393–422.

**Grime, J.P., Mason, G., Curtis, A.V., Rodman, J., Band, S.R., Mowforth, M.A.G., Neal, A.M. & Shaw, S. (1981).** A comparative study of germination characteristics in a local flora. *Journal of Ecology*, **69**, 1017–1059.

**Grime, J.P., Crick, J.C. & Rincon, E.R. (1986).** The ecological significance of plasticity. In *Plasticity in Plants* (Ed. by D.H. Jennings & A.J. Trewavas), pp. 5–19. Company of Biologists, Cambridge.

**Grubb, P.J., Kelly, D. & Mitchley, J. (1982).** The control of relative abundance in communities of herbaceous plants. In *The Plant Community as a Working Mechanism* (Ed. by E. Newman), pp. 77–97. *Special Publication series of the British Ecological Society No. 1*. Blackwell Scientific Publications, Oxford.

**Gupta, P.L. & Rorison, I.H. (1975).** Seasonal differences in the availability of nutrients down a podzolic profile. *Journal of Ecology*, **63**, 521–534.

**Hewitt, E.J. (1966).** *Sand and Water Culture Methods used in the Study of Plant Nutrition*, 2nd edn, Commonwealth Agricultural Bureaux Technical Communication 22. Farnham Royal.

**Rorison, I.H. & Gupta, P.L. (1974).** The growth of seedlings in response to variable phosphorus supply. In *Plant Analysis and Fertilizer Problems* (Ed. by J. Wehrmann), pp. 378–382. German Society of Plant Nutrition, Hanover.

**Taylor, A.A., De-Felice, J. & Havill, D.C. (1982).** Seasonal variation in nitrogen availability and utilization in an acidic and calcareous soil. *New Phytologist*, **92**, 141–152.

**Viets, F.G. (1965).** The plant's need for and use of nitrogen. In *Soil Nitrogen* (Ed. by W.V. Bartholomew & F.E. Clark), pp. 543–554, Agronomy Series 10. Academic Press, London.

**Wiersum, L.K. (1958).** Density of root branching as affected by substrate and separate ions. *Acta Botanica Neerlandica*, **7**, 174–190.

**Woods, D.B. & Turner, N.C. (1971).** Stomatal response to changing light by four tree species of varying shade tolerance. *New Phytologist*, **70**, 77–84.

# Extraction of potential allelochemicals and their effects on root morphology and nutrient contents

D. VAUGHAN AND B.G. ORD
*Plants Division, The Macaulay Land Use Research Institute, Craigiebuckler, Aberdeen AB9 2QJ, UK*

## SUMMARY

1 The role of allelopathy in influencing the composition of plant communities is assessed.
2 Phenolic acids are often alleged to have allelopathic effects and examples are discussed of the phytotoxicity produced by these acids.
3 The release of phenolic acids in root exudates is reported together with the details of equipment designed to extract these substances under axenic conditions.
4 Externally available phenolic acids can modify root morphology, an effect which is concentration dependent and related to N availability.
5 Because of the importance of roots in the absorption of minerals from the soil solution, the influence of phenolic acids (ferulic, vanillic, *p*-coumaric, *p*-hydroxybenzoic and cinnamic acids) on the mineral composition of roots and shoots is reported in relation to the major elements Ca, K, Mg and Na and the trace elements Cu, Mn and Zn.
6 The effect of phenolic acids on the uptake of $^{36}Cl$, $^{22}Na$ and $^{45}Ca$ and on protein synthesis in excised pea root tissues is reported.
7 The mode of action of phenolic acids is discussed in relation to root growth, metabolism and nutrient uptake.

## INTRODUCTION

### *The phenomenon of allelopathy*

Many factors are involved in either changing, or maintaining in equilibrium, the species composition of plant communities. These factors include *competition* for resources, which may be limiting at certain times, such as water, space, nutrients or light energy (Donald 1963), or else *interference* due to allelopathy (Lovett & Levitt 1981; Rice 1984; Muller, 1969). There is often difficulty in distinguishing unequivocally between these interacting factors (Bowman & Kirkpatrick 1986). The animal component can impose an additional factor, especially when there is a herbivore capable of eating one species thus giving an advantage to another.

In this paper, only factors which are broadly related to the phenomenon of allelopathy are considered. The term allelopathy was suggested by Molisch (1937) to describe the biochemical interactions among plants, including microbes. These interactions included both stimulations and inhibitions, but most research into allelopathy has concentrated on the inhibitory factors. Because of this, and the derivation of the term itself (from the Greek, *allelon* — of each other, and *pathos* — to suffer), allelopathy and inhibition are often regarded as being synonymous. Recently, Rice (1986) has attempted to correct this imbalance by reporting examples of plant/plant, plant/microbial and microbial/microbial stimulations.

Borner (1960) has pointed out that the concept of allelopathy is not new. Indeed de Candolle (1832) suggested that organic substances are released by plant roots which subsequently affect the growth of the same or different species (the toxic theory). However, one of the best documented examples of allelopathy is the release of juglone (5-hydroxy-1,4-naphthoquinone) from black walnut (*Juglans nigra*) reported by Davis (1928). In this case, soil in the vicinity of these trees is toxic to tomato, potato and alfalfa. The known examples of allelopathy are only the most obvious cases which stand out from the matrix of more general and less obvious influences (Whittaker 1970). In natural plant communities allelopathy may be regarded as part of the mechanism that maintains relative stability. Many of the responses involving this process are, therefore, subtle having developed during evolution of the community (Lovett & Levitt 1981). Such possibilities have prompted Putnam & Tang (1986) to speculate on the permanent long-term impact of a 20% growth reduction caused by allelopathy during a 1 week growth period early in the life of a plant. While the importance of allelopathy in agroecosystems has become increasingly recognized in recent years (Rice 1983; Putnam 1985; Putnam & Duke 1978), their role in natural plant communities is the best documented (Muller 1966).

## *Substances involved in allelopathy*

Allelochemicals belong to a group of substances classified as secondary metabolites (Whittaker & Feeney 1971; Swain 1977). It has been suggested (Mandava 1985) that these substances belong to diverse major groups including fatty acids and

TABLE 1. Some possible broad groupings of allelochemicals

| Broad grouping | Individual substances |
|---|---|
| Aliphatic compounds | Methanol, ethanol, *n*-propanol and *n*-butanol; crotonic, oxalic, formic, butyric, lactic, acetic and succinic acids |
| Unsaturated lactones | Patullin, psilotin, psilotinin and protoanemonin |
| Fatty acids and lipids | Dihydrostearic acid |
| Cyanogenic glycosides | Amygdalin and dhurrin |
| Terpenoids | α-pinene, β-pinene, camphor and cineole; caryophyllene, bisabolone, chamazulene, arbusculin, achillin and viscidulin *c* |
| Aromatic substances | Phenols, phenolic acids, coumarins, flavonoids, quinones and tannins |

TABLE 2. Some phytotoxic aromatic substances often implicated in allelopathy

| Broad grouping | Individual components |
|---|---|
| Simple phenols | Hydroquinone (its glycoside, arbutin) |
| Phenolic acids | Benzoic series<br>Cinnamic series |
| Coumarins | Scopoletin, scopolin, esculetin, esculin, methylesculin |
| Quinones | Juglone (5-hydroxy-1,4-naphthoquinone) |
| Flavonoids | Phlorizin, quercetin, myricetin |
| Tannins | Condensed<br>Hydrolysed |

lipids, cyanogenic glycosides, terpenoids, unsaturated lactones, aliphatic and phenolic acids, alcohols, coumarins, flavenoids, quinones and tannins (Table 1). Indeed, allelochemicals are probably normal constituents of the environment of land plants (Whittaker 1970). Swain (1977) has considered that the function of such substances is to act as chemical signals in ecosystems. In this paper emphasis is given to phytotoxins which are classified as aromatic substances (Table 2). They are released in a number of ways such as leachates of water-soluble components from leaves and stems of living plants, or from plant litter and residues, or as exudates from roots or released as volatile substances from plants and microbes.

## *Phenolic acids and allelopathy*

Among the most frequently identified allelochemicals are the phenolic acids related to benzoic and cinnamic acids (Table 3). Thus, for example, cinnamic acid is excreted by *Parthenium argentatum* (Bonner 1946) and bracken *Pteridium aquilinum* (Gliessman & Muller 1972). Chou & Lin (1976) showed that decomposing rice straw is toxic to lettuce seedlings. Vanillic, ferulic, *p*-hydroxybenzoic and *p*-coumaric acids were identified as being implicated in the response. These phenolics have also been isolated from residues of *Zea mays, Triticum aestivum, Avena sativa*

TABLE 3. Some phenolic acids alleged to be allelochemicals

| Benzoic series | Cinnamic series |
|---|---|
| Benzoic | Cinnamic |
| Vanillic | *p*-Coumaric |
| *p*-Hydroxybenzoic | Sinapic |
| Syringic | Caffeic |
| Protocatechuic | Ferulic |
| Gallic | Chlorogenic |
| Ellagic | |
| Salicylic | |
| Sulphosalicylic | |

and *Sorghum bicolor* (Guenzi & McCalla 1966) and associated with bracken (Glass 1976). Phenolic acids are found widely in the soil solution (Walters 1917; Bohm & Tryon 1967; Hartley & Whitehead 1985) and their concentrations are often linked to the dominant plant species (Whitehead 1964; Wang, Yang & Chuang 1967; Shindo, Ohta & Kuwatsuka 1978; Whitehead, Dibb & Hartley 1982; Kuiters & Denneman 1987).

A substantial literature shows that phenolic acids inhibit the growth, for example, of *Sorghum bicolor, Raphanus sativus, Glycine max, Helianthus annuus, Nicotiana tabacum* and *Cucumis sativus* (Bell & Koeppe 1972; Einhellig & Rasmussen 1979; Patterson 1981; Blum, Weed & Dalton 1987). In forestry, problems associated with reafforestation, or the regeneration of natural woodlands are often linked to the presence of phenolic acids derived from former tree species or herbaceous vegetation (Fisher 1980; Becker & Drapier 1984).

### *Proof of allelopathy*

Despite these observations, little critical work has been done on the origin of phenolic acids found in the soil, particularly those derived from the exudates of higher plants, or on their effects on root morphology and important physiological processes such as mineral uptake. This paper concentrates on those aspects which are vital to maintain or enhance plant growth. It should be emphasized that the role of phenolic acids in allelopathy is frequently *assumed* rather than proved. Thus Glass (1976) referred to the allelopathic potential of phenolic acids. Seldom is there any attempt to establish allelopathy conclusively by following a specific protocol as suggested by Putnam & Tang (1986). For this reason phenolic acids, which most frequently produce inhibitory effects, are referred to subsequently in our paper as phytotoxins.

## MATERIALS AND METHODS

### *Germination and growth of seedlings*

All experiments were carried out under axenic conditions to eliminate complications arising from the presence of micro-organisms. Seeds of *Pisum sativum* L. (cv. Meteor) were sterilized with 1% (v/v) bromine water, rinsed in sterile water and germinated in previously autoclaved moist vermiculite for 2 days at 22 °C (Vaughan, DeKock & Cusens 1974). Seedlings with radicles 25–35 mm long were transferred to stainless-steel grids (Fig. 1). The grids, 85 mm diameter and containing twelve holes of 1 mm diameter per 100 $mm^2$, were supported on three stainless-steel legs each of height 75 mm (Vaughan & Linehan 1976). The radicles were grown in previously autoclaved standard Hoagland nutrient solution (Hoagland & Arnon 1950). Seeds of wheat (*Triticum aestivum* L. var. Capelle) were sterilised as described above but germinated directly on the metal grids (Vaughan & Linehan 1976).

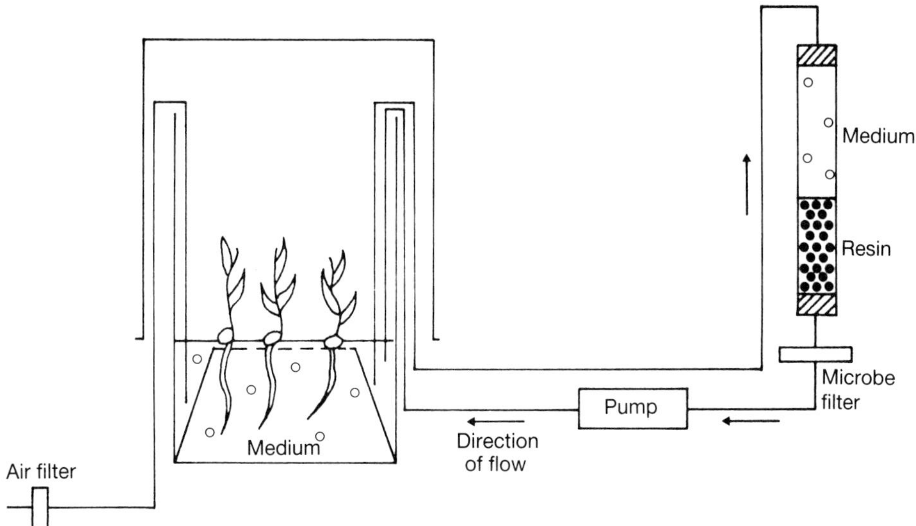


FIG. 1. Diagram of apparatus used to remove phenolic acids from root exudates using XAD-4 resin.

*Extraction and estimation of phenolic acids*

The solution was pumped through an Amberlite XAD-4 resin (BDH Ltd, Poole, Dorset) column (40 mm × 20 mm diameter) to remove exuded phenolic acids before being returned to the main vessel (Fig. 1). This vessel was covered by an inverted, autoclaved beaker. Because of problems arising from the invasion of micro-organisms (Vaughan, Jones & Ord 1988), a 0·2 μm Millipore filter was inserted in the return system just below the resin column. Phenolic acids were removed from the XAD-4 resin with three bed volumes of HPLC grade methanol and evaporated to dryness under reduced pressure. The residue was dissolved in 10% (v/v) aqueous solution of acetic acid containing 10% (v/v) methanol. Samples, and standards, of phenolic acids were analysed on a Kontron HPLC system using a 250 mm × 4·6 mm diameter Spherisorb ODS 2–5 μm column (Jones Chromatograph Ltd, Hengoed, Glamorgan) with 10% acetic acid as mobile phase A and methanol as mobile phase B. A stepwise dilution was performed starting with 10% B and increasing to 100% B over 40 min at a flow rate of 1·5 ml $min^{-1}$ with UV detection at 254 nm. All XAD-4 resin was initially treated for 24 h with methanol and a further 24 h with acetonitrile. Untreated resin was contaminated with impurities and unsuitable for use.

Phenolic acids in the soil solution, or in calcium hydroxide extracts, were measured as described by Hartley & Whitehead (1985).

*Measurements of growth and mineral contents*

To test the effects of *added* phenolic acids on the growth and mineral contents of plants, the Hoagland nutrient solution was supplemented with the appropriate acid. Except in cases where the concentrations of the added phenolic acids were lowered to a predetermined level by adjusting the flow rate, the XAD-4 resin column was not incorporated into the system. Changes in root morphology were recorded photographically. Fresh weights of roots and shoots were determined after blotting the surfaces dry, and the dry weights were measured after heating in an oven at 80 °C for 24 h. To determine ash contents, the oven-dried material was ashed at 450 °C for 6 h and the ash dissolved in 6 M HCl by heating gently. After evaporation to dryness, the ash was dissolved in 6 M HCl (final concentration adjusted to 0.1 M), filtered and analysed by atomic absorption spectroscopy.

### *Measurements of radioactivity*

For studies involving radioactive substances, pea seedlings with radicles 25–35 mm long were placed on the metal grids and grown for 24 h in water with or without phenolic acids. The radicles were rinsed with water and the apical 10 mm excised with a razor blade and incubated in the appropriate radioactive solution containing 2% (w/v) sucrose as an energy source. The uptakes of chloride (as $K^{36}Cl$), sodium (as $^{22}NaCl$) and calcium (as $^{45}CaCl_2$) were measured in 15 ml solution over 2 h at 25 °C as described previously (Vaughan, DeKock & Cusens 1974).

Uptake and incorporation of uniformly labelled $^{14}C$-L-leucine into proteins were determined by incubating groups of ten, 0–10 mm excised root segments in a medium containing 2% sucrose and 10 μM labelled amino acid (activity 3700 Bq $ml^{-1}$) for 2 h at 25 °C as described by Vaughan & Cusens (1973).

Unless otherwise stated, the results are the mean values of at least three separate experiments with triplicate treatments in each experiment.

## RESULTS AND DISCUSSION

### *Occurrence of phenolic acids*

All the phenolic acids encountered in this study were related to either benzoic acid (*p*-hydroxybenzoic, syringic and vanillic acids) or to cinnamic acid (caffeic, *p*-coumaric, ferulic and sinapic acids). Ferulic, vanillic, *p*-coumaric and *p*-hydroxybenzoic acids were present in wheat root and pea root exudates and in the soil solution (Table 4). Additionally, caffeic and syringic acids were also detected in wheat root exudates, but only under axenic conditions. The data have not been given for phenolic acid concentrations because root exudates also contain enzymes such as phenolase which can differentially utilize phenolic acids as substrates. However, 5 and 0·4 mg, respectively, for *p*-hydroxybenzoic and *p*-coumaric acids have been obtained from 1 g fresh weight of pea roots during a 14-day culture period. The technique of trapping phenolic acids on an XAD-4 resin has been used by other workers but not under axenic conditions. Using this technique, Tang & Young (1982) found benzoic, cinnamic, syringic, sinapic and ferulic acids in root exudates of Bigalta limpograss (*Hemathria altissima*).

Table 4. Phenolic acids in wheat root exudates and in the soil solution

| Phenolic acid | Exudate* | Soil solution[†] |
|---|---|---|
| *p*-OH benzoic | + | 1150 |
| Vanillic | + | 190 |
| *p*-Coumaric | + | 65 |
| Ferulic | + | 15 |
| Syringic | + | — |
| Caffeic | + | — |

* Varying amounts of phenolase and peroxidase found in exudates, so figures are of limited value at this stage.
[†] Values are expressed as nM and extracted from a Countesswells series soil of 25.1% moisture.

The precise amount of phenolic acids extracted from the soil is dependent on the extractant and the pH (Hartley & Whitehead 1985). Thus, typically for *p*-hydroxybenzoic acid, calcium hydroxide at pH 11·1 removes 55–65 $\mu g\,g^{-1}$ dry agricultural top soil (Countesswells series, from the grounds of the Macaulay Institute, National grid reference NJ 905 045), whereas aqueous extraction at the soil pH of 5·4 removes 70–80 $ng\,g^{-1}$ dry soil. The water extraction values, given in Table 4, are regarded as being a reliable guide to their availability for plant uptake because substantial amounts of phenolic acids are adsorbed differentially by soil clay minerals (Huang *et al.* 1977; Sparling, Ord & Vaughan 1981).

### *Root growth and morphology*

Einhellig (1986) has emphasized that cell division and elongation are essential phases in root development and that potential allelochemicals, such as coumarin, scopelitin and umbelliferone, can modify these events. Indeed the most obvious visual effect of phenolic acids is on overall root growth and morphology. All phenolic acids tested inhibited the increase in fresh weights of roots, the precise effect being dependent on the phenolic acid, and its concentration. Effects of phenolic acids on shoot growth were generally much less than on the roots. Typical values are given in Table 5 for three phenolic acids. At the highest concentrations used (1 mM) the most inhibitory phenolic acid, assessed by increases in fresh weight, was vanillic acid and the least effective was *p*-hydroxybenzoic acid. At a concentration of 0·5 mM, *p*-coumaric acid was more inhibitory than vanillic acid. A different result was obtained by measuring the lengths of the main root and showed that *p*-hydroxybenzoic acid was as an effective inhibitor as vanillic and *p*-coumaric acids.

The effects of 0·5 mM concentrations of *p*-hydroxybenzoic, ferulic and *p*-coumaric acids on pea root morphology after 14 days of continuous growth are shown in Fig. 2. The three phenolic acids all inhibited the growth of the main root and the length of the laterals. No lateral roots were produced on roots cultured in

TABLE 5. Influence of some phenolic acids on the growth of peas after 14 days of culture

| Treatment | Shoot (mg fresh wt $plant^{-1}$) | Root (mg fresh wt $plant^{-1}$) | Length of main root (cm) |
|---|---|---|---|
| Control | 620 ± 12·8 | 650 ± 12·7 | 10·8 ± 0·7 |
| Vanillic | | | |
| 1 mM | 400 ± 9·3 | 230 ± 5·7 | 3·2 ± 0·2 |
| 0·5 mM | 610 ± 13·1 | 510 ± 10·9 | 7·2 ± 0·5 |
| 0·1 mM | 665 ± 14·2 | 650 ± 13·6 | 8·9 ± 0·5 |
| *p*-Coumaric | | | |
| 1 mM | 550 ± 12·4 | 310 ± 8·7 | 3·4 ± 0·3 |
| 0·5 mM | 550 ± 10·3 | 450 ± 9·4 | 7·8 ± 0·5 |
| 0·1 mM | 580 ± 11·7 | 650 ± 11·3 | 10·1 ± 0·8 |
| HBA* | | | |
| 1 mM | 650 ± 13·2 | 570 ± 7·2 | 3·7 ± 0·2 |
| 0·5 mM | 670 ± 12·9 | 630 ± 11·9 | 7·2 ± 0·6 |
| 0·1 mM | 690 ± 13·7 | 690 ± 11·7 | 9·9 ± 0·8 |

Values are the mean ± S.E. for twelve samples per treatment.
* *p*-Hydroxybenzoic acid.

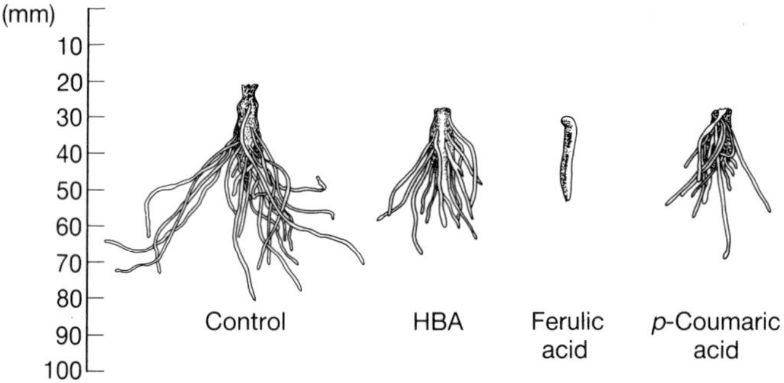


FIG. 2. Morphology of pea roots grown continuously for 14 days in Hoagland nutrient solution with or without 0·5 mM concentrations of *p*-hydroxybenzoic (HBA), ferulic or *p*-coumaric acids.

ferulic acid. Caffeic, syringic and vanillic acids also produced a marked effect on root morphology at 0·5 mM concentrations (Fig. 3). The most effective was syringic acid which reduced the length of the main root and the number and size of the laterals. Vanillic acid inhibited growth of the main root and the number of laterals formed. Caffeic acid inhibited to a lesser extent the growth of the main stem and the laterals but always produced a black main root and lateral root tips.

The phenolic acid must be present continuously at the appropriate concentration to exert its maximum effect. The effect on root morphology of transferring pea seedlings from a nutrient solution to a medium containing 0·5 mM syringic acid on

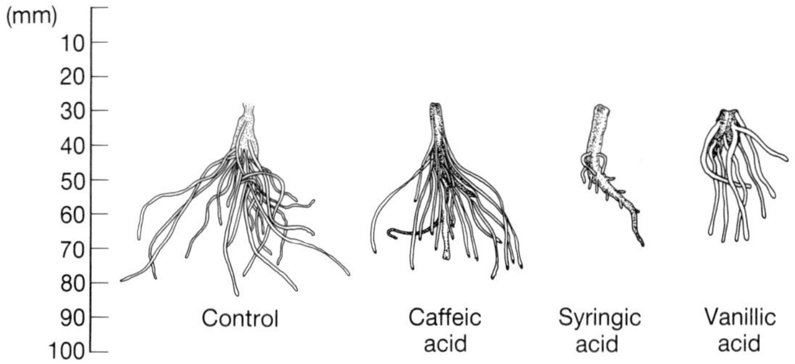


FIG. 3. Morphology of pea roots grown continuously for 14 days in Hoagland nutrient solution with or without 0·5 mM concentrations of caffeic, syringic or vanillic acids.

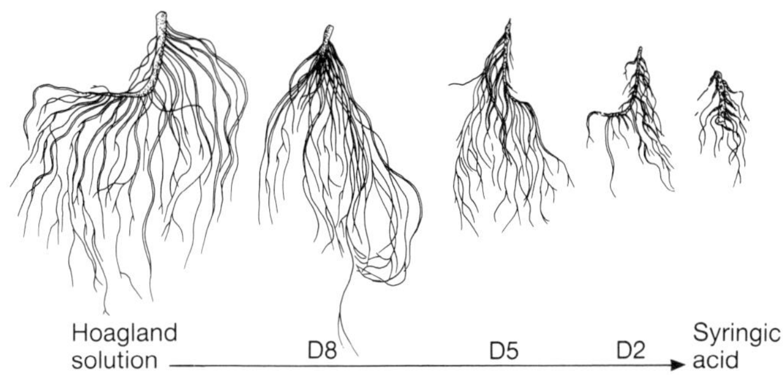


FIG. 4. Effect on root morphology of transferring pea seedlings from media containing Hoagland nutrient solution to nutrient media supplemented with 0·5 mM syringic acid on days 2, 5 and 8 of a 14-day culture period.

days 2, 5 and 8 of the 14-day incubation period is shown in Fig. 4. The effectiveness of the phenolic acid is diminished with increasing transfer time, but even after an 8 day transfer, the morphological effects are evident particularly on the length of the main root. The converse is also true when the seedlings are transferred from a nutrient medium containing the syringic acid to a nutrient medium alone. The influence of these transfers on changes in fresh weight are given in Fig. 5 and shows that the effectiveness of the syringic acid depends on the continuous presence of the acid. This was also the case for all the other phenolic acids tested and is illustrated by the changes in fresh weights of pea roots subjected to transfers to and from nutrient solutions with and without 0·5 mM caffeic acid (Fig. 6). The caffeic acid results were similar to those described for syringic acid.

In common with most other studies described in the literature, we have used phenolic acids at much higher concentrations than normally encountered in the soil solution. Typically the soil solution contains concentrations of phenolic acids in the order of 1–2 μM (Whitehead 1964; Glass 1976) (Table 4). However phenolic

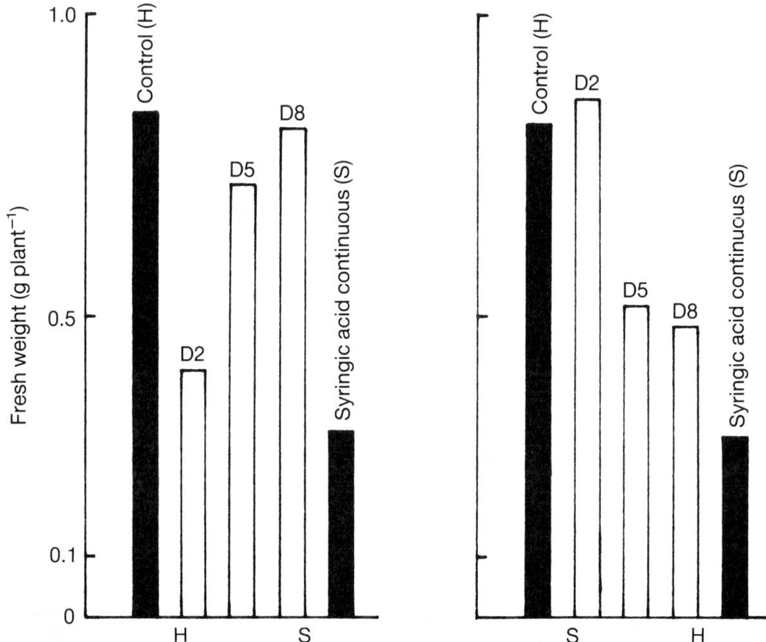


FIG. 5. Effect on the increase in fresh weight of transferring pea seedlings between nutrient solutions (H) with or without 0·5 mM syringic acid (S), at intervals of 2, 5 or 8 days of a 14-day culture period.

acids can exert inhibitory effects at even lower concentrations under conditions where one or more nutrients may be limiting. In Hoagland nutrient solution containing only 10% of its normal nitrate content, 1 μM concentrations of vanillic, *p*-hydroxybenzoic and *p*-coumaric acids all inhibited growth of roots and modified their morphology (Fig. 7). Additionally for vanillic and *p*-coumaric acids, the number of lateral roots were decreased and the fresh weight reduced. In this context, Stowe & Osborn (1980) showed that the phytotoxicity of some phenolic acids could be lowered by enhancing N and P availability. Modification of the phytotoxic thresholds of phenolic acids by altering the nutrient status of a growth medium is well established and first reported by Schreiner & Skinner (1912). More recently Glass (1976) showed that the addition of mineral nutrients decreased the inhibition of bracken growth produced by a mixture of phenolic acids.

Several mechanisms have been put forward to account for the influence of phenolic acids on plant growth. Accumulated evidence from the past 30 years shows that one likely mechanism of phenolic acid action operates by altering the balance of hormones involved with regulating plant growth. Henderson & Nitsch (1962) reported that chlorogenic and caffeic acids prevented the disappearance of IAA in *Avena* and act synergistically with IAA. Tomaszewski & Thimann (1966) showed that sinapic and ferulic acids also synergize IAA-induced growth in *Avena* coleoptiles by counteracting IAA-decarboxylation. The destruction of IAA can

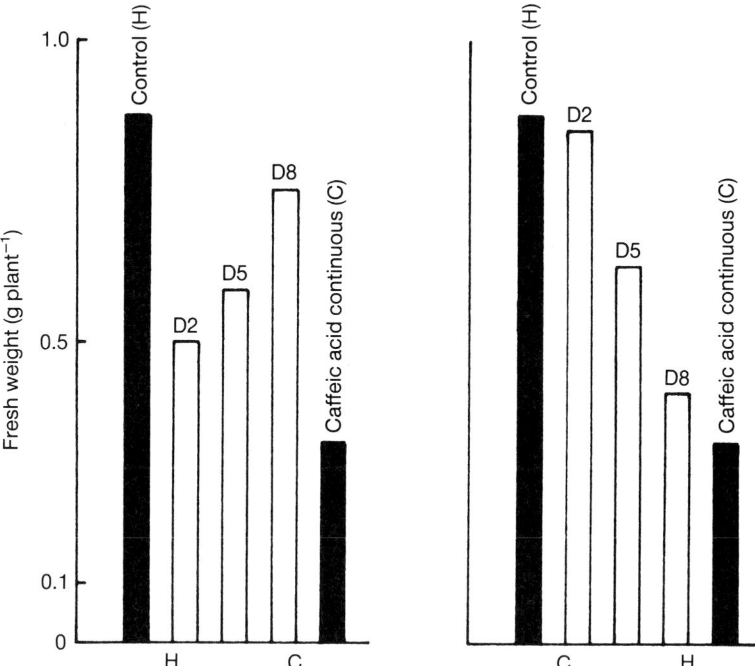


FIG. 6. Effect on the increase in fresh weight of pea roots after 14 days of transferring pea seedlings between nutrient solutions (H), with or without 0·5 mM caffeic acid (C), at intervals of 2, 5 or 8 days of a 14-day culture period.

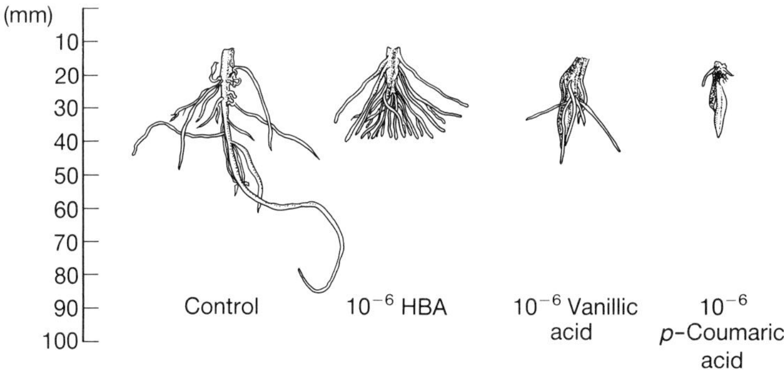


FIG. 7. Effect of 1 μM concentrations of *p*-hydroxybenzoic, vanillic and *p*-coumaric acids on the morphology of pea roots grown on a Hoagland nutrient solution containing a low N concentration (1 mM) for 14 days.

be brought about by the action of IAA oxidase so that inhibitions of this enzyme should lead to higher auxin levels. Rabin & Klein (1957) have demonstrated that in peas, chlorogenic acid was an inhibitor of IAA oxidase, an observation confirmed by Sondheimer & Griffin (1960). It has been reported, mainly for aerial parts of excised plant tissues, that monohydroxy phenolic acids are cofactors of IAA oxidase, resulting in increased rates of destruction of auxin and subsequent decreased growth, while dihydroxy phenolic acids inhibit IAA decarboxylation resulting in enhanced or maintained growth (Gortner & Kent, 1958; Henderson & Nitsch 1962; Zenk & Muller 1963; Tomaszewski & Thimann 1966). These observations are not always applicable to the roots of intact plants where most phenolic acids are inhibitory. Einhellig (1986) has emphasized that additional work is required to clarify the manner in which phytotoxic inhibition of growth is related to hormonal activity, although in intact plants it is probably more complex than simply an interaction with auxin.

Another method by which phenolic acids could influence growth is by altering changes in the plant–water balance (Einhellig 1986). Thus *p*-coumaric and ferulic acids lowered the leaf water potential of grain sorghum and soya-bean within one day of treatment. This difference resulted from reductions of both osmotic and turgor pressure. Patterson (1981) reported that the growth of soya-beans was inhibited by caffeic, ferulic and gallic acids and this inhibition was accompanied by reductions in water potential.

Any substance which interferes with energy production in a tissue would be expected to inhibit overall growth. This has been emphasized by Demos *et al.* (1975) who showed that compounds which alter respiration, or uncoupling responses in isolated mitochondria, also inhibit hypocotyl growth in mung beans and may reflect a mechanism of action for these naturally occurring growth inhibitors.

Of the other mechanisms which may be put forward to account for the influence of phenolic acids on plant growth, one is of particular relevance, in that it emphasizes that phenolics may be incorporated and then form an integral part of the cell wall structure. DeKock & Vaughan (1975) reported that $^{14}$C-cinnamic acid is taken up actively and incorporated irreversibly, much as ferulic acid, into the cell walls of excised pea roots. Whitmore (1974) showed that in the cell walls of *Avena* coleoptiles, ferulic acid is bonded to carbohydrate, probably via an ester linkage. An increase in the incorporation of phenolic acids into cell walls could lead to an increase in wall rigidity by a cross-linking of the diferulic acid type (DeKock & Vaughan 1975), or by forming quinonoid structures which link through a calcium chelate bridge as postulated by Painter & Neukom (1968).

Under field conditions phenolic acids may influence one or more growth mechanisms but such effects are probably constant for a particular species and a given phenolic acid. Phenolic acids may act additively, or even synergistically, on plant growth (Einhellig 1986) because each may affect a different mechanism.

## *Mineral contents of plants*

The dramatic effects produced by phenolic acids on root growth and morphology suggest that these substances can also have substantial effects on the uptake of minerals. Such effects have been reported but very few studies have tested directly the influence of specific potential allelochemicals on mineral uptake by plants

(Balke 1985). Among the recent reports, Kobza & Einhellig (1987) have shown that ferulic acid can reduce the concentrations of P, K and Mg in plants, an effect which is concentration dependent. Similarly Alsaadawi, Al-Hadithy & Arif (1986) have reported that *Vigna sinensis* L. (cowpeas) inhibited by syringic, caffeic and protocatechuic acids contained less N, P, K, Fe and Mo than the untreated controls. Caffeic acid also altered the mineral content of *Argyrodendron trifoliatum* (Chandler & Goosem 1982). At a concentration of 10 $\mu g\,ml^{-1}$, caffeic acid increased the contents of Zn, Mn and $PO_4$, but at a higher concentration (50 $\mu g\,ml^{-1}$) Mn and PO4 contents were decreased compared with their values at 10 $\mu g\,ml^{-1}$. Such results indicate that the mineral contents of plants can be altered under conditions where allelopathy (phytotoxicity) may exist.

In this paper the direct influence of phenolic acids on the mineral composition of plants, mainly peas, under axenic conditions is reported. Vanillic and *p*-hydroxybenzoic acids have a substantial effect on the Mn and Zn, and to a lesser extent the Cu contents of shoots and roots (Table 6). At the highest concentration used (1 mM), both phenolic acids markedly inhibited Mn and Zn. This effect was much more noticeable in the roots than in the shoots. The effect was concentration dependent and at the lowest concentration used (0·1 mM) *p*-hydroxybenzoic acid even enhanced the Mn content of the roots.

The enhancement of the Mn content in roots at lower concentrations of both vanillic and *p*-hydroxybenzoic acids can be seen clearly when the results of Table 6 are recalculated on a dry weight basis (Table 7). Again the difference in response between the shoots and the roots is evident and there is some suggestion that lower concentrations of the phenolics enhance the Zn content.

Phenolic acids also have a substantial effect on the Na, K, Ca and Mg concentrations of pea roots (Table 8). On a 'per root' basis the contents of all these cations

TABLE 6. Influence of vanillic and *p*-hydroxybenzoic (HBA) acids on the copper, manganese and zinc contents of pea shoot and roots after 14 days continuous culture in Hoagland nutrient solution

| | Shoots | | | Roots | | |
|---|---|---|---|---|---|---|
| Treatment | Cu | Mn | Zn | Cu | Mn | Zn |
| | | | ($\mu g$ plant$^{-1}$) | | | |
| None (control) | 1·1 | 3·8 | 4·8 | 0·9 | 12·9 | 3·0 |
| Vanillic | | | | | | |
| 1 mM | 0·9 | 1·7 | 2·8 | 0·6 | 1·2 | 1·5 |
| 0·5 mM | 0·7 | 2·2 | 2·9 | 0·6 | 3·1 | 2·1 |
| 0·1 mM | 1·2 | 3·6 | 4·9 | 0·8 | 10·6 | 2·7 |
| HBA* | | | | | | |
| 1 mM | 0·7 | 1·9 | 2·1 | 0·4 | 3·1 | 1·5 |
| 0·5 mM | 0·9 | 3·9 | 4·6 | 0·6 | 13·0 | 2·3 |
| 0·1 mM | 1·2 | 3·6 | 4·4 | 0·7 | 16·9 | 3·0 |

* *p*-Hydroxybenzoic acid.

TABLE 7. Copper, manganese and zinc contents of pea shoots and roots after 14 days in Hoagland nutrient solution expressed on a dry weight basis

| | Shoots | | | Roots | | |
|---|---|---|---|---|---|---|
| Treatment | Cu | Mn | Zn | Cu | Mn | Zn |
| | | | (μg 100 mg$^{-1}$ dry wt) | | | |
| None (control) | 1·2 ± 0·08 | 5·3 ± 0·3 | 6·5 ± 0·4 | 2·8 ± 0·2 | 40·9 ± 3·2 | 9·7 ± 0·7 |
| Vanillic | | | | | | |
| 1 mM | 1·7 ± 0·1 | 3·3 ± 0·2 | 5·6 ± 0·2 | 4·7 ± 0·3 | 5·7 ± 0·3 | 7·8 ± 0·6 |
| 0·5 mM | 1·4 ± 0·08 | 4·5 ± 0·3 | 5·6 ± 0·3 | 2·8 ± 0·2 | 15·3 ± 1·2 | 9·8 ± 0·8 |
| 0·1 mM | 1·8 ± 0·11 | 5·2 ± 0·13 | 7·1 ± 0·4 | 3·6 ± 0·2 | 46·2 ± 2·9 | 11·9 ± 1·1 |
| HBA | | | | | | |
| 1 mM | 1·5 ± 0·09 | 3·8 ± 0·3 | 4·3 ± 0·3 | 1·7 ± 0·2 | 16·2 ± 1·2 | 7·6 ± 0·5 |
| 0·5 mM | 1·6 ± 0·1 | 5·3 ± 0·2 | 6·2 ± 0·5 | 2·4 ± 0·2 | 52·8 ± 3·1 | 8·0 ± 0·5 |
| 0·1 mM | 1·6 ± 0·08 | 6·0 ± 0·4 | 7·4 ± 0·4 | 2·8 ± 0·2 | 56·0 ± 2·9 | 12·3 ± 0·9 |

Values are the mean ± S.E. for nine samples per treatment.

TABLE 8. Influence of vanillic, *p*-coumaric and *p*-hydroxybenzoic (*p*-HBA) acids on the sodium, potassium, calcium and magnesium contents of pea roots after 7 days continuous culture in Hoagland nutrient solution

| Treatment | Dry weights (mg root$^{-1}$) | Ca | Na | K | Mg |
|---|---|---|---|---|---|
| | | | (μg root$^{-1}$) | | |
| Control | 16·7 ± 1·2 | 290 | 220 | 5900 | 310 |
| Vanillic | | | | | |
| 1 mM | 1·3 ± 0·1 | 50 | 30 | 210 | 15 |
| 0·5 mM | 9·7 ± 0·6 | 220 | 160 | 3000 | 150 |
| 0·1 mM | 16·8 ± 1·4 | 300 | 250 | 5950 | 320 |
| *p*-Coumaric | | | | | |
| 1 mM | 1·5 ± 0·2 | 60 | 30 | 240 | 10 |
| 0·5 mM | 7·2 ± 0·4 | 170 | 120 | 2070 | 25 |
| 0·1 mM | 15·3 ± 1·3 | 280 | 260 | 5300 | 210 |
| HBA | | | | | |
| 1 mM | 6·5 ± 0·3 | 140 | 120 | 1110 | 80 |
| 0·5 mM | 14·7 ± 0·9 | 320 | 220 | 4750 | 250 |
| 0·1 mM | 17·8 ± 1·4 | 320 | 250 | 6690 | 340 |

Values for dry weights are the mean ± S.E. for six samples per treatment.

were lower in the presence of high concentrations of phenolic acids, with lower concentrations having progressively less effect. These results can be related to the reduced growth. The precise effect depends as much on the phenolic acid as it does on its concentration. With the exception of a stimulation of K content by (0·1 mM)

TABLE 9. Sodium, potassium, calcium and magnesium contents of pea roots after 7 days in Hoagland nutrient solution expressed on a dry weight basis

| Treatment | Ca | Na | K | Mg |
|---|---|---|---|---|
| | (mg g$^{-1}$ dry wt) | | | |
| Control | 1·7 ± 0·1 | 1·3 ± 0·1 | 35·3 ± 2·9 | 1·8 ± 0·2 |
| Vanillic | | | | |
| 1 mM | 5·7 ± 0·4 | 3·4 ± 0·2 | 20·6 ± 1·7 | 1·2 ± 0·1 |
| 0·5 mM | 2·3 ± 0·2 | 1·7 ± 0·1 | 31·0 ± 1·8 | 1·5 ± 0·2 |
| 0·1 mM | 1·8 ± 0·1 | 1·5 ± 0·2 | 35·4 ± 2·1 | 1·9 ± 0·2 |
| *p*-Coumaric | | | | |
| 1 mM | 4·6 ± 0·3 | 2·4 ± 0·2 | 20·0 ± 1·6 | 1·0 ± 0·1 |
| 0·5 mM | 2·4 ± 0·2 | 1·7 ± 0·1 | 28·7 ± 1·7 | 1·3 ± 0·2 |
| 0·1 mM | 1·8 ± 0·1 | 1·7 ± 0·2 | 34·4 ± 1·9 | 1·4 ± 0·2 |
| HBA | | | | |
| 1 mM | 2·2 ± 0·2 | 1·9 ± 0·2 | 17·0 ± 1·2 | 1·3 ± 0·2 |
| 0·5 mM | 2·2 ± 0·1 | 1·5 ± 0·2 | 32·3 ± 1·7 | 1·7 ± 0·4 |
| 0·1 mM | 1·8 ± 0·2 | 1·4 ± 0·2 | 37·6 ± 2·4 | 1·9 ± 0·3 |

Values are the mean ± S.E. for six samples per treatment.

*p*-hydroxybenzoic acid, the lower concentrations of all acids had no significant effect on the content of the cations. A different conclusion may be reached when the cation contents are expressed on a dry weight basis, when the inhibitions for Na, K and Mg contents are clearly less well defined (Table 9). Indeed, in this case, Ca and Na contents were enhanced in the presence of high concentrations of the phenolic acids.

The data for Ca is interesting in that the final content of this cation in root cells results from two processes, *viz.* a passive influx and an active efflux (Macklon 1984). When the cell membrane becomes leaky, the control of Ca accumulation is lost resulting in a higher content as the cells reach equilibrium with the external medium. Several workers (Glass, 1974; Balke 1985) have reported that a range of phenolic acids can alter membrane permeability. It is, therefore, likely that the higher Ca contents measured in the roots of plants grown in phenolic acids (Table 9) is the result of an effect of these acids on the integrity of the cell membranes. Such an effect is not reflected when the results are expressed on a per root basis (Table 8) because the roots are much smaller in the highest phenolic treatments than in the controls.

There is little doubt that phenolic acids can influence the nutrient contents of pea seedlings. However, the precise effect of these phenolics on nutrient absorption may be caused by other factors such as translocation, or even by the nutrient reserves held in the large cotyledons. The use of excised roots eliminates these other factors thus allowing nutrient absorption to be studied directly (Balke 1985).

*Uptake of nutrients by excised roots*

Previous workers, using concentrations in the range used by us, have shown that phenolic acids usually inhibit the uptake of ions into excised roots. Glass (1973, 1974) reported that several phenolic acids inhibited the uptake of $K^+$ and $PO_4^{3-}$ into barley. Similarly McClure, Gross & Jackson (1978) have shown that ferulic acid inhibited $PO_4^{3-}$ uptake into *Glycine max*. Most studies on mineral uptake into excised roots have been conducted with a limited number of ions such as $K^+$, $PO_4^{3-}$, $Cl^-$ and $NO_3^-$. A further limitation in such studies has been the small number of plants tested which have generally been confined to the monocotyledons (Balke 1985).

In our study, the uptakes of radioactively labelled $^{36}Cl$, $^{22}Na$ and $^{45}Ca$ were measured over 2 h using 1-cm-long root segments excised from pea seedlings previously grown in the appropriate phenolic acid for 24 h. At 20 °C, $^{36}Cl$ uptake was inhibited progressively by increasing concentrations of vanillic and *p*-coumaric acids (Table 10). In contrast, *p*-hydroxybenzoic acid had no inhibitory effect. At 2–4 °C, the uptakes of $^{36}Cl$ into the roots was only 10–15% that at 20 °C and the phenolic acids were without effect on this uptake. It is, therefore, likely that the phenolic acids inhibit chloride uptake by influencing processes dependent on active metabolism.

$^{22}Na$ uptakes were inhibited by high concentrations of vanillic and *p*-coumaric acids whereas lower concentrations of both phenolics resulted in enhanced uptakes (Table 10). *p*-Hydroxybenzoic acid had virtually no effect on Na uptake. $^{45}Ca$ uptake was the least affected by phenolics, although lower concentrations of

TABLE 10. Effect of some phenolic acids in the culture medium on $^{36}Cl$, $^{45}Ca$ and $^{22}Na$ uptakes by 0–10 mm excised pea root segments after 1 h

| Treatment | $^{36}Cl$ | $^{45}Ca$ Bq segment$^{-1}$ | $^{22}Na$ |
|---|---|---|---|
| Control | 11·5 ± 0·7 | 35·4 ± 2·3 | 17·3 ± 1·2 |
| Vanillic | | | |
| 1 mM | 6·2 ± 0·3 | 39·4 ± 2·9 | 11·9 ± 0·8 |
| 0·5 mM | 10·7 ± 0·4 | 41·3 ± 2·8 | 20·3 ± 1·4 |
| 0·1 mM | 10·6 ± 0·6 | 46·3 ± 3·2 | 25·2 ± 1·7 |
| *p*-Coumaric | | | |
| 1 mM | 6·5 ± 0·4 | 37·3 ± 2·8 | 12·3 ± 0·1 |
| 0·5 mM | 11·6 ± 0·6 | 36·2 ± 2·7 | 23·7 ± 1·5 |
| 0·1 mM | 13·3 ± 0·9 | 39·9 ± 2·7 | 25·4 ± 1·6 |
| HBA | | | |
| 1 mM | 12·3 ± 0·8 | 37·8 ± 2·9 | 16·4 ± 0·9 |
| 0·5 mM | 12·2 ± 0·7 | 40·4 ± 2·8 | 18·9 ± 1·4 |
| 0·1 mM | 11·8 ± 0·8 | 41·0 ± 3·0 | 18·3 ± 1·4 |

Values are the mean ± S.E. for six samples per treatment.

TABLE 11. Effect of some phenolic acids (0·5 mM) in the assay medium only on $^{36}Cl$, $^{45}Ca$ and $^{22}Na$ uptakes *per se* by 0–10 mm excised pea root segments after 1 h

| Treatment | $^{36}Cl$ | $^{45}Ca$ (Bq segment$^{-1}$) | $^{22}Na$ |
|---|---|---|---|
| Control | 8·3 ± 0·5 | 45·1 ± 3·1 | 18·8 ± 1·1 |
| Vanillic | 5·0 ± 0·3 | 39·6 ± 3·0 | 16·1 ± 1·1 |
| *p*-Coumaric | 3·5 ± 0·3 | 46·0 ± 3·0 | 14·9 ± 1·0 |
| Cinnamic | 2·2 ± 0·2 | 44·4 ± 2·9 | 12·1 ± 1·0 |
| HBA | 6·9 ± 0·5 | 43·7 ± 2·8 | 23·2 ± 1·4 |
| Ferulic | 3·9 ± 0·2 | 46·9 ± 3·2 | 11·3 ± 0·7 |

Values are the mean ± S.E. for six samples per treatment.

vanillic, *p*-coumaric and *p*-hydroxybenzoic acids produced varying degrees of stimulation. No phenolic acid inhibited Ca uptake.

The data of Table 10 illustrate that the effect of phenolic acids on ion uptake depends on the individual phenolic acid, its concentration and the ion under investigation. The data do not enable a distinction to be made between an effect of a phenolic acid on the development of an ion uptake mechanism or on the operation of that mechanism *per se*. The latter aspect can be assessed readily by growing the roots in the absence of a phenolic acid but subsequently incorporating the phenolic material into the uptake assay medium.

All phenolic acids investigated inhibited $^{36}Cl$ uptake *per se*, the extent depending on the phenolic acid (Table 11). $^{22}Na$ was also generally inhibited, but the relative effectiveness of each phenolic acid was different than for $^{36}Cl$ uptake. One phenolic acid, *p*-hydroxybenzoic acid, stimulated $^{22}Na$ uptake significantly. Generally the phenolic acids had little effect on $^{45}Ca$ uptake *per se*, although vanillic acid was inhibitory. The response of Ca to phenolic acids may reflect the lack of an active uptake mechanism for this cation in pea roots (Macklon 1984).

In order to produce the maximum effects on $^{36}Cl$ and $^{22}Na$ uptakes, the phenolic acid must be in continuous contact with the excised root. In this connection, the inhibitions of K and $PO_4^{3-}$ in barley roots (Glass 1973, 1974) and $PO_4^{3-}$ in soyabean (McClure, Gross & Jackson 1978) were non-competitive and reversed when the phenolic acids were removed.

The question now arises as to the mode of action of phenolic acids in influencing nutrient uptake.

### *Mode of action of phenolic acids*

The mechanisms by which phenolic acids could influence ion uptake into plant roots is outlined in Fig. 8. With the exception of an effect produced by chelation of a cation by a phenolic acid in the external medium (Van der Werff 1981), all the mechanisms imply an effect on active mineral absorption (Balke 1985). Among the main mechanisms are:

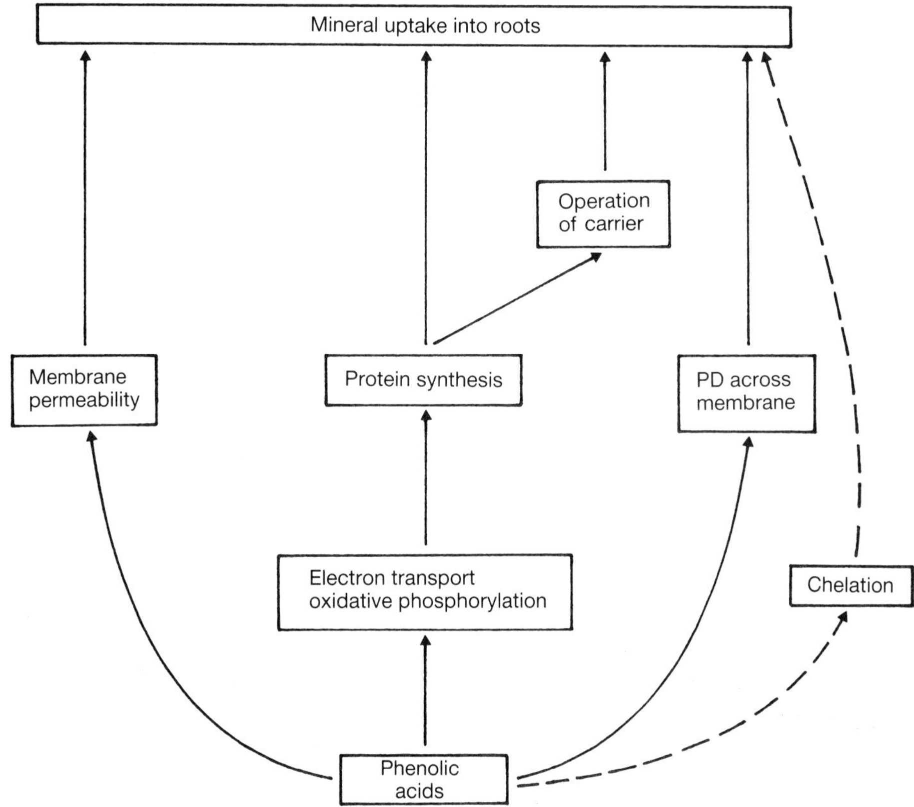


FIG. 8. Schematic diagram for the likely effects of phenolic acids on mineral uptake by plant roots.

1 Alteration of membrane permeability. There is now substantial evidence to support the view that phenolic acids can alter membrane permeability and hence inhibit mineral absorption. Glass (1974) noted that the inhibitory effect of a particular phenolic acid correlated well with its lipid solubility. Balke (1985) has discussed in some detail the two hypotheses proposed to explain how phenolic acids can directly increase membrane permeability. Firstly, phenolics solubilize into cellular membranes and thus cause a 'loosening' of the membrane structure so that ions can leak in and out readily (Glass 1973, 1974; Glass & Dunlop 1974). Secondly to account for the selective effect on the permeability of artificial phospholipid membranes, or those in the neurones of *Navanax*, towards $K^+$ and $Cl^-$, it was proposed that the anions of organic acids adsorb to membranes to produce either a negative surface potential (McLaughlin 1973), or an increase in the ionic field strength of the membrane (Levitan & Barker 1972). It still remains to be determined if one or both hypotheses are generally correct for higher plants, although the first has the advantage of being derived from data obtained using barley.

2 Effect on electrical potential (PD). The electrical potential across membranes

is important because without it ions would distribute only according to their chemical concentrations and usually would leak out of the cells to reach equilibrium with the ions in the external medium. Phenolic acids can influence the electrical potential across plant membranes. Glass & Dunlop (1974) have shown that at pH 7·2, salicylic acid quickly depolarized the electrical potential in the epidermal cells of barley roots. When the acid was removed there was a much slower recovery of the electrical potential. The precise effect of the phenolic acid is dependent on the pH of the external medium. Balke (1985) has reported that for the epidermal cells of oat roots, salicylic acid produced a dramatic depolarization at pH 4·5, but had no effect at pH 6·5. These data are related to the influence of salicylic acid on $K^+$ absorption in oat roots (Harper & Balke 1981). The ability of several benzoic and salicylic acid derivatives to increase the electrical potentials in some systems is related to their octanol/water partition coefficients and pKa values (Levatin & Barker 1972). These workers suggested that the organic acid anions bound to the membranes producing changes in the electrical potential. Balke (1985) proposed that electrical changes across membranes could result from inhibition of ATPase activity or from reductions in the substrate ATP.

3 Effect on respiration and oxidative phosphorylation. Current evidence suggests that phenolic acids do not affect mineral uptake by influencing ATPase enzymes (Balke 1985). They can, however, decrease the ATP content of tissues. Koch & Wilson (1977) reported that gentisic, ferulic and *p*-coumaric acids reduced the rate of respiration and the ADP : O ratios in mung bean. They suggested that phenolic acids altered the integrity of both inner and outer mitochondrial membranes. An effect of phenolics on membranes is also suggested by Balke (1985) who reported that salicylic acid decreased the ATP content of oat roots in a pH-dependent manner. Additionally phenolic acids may uncouple oxidative phosphorylation. In this context, Demos *et al.* (1975) showed that phenolic substances uncouple oxidative phosphorylation in mung bean mitochondria. Mitochondria are the main source of ATP production in tissues without chlorophyll so that an effect on ATP production will influence protein synthesis and growth. The effect of phenolic acids on ATP production is dependent on the individual phenolic acid, its concentration and the plant species under investigation. Thus, substituted cinnamic acids did not affect ATP production in mitochondria of cucumber hypocotyls, maize coleoptiles or pea roots (Balke 1985).

4 Protein synthesis. Ferulic acid inhibits protein synthesis in yeast cells and lettuce (van Sumere *et al.* 1972; Cameron & Julian 1980). Many ions are taken up actively by a mechanism dependent on protein synthesis. It is, therefore, surprising that this aspect has been neglected in investigations involving the influence of phenolic acids on ion uptake into roots. In this current investigation using excised pea root segments it has been found that all phenolic acids tested at a concentration of 0·5 mM, represented in Table 12 by vanillic, *p*-coumaric and *p*-hydroxybenzoic acids, inhibited the uptake and incorporation into proteins of $^{14}C$-leucine. However, these effects are not reflected in the $^{36}Cl$ uptake values which are unaffected by these concentrations of phenolic acids. Furthermore,

TABLE 12. Influence of some phenolic acids on the uptake and incorporation of proteins of $^{14}C$-leucine, and the uptake of $^{36}Cl$, into 0–10 mm excised pea root segments after 2 h

| Treatment | Total leucine uptake | Incorporation into proteins (Bq segment$^{-1}$) | $^{36}Cl$ uptake |
|---|---|---|---|
| Control | 266 ± 7·9 | 60·1 ± 4·2 | 11·2 ± 0·6 |
| 1 μg ml$^{-1}$ CHM* | 30·3 ± 2·1 | 3·9 ± 0·2 | 2·5 ± 0·2 |
| 0·5 mM VAN† | 32·9 ± 2·1 | 3·6 ± 0·2 | 10·9 ± 0·8 |
| CHM + VAN | 29·6 ± 2·2 | 3·6 ± 0·3 | 2·4 ± 0·1 |
| 0·5 mM *p*-Coumaric | 39·8 ± 2·7 | 4·9 ± 0·4 | 11·4 ± 0·8 |
| CHM + *p*-Coumaric | 23·6 ± 1·3 | 4·7 ± 0·3 | 2·8 ± 0·2 |
| 0·5 mM HBA‡ | 52·3 ± 3·2 | 8·7 ± 0·4 | 1·6 ± 0·1 |
| CHM + HBA | 35·7 ± 2·5 | 4·1 ± 0·3 | 3·1 ± 0·2 |

Values are the mean ± S.E. for six samples per treatment.
* Cycloheximide.
† Vanillic acid.
‡ *p*-Hydroxybenzoic acid.

cycloheximide, an inhibitor of protein synthesis, when used at a concentration which produced an effect on protein synthesis similar to that produced by 0·5 mM phenolic acids also markedly inhibited $^{36}Cl$ uptake. Cycloheximide still produced this response when used together with a phenolic acid (Table 12). Similar results to those obtained with cycloheximide were also produced using other 'specific' inhibitors of protein synthesis such as puromycin and D-threo-chloramphenicol. Hence it is unlikely that for pea roots, the phenolic acids inhibit Cl uptake, or other active mineral uptake, by interfering with protein synthesis. Because the ATP content of pea roots is unaffected by phenolic acids (Balke 1985), it is likely that the major effect of phenolic acids in this species is at the membrane level either on changes in permeability and/or electrical potential. Experiments are now underway to investigate this aspect of the work.

## GENERAL CONCLUSION

There is no doubt that some phenolic acids can influence root growth and morphology and mineral uptakes. The precise effects depend on factors such as the chemical and physical properties of the individual phenolic acid, its concentration, the pH and nutrient status of the external medium and the species under investigation. The growth mechanism affected may also depend on a particular phenolic acid so that a mixture of phenolic acids may interfere with several mechanisms. All these factors must be considered where a number of different phenolic acids or other phytotoxic substances occur.

Phenolic acids are probably involved in many cases of allelopathy and when considered together with a wide range of other potential allelochemicals offer a vast scope of external conditions and internal plant mechanisms under which growth inhibition can occur.

## REFERENCES

**Alsaadawi, I.S., Al-Hadithy, S.M. & Arif, M.B. (1986).** Effects of three phenolic acids on chlorophyll content and ion uptake in cowpea seedlings. *Journal of Chemical Ecology*, **12**, 221–227.

**Balke, N.E. (1985).** Effects of allelochemicals on mineral uptake and associated physiological processes. *The Chemistry of Allelopathy* (Ed. by A.C. Thompson), pp. 163–178. American Chemical Society, Washington.

**Becker, M. & Drapier, J. (1984).** Rôle de l'allelopathie dans les difficultés de régénération du sapin (*Abies alba* Mill). 1. Propriétés phytotoxiques des hydrosolubles d'aigulles de sapin. *Acta Ecologica Ecologia Plantarum*, **5**, 347–356.

**Bell, D.T. & Koeppe, D.E. (1972).** Noncompetitive effects of giant foxtail on the growth of corn. *Agronomy Journal*, **64**, 321–325.

**Blum, U., Weed, S.B. & Dalton, B.R. (1987).** Influence of various soil factors on the effects of ferulic acid on leaf expansion of cucumber seedlings. *Plant and Soil*, **98**, 111–130.

**Bohm, B.A. & Tryon, R.M. (1967).** Phenolic compounds in ferns. 1. A survey of some ferns for cinnamic acid and benzoic acid derivatives. *Canadian Journal of Botany*, **45**, 585–593.

**Bonner, J. (1946).** Further investigation of toxic substances which arise from guayule plants: relation of toxic substances to the growth of guayule in soil. *Botanical Gazette*, **107**, 343–357.

**Borner, H. (1960).** Liberation of organic substances from higher plants and their role in the soil sickness problem. *Botanical Review*, **26**, 393–424.

**Bowman, D.M.J.S. & Kirkpatrick, J.B. (1986).** Establishment, suppression and growth of *Eucalyptus delegatensis* in multiaged forests. III. Intraspecific allelopathy, competition between adult and juvenile for moisture and nutrients, and frost damage to seedlings. *Australian Journal of Botany*, **34**, 81–94.

**Cameron, H.J. & Julian, G.R. (1980).** Inhibition of protein synthesis in lettuce (*Latuca sativa* L.) by allelopathic compounds. *Journal of Chemical Ecology*, **6**, 989–995.

**Chandler, G. & Gossem, S. (1982).** Aspects of rain forest regeneration. III. The interaction of phenols, light and nutrients. *New Phytologist*, **92**, 369–380.

**Chou, C.H. & Lin, H.J. (1976).** Auto-intoxication mechanism of *Oryza sativa*. 1. Phytotoxic effects of decomposing rice residues in soil. *Journal of Chemical Ecology*, **2**, 353–367.

**Davis, E.F. (1928).** The toxic principle of *Juglans nigra* as identified with synthetic juglone, and its toxic effects on tomato and alfalfa plants. *American Journal of Botany*, **15**, 620–628.

**De Candolle, M.A. (1832).** *Physiologie Vegetale*, **III**, 1478–1484.

**DeKock, P.C. & Vaughan, D. (1975).** Effects of some chelating and phenolic substances on the growth of excised pea root segments. *Planta*, **126**, 187–195.

**Demos, E.K., Woolvine, M., Wilson, M. & McMillan, R.H. (1975).** The effects of ten phenolic compounds on hypocotyl growth and mitochondrial metabolism of mung bean. *American Journal of Botany*, **62**, 97–102.

**Donald, C.M. (1963).** Competition among crop and pasture plants. *Advances in Agronomy*, **15**, 1–118.

**Einhellig, F.A. (1986).** Mechanisms and modes of action of allelochemicals. *The Science of Allelopathy* (Ed. by A.R. Putnam & C.S. Tang), pp. 171–188. Wiley, New York.

**Einhellig, F.A. & Rasmussen, J.A. (1978).** Synergistic inhibitory effects of vanillic and *p*-hydroxybenzoic acids on radish and grain sorghum. *Journal of Chemical Ecology*, **4**, 425–436.

**Einhellig, F.A. & Rasmussen, J.A. (1979).** Effects of three phenolic acids on chlorophyll content and growth of soybean and grain sorghum. *Journal of Chemical Ecology*, **5**, 815–825.

**Fisher, R.F. (1980).** Allelopathy: a potential cause of regeneration failure. *Journal of Forestry*, **6**, 346–350.

**Glass, A.D.M. (1973).** Influence of phenolic acids on ion uptake — I. Inhibition of phosphate uptake. *Plant Physiology*, **51**, 1037–1041.

**Glass, A.D.M. (1974).** Influence of phenolic acids upon ion uptake — III. Inhibition of potassium absorption. *Journal of Experimental Botany*, **25**, 1104–1113.

**Glass, A.D.M. (1976).** The allelopathic potential of phenolic acids associated with the rhizosphere of *Pteridium aquilinum*. *Canadian Journal of Botany*, **54**, 2440–2444.

**Glass, A.D.M. & Dunlop, J. (1974).** Influence of phenolic acids on ion uptake — IV. Depolarisation of membrane potentials. *Plant Physiology*, **54**, 855–858.

**Gliessman, S.R. & Muller, C.H. (1972).** The phytotoxic potential of bracken, *Pteridium aquilinum* (L.) Kuhn. *Madrono*, **21**, 299–304.

**Gortner, W.A. & Kent, M.J. (1958).** The co-enzyme requirement and enzyme inhibitors of pineapple indoleacetic acid oxidase. *Journal of Biological Chemistry*, **233**, 731–735.

**Guenzi, W.D. & McCalla, T.M. (1966).** Phytotoxic substances extracted from soil. *Soil Science Society of America Proceedings*, **30**, 214–216.

**Harper, J.R. & Balke, N.E. (1981).** Characterization of the inhibition of K absorption into oat roots by salicylic acid. *Plant Physiology*, **68**, 1349–1353.

**Hartley, R.D. & Whitehead, D.C. (1985).** Phenolic acids in soils and their influence on plant growth and soil microbial processes. *Developments in Plant and Soil Sciences* (Ed. by D. Vaughan & R.E. Malcolm), Vol 16, pp. 109–149. Martinus Nijhoff, Dordrecht.

**Henderson, J.H.M. & Nitsch, J.P. (1962).** Effect of certain phenolic acids on the elongation of *Avena* first nodes in the presence of auxins and tryptophan. *Nature (London)*, **195**, 780–782.

**Hoagland, D.R. & Arnon, D.I. (1950).** The water culture method for growing plants without soil. California Agricultural Experimental Station Circular, **347**.

**Huang, P.M., Wang, T.S.C., Wang, M.K., Wu, M.H. & Hsu, N.W. (1977).** Retention of phenolic acids by non-crystalline hydroxy-aluminium and iron compounds and clay minerals of soils. *Soil Science*, **123**, 213–220.

**Kobza, J. & Einhellig, F.A. (1987).** The effects of ferulic acid on the mineral nutrition of grain sorghum. *Plant and Soil*, **98**, 99–109.

**Koch, S.J. & Wilson, R.H. (1977).** Effects of phenolic acids in hypocotyl growth and mitochondrial respiration in mung bean (*Phaseolus aureus*). *Annals of Botany*, **41**, 1091–1092.

**Kuiters, A.T. & Denneman, A.J. (1987).** Water-soluble phenolic substances in soils under several coniferous and deciduous tree species. *Soil Biology & Biochemistry*, **19**, 765–769.

**Levatin, H. & Barker, J.L. (1972).** Membrane permeability: cation selectivity reversibility by salicylate. *Science*, **178**, 63–64.

**Lovett J.V. & Levitt, J. (1981).** Allelochemicals in a future agriculture. *Biological Husbandry: a Scientific Approach to Organic Farming* (Ed. by B. Stonehouse), pp. 169–180. Butterworths, London.

**Macklon, A.E.S. (1984).** Calcium fluxes at plasmalemma and tonoplast. *Plant, Cell and Environment*, **7**, 407–413.

**Mandava, N.B. (1985).** Chemistry and biology of allelopathic agents. *The Chemistry of Allelopathy* (Ed. by A.C. Thompson), pp. 33–54. American Chemical Society, Washington.

**McClure, P.R., Gross, H.D. & Jackson, W.A. (1978).** Phosphate absorption by soybean varieties: the influence of ferulic acid. *Canadian Journal of Botany*, **56**, 764–767.

**McLaughlin, S. (1973).** Salicylates and phospholipid bilayer membranes. *Nature (London)*, **243**, 234–236.

**Molisch, H. (1937).** *Der Einfluss einer Pflanze auf die andere — Allelopathie*. Gustav Fischer Verlag, Jena.

**Muller, C.H. (1966).** The role of chemical inhibition (allelopathy) in vegetational composition. *Torrey Botanical Club Bulletin*, **93**, 332–351.

**Muller, C.H. (1969).** Allelopathy as a factor in ecological process. *Vegetatio*, **18**, 348–357.

**Painter, T.J. & Neukom, H. (1968).** The mechanism of oxidation gelation of a glycoprotein from wheat flour. Evidence from a model system based on caffeic acid. *Biochimica et Biophysica Acta*, **158**, 363–387.

**Patterson, D.T. (1981).** Effects of allelopathic chemicals on growth and physiological processes of soybean (*Glycine max*). *Weed Science*, **29**, 53–59.

**Putnam, A.R. (1985).** Allelopathic research in agriculture. *The Chemistry of Allelopathy* (Ed. by A.C. Thompson), pp. 1–8. American Chemical Society, Washington.

**Putnam, A.R. & Duke, W.B. (1978).** Allelopathy in agroecosystems. *Annual Review of Phytopathology*, **16**, 431–451.

**Putnam, A.R. & Tang, C.S. (1986).** Allelopathy: state of the science. *The Science of Allelopathy* (Ed. by A.R. Putnam & C.S. Tang), pp. 1–19. Wiley, New York.

**Putnam, A.R. & Weston, L.A. (1986).** Adverse impacts of allelopathy in agricultural systems. *The*

*Science of Allelopathy* (Ed. by A.R. Putnam & C.S. Tang), pp. 43–56. Wiley, New York.

**Rabin, R.S. & Klein, R.M. (1957).** Chlorogenic acid as a competitive inhibitor of indoleacetic acid oxidase. *Archives of Biochemistry and Biophysics*, **70**, 11–15.

**Rice, E.L. (1983).** *Pest Control with Nature's Chemicals*. University of Oklahoma Press, Norman, 224 pp.

**Rice, E.L. (1984).** *Allelopathy*, 2nd edn. Academic Press, New York.

**Rice, E.L. (1986).** Allelopathic growth stimulation. *The Science of Allelopathy* (Ed. by A.R. Putnam & C.S. Tang), pp. 23–42. Wiley, New York.

**Schreiner, O. & Skinner, J.J. (1912).** The toxic action of organic compounds as modified by fertilizer salts. *Botanical Gazette (Chicago)*, **54**, 31–48.

**Shindo, H., Ohta, S. & Kuwatsuka, S. (1978).** Behaviour of phenolic substances in the decaying process of plants — IX. Distribution of phenolic acids in paddy fields and forests. *Soil Science and Plant Nutrition*, **24**, 233–243.

**Sondheimer, E. & Griffin, D.H. (1960).** Activation and inhibition of indoleacetic acid oxidase activity from peas. *Science*, **131**, 672.

**Sparling, G.P., Ord, B.G. & Vaughan, D. (1981).** Changes in microbial biomass and activity in soils amended with phenolic acids. *Soil Biology & Biochemistry*, **13**, 455–460.

**Stowe, L.G. & Osborn, A. (1980).** The influence of nitrogen and phosphorus levels on the phytotoxicity of phenolic compounds. *Canadian Journal of Botany*, **58**, 1149–1153.

**Swain, T. (1977).** Secondary compounds as protective agents. *Annual Review of Plant Physiology*, **28**, 479–501.

**Tang, C.S. & Young, C.C. (1982).** Collection and identification of allelopathic compounds from the undisturbed root system of Bigatta Limpograss (*Hemarthria altissima*). *Plant Physiology*, **69**, 155–160.

**Tomaszewski, M. & Thimann, K.V. (1966).** Interactions of phenolic acids, metallic ions and chelating agents on auxin-induced growth. *Plant Physiology*, **41**, 1443–1454.

**Van der Werff, M.M. (1981).** *Ecotoxicity of heavy metals in aquatic and terrestrial higher plants*. Ph.D. thesis, Vrije Universiteit, Amsterdam.

**Van Sumere, C.F., Cottenie, J., De Greef, J. & Kint, J. (1972).** Biochemical studies in relation to the possible germination regulatory role of naturally occurring coumarin and phenolics. *Recent Advances in Phytochemistry*, **4**, 165–221.

**Vaughan, D. & Cusens, E. (1973).** Effect of hydroxyproline on the growth of excised root segments of *Pisum sativum* under aseptic conditions. *Planta*, **112**, 243–252.

**Vaughan, D. & Linehan, D.J. (1976).** The growth of wheat plants in humic acid solutions under axenic conditions. *Plant and Soil*, **44**, 445–449.

**Vaughan, D., DeKock, P.C. & Cusens, E. (1974).** Effects of hydroxyproline and other amino acid analogues on the growth of pea root segments. *Physiologia Plantarum*, **30**, 255–259.

**Vaughan, D., Jones, D. & Ord, B.G. (1988).** Soil phenolic acids — their origin and utilisation by soil fungi. *Journal of the Science of Food and Agriculture*, **52**, 289–299.

**Walters, E.H. (1917).** The isolation of *p*-hydroxybenzoic acid from soil. *American Chemical Society Journal*, **39**, 1778.

**Wang, T.S.C., Yang, T. & Chuang, T. (1967).** Soil phenolic acids as plant growth inhibitors. *Soil Science*, **103**, 239–246.

**Whitehead, D.C. (1964).** Identification of *p*-hydroxybenzoic, vanillic, *p*-coumaric and ferulic acids in soils. *Nature (London)*, **202**, 417–418.

**Whitehead, D.C., Dibb, H. & Hartley, R.D. (1982).** Phenolic compounds in soil as influenced by the growth of different plant species. *Journal of Applied Ecology*, **19**, 579–588.

**Whitmore, F.W. (1974).** Phenolic acids in wheat coleoptile cell walls. *Plant Physiology*, **53**, 728–731.

**Whittaker, R.H. (1970).** The biochemical ecology of higher plants. *Chemical Ecology* (Ed. by E. Sondheimer & J.R. Simeone), pp. 43–70. Academic Press, New York.

**Whittaker, R.H. & Feeny, P.P. (1971).** Allelochemics: chemical interactions between species. *Science*, **171**, 757–770.

**Zenk, M.H. & Muller, G. (1963).** *In vivo* destruction of exogenously applied indolyl-3-acetic acid as influenced by naturally occurring phenolic acids. *Nature (London)*, **200**, 761–763.

# Hydraulic lift: ecological implications of water efflux from roots

M.M. CALDWELL, J.H. RICHARDS AND W. BEYSCHLAG*
*Department of Range Science and the Ecology Center, Utah State University, Logan, Utah 84322–5230, USA*

## SUMMARY

1 Field measurements of diel fluctuations in soil water potential, transpiration manipulation experiments, and deuterium labelling support the hypothesis that water absorbed by deep roots of *Artemisia tridentata* shrubs in moist soil moves through the roots, is released in the upper soil profile at night, and is stored there until it is resorbed by roots the following day. This phenomenon is termed hydraulic lift.

2 The deuterium-labelling experiments also show the potential for parasitism of the water stored in the upper soil layers by neighbouring plant roots. Although the quantitative significance of interspecific water transfer remains to be explored, water parasitism may be significant for survival of shallow-rooted species during drying cycles.

3 Under conditions where only the sparse, deep roots of plants have access to moist soil, hydraulic lift increases the effectiveness of water absorption by deep roots as indicated by reductions of 25–50% in transpiration of *Artemisia tridentata* on days following experimental circumvention of hydraulic lift.

4 The upper soil layers serve as a capacitor in the soil–plant–atmosphere continuum during hydraulic lift. The plant tissues of *Artemisia* appear not to play a large role in this water storage at night. The status of the short-term water reservoir in the upper soil layers has an important bearing on diurnal plant gas exchange patterns. This reservoir may also serve as a storage buffer for periods longer than the diel cycles.

5 Self-irrigating roots associated with hydraulic lift should prolong activity of fine roots and their associated micro-organisms during drying cycles. Flushing of localized depletion zones near roots by water efflux may also facilitate nutrient uptake in drying soils.

## EVIDENCE OF ROOT WATER EFFLUX

Roots function as water-absorbing organs and yet often find themselves in situations where the soil water potential, $\Psi_s$ is more negative than the root xylem

* Permanent address: Lehrstuhl für Botanik II, Universität Würzburg, Mittlerer Dallenbergweg 64, D-8700 Würzburg, Germany.

potential, $\Psi_r$. Whether or not a significant efflux of water from the roots occurs under these circumstances has been somewhat controversial (Passioura 1988). Some laboratory experiments have demonstrated a considerable efflux of water from roots in dry soil, when another portion of the root system was amply supplied with moisture (van Bavel & Baker 1985; Baker & van Bavel 1986) while other laboratory experiments have failed to show this (Dirksen & Raats 1985).

There is some evidence that roots may develop a greatly reduced radial permeability to water in drying soil and, thus, curtail water efflux. When roots of *Agave deserti* and *Ferocactus acanthodes* were subjected to air drying (equivalent to a water potential of −93·6 MPa), their radial permeability to water dropped by an order of magnitude within 3–6 h and after rewetting exhibited the original permeability level within 6 h (Nobel & Sanderson 1984). The extreme conditions that induced these 'rectifier-like' properties may be approached in the very shallow soil layers of deserts. More often, however, roots are subjected to much less severe drying as water is slowly depleted from the soil profile, usually progressing from shallow to deeper soil layers. Under such conditions, changes in radial permeability may also occur (Passioura 1988); however, if water flows from the roots each night into the drying soil of the upper profile, this may delay reductions in radial permeability. Thus, this self-irrigating feature of roots may then prolong the very phenomenon of water efflux.

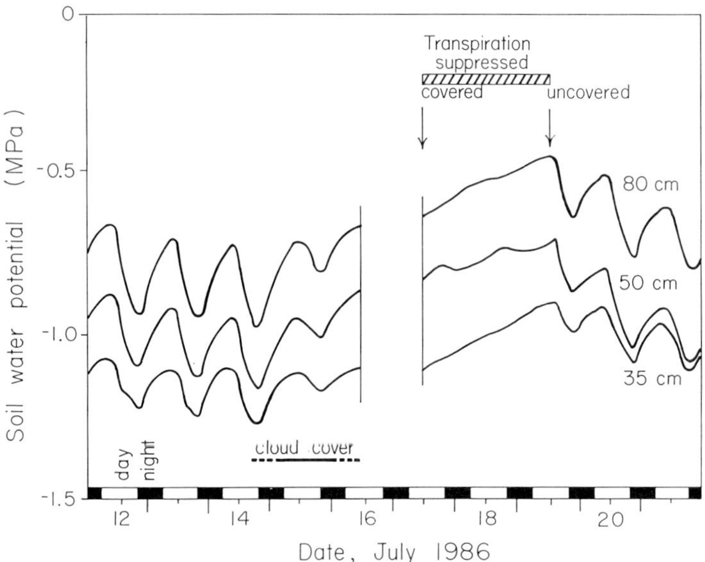


FIG. 1. Time course of $\Psi_s$ at three soil depths during a drying cycle and a transpiration–suppression experiment in which *Artemisia tridentata* shrubs were covered with opaque plastic bags for a 3 day period to curtail transpiration (from Richards & Caldwell 1987). Periods of cloud cover and the transpiration suppression are indicated. Decreases in $\Psi_s$ were lessened during cloudy weather and were totally absent when transpiration was suppressed. Missing data on 16–17 July are the result of a power failure.

Little evidence has accrued from field observations that can be used to test the notion of root water efflux. Field measurements of plant xylem potentials and plant and soil osmotic potentials of *Prosopis tamarugo* in the Atacama Desert, an area of essentially no precipitation but abundant ground water, led to the conjecture that water absorbed by deep roots from the ground water flowed out of roots into the shallower soil layers (Mooney *et al.* 1980). This flow was postulated to be responsible for the comparatively moist soil and an anomalous mat of dense roots in the upper soil layers.

Measurements of diel $\Psi_s$ fluctuations provided the first field observations of nocturnal increases of $\Psi_s$ considered to reflect a significant efflux of water from roots (Richards & Caldwell 1987). These observations involved *Artemisia tridentata* shrubs growing in a dry-summer, semi-arid environment and the fluctuations became apparent at several depths in the upper metre of the soil profile during long drying cycles (Fig. 1). The nocturnal increases of $\Psi_s$ represented water accumulation that is several orders of magnitude greater than could be attributed to liquid- and vapour-phase water movement from deeper to shallower layers through the soil itself (Richards & Caldwell 1987). The $\Psi_s$ fluctuations could be manipulated by covering the shrubs to suppress transpiration (Fig. 1) or by nighttime illumination of the shrubs (Caldwell & Richards 1989). These manipulations indicated that

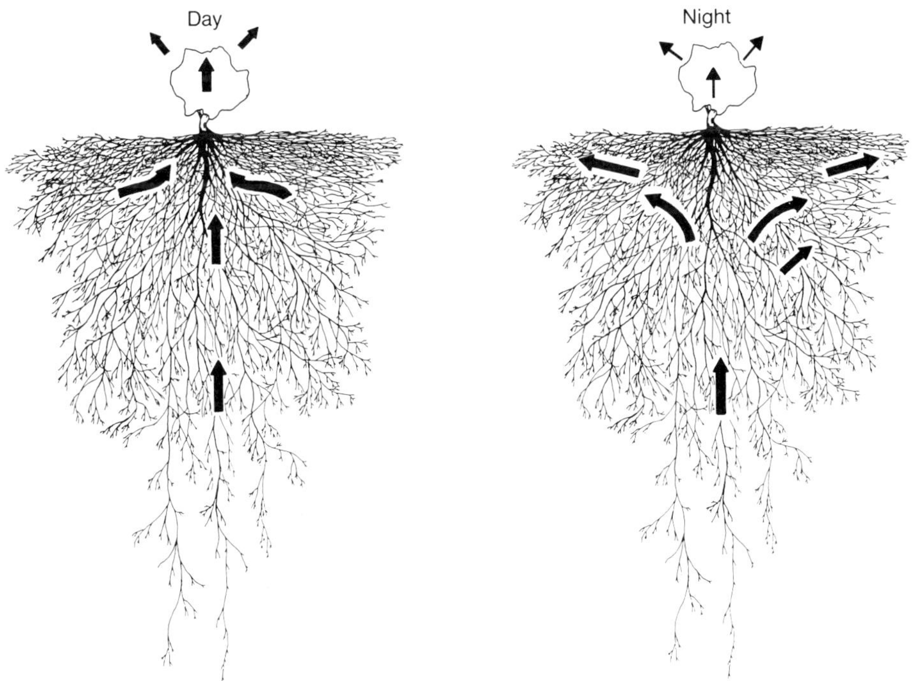


FIG. 2. Pattern of water movement according to the hydraulic lift hypothesis during day and night periods (from Caldwell 1988).

transpiration was clearly responsible for the diurnal decreases in $\Psi_s$. The nocturnal increases were postulated to result from upward transport by roots and efflux into the soil since no other avenue of water transport between soil layers appeared to be quantitatively feasible.

A postulated scheme of water movement shown in Fig. 2 involves water absorption by deep roots in moist soil and transport to the upper layer roots. With negligible transpiration at night, the prevailing gradient would be from the roots to the soil in the relatively dry upper layers of the soil profile. Thus, water would be lifted from deeper to upper soil layers using the root system as a bridge. During the day, the water released into the upper soil layers the previous night would be absorbed by the roots in the upper soil layers. This water, along with water absorbed by the deep roots, would flow into the transpiration stream of the shoot. The overall process is called hydraulic lift.

Additional support for this hypothesis came from deuterium-labelling experiments designed to determine if deuterium taken up by deep roots of *Artemisia* shrubs would appear in the transpiration stream of neighbouring plants (Fig. 3). The ends of individual intact roots of the *Artemisia* shrubs were freed from the soil

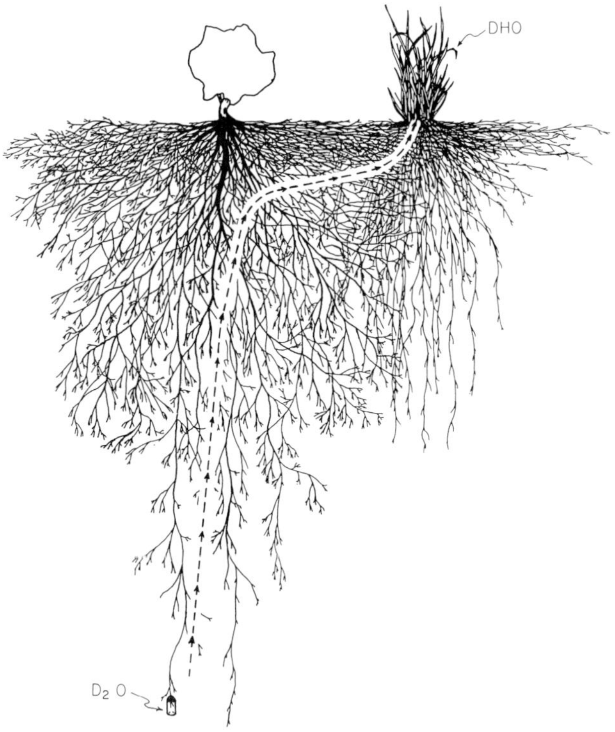


FIG. 3. Deuterium-labelling experiment designed to determine if water absorbed by deep roots of *Artemisia tridentata* would appear in the stem water of neighbouring *Agropyron desertorum* tussock grasses. Intact deep roots of *Artemisia tridentata* were immersed in vials of $D_2O$ and the DHO content of stem water in the grasses was monitored. The presumed pathway of deuterium movement is indicated.

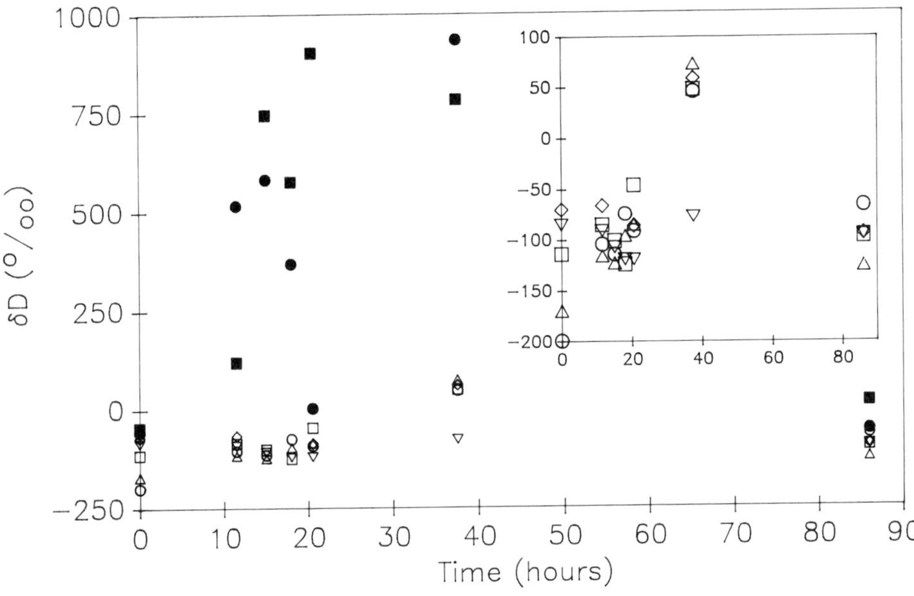


FIG. 4. Deuterium content of stem water as δD (per ml deviation from Standard Mean Ocean Water) of individual *Artemisia* shrubs (closed symbols) and neighbouring *Agropyron* tussock grasses (open symbols) following the initiation of $D_2O$-labelling of deep *Artemisia* roots (from Caldwell & Richards 1989). The inset shows only the *Agropyron* values plotted with an expanded y-axis scale. Each point represents δD from a pooled sample of five stems from a single plant. Each symbol type represents an individual plant sampled sequentially. The $D_2O$-labelling began at 1930 h solar time and lasted for 37 h. The range of background δD (in part due to some deuterium enrichment which occurred because water evaporated from the stems) is indicated by the values at time zero. All but one of the tussock grasses exhibited a similar pulse of HDO approximately 24 h following the large pulse which appeared in the shrubs. The one tussock grass that did not exhibit the HDO pulse was the furthest of the neighbouring grasses from the shrubs which absorbed the bulk of the $D_2O$.

wall of a deep trench excavated adjacent to the shrubs and placed in vials of $D_2O$. Stem water of the shrubs and of neighbouring perennial tussock grasses was then monitored over a 4 day period (Fig. 4). Not surprisingly, a large pulse of deuterated water, HDO, appeared in the stem water of the shrubs soon after the initiation of root $D_2O$ labelling. Of particular interest, however, was a delayed and weaker pulse of HDO which appeared in most of the neighbouring tussock grasses. This is clearly consistent with the hydraulic lift hypothesis. Alternative pathways of deuterium movement, such as by interspecific root grafts or mycorrhizal hyphae common to both shrubs and grasses, are possible, but were considered unlikely due to the magnitude and consistency of the HDO pulse in the grasses (Caldwell & Richards 1989).

## WATER PARASITISM

The potential for water parasitism by the grasses was also shown by these labelling experiments which raises an intriguing issue in plant competition. Clustering of different plant species in water-limited environments has been attributed to several factors including shading protection, increased infiltration of precipitation, and animal-facilitated dissemination of plant propagules (West 1989). Increased soil fertility associated with clumps of vegetation has also been advanced as a reason contributing to contagious distributions of vegetation, especially in water-limited environments where soil fertility is often low. Such 'islands of fertility' are thought to exist because wind-blown organic debris is trapped by the 'islands' of plants and litter deposition and nutrient cycling are concentrated in these islands. The increased organic matter in the 'island' soils also increases soil moisture holding capacity and moisture infiltration which further promote favourable growing conditions (West 1989). Hydraulic lift might contribute to the vegetation island phenomenon as shallow-rooted plants cluster near deep-rooted species to take advantage of the water they release into the shallow soils. Numerous reports have appeared suggesting that the close proximity of shrubs in water-limited environments can result in substantially increased production of herbaceous species (e.g. Garcia-Moya & McKell 1970; Halvorson & Patten 1975; Barth & Klemmedson 1978; Rumbaugh, Johnson & Van Epps 1982). There are also observations of moist soil occurring near shrubs and trees in desert areas. For example, Bergström & Skarpe (personal communication) reported greatly increased soil moisture within a metre of *Acacia mellifera* shrubs in the Kalahari Desert. The moisture may have been due to release from the shrub roots. Although such accounts do not exclude alternative explanations, they are consistent with the hydraulic lift hypothesis and suggest that the opportunity for water parasitism exists in water-limited environments. If water parasitism is prevalent in nature, the consequences for plant associations, distribution and competitive relationships are profound. Practical implications of water parasitism for mixed species associations in agronomy, forestry and range management should also be pursued.

Yet, the question remains as to the amount of water that might be parasitized. The deuterium-labelling experiment described earlier (Fig. 3) clearly indicates the potential for parasitism, but does not show the net quantity of water which passed from the shrubs to the grasses or how important the parasitized water is for the water economy of the grasses. In these experimental plots, the shrub and grass roots mix extensively and are often within millimetres of each other as revealed by detailed mapping of individual root distributions from frozen soil blocks (Fig. 5). Similar excavations and mapping of root distributions around soil psychrometers revealed that root density was similar to that in the bulk soil and roots had not aggregated around the psychrometers. Thus, the increases of $\Psi_s$ at night are considered to reflect a general increase in bulk $\Psi_s$ and not just a localized wetting near the root surface. Although the grass roots should be sufficiently close to the shrub roots to absorb water released by them, the shrub roots would still likely be more effective in resorbing the water they had released the previous night.

Water parasitism may be particularly beneficial for plants, such as establishing seedlings, whose roots do not extend to moist soil layers and, therefore, cannot conduct hydraulic lift themselves. Such a situation was simulated in laboratory

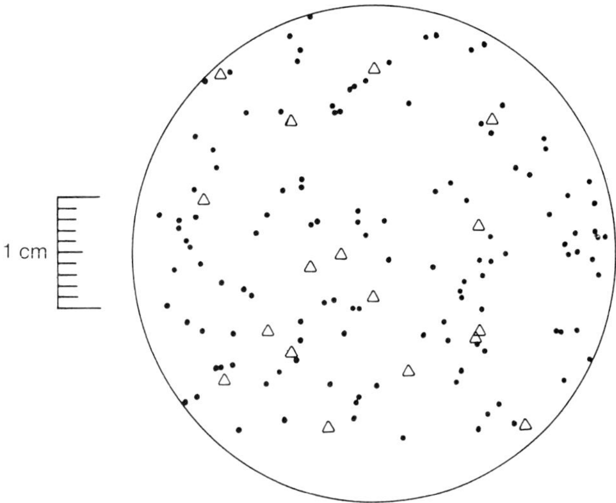


FIG. 5. Distribution of individual fine roots in a cross-section of soil at 10 cm depth in the interspace between *Artemisia* (triangles) and *Agropyron* (circles) plants (adapted from Caldwell *et al.* 1987).

experiments of Corak, Blevins & Pallardy (1987) in which alfalfa roots extended throughout two vertically stacked soil-filled plastic tubes. The alfalfa roots also spanned a 5 cm air gap between the two tubes. Maize roots were restricted to the upper tube which received no water. Tritium-labelling experiments showed that $^3H_2O$ placed in the soil of the lower tube appeared in the maize shoot tissue in large quantity. The maize plants also exhibited more positive water potentials and greater survival compared to those in an identical tube system without alfalfa roots. Yet, the amount of water transferred was not sufficient to sustain net growth of the maize nor detectably greater transpiration rates than in the system without alfalfa roots. Alfalfa is a species which did not exhibit detectable water efflux in other laboratory experiments (Dirksen & Raats 1985). Thus, choice of another species to grow with the maize, might have led to greater water transfer. In any case, it is clear that the quantitative significance of water parasitism in nature awaits exploration.

## HYDRAULIC BENEFITS

Water efflux from the root system in the upper soil layers may result in forfeiture of some water to neighbouring plants or even to evaporation if efflux occurs close to the soil surface. Yet, these losses may be offset by an advantage that can obtain from hydraulic lift under some circumstances.

Most of the fine root length of plants is typically found in the upper part of the soil profile where most of the organic matter and mineral nutrients reside (Richards 1986). Accordingly, soil moisture is usually depleted first in the upper layers of the profile. Thus, during prolonged drying cycles, the bulk of the root length is in the drier, upper soil layers and comparatively few roots access the moister soil in deeper layers. The sparsity of roots in these deeper soil layers can limit rates of water uptake (Passioura 1988). We hypothesized that hydraulic lift could increase the total amount of water absorbed from deeper soil layers because the deeper roots continue to absorb and transport water at night and this water is stored in the upper soil layers (Fig. 2). During the day, uptake of stored water by the abundant shallow roots should supplement water uptake by deep roots and result in higher rates of water supply to the shoot than would otherwise occur from deep roots alone. The 24 h uptake of water by the deep roots under these circumstances would increase their effectiveness considerably.

This hypothesis was tested by measuring whole-plant water transport capacity following experimental circumvention of hydraulic lift (Caldwell & Richards 1989). In the field, entire shoots of *Artemisia tridentata* shrubs were placed in gas exchange cuvettes in which temperature, humidity and $CO_2$ concentrations were controlled. These cuvettes, in turn, were placed under lamps in an opaque tent erected over the shrubs. Thus, the shrubs only received artificial illumination. Whole-plant transpiration with the cuvettes programmed for high evaporation demand conditions typical of midsummer weather in the area was taken as an index of the capacity of the plants to absorb and transport water to the atmosphere. During the night periods, the cuvettes were programmed for cool temperatures typical of this time of year. Occasionally, hydraulic lift was circumvented by nighttime illumination at these cool temperatures. Thus, stomates were open due to the illumination, but the plants were not consuming large amounts of water. Soil psychrometric data indicated that this treatment resulted in the anticipated circumvention of hydraulic lift because $\Psi_s$ did not increase when the plants were illuminated at night (Caldwell & Richards 1989).

Whole-plant transpiration on the day following circumvention of hydraulic lift was greatly reduced (Fig. 6), which is consistent with the hypothesis. This effect was largely reversible as indicated by increased daytime transpiration following a subsequent normal dark period. Over the course of the experiment, a day-to-day decrease in transpiration occurred due to the normal drying of the soil profile. Thus, the effects of treatment or recovery were based on deviations from a regression of transpiration through time.

In another experiment, a similar protocol was employed, but a 1·5 h test period with low $CO_2$ in the cuvettes was imposed early in the day to assess maximum whole-plant water conductance without normal stomatal control. The low $CO_2$ forced the stomates to open further than in normal air resulting in a two-fold increase in water vapour conductance. Maximum whole-plant conductance monitored over several days exhibited the same pattern of response to circumvention of hydraulic lift and recovery as did the whole-plant transpiration data (Caldwell & Richards 1989).

Over the course of replicate experiments, the reductions of transpiration following circumvention of hydraulic lift ranged between 25% and 50% (and averaged

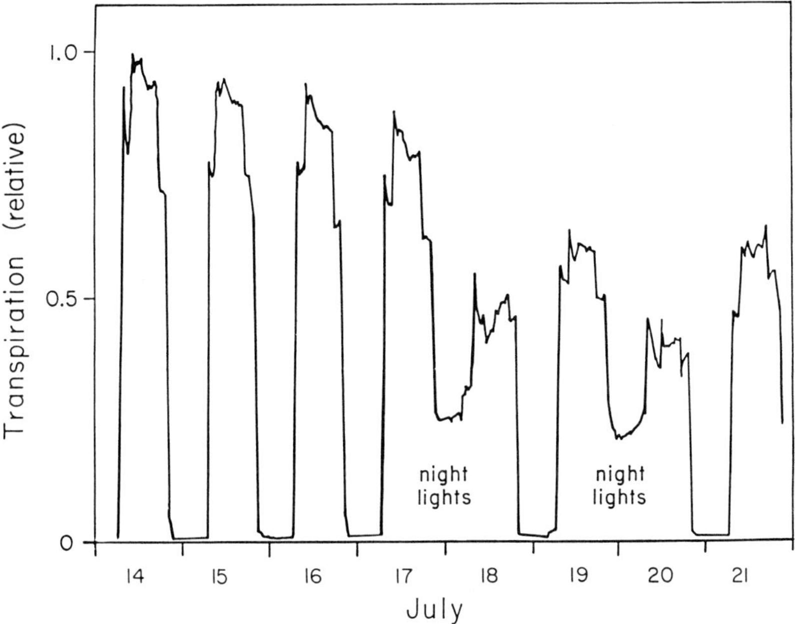


Fig. 6. Transpiration of an *Artemisia* shrub in controlled cuvette conditions of high evaporation demand during the day (reaching 34°C with a vapour pressure deficit of 4700 Pa during the middle hours of the day) and cool night temperatures (12·5°C) for eight days (from Caldwell & Richards 1989). Illumination was only by lamps. Treatment periods of 'nighttime' illumination are also shown.

34%). If hydraulic lift increases transpiration on a landscape scale by the magnitude indicated in these experiments, this process has an important influence on the partitioning of water between evapotranspiration and subsurface flow in arid areas. *Artemisia tridentata* is an important shrub of the Great Basin of North America — a large province of mountain ranges separated by closed basins with shrub steppe vegetation. There is no drainage from the Great Basin and water ultimately collects and evaporates from the basins. While the basins receive only modest annual precipitation (on the order of 200–400 mm), the mountains receive much more. The water draining from the mountains flows into extensive aquifers underlying most of the basins. Since this is a closed system with internal drainage, these aquifers discharge in springs and playas in the lower parts of the basins where the water eventually evaporates (Eakin, Price & Harrill 1967) (Fig. 7). The depth of the soil moisture associated with the aquifers varies with location in the basins and tends to be closer to the surface lower in the basins. Depth to this water is important for the distribution of species in the basins. Conversely, the vegetation composition determines how much of the subsurface water is tapped before it reaches the playas and springs. While it is obvious that deeply rooted plants will tap this moisture, hydraulic lift may substantially increase the effectiveness of water uptake from subsurface moisture. Thus, hydraulic lift, if operating effectively, may result in much less water accumulation in the aquifers of the low basin elevations.

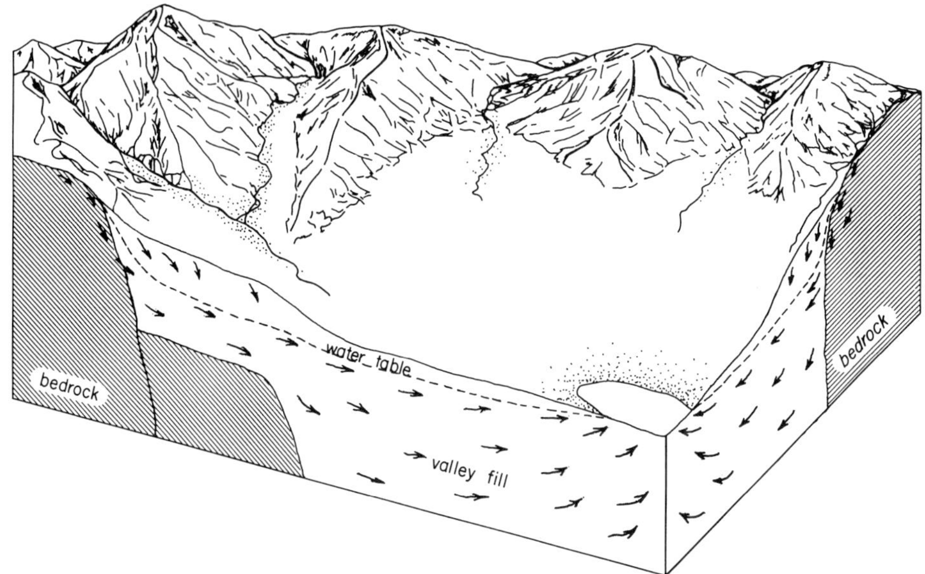


FIG. 7. Patterns of subsurface water flow in aquifers of a valley in the Great Basin (adapted from Eakin, Price & Harrill 1967). Much of the recharge of the basin subsurface water comes from downslope flow from the mountains. In these closed basins of internal drainage, water ultimately surfaces in springs or playas where it evaporates. The dashed line indicates the water table.

## STORAGE OF WATER

With hydraulic lift, the upper soil layers serve as a capacitance in the soil–plant–atmosphere continuum. Yet, does the plant itself play a significant role in the water storage at night? The comparatively rapid reversal of $\Psi_s$ trends in the early evening (Fig. 1) suggest that water flows readily into the soil when the stomates close. This reversal of $\Psi_s$ was even more rapid in experiments conducted under artificial illumination when there was an abrupt transition from light to dark (Caldwell & Richards 1989).

As an indirect test of plant capacitance, we measured water vapour efflux from intact roots of *Artemisia tridentata* as transpiration rates of the shrub were rapidly altered. The ends of intact roots of the shrub were freed from the soil wall of a trench excavated adjacent to the shrub and placed in a leaf gas exchange cuvette (Bingham & Coyne 1977). The cuvette was operated with an air temperature of 20 °C and 90% relative humidity, $\Psi_{air}$ = 14·1 MPa (greater humidity would have resulted in condensation problems and loss of measurement sensitivity). When the plant was actively transpiring, the roots exhibited water efflux at a steadily declining rate which probably reflected water loss from the root at a rate exceeding the supply to the root. This is indicated as the slope in Fig. 8. When the shrub was covered with opaque plastic to curtail transpiration, there was a decrease in the rate of decline of water vapour efflux within 10 min. This rate of change (slope)

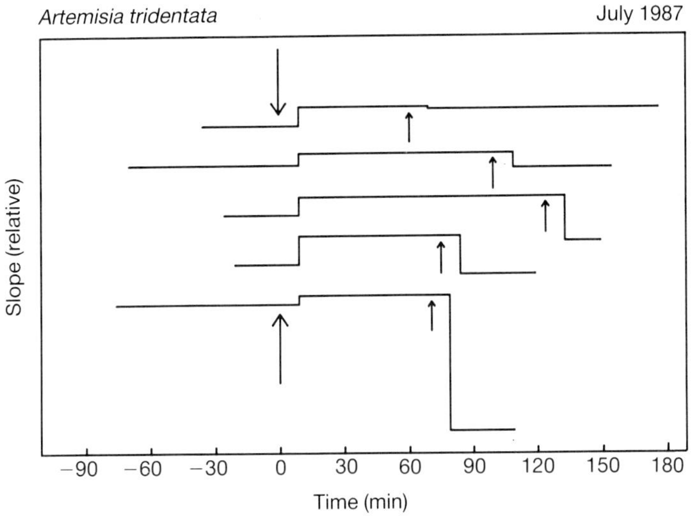


FIG. 8. Rate of change of water vapour efflux from *Artemisia tridentata* roots held in a small gas exchange cuvette as the transpiration rate of the shoot was manipulated. Root water vapour efflux was measured when the shoot was under conditions of high evaporative demand in the field. The efflux rate exhibited a steadily declining rate indicated as relative slope. The shrub was covered with opaque plastic at time zero indicated by large arrows. After a step change to a new steady rate of efflux decline, the shrub was uncovered at the times indicated by the small arrows. The results of five replicated experiments are shown. In the different experiments, the shrubs were uncovered at different times of day which contributed to the variation among experiments in the new slope achieved following uncovering of the plants.

then persisted until the shrub was uncovered, in which case another step change in the slope occurred within 10 min. These rapid changes to different and steady slopes are consistent with the notion that most of the water storage is occurring in the soil and not in the plant. If the plant tissues were playing a large role in the capacitance, a slow and more gradual change in root water vapour efflux following changes in transpiration would be expected as water storage in the shoot and root tissues would buffer the change in water supply to the 'transpiring' root. Similarly, if a significant portion of the hydraulic lift water were stored in the plant tissues, with the onset of darkness one would expect a pronounced lag in the rise of $\Psi_s$ as plant tissues were first recharged with water before a significant amount of water flowed into the soil. Thus, for *Artemisia tridentata*, the capacitance in the soil–plant–atmosphere continuum appears to occur in the soil.

The status of the temporary soil moisture reservoir in the upper soil layers stemming from hydraulic lift has a bearing on the diurnal course of shoot gas exchange. When weather is conducive for high transpiration rates, the daily pattern of leaf conductance, $g$, and transpiration is considered to reflect atmospheric factors and the progression of leaf water potential, $\Psi_l$. The $\Psi_l$ in turn, characteristically decreases during the day and then increases overnight as shoot $\Psi$ comes into equilibrium with $\Psi_s$. (The $\Psi_s$ is usually thought to only gradually decrease from day to day.) The divergence of $\Psi_l$ from $\Psi_s$ during the day is considered to reflect plant water loss rates that exceed the rate of absorption by the plant (Slatyer 1967). However, when hydraulic lift is operating, diel changes in $\Psi_s$ are superimposed on this pattern. The diurnal change in *g* and transpiration, thus, must (in addition to atmospheric factors) also reflect the decrease in $\Psi_s$ during the day as stored water in the upper soil layers is depleted.

Not only are the upper soil layers a site of diel water storage, but they may also serve as a buffer for several days. The potential magnitude of the water storage buffer, as indicated by the transpiration-suppression experiments (Fig. 1), is much larger than the normal diel storage (Richards & Caldwell 1987). During periods of cloudy weather when transpiration is reduced, storage can exceed the daily use of water (Fig. 1). During midsummer in the Great Basin, heavy precipitation events are comparatively rare, but alternating periods of cloudy and sunny weather are common. Following cloudy periods, even without precipitation, the shrubs should have more water available than following several days of clear warm weather.

## IMPLICATIONS FOR THE RHIZOSPHERE

As the upper layers of the soil are depleted of water during a drying cycle, self-irrigating roots associated with hydraulic lift should prolong life of the fine roots. Small roots in upper soil layers may be quite susceptible to even moderate drying. [Deans (1979) reported that fine root death of *Picea sitchensis* could start when $\Psi_s$ dropped to only −10 kPa.] Thus, hydraulic lift may be of importance not only in environments that experience pronounced drying cycles, but also in ecosystems where only transient drying of the upper soil layers occurs. Since fine-root turnover is estimated to represent a large fraction of primary productivity in many ecosystems (Caldwell 1979), the importance of hydraulic lift in mitigating fine-root death may be far-reaching.

The daily efflux of water from the upper roots should also help to prolong the activity of mycorrhizae and other soil micro-organisms. This would facilitate continued nutrient uptake and other processes such as mineralization.

As water and nutrients move towards the root from the surrounding cylinder of soil, localized depletions of water and nutrients can develop next to the root. Localized mineral ion depletions develop if the rate of diffusion of the ions is slow relative to the uptake by the root (Nye & Tinker 1977). These depletions can result in reduced uptake by the root because the ion concentration at the root surface can be considerably lower than in the bulk soil. Localized depletions of water are usually thought to be of modest proportion unless the soil is quite dry (Newman 1974). However, mineral ion diffusivity is reduced even with small reductions of $\Psi_s$ (Nye & Tinker 1977). Thus, even small localized reductions of $\Psi_s$ near the root surface could reduce mobility and uptake of nutrient ions. Flushing of the depletion zones due to hydraulic lift, even if very limited, might facilitate uptake of nutrient ions.

## CONCLUDING THOUGHTS

The study of hydraulic lift under field conditions has hardly begun and, therefore, little can be said of its generality. Nevertheless, the numerous implications for plant–water relations, the maintenance of fine roots and facilitation of mineral nutrient uptake, competitive and parasitic interactions among neighbouring plants, facilitation of water uptake by deep roots and the partitioning of water between evapotranspiration and subsurface flow in arid lands suggest that further study is well warranted.

## ACKNOWLEDGEMENTS

The research dealing with *Artemisia tridentata* was supported by the National Science Foundation (BSR 8705492) and the Utah Agricultural Experiment Station. The support of W. Beyschlag by the Alexander von Humboldt-Stiftung during his stay at Utah State University is also gratefully acknowledged.

## REFERENCES

**Baker, J.M. & van Bavel, C.H.M. (1986).** Resistance of plant roots to water loss. *Agronomy Journal*, **78**, 641–644.

**Barth, R.C. & Klemmedson, J.O. (1978).** Shrub-induced spatial patterns of dry matter, nitrogen and organic carbon. *Soil Science Society of America Journal*, **42**, 804–809.

**Bingham, G.E. & Coyne, P.I. (1977).** A portable temperature-controlled, steady-state porometer for field measurements of transpiration and photosynthesis. *Photosynthetica*, **11**, 148–160.

**Caldwell, M.M. (1979).** Root structure: the considerable cost of belowground function. In *Topics in Plant Population Biology* (Ed. by O.T. Solbrig, S. Jain, G.B. Johnson & P.H. Raven), pp. 408–427. Columbia University Press, New York.

**Caldwell, M.M. (1988).** Plant root systems and competition. In *Proceedings of the XIV International Botanical Congress* (Ed. by W. Greuter & B. Zimmer), pp. 385–404. Koeltz, Königstein.

**Caldwell, M.M. & Richards, J.H. (1989).** Hydraulic lift: water efflux from upper roots improves effectiveness of water uptake by deep roots. *Oecologia*, **79**, 1–5.

**Caldwell, M.M., Richards, J.H., Manwaring, J.H. & Eissenstat, D.M. (1987).** Rapid shifts in phosphate acquisition show direct competition between neighbouring plants. *Nature*, **327**, 615–616.

**Corak, S.J., Blevins, D.G. & Pallardy, S.G. (1987).** Water transfer in an alfalfa/maize association: Survival of maize during drought. *Plant Physiology*, **84**, 582–586.

**Deans, J.D. (1979).** Fluctuations of the soil environment and fine root growth in a young Sitka spruce plantation. *Plant and Soil*, **52**, 195–208.

**Dirksen, C. & Raats, P.A.C. (1985).** Water uptake and release by alfalfa roots. *Agronomy Journal*, **77**, 621–626.

**Eakin, T.E., Price, D. & Harrill, J.R. (1967).** Summary appraisals of the nation's ground-water resources — Great Basin Region. *USGS Professional Paper 813 G*, Washington, D.C., p. G37.

**Garcia-Moya, E. & McKell, C.M. (1970).** Contribution of shrubs to the nitrogen economy of a desert wash plant community. *Ecology*, **51**, 81–88.

**Halvorson, W.L. & Patten, D.T. (1975).** Productivity and flowering of winter ephemerals in relation to Sonoran Desert shrubs. *American Midland Naturalist*, **93**, 311–319.

**Mooney, H.A., Gulman, S.L., Rundel, P.W. & Ehleringer, J. (1980).** Further observations on the water relations of *Prosopis tamarugo* of the northern Atacama Desert. *Oecologia*, **44**, 177–180.

**Newman, E.I. (1974).** Root and soil water relations. In *The Plant Root and its Environment* (Ed. by E.W. Carson), pp. 363–440. University Press of Virginia, Charlottesville.

**Nobel, P.S. & Sanderson, J. (1984).** Rectifier-like activities of roots of two desert succulents. *Journal of Experimental Botany*, **35**, 727–737.

**Nye, P.H. & Tinker, P.B. (1977).** *Solute Movement in the Soil–Root System*. University of California Press, Berkeley, 342 pp.

**Passioura, J.B. (1988).** Water transport in and to roots. *Annual Review of Plant Physiology and Molecular Biology*, **39**, 245–265.

**Richards, J.H. (1986).** Root form and depth distribution in several biomes. In *Mineral Exploration: Biological Systems and Organic Matter* (Ed. by D. Carlisle, W.L. Berry, I.R. Kaplan & J.R. Watterson), pp. 83–97. Prentice-Hall, Englewood Cliffs, NJ.

**Richards, J.H. & Caldwell, M.M. (1987).** Hydraulic lift: substantial nocturnal water transport between soil layers by *Artemisia tridentata* roots. *Oecologia*, **73**, 486–489.

**Rumbaugh, M.D., Johnson D.A. & Van Epps, G.A. (1982).** Forage yield and quality in a Great Basin shrub, grass, and legume pasture experiment. *Journal of Range Management*, **35**, 604–609.

**Slatyer, R.O. (1967).** *Plant–Water Relationships*. Academic Press, New York, 347 pp.

**Van Bavel, C.H.M. & Baker, J.M. (1985).** Water transfer by plant roots from wet to dry soil. *Naturwissenschaften*, **70**, 606–607.

**West, N.E. (1989).** Spatial pattern — functional interactions in shrub-dominated plant communities. In *The Biology and Utilization of Shrubs* (Ed. by C.M. McKell), pp. 283–305. Academic Press, New York.

# The role of root-inhabiting fungi in mixed-species plant communities

E.I. NEWMAN
*Department of Botany, University of Bristol, Bristol BS8 1UG, UK*

## SUMMARY

1 The paper discusses the influence of root pathogens and mycorrhizas in plant communities, paying special attention to their relationship to plant–plant interactions and the species composition of the vegetation.
2 *Phytophthora cinnamomi* in Australian forests is taken as an example of a root pathogen that can have substantial effects on the species composition of native vegetation. In recent decades it has caused conspicuous death of *Eucalyptus* and other species in forests in western and south-eastern Australia. In Victoria marked changes in the species composition of open *Eucalyptus* woodland caused by *P. cinnamomi* epidemic have been documented over 20 years, and there are signs that an equilibrium between fungus and vegetation may be developing in some areas.
3 A pot experiment showed VA-mycorrhiza having little effect on the growth of *Lolium perenne* and *Plantago lanceolata* when they were separate, but greatly increasing the ability of *P. lanceolata* to continue growth when in the presence of *L. perenne*.
4 When leaves of dicotyledonous tree species are compared, ectomycorrhizal species usually have a higher N : P ratio than VA-mycorrhizal species. This could be because of greater ability by ectomycorrhizal species to obtain nitrogen from proteins in litter. The leaf litter of ectomycorrhizal trees usually decomposes more slowly than that of VA-mycorrhizal trees. These differences may have important implications when species of the two types grow together.
5 Although the paper confines its discussion largely to fungi of low host-specificity, the examples covered are sufficient to illustrate major effects that root fungi can have in plant communities.

## INTRODUCTION

Fungi that live within plant roots range from strongly beneficial mycorrhizal species, through some that have little or no effect on the host plant, to others that are severe pathogens. There has been much study of such fungi and their interrelations with the host plant. Most of this research has involved plants growing singly, for example in pots, or in single-species stands, for example of crop species or forestry plantations. However, much of the world's land surface is covered by vegetation comprising several or many plant species. This raises questions about fungus–plant interactions that have been much less studied, in particular the two related questions: (i) how does the structure and species composition of the vegetation affect the spread of the fungi and their abundance in the plants? (ii) how do the fungi influence the balance between plant species and hence the composition and structure of the vegetation? This paper addresses these two questions. To illustrate the potential implications of root pathogens in natural vegetation one example is taken, *Phytophthora cinnamomi* in Australian forests. The section on mycorrhiza considers how they can influence relations between plant species, firstly where the species have the same type of mycorrhiza and then where the type of mycorrhiza differs between coexisting plant species.

## THE ROOT PATHOGEN *PHYTOPHTHORA CINNAMOMI* IN AUSTRALIAN FORESTS

*Phytophthora cinnamomi* is a fungus which infects the roots of many plant species, most of them woody species. It causes root rot, accompanied by a variety of shoot symptoms (Zentmyer 1980). In trees there may be die-back of individual branches which gradually extends, so death of the tree is slow. *P. cinnamomi* has been found in every continent except Antarctica (Zentmyer 1985), and in native vegetation as well as orchards and plantations. In south-eastern and south-western Australia epidemics of *P. cinnamomi* have caused conspicuous patches of dying plants in native woodland and sclerophyll forest dominated by some *Eucalyptus* species. This provides an excellent opportunity to study a root pathogen having a major effect on native vegetation. The biology of *P. cinnamomi* in Australian forests has been reviewed by Weste & Marks (1987). I here confine my attention to the relationship between the fungus and the species composition of the vegetation. The two primary, interacting questions are: (i) how does the fungus affect the composition of the vegetation? (ii) how does the composition of the vegetation affect the spread and severity of the disease?

### *Western Australia*

*Eucalyptus marginata* (jarrah) dominates the largest area of forest in Western Australia. Besides its importance as timber, this open sclerophyll forest covers the catchment area that provides much of the water supply for the Perth metropolitan area. Death of patches of the forest ('jarrah dieback') were first reported in the 1920s, and various possible causes were suggested. Following research in the 1960s (Podger 1972) it became widely accepted that *Phytophthora cinnamomi* was the major cause. At that time the attitude in the Forests Department of Western Australia was, according to Abbott & Loneragan (1986), that 'the forest was thought to be doomed'. It had been noted that cutting areas of jarrah forest led to increased salinity in the run-off water, and it was suggested that jarrah dieback might in this way imperil Perth's water supply (Shea *et al.* 1975). This was seen as a challenge in applied ecology, to devise a system of forest management that would control the spread of the disease.

Podger (1972) recorded in dieback areas the apparent susceptibility of various plant species to the fungus. *Eucalyptus calophylla*, which is codominant with *E. marginata* in some parts of the forest, was less affected, though not totally immune, a difference further investigated by Malajczuk, McComb & Parker (1977). Many of the shrub species were susceptible, some more susceptible than *E. marginata*, but some appeared unaffected, as did most members of the Restionaceae and Cyperaceae. Thus *P. cinnamomi* might be expected to cause a marked change in the species composition of the understorey, and perhaps to increase the abundance of *E. calophylla* relative to *E. marginata* (jarrah). Unfortunately, only limited records have been made of the vegetation, and of the growth and mortality of jarrah, in infected and uninfected areas to allow comparison. Podger (1972) found that complete death of jarrah trees at infected sites was slow: he reported 2–5% dying per year at three infected sites. Surviving jarrah trees in dieback areas did not have slower girth increment than similar-sized trees in uninfected areas in a 5 year study by Davison & Tay (1988). Davison (1988) has suggested that *P. cinnamomi* may favour jarrah in some areas by killing susceptible shrubs which were competing for water with the jarrah trees.

Since the 1950s the Forests Department has burnt each area of jarrah forest in spring about every 5–7 years. It was suggested (Shea, McCormick & Portlock 1979) that a change in burning regime could, by altering the composition of the understorey, decrease the severity of attack by *P. cinnamomi*: the preponderance of proteaceous species, many of which are very susceptible to *P. cinnamomi*, could be replaced by a preponderance of leguminous shrubs, some of which are less susceptible. This interesting suggestion, using manipulation of the species composition of vegetation to control a pathogen, was not adopted, and evidence that it would work is still very limited.

### *South-eastern Australia*

Weste and co-workers have made detailed studies of *Phytophthora cinnamomi* epidemics in open *Eucalyptus* forests in Victoria. The most detailed published evidence is on areas in the Brisbane Ranges, about 60 km south-west of Melbourne (Weste & Taylor 1971; Weste, Cooke & Taylor 1973; Dawson, Weste & Ashton 1985; Dawson & Weste 1985; Weste 1986). The dominant trees are *Eucalyptus obliqua, E. dives* and *E. macrorhyncha*, over a shrubby understorey. Although the disease was almost certainly present in the area before 1962, in 1970 it was still confined to patches of forest, mostly near roads, totalling about 1% of the area surveyed. By 1980–81 the infection had spread substantially.

Permanent quadrats were used to record changes in plant species in areas infected by *P. cinnamomi* and in uninfected areas. In 1975 Weste (1986) recorded vegetation in a 21 × 30 m plot at each of three sites: a control site that was free of

TABLE 1. Data from a permanent plot at each of three sites in *Eucalyptus* woodland in the Brisbane Ranges, Victoria (Weste 1986)

| *Phytophthora* infection | Absent | | Recent | | Old | |
|---|---|---|---|---|---|---|
| | 1975 | 1985 | 1975 | 1985 | 1975 | 1985 |
| *Phytophthora cinnamomi* isolated from soil | − | − | + | + | + | − |
| Bare ground (%) | 20 | 20 | 10 | 40 | 40 | 60 |
| Number of plant species† | 30 | 30 | 18 | 24 | 24 | 28 |
| | Number of individuals per 360 m² | | | | | |
| *Eucalyptus*, all species | 39 | 49 | 43 | 23 | 28 | 16 |
| *Isopogon ceratophyllus* (P) | 43 | N.S. 52 | 327 | *** 0 | 0 | 0 |
| *Xanthorrhoea australis* (L) | 153 | * 174 | 162 | *** 9 | 0 | 0 |
| *Banksia marginata* (P) | 4 | N.S. 1 | 70 | N.S. 64 | 38 | N.S. 75 |
| *Hakea sericea* (P) | 35 | N.S. 41 | 4 | *** 20 | 0 | 0 |
| *Lomandra filiformis* (L) | 0 | N.S. 13 | 0 | *** 53 | 73 | *** 229 |

† In 630 m².
Plant families: (P) = Proteaceae, (L) = Liliaceae or closely related. Significance of difference 1975–85: *** $P < 0{\cdot}001$, * $P < 0{\cdot}05$, N.S. not significant.

the pathogen, a site on a boundary where the pathogen was invading, and an old diseased site where the pathogen had been present for at least 6 years. The same plots were recorded again 10 years later, and Table 1 summarizes some of the results. Although there were some differences between the plots not attributable to the pathogen, the four right-hand columns may be considered to represent approximately a time sequence. Bare ground increased substantially in the infected areas. However, the number of plant species was recovering after the initial fall, mainly because new, pathogen-resistant species were invading. Results for some of the individual species have been chosen to illustrate the differences in response: some species decreased markedly, some increased markedly and some changed little. Some shrub species, such as *Isopogon ceratophyllus* and *Xanthorrhoea australis* are extremely susceptible to *Phytophthora* and were virtually eliminated from the infected plots. A substantial proportion of the *Eucalyptus* trees died, but in 1985 there were still some left even in the old infected site.

There were several signs that by 1985 the severity of pathogen effects were decreasing in the old infected site. Firstly, Weste was not able to isolate *P. cinnamomi* from the old site from 1981 onwards. Secondly, *Eucalyptus* species on the recently infected site showed extensive die-back symptoms in their crowns, but in the old site by 1985 all remaining trees appeared healthy and had vigorous intact crowns free from die-back. Thirdly, five species that had been eliminated from the old disease site later reappeared as a few seedlings. Whether they would survive to become permanently established these results do not show, but further information on this comes from the parallel study in the same area by Dawson, Weste & Ashton

(1985). They recorded the distribution of four understorey species within a 48 × 48 m permanent plot in 1962, 1974 and 1981. The pathogen evidently invaded the plot from the north-west corner and moved slowly across it, as indicated by progressive dying out of susceptible plant species and their replacement by non-susceptible species. However, one of the susceptible species, *Xanthorrhoea australis*, later reinvaded and individuals survived for several years at least. And one of the non-susceptible species, *Hakea sericea*, which invaded from the north-west, had by the end declined in abundance there below its maximum. Isolation tests found *P. cinnamomi* less abundant in the soil in 1982 than in 1974.

These results record the changes in the vegetation as the fungal epidemic passed through. They show the eucalypt overstorey decreasing but not disappearing, resulting in a more open canopy. The understorey composition changed markedly, as susceptible species disappeared or greatly decreased, and non-susceptible species increased or invaded. The results also suggest a new equilibrium vegetation that may establish. The pathogen evidently decreased markedly after the epidemic; this may well be partly because of the decreased abundance of susceptible hosts, but other causes, such as increase in soil micro-organisms and micro-fauna that can attack the fungus, are also possible. This reduction in the fungus may in turn be what allows survival or reinvasion of susceptible plant species. We can speculate that a new plant community will stabilize, very different in composition from the original but containing some susceptible species in much reduced quantities. The fungus may well remain, but at a reduced level. This is suggested by the fact that it has been found in native and apparently undisturbed forests elsewhere, e.g. in Queensland (Brown 1976) and Papua–New Guinea (Arentz & Simpson 1986). Such an equilibrium could later be upset by special conditions, for example unusual weather. The epidemic in the Brisbane Ranges became more severe after an unusually wet year in 1971 (Weste, personal communication). Higher mortality of jarrah in Western Australia appears to be associated with years of particular weather conditions (Davison 1988).

The final result of the invasion of Australian sclerophyll forests by *Phytophthora cinnamomi* may thus be markedly altered forests, but in which the disease symptoms and the fungus itself are rare and inconspicuous. One may then ask, if we had not observed the epidemic would we have realized the major effect of the fungus? And, are there many other sorts of native vegetation whose composition is greatly affected by a pathogen without our realizing it, because we were not present to observe the original invasion?

## MYCORRHIZAS

### *Introduction*

Although there has been considerable interest in the role of mycorrhizas in non-crop ecosystems, this subject is very difficult to investigate experimentally. Because mycorrhizal inoculum is present in virtually all soils, most experiments have involved sterilizing soil and then growing plants in this bare soil devoid of older plants; a mixed-age, mixed-species community is difficult to simulate in mycorrhiza experiments. One reason why mycorrhizas may influence interactions between plants is that plants of different species can be linked by mycorrhizal hyphae. This has so far been demonstrated for only a few species pairs (Newman 1988), but may be widespread. It could have profound effects on competitive relations between plants because they are drawing nutrients from a common mycelial network rather than separately extracting it from the soil. Mineral nutrients and organic compounds could pass from one plant to another, thus perhaps reducing competitive dominance; and nutrients could pass from a dying to a living root without ever entering the soil solution. I have reviewed available evidence on the functioning of these mycorrhizal links (Newman 1988), the principal conclusion being that the evidence is at present inadequate on virtually every aspect. The discussion is not repeated here, but the potential effects of mycorrhizal links between plants have implications in the topics that follow.

In the following section vesicular–arbuscular mycorrhiza is abbreviated to VAM and ectomycorrhiza to ECM.

## *Competition and mycorrhizas*

There have been few experiments on the effect of mycorrhizas when two or more species are growing together. Grime *et al.* (1987) grew the grass *Festuca ovina* in trays of nutrient-poor sand, either VA-mycorrhizal or non-mycorrhizal. Seeds of twenty grassland species were sown into each tray. In the mycorrhizal trays *F. ovina* grew less well but seedlings of many of the other species grew better. These results suggest that mycorrhizas promote coexistence of species and hence diversity in this type of grassland. One question raised is, would similar effects of mycorrhiza on growth have been found if the species had each been grown separately?

Ramos (1987) conducted an experiment involving the grass *Lolium perenne* and the dicotyledonous herb *Plantago lanceolata* which often grows with it. Unlike the experiment of Grime *et al.* (1987) the plants were all sown at the same time, either two *Lolium* per pot of soil, two *Plantago* or one of each. In competition experiments of this type, it is possible for the final plant weight and nutrient content to be substantially influenced by early events (e.g. spread of the canopy) before interaction between the plants has become intense. To ensure that mycorrhizas operated only while interactions between the plants were occurring, Ramos grew all plants initially in heat-sterilized soil. After 3 months the sizes of the shoots made it clear that competition between the species was occurring, and half of the pots were then inoculated with VAM. The percentage of root length infected reached an approximate plateau by 5 months, and Fig. 1 shows the relative growth rates (calculated from shoot weight only) for the next 2 months. If the plants were non-mycorrhizal, the two species had similar growth rates when separate; yet when they were in the same pot *Lolium* grew well but *Plantago* scarcely at all. In the mycorrhizal pots again the two species grew about equally fast if separate, but here they were also

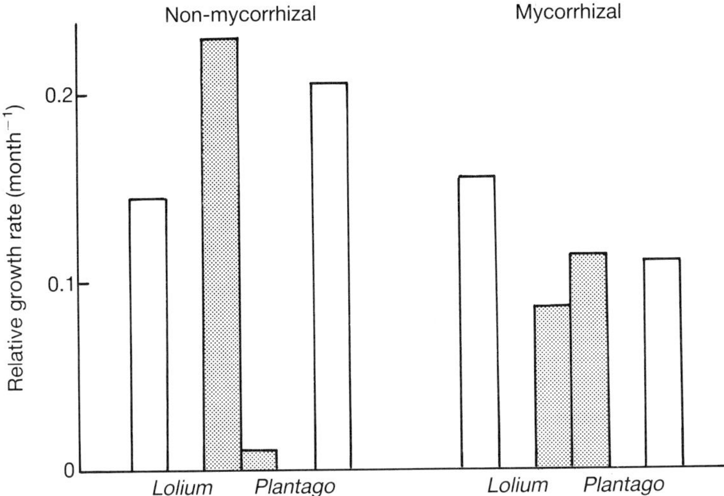


FIG. 1. Relative growth rate of shoots of *Lolium perenne* and *Plantago lanceolata* growing in pots of nutrient-deficient soil. Open bars, each species in separate pot; shaded bars, both species in same pot. Data of Ramos (1987).

well balanced if growing together. These results are supported by earlier experiments with the same species on the same soil (Newman & Ritz 1986).

It would be interesting to know the mechanisms by which the plants interact to produce the results in Fig. 1. Heap & Newman (1980) showed that these two species when grown on this soil form mycorrhizal links between them, which opens the possibility for special interactions of the types mentioned above. One clear conclusion from Ramos' results is that experiments on the response of a species to mycorrhizal infection which involve only the plant growing on its own are not necessarily adequate to predict the influence of mycorrhizas in many species vegetation in the field.

## *The role of mycorrhizas in vegetation: evidence from the field*

Plants in pots are in many respects inadequate simulators of most of the vegetation we see outdoors. However, there are grave technical barriers to conducting experiments in undisturbed vegetation to elucidate the role of mycorrhizas. What is required is a way of obtaining areas of undisturbed vegetation identical in every way except that some have normal mycorrhizal infection while others have none. This could be achieved if a chemical could be applied to the soil which killed the mycorrhizal fungi but had no effect on any other organism; but existing fungicides seem not to approach the desired specificity. Ingham *et al.* (1986) and Fitter (1986) have applied fungicides to natural vegetation with the aim of reducing mycorrhizal infection. The results of both studies were difficult to interpret. In the experiment of Ingham *et al.* some non-target organisms were altered at least as much as VA-mycorrhizal fungi; it is not known whether this occurred in Fitter's experiments.

Fitter (1985) reviewed some field experiments on the response of plants to mycorrhizal infection. He concluded that the responses 'tended to be . . . much smaller in magnitude than in comparable pot experiments' and that the results showed 'low effectiveness of VA fungi as P uptake systems in field conditions'. In my view these conclusions by Fitter are not justified by the evidence. Of the seven papers he cited, one (Sparling & Tinker 1978) describes only pot experiments. The remainder describe field experiments, in which plants were sown into unsterilized soil either with or without added mycorrhizal inoculum. In all experiments the uninoculated plants became infected with indigenous mycorrhizal fungi; sometimes the additional inoculation increased the total infection, sometimes it did not. Most pot experiments on effects of mycorrhiza involve a treatment in which all mycorrhizal inoculum has been removed by treating the soil, and the plants fail to become mycorrhizal. In my view, the differences between results from pot and field experiments arise because of the difficulty in producing in the field a non-mycorrhizal control, not through any of the more fundamental reasons suggested by Fitter.

As an alternative to experiments, one can obtain indirect evidence on the role of mycorrhizas in the field by observation and measurement. For example, the occurrence of non-mycorrhizal plants in certain habitats can indicate conditions under which mycorrhizas are not essential. Surprisingly, non-mycorrhizal plants can occur in sites that are likely to be very deficient in available nutrients, for example in the 'yellow' phase of sand dunes (Ernst, van Duin & Oolbekking 1984). But such sites are associated with disturbance and hence reduced intensity of competition, suggesting that mycorrhizas may be involved in competitive ability. This is supported by some other field observations. Gay *et al.* (1982) noted that some small plants of British chalk grassland are non-mycotrophic (i.e. never form mycorrhiza), but these species are not 'turf-compatible', i.e. are found only in locally disturbed sites.

Generalizations can be made about the geographical distribution of VAM and ECM species in relation to climate and associated soils (Read 1984; Högberg 1986a), which can suggest differences in the functioning of these two types of mycorrhizas. Here I want instead to consider situations where VAM and ECM plants grow together. VAM and ECM trees grow together in many temperate forests, and in some savannas and tropical forests (Högberg 1986b; Alexander & Högberg 1986). In other forests where the tree layer is entirely of ECM species the undergrowth is often predominantly VAM. Where plants of these two mycorrhizal types are growing together there are two sets of plants each of which can potentially be linked underground; but between the two sets links cannot occur. It is relevant to compare ECM and VAM plants, to find out whether they differ consistently in characters related to nutrient acquisition and cycling, and how this might influence relations between them.

## *Nutrient concentrations in ECM and VAM plants*

Gerloff, Moore & Curtis (1964) collected shoot material of over 200 species growing in various native vegetation types in Wisconsin, USA, and analysed them for concentrations of fifteen elements. In Table 2 the species have been grouped according to the type of mycorrhiza they form, where this is known. The results for micro-nutrients were variable, but Mn showed a highly significant tendency for ericoid species to have the highest concentration and VAM species the lowest. In contrast, all macro-nutrients except nitrogen had mean concentrations in the order VAM > ECM > ericoid; however, the only significant differences were for phosphorus and potassium. For nitrogen, VAM and ECM species were on average very similar (Table 2). It could be suggested that the mean concentration of P and K was low in ECM species because some of them were conifers, and was high in VAM species because many of them were herbs. If dicotyledonous woody plants only are compared, K concentrations of ECM and VAM species become much more similar, but the difference in P concentration between ECM and VAM species is maintained (Table 2). So among these Wisconsin species there is a tendency for VAM species to have higher shoot concentration of phosphorus but not of nitrogen.

Another possible explanation for these results is that they reflect different distributions of the species within Wisconsin, ECM species tending to grow where phosphorus is more limiting. To test this we need to compare ECM and VAM dicotyledonous woody plants growing at the same site. Within the data of Gerloff, Moore & Curtis (1964) there are three sites from which they collected the required material (but of few species from each site); at each site the VAM species maintained the higher P concentration. Among other comparisons at a single site

TABLE 2. Mean nutrient concentration in shoot material of native plants of Wisconsin analysed by Gerloff, Moore & Curtis (1964), grouped according to their normal mycorrhizal type. Species that have been recorded with more than one type of mycorrhiza have been omitted. Stars indicate a statistically significant difference between the figures immediately above and below (*** $P<0{\cdot}001$, ** $P<0{\cdot}01$). The bottom line indicates the overall statistical significance by analysis of variance

| Mycorrhiza type | Number of species | Concentration (mg g$^{-1}$) | | | |
|---|---|---|---|---|---|
| | | N | P | K | Mn |
| Ericoid | 6 | 10·1 | 1·1 | 5·1 | 1·38 |
| | | | | | *** |
| ECM | 15 (9)† | 14·2 (16·1) | 1·5 (1·6) | 6·7 (8·3) | 0·58 (0·65) |
| | | | *** | *** | ** |
| VAM | 50 (13) | 14·6 (16·0) | 2·6 (2·6) | 21·9 (11·2) | 0·23 (0·40) |
| Statistical significance | | N.S. | $P<0{\cdot}001$ | $P<0{\cdot}001$ | $P<0{\cdot}001$ |

† Figures in brackets are for dicotyledonous woody species only.

TABLE 3. Nutrient concentrations in leaves of trees at five savanna sites in Tanzania, grouped into ectomycorrhizal (ECM) and VA-mycorrhizal (VAM) species. Calculated from data of Högberg (1986b). + indicates that at each site the element had a consistently higher concentration in species of one mycorrhizal type than the other type

| | Number of species | Concentration ($mg\ g^{-1}$) N | P | K | Ca | Mg |
|---|---|---|---|---|---|---|
| ECM | 4 | 19·9 | 1·65 | 6·1 | 19·1 | 2·9 |
| | | + | | + | + | |
| VAM | 4 | 15·5 | 1·56 | 10·1 | 12·3 | 3·1 |

within USA one also showed the higher P concentration in VAM trees but two others did not (Chapin & Kedrowski 1983; Whittaker *et al.* 1979; Boerner 1984). Högberg (1986b) presented data on mycorrhizal status and foliar nutrient concentrations in deciduous trees at five sites in Tanzanian savanna. Table 3 shows the mean values, excluding the species that are potentially N fixers. N and Ca concentrations were higher in ECM species and K in VAM species, and this was consistent among the species at each site. P and Mg showed little difference between ECM and VAM plants.

Thus there does not appear to be any consistent difference between ECM and VAM species in their foliar concentration of either nitrogen or phosphorus. However, there is a tendency for the ratio of N:P to be higher in ECM species. Table 4 shows that in two studies the difference between the mean ratios was large; and in three others it was smaller but in the same direction. Research by Abuzinadah & Read (1986a,b) and Abuzinadah, Finlay & Read (1986) suggests a possible explanation of these differences in N:P ratio. They found that several ECM fungi confer on several tree species the ability to use protein as a nitrogen source. This would give ECM plants the potential to obtain nitrogen from the protein of litter before free-living decomposer organisms have broken it down to simple soluble compounds. By giving ECM plants an additional source of N not available to VAM plants, this could raise their N:P ratio. Whether the ratio was altered by higher N concentration or lower P concentration would depend on which of these elements was the more limiting to growth at that site.

TABLE 4. Mean ratio of nitrogen concentration: phosphorus concentration for leaves of dicotyledonous trees growing at the same sites. Species with nitrogen-fixing symbioses are excluded

| | N:P ratio | | Number of species | | |
|---|---|---|---|---|---|
| Site | ECM | VAM | ECM | VAM | Reference |
| Wisconsin | 14·0 | 7·3 | 4 | 4 | Gerloff *et al.* (1964) |
| Tennessee | 21·0 | 14·8 | 3 | 2 | Chapin & Kedrowski (1983) |
| New Hampshire | 13·2 | 12·2 | 2 | 1 | Whittaker *et al.* (1979) |
| Ohio | 13·6 | 11·3 | 3 | 1 | Boerner (1984) |
| Tanzania | 12·8 | 11·7 | 4 | 4 | Högberg (1986b) |

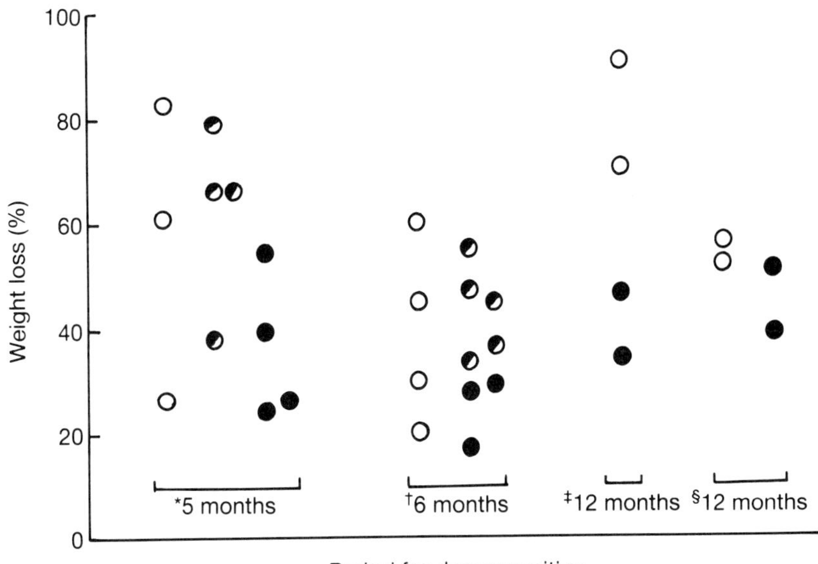


FIG. 2. Percentage loss of dry weight during decomposition of senescent leaves of tree species from northern temperate deciduous forests. ○ VAM species, ● ECM species, ◐ can be either VAM or ECM. Source of data: *Bocock (1964); †Broadfoot & Pierre (1939); ‡Witkamp (1966); §MacLean & Wein (1978). The time allowed for decomposition is shown for each data set.

### *Rate of decomposition of leaves*

Figure 2 summarizes results from four studies of decomposition rates of senescent leaves from deciduous dicotyledonous trees of northern temperate forests. While there is substantial variation between species, there is a clear tendency for the fastest weight loss at each site to be by species that are always or sometimes VAM. A median test showed that the tendency of VAM species to decompose faster than ECM species was significant at $P<0{\cdot}01$. If ECM conifers were included the difference would be even greater. What properties of the leaves caused these differences in decomposition rate is uncertain. In the studies by Broadfoot & Pierre (1939) and Bocock (1964) the correlation between initial nitrogen concentration and decomposition rate was weak. In freshly fallen leaves of North American dicotyledonous tree species there was no consistent difference between the three ECM species and twelve VAM species in concentration of phenolics (Neuhauser & Hartenstein 1978).

The picture emerging is that ECM species tend to have more slowly decomposing litter than VAM species. This difference may be enhanced by a direct effect of ECM fungi on decomposition: Gadgil & Gadgil (1975) found that leaves of *Pinus radiata* lost weight more slowly when closely associated with mycorrhizas of *P. radiata* than when associated with the same species non-mycorrhizal or when separated from the plant. Some ECM have increased ability to acquire nitrogen from insoluble organic sources (see above); and some ECM fungi produce hyphal strands which allow spread of the fungus through the litter. In contrast, although VAM fungi have been shown to translocate phosphorus several centimetres in artificial conditions, it is likely that in soil most of their uptake is within a few millimetres of the root (Owusu-Bennoah & Wild 1979). Thus ECM plants seem better equipped to obtain nutrients from litter, and to be associated with litter that decomposes more slowly. We should not assume, however, that all ECM fungi are the same in these respects. Abuzinadah & Read (1986a,b) found marked differences between ECM fungal species in proteinase activity; and Reddell & Malajczuk (1984) found that two mycorrhiza types of *Eucalyptus marginata* were predominant in litter while another type predominated in mineral soil.

On the basis of the limited evidence, I now speculate on the relations between the plants in communities that include both ECM and VAM species. Because host-specificity of VAM fungi is low, all VAM plants may well be linked into a common mycorrhizal–mycelial system (Newman 1988). Whether this also happens for the ECM plants will depend on the specificity of the ECM fungi present. If there are essentially two mycorrhizal systems in a community, the fungal mycelium of each will be taking up mineral nutrients which are then shared among the attached plants. What determines this allocation is unknown, but we may expect the competitive relations among linked species to be different from relations between species having different mycorrhizal types. This will be enhanced if shaded plants gain organic compounds from other plants via hyphal links (Read, Francis & Finlay 1985). When roots die, nutrients in them may be preferentially transferred to other plants sharing the same mycorrhiza (Newman 1988). It seems that allocation of nutrients from dying above-ground parts may also differ between mycorrhizal types. The litter from ECM plants, decomposing more slowly, may be more effectively tapped for nutrients by some ECM fungi, because of their hyphal strands and proteinase activity. The more rapidly decomposing litter of VAM species will provide more nutrients in soluble inorganic form, which may more rapidly leach to the proximity of roots where VAM fungi can take them up. One may thus speculate that competition for nutrients between these two groups of plants, ECM and VAM, is sometimes relatively weak because each group has a substantially separate nutrient cycling process.

## CONCLUDING REMARKS

The role of root-inhabiting fungi in plant communities is clearly a large topic, and this paper has dealt with only some aspects of it. The examples given have been sufficient to illustrate how in plant communities a pathogenic or mycorrhizal fungus can influence some plant species differently from others, and hence can alter the balance between species. The results suggest that root fungi may be having major effects on the structure and species composition of vegetation. A common feature of the examples considered was low specificity. *Phytophthora cinnamomi* can infect a large number of plant species but not all. Mycorrhizal effects were discussed mostly on the basis of two groups of plants living together, VAM-infected and ECM-infected. In fact some ECM fungi have limited host range, as do some root pathogens, and the implications of this for natural vegetation deserve further study. Where there is low specificity, and hence the plants are viewed as belonging to two or a few groups (*Phytophthora* susceptible or resistant, VAM- or ECM-infected), this implies low niche separation or few barriers to interaction, e.g. in transfer of infection or formation of mycorrhizal links. This contrasts, on the one hand with a very genetically uniform crop monoculture which would all be potentially available for attack by one fungal pathogen species, and the other extreme possibility of every plant species in a species-rich forest having a separate specific foliar pathogen. The relations between plants and fungi may have influences on the structure and functioning of ecosystems more profound and far-reaching than we can at present envisage.

## ACKNOWLEDGEMENTS

My visits to Australian forests were made possible by a University of Western Australia 75th Anniversary Fellowship and financial assistance from the Royal Society. The research by Dr M.I.R.F. Ramos was supported by the Gulbenkian Foundation. I thank Dr E.M. Davison and Dr G. Weste for useful information about *Phytophthora cinnamomi* and for comments on an earlier version of this paper and Dr F.S. Chapin for drawing my attention to the data of Gerloff, Moore & Curtis (1964).

## REFERENCES

**Abbott, I. & Loneragan, O. (1986).** *Ecology of Jarrah (Eucalyptus marginata) in the Northern Jarrah Forest of Western Australia.* Department of Conservation and Land Management, Perth.

**Abuzinadah, R.A. & Read, D.J. (1986a).** The role of proteins in the nitrogen nutrition of ectomycorrhizal plants. I. Utilization of peptides and proteins by ectomycorrhizal fungi. *New Phytologist*, **103**, 481–493.

**Abuzinadah, R.A. & Read, D.J. (1986b).** The role of proteins in the nitrogen nutrition of ectomycorrhizal plants. III. Protein utilization by *Betula*, *Picea* and *Pinus* in mycorrhizal association with *Hebeloma crustuliniforme*. *New Phytologist*, **103**, 507–514.

**Abuzinadah, R.A., Finlay, R.D. & Read, D.J. (1986).** The role of proteins in the nitrogen nutrition of ectomycorrhizal plants. II. Utilization of protein by mycorrhizal plants of *Pinus contorta*. *New Phytologist*, **103**, 495–506.

**Alexander, I.J. & Högberg, P. (1986).** Ectomycorrhizas of tropical angiospermous trees. *New Phytologist*, **102**, 541–549.

**Arentz, F. & Simpson, J.A. (1986).** The distribution of *Phytophthora cinnamomi* in Papua New Guinea and notes on its origin. *Transactions of the British Mycological Society*, **87**, 289–295.

**Bocock, K.L. (1964).** Changes in the amounts of dry matter, nitrogen, carbon and energy in decomposing woodland leaf litter in relation to the activities of the soil fauna. *Journal of Ecology*, **52**, 273–284.

**Boerner, R.E.J. (1984).** Foliar nutrient dynamics and nutrient use efficiency of four deciduous tree species in relation to site fertility. *Journal of Applied Ecology*, **21**, 1029–1040.

**Broadfoot, W.M. & Pierre, W.H. (1939).** Forest soil studies: I. Relation of rate of decomposition of tree leaves to their acid–base balance and other chemical properties. *Soil Science*, **48**, 329–348.

**Brown, B.N. (1976).** *Phytophthora cinnamomi* associated with patch death in tropical rain forests in Queensland. *Australian Plant Pathology Society Newsletter*, **5**, 1–4.

**Chapin, F.S. & Kedrowski, R.A. (1983).** Seasonal changes in nitrogen and phosphorus fractions and autumn retranslocation in evergreen and deciduous taiga trees. *Ecology*, **64**, 376–391.

**Davison, E.M. (1988).** The role of waterlogging and *Phytophthora cinnamomi* in the decline and death of *Eucalyptus marginata* in Western Australia. *GeoJournal*, **17**, 239–244.

**Davison, E.M. & Tay, F.C.S. (1988).** Annual increment of *Eucalyptus marginata* trees on sites infested with *Phytophthora cinnamomi*. *Australian Journal of Botany*, **36**, 101–106.

**Dawson, P. & Weste, G. (1985).** Changes in the distribution of *Phytophthora cinnamomi* in the Brisbane Ranges National Park between 1970 and 1980–81. *Australian Journal of Botany*, **33**, 309–315.

**Dawson, P., Weste, G. & Ashton, D. (1985).** Regeneration of vegetation in the Brisbane Ranges after fire and infestation by *Phytophthora cinnamomi*. *Australian Journal of Botany*, **33**, 15–26.

**Ernst, W.H.O., van Duin, W.E. & Oolbekking, G.T. (1984).** Vesicular–arbuscular mycorrhiza in dune vegetation. *Acta Botanica Neerlandica*, **33**, 151–160.

**Fitter, A.H. (1985).** Functioning of vesicular–arbuscular mycorrhizas under field conditions. *New Phytologist*, **99**, 257–265.

**Fitter, A.H. (1986).** Effect of benomyl on leaf phosphorus concentration in alpine grasslands: a test of mycorrhizal benefit. *New Phytologist*, **103**, 767–776.

**Gadgil, R.L. & Gadgil, P.D. (1975).** Suppression of litter decomposition by mycorrhizal roots of *Pinus radiata*. *New Zealand Journal of Forest Science*, **5**, 33–41.

**Gay, P.E., Grubb, P.J. & Hudson, H.J. (1982).** Seasonal changes in the concentrations of nitrogen, phosphorus and potassium, and in the density of mycorrhiza, in biennial and matrix-forming perennial species of closed chalkland turf. *Journal of Ecology*, **70**, 571–593.

**Gerloff, G.C., Moore, D.G. & Curtis, J.T. (1964).** Mineral content of native plants of Wisconsin. *Wisconsin Experimental Station Research Report*, **14**, 3–27.

**Grime, J.P., Mackey, J.M.L., Hillier, S.H. & Read, D.J. (1987).** Floristic diversity in a model system using experimental microcosms. *Nature*, **328**, 420–422.

**Heap, A.J. & Newman, E.I. (1980).** Links between roots by hyphae of vesicular–arbuscular mycorrhizas. *New Phytologist*, **85**, 169–171.

**Högberg, P. (1986a).** Soil nutrient availability, root symbioses and tree species composition in tropical Africa: a review. *Journal of Tropical Ecology*, **2**, 359–372.

**Högberg, P. (1986b).** Nitrogen-fixation and nutrient relations in savanna woodland trees (Tanzania). *Journal of Applied Ecology*, **23**, 675–688.

**Ingham, E.R., Trofymow, J.A., Ames, R.N., Hunt, H.W., Morley, C.R., Moore, J.C. & Coleman, D.C. (1986).** Trophic interactions and nitrogen cycling in a semi-arid grassland soil. II. System responses to removal of different groups of soil microbes or fauna. *Journal of Applied Ecology*, **23**, 615–630.

**Malajczuk, N., McComb, A.J. & Parker, C.A. (1977).** Infection by *Phytophthora cinnamomi* Rands of roots of *Eucalyptus calophylla* R. Br. and *Eucalyptus marginata* Donn. ex Sm. *Australian Journal of Botany*, **25**, 483–500.

**MacLean, D.A. & Wein, R.W. (1978).** Weight loss and nutrient changes in decomposing litter and forest floor material in New Brunswick forest stands. *Canadian Journal of Botany*, **56**, 2730–2749.

**Neuhauser, E.F. & Hartenstein, R. (1978).** Phenolic content and palatability of leaves and wood to soil isopods and diplopods. *Pedobiologia*, **18**, 99–109.

**Newman, E.I. (1988).** Mycorrhizal links between plants: their functioning and ecological significance. *Advances in Ecological Research*, **18**, 243–270.

**Newman, E.I. & Ritz, K. (1986).** Evidence on the pathways of phosphorus transfer between vesicular–arbuscular mycorrhizal plants. *New Phytologist*, **104**, 77–87.

**Owusu-Bennoah, E. & Wild, A. (1979).** Autoradiography of the depletion zone of phosphate around onion roots in the presence of vesicular–arbuscular mycorrhiza. *New Phytologist*, **82**, 133–140.

**Podger, F.D. (1972).** *Phytophthora cinnamomi*, a cause of lethal disease in indigenous plant communities in Western Australia. *Phytopathology*, **62**, 972–981.

**Ramos, M.I.R.F. (1987).** *Studies of interactions between grassland plants*. Ph.D. thesis, University of Bristol.

**Read, D.J. (1984).** The structure and function of the vegetative mycelium of mycorrhizal roots. In *The*

*Ecology and Physiology of the Fungal Mycelium* (Ed. by D.H. Jennings & A.D.M. Rayner), pp. 215–240. Cambridge University Press, Cambridge.

**Read, D.J., Francis, R. & Finlay, R.D. (1985).** Mycorrhizal mycelia and nutrient cycling in plant communities. In *Ecological Interactions in Soil* (Ed. by A.H. Fitter), pp. 193–217. Blackwell Scientific Publications, Oxford.

**Reddell, P. & Malajczuk, N. (1984).** Formation of mycorrhizae by jarrah (*Eucalyptus marginata* Donn ex Smith) in litter and soil. *Australian Journal of Botany*, **32**, 511–520.

**Shea, S.R., Hatch, A.B., Havel, J.J. & Ritson, P. (1975).** The effect of changes in forest structure and composition on water quality and yield from the northern jarrah forest. *Proceedings of the Ecological Society of Australia*, **9**, 58–73.

**Shea, S.R., McCormick, J. & Portlock, C.C. (1979).** The effect of fires on regeneration of leguminous species in the northern jarrah (*Eucalyptus marginata* Sm) forest of Western Australia. *Australian Journal of Ecology*, **4**, 195–205.

**Sparling, G.P. & Tinker, P.B. (1978).** Mycorrhizal infection in Pennine grassland. II. Effects of mycorrhizal infection on the growth of some upland grasses on $\gamma$-irradiated soils. *Journal of Applied Ecology*, **15**, 951–958.

**Weste, G. (1986).** Vegetation changes associated with invasion by *Phytophthora cinnamomi* of defined plots in the Brisbane Ranges, Victoria, 1975–1985. *Australian Journal of Botany*, **34**, 633–648.

**Weste, G.M. & Taylor, P. (1971).** The invasion of native forest by *Phytophthora cinnamomi*. I. Brisbane Ranges, Victoria. *Australian Journal of Botany*, **19**, 281–294.

**Weste, G. & Marks, G.C. (1987).** The biology of *Phytophthora cinnamomi* in Australasian forests. *Annual Review of Phytopathology*, **25**, 207–229.

**Weste, G., Cooke, D. & Taylor, P. (1973).** The invasion of native forest by *Phytophthora cinnamomi*. II. Post-infection vegetation patterns, regeneration, decline in inoculum, and attempted control. *Australian Journal of Botany*, **21**, 13–29.

**Whittaker, R.H., Likens, G.E., Bormann, F.H., Eaton, J.S. & Siccama, T.G. (1979).** The Hubbard Brook ecosystem study: forest nutrient cycling and element behavior. *Ecology*, **60**, 203–220.

**Witkamp, M. (1966).** Decomposition of leaf litter in relation to environment, microflora, and microbial respiration. *Ecology*, **47**, 194–201.

**Zentmyer, G.A. (1980).** *Phytophthora cinnamomi* and the diseases it causes. *Monograph 10, American Phytopathological Society*. St Paul, Minnesota.

**Zentmyer, G.A. (1985).** Origin and distribution of *Phytophthora cinnamomi*. In *Ecology and Management of Soilborne Plant Pathogens* (Ed. by C.A. Parker, A.D. Rovira, K.J. Moore & P.T.W. Wong), pp. 71–72. American Phytopathological Society, St Paul, Minnesota.

# Effects of root herbivory on vegetation dynamics

V.K. BROWN AND A.C. GANGE
*Imperial College at Silwood Park, Ascot, Berkshire SL5 7PY, UK*

## SUMMARY

1 Relatively few orders of insects contain representatives which feed on plant roots. Even though many species of root-feeding insects are pests of agricultural importance, they have largely been ignored by ecologists.
2 Root-feeding insects are dependent on the moisture and nutrient content of soils. Supplying plants with increased water or nutrients can mitigate the effects of herbivory.
3 Because of long life cycles, populations of these insects tend to build up slowly with time. As a result, the consequences of root herbivory are most clearly seen in long-lived crops or well-established plant communities.
4 In pot trials, losses of yield due to root feeding may be up to 70%. More realistic measurements come from field experiments where 30–40% crop loss can occur.
5 Manipulation experiments, involving appropriate insecticides, have demonstrated that root-feeding insects have dramatic effects on the establishment and development of natural plant communities. Reducing herbivore numbers results in significant increases in plant-species richness, since they are important agents of seedling mortality. Perennial forbs are particularly affected by root herbivory, with both species richness and cover abundance being increased as a result of its reduction.
6 Feeding by these insects may have important direct consequences for mycorrhizal infection of roots and indirect consequences for foliar-feeding insects. It is suggested that subterranean herbivores play a previously unrecognized, but vital, role in plant community ecology.

## INTRODUCTION

During the last decade, there has been a surge of interest in insect–plant interactions and one aspect, now receiving considerable attention, is herbivory. Thus, authors have examined the effect of insects on the growth and reproduction of single plant species (e.g. Whittaker 1982; Kinsman & Platt 1984) or whole communities (e.g. Brown, Gange & Gibson 1988). Attempts have also been made to relate these effects to plant population changes (Bentley & Whittaker 1979; Gange *et al.* 1989) and to the process of plant community development, or succession (Brown, Jepsen & Gibson 1988). A feature common to these studies is that they deal almost exclusively with foliar-feeding insects. The role played by root-feeding insects has been largely ignored, a fact emphasized in Crawley's (1983) review of herbivory. Of over 1000 references in this book, less than 2% are concerned with root-feeding insects.

Although ecologists have largely ignored subterranean insect feeders, the agricultural literature often documents the effects of root herbivory on plants, since many root-feeding insects are important crop pests. Indeed, root-feeding insects can be so destructive that they have been used as agents in the biological control of weeds. Among the most successful introductions are *Sphenoptera jugoslavica* Obenb. for the control of diffuse and spotted knapweed, *Centaurea diffusa* Lam. and *C. maculosa* s. *lat.* Lam. respectively (Harris & Myers 1981), and *Longitarsus jacobaeae* Wat. for ragwort, *Senecio jacobaea* L. (Harris, Wilkinson & Myers 1981).

In this paper, we first outline the types of insects commonly found feeding on plant roots and review the factors which affect their feeding. We then describe the effects these insects have on plant growth and the mechanisms whereby a plant may compensate. The role of root herbivory in natural plant communities is then illustrated in detail using data from current field experiments at Silwood Park, Berkshire, UK, which have been specifically designed for this purpose.

## REVIEW OF PREVIOUS STUDIES

### *Root-feeding insects*

Root-feeding insects are mainly restricted to six orders: Coleoptera, Diptera, Hemiptera, Isoptera, Lepidoptera and Orthoptera (Table 1). With the exception of Hemiptera, Isoptera and Orthoptera, it is only the immature stages which feed on roots. Generally, very little is known about the behaviour of root-feeding insects in the field, although facts such as the amount of movement and depth of feeding have important implications for their effects on vegetation dynamics. On a worldwide scale, the most economically important species are chewing Coleoptera larvae (Hill 1987). For example, larvae of the weevil, *Sitona discoideus* Gyllenhal, caused up to 43% yield loss of lucerne in New Zealand (Goldson *et al.* 1985), while in the eastern United States, damage caused by *Popillia japonica* Newman represented an average loss of \$4·5 million per state in 1981 (McDavid 1981). A striking example of the effects of pasture scarabs on grass growth and sheep grazing was provided by Roberts & Ridsdill-Smith (1979). At low stocking rates, grass grubs (*Sericesthis geminata* Boisd.) were so abundant (150 $m^{-2}$) that sheep growth was similar to that at high stocking rates. In the UK, the major root-chewing crop pests include the carrot fly, *Psila rosae* (F.), and the cabbage root fly, *Delia radicum* L. (Hill 1987). Losses of up to 32% (Clements 1984) have been estimated as a result of damage, mainly by larvae of Scarabaeidae, Tipulidae and Noctuidae to permanent pasture which covers 5 million ha in England and Wales (Clements & Bentley 1983; Henderson & Clements 1974).

Root-sucking insects are also important agricultural pests and can reach very high densities in soil. Dunn (1959) recorded up to 64 500 individuals $m^{-2}$ of the lettuce root aphid, *Pemphigus bursarius* L., and similar densities for other aphids have been recorded by Dixon (1971). The grape phylloxera, *Daktulosphaira vitifolii*

TABLE 1. Examples of the major insect groups containing species which feed on plant roots for part or all of their life cycle

| Order | Family | Examples | Host |
|---|---|---|---|
| Coleoptera | Chrysomelidae | *Diabrotica virgifera* | Maize |
| | Curculionidae | *Sitona discoideus* | Lucerne (*M. sativa*) |
| | | *S. hispidulus* | Alfalfa (*M. sativa*) |
| | Elateridae | *Agriotes* spp. | Pasture grasses |
| | Scarabaeidae | *Costelytra zealandica* | Pasture grasses |
| | | *Popillia japonica* | Wide range of grasses and forbs |
| Diptera | Anthomyiidae | *Delia radicum* | Brassicas |
| | Psilidae | *Psila rosae* | Carrots, etc. |
| | Tipulidae | *Tipula maxima* | Pasture grasses |
| Hemiptera S.O. Homoptera | Aphididae | *Pemphigus bursarius* | Lettuce |
| | Cercopidae | *Tomaspis* spp. | Sugarcane |
| | Cicadidae | *Magicicada* spp. | Trees |
| | Pseudococcidae | *Planococcus citri* | Coffee, citrus |
| | Phylloxeridae | *Daktulosphaira vitifolii* | Vines |
| Isoptera | Termitidae | *Macrotermes* spp. | Grasses |
| Lepidoptera | Hepialidae | *Hepialus humuli* | Pasture grasses |
| | Noctuidae | *Noctua pronuba* | Wide range of forbs and grasses |
| Orthoptera | Gryllotalpidae | *Gryllotalpa* spp. | Grasses |

(Fitch), is a very serious pest of vines in the USA and Europe. Here, the 'primary host' is the aerial part of the vine and the 'secondary host' is the root system (Bartlett 1984).

Apart from yield losses caused by root-feeding, insects may act as disease vectors. Gilbertson *et al.* (1986) have shown that feeding by larvae of a chrysomelid, *Diabrotica virgifera* Leconte, can transmit *Fusarium* stalk rot, while Witcosky & Hansen (1985) found that three species of beetle were the vectors of black-stain root disease of Douglas fir.

## *Factors affecting root herbivory*

### *Species and age of host plant*

Most of the root-feeding species which have been studied are not generalist herbivores, but show distinct host-plant preference. Radcliffe (1970, 1971d), working with the scarabaeid, *Costelytra zealandica* White, demonstrated a larval preference for white clover (*Trifolium repens* L.) over a variety of pasture grasses. Clover was more susceptible to damage in monoculture than in a mixed sward. Similar preferences and damage susceptibility were recorded by Prestidge, Van der Zijpp & Badan (1985). Aeschlimann (1980) found that annual medics (*Medicago* spp.) were more susceptible to attack by *S. discoideus* larvae than lucerne (*Medicago sativa* L.) when the different species were grown together. Kard & Hain (1987) reported that root damage to Fraser fir trees by scarab larvae was greatest when the trees were planted in pastures supporting grasses and herbs. The pasture was attractive to ovipositing adults and preferred by the larvae. When the grasses and herbs were eliminated, root damage to the trees was greatly reduced. The effect of plant density on the activity of root feeders was investigated by Radcliffe (1971c) who found no interactions between the feeding of *C. zealandica* and the density of two grass species.

Compared with many crops, where the effects of root-feeding insects are manifest rapidly, grass is a long-lived crop where pest numbers build up over a number of years (Clements 1984). Experiments have shown that where insecticides are applied to established grassland, the increased yield was greater than that on a newly-established sward (Henderson & Clements 1974). Cranshaw (1985), working with *Sitona hispidulus* (F.) on alfalfa, found no infestation during the first 2–3 months of stand establishment, but after this increased root attack occurred with increasing stand age. However, Goldson & French (1983) found the opposite effect with *S. discoideus* on lucerne. Densities of larvae at five sites declined by as much as 40% per year. They attributed this to very young *S. discoideus* larvae feeding on root nodules. As plants age, the nodules are located deeper in the soil and thus are less accessible to the larvae.

### *Soil moisture and nutrient content*

Root-feeding insects are greatly affected by soil moisture. In laboratory experiments, Radcliffe (1971b) found that *C. zealandica* larvae survived better in dry soil (57% moisture capacity — MC) than wet (90% MC). Soil moisture also affects the depth at which insects feed and thus the roots available as a source of food. In dry conditions, plants may suffer drought stress and this combined with enhanced insect survival will lead to increased damage and reduced yield. Ladd & Buriff (1979) grew Kentucky Blue Grass, *Poa pratensis* L., in pots and subjected them to root feeding at different densities of *P. japonica* larvae. At 60% field capacity, the reduction in yields was 12–35% greater than at 90% (Fig. 1). Similarly, Godfrey & Yeargan (1985) found increased soil moisture levels resulted in decreased larval survival of *S. hispidulus* and thus less damage and higher yields. In field situations, studies have shown that there are critical levels of soil moisture below which young larvae of root-feeding insects fail to establish. Quinn & Hower (1986) found that the number of first and second instar *S. hispidulus* was strongly correlated with soil moisture and thus nodule development of lucerne. In New Zealand, populations of *S. discoideus* were strongly affected by moisture conditions in September–October, the peak time of oviposition and larval establishment (Goldson, Frampton & Jamieson 1986). Larval effects on yield were subsequently related to soil moisture, with a damage threshold to the crop of 1200 larvae $m^{-2}$ in the dry season and 2100 larvae $m^{-2}$ in the wet season (Goldson *et al.* 1985). Root-sucking insects are also sensitive to soil moisture and inflict most damage in dry soils (Anon 1984; Bartlett 1984). However, plant root systems will tend to be better developed in drier soils,

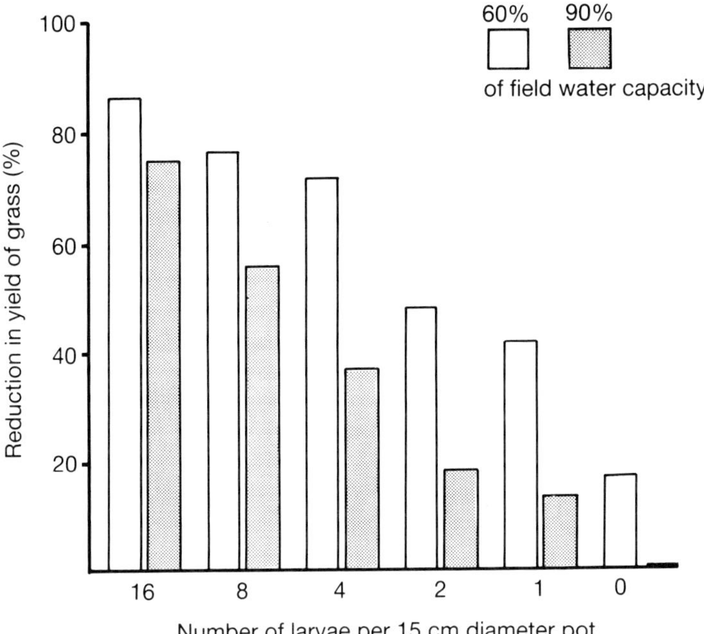


FIG. 1. Reduction of foliage yield caused by *Popillia japonica* larval feeding at two field water capacities and at different larval densities. Control is 0 larvae, 90% field capacity (from Ladd & Buriff 1979).

thus providing a greater availability of food for insects and thereby supporting larger populations.

Davidson (1979) reported that manual root pruning at a constant depth of grass and clover affected a higher proportion of roots when plants were fertilized and suggested that this was due to the shallower root system. In unfertilized plants, roots penetrated below the level of cutting and therefore suffered less damage. However, for the root feeder (*C. zealandica*), Prestidge, Van der Zijpp & Badan (1985) found no interaction between feeding and fertilizer application. It has been suggested that the addition of fertilizers may allow plants to overcome root damage by insects (Radcliffe 1970, 1971a; Wiseman, Leach & McMillan 1973). More work needs to be done on this subject especially as root-feeding larvae appear to be relatively immobile and feed at a constant depth (Davidson 1979).

## *Effects of root herbivory on plant growth*

### *Yield loss*

Since most studies of root herbivory focus on insects of agricultural importance, the effects are measured in terms of yield. Reported yield losses are highly variable, due to the range of plant and insect species, the different densities of insects, and the range of experimental conditions employed.

In pot trials, Ladd & Buriff (1979) demonstrated the effect of *P. japonica* larvae on *P. pratensis* (Fig. 1). In a series of experiments, Radcliffe (1971a–d) showed the effect of *C. zealandica* on a range of grasses and clovers, and recorded yield losses of up to 70%. More realistic measurements may, however, be obtained from field trials where natural populations of root-feeding insects are reduced by insecticides. For example, Henderson & Clements (1974) recorded increases in the yield of established pasture by up to 30% in any 1 year as a result of the application of insecticides. In another experiment, yields were increased by up to 32% (Henderson & Clements 1977). An indication that yield loss resulted from low levels of insect herbivory rather than a direct effect of pesticide application was provided by Clements, Bentley & Nuttall (1987). In pastures where pests were uncommon there was no increase in yield after pesticide application, whereas when pests were common yields were increased by up to 33%.

A common feature of these experiments involving root feeding and plant growth is the existence of damage thresholds below which insects have no effect and above which yield is impaired. However, these thresholds will vary according to the insect/plant system involved. Davidson (1979) reported that up to 50% of the roots of pasture grasses and clover could be damaged before significant shoot loss occurred, while for two species of Coleoptera larvae, feeding on three grass species, the threshold approached 60% (Fig. 2). Fleming (1972) considered that there were also threshold densities of *P. japonica* larvae before significant loss of yield occurred and a similar situation has been reported for *S. discoideus* on lucerne (Goldson *et al.* 1985).

### *Drought stress*

Few experiments have measured leaf water content during root feeding. However, Ridsdill-Smith (1977) found that root feeding by the scarabaeid, *Sericesthis nigrolineata* (Boisd.), significantly reduced relative water content of leaves of *Lolium perenne* causing a form of drought stress. A similar effect was recorded by Goldson, Bourdot & Proffitt (1987) who found that severe damage by *S. discoideus* larvae caused a form of drought-related dormancy in lucerne. Unfortunately, neither of these experiments were conducted under different watering regimes. However, it seems likely that reports of the effects of root herbivory being offset by increasing soil moisture levels (e.g. Radcliffe 1971d) may be explained by the plants overcoming drought stress when supplied with ample water.

### *Plant fitness*

Root-feeding insects can depress plant reproduction, both as a consequence of reductions in the rate of nutrient and water uptake and carbohydrate supply for seed production. However, few instances of effects of root herbivory on plant reproduction have been reported. Harris & Myers (1981) found that larvae of the beetle *S. jugoslavica* reduced seed production of diffuse knapweed by up to 42%,

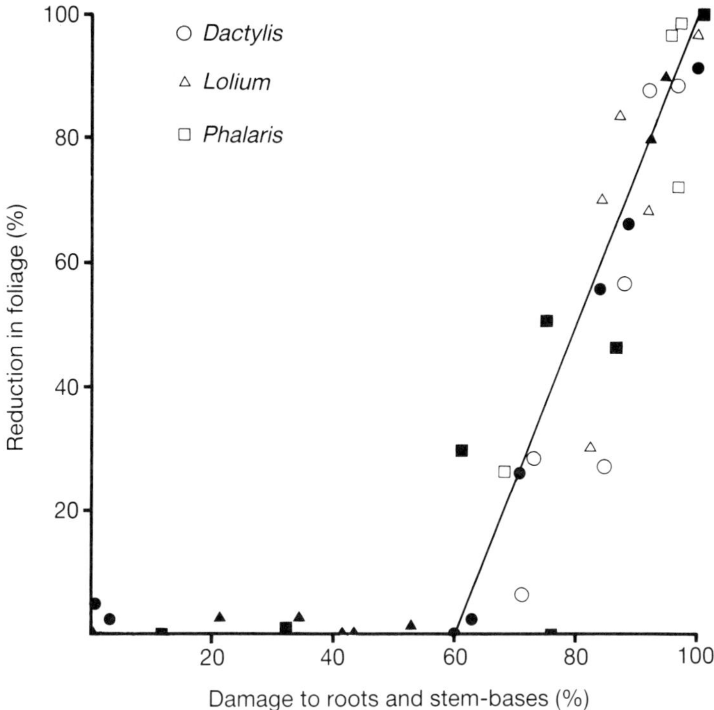


FIG. 2. Relationship between root damage and reduction of foliage of three grass species caused by two species of Coleopteran larvae (from Davidson 1979). The two insects are represented by open and solid symbols.

when introduced as a biological control agent for the weed and recently, Brown & Gange (1989) have found increased reproduction of annual forbs (non-graminaceous species) in a ruderal community when the levels of subterranean insects were reduced (see p. 463).

Root-feeding insects may also have considerable effects on seedlings after germination, but before emergence above the soil surface. Clements & Bentley (1983) found that applications of soil insecticide significantly increased the establishment of sown *T. repens* where tipulid larvae were a common pest. The establishment of alfalfa seedlings was reduced by 44%, 22 days after seeding by *S. hispidulus* (Godfrey, Legg & Yeargan 1986). More recently, Brown & Gange (1989) have demonstrated increased plant species richness in a 2-year-old natural community when root herbivory was reduced. They suggested that this was due to increased seedling establishment (see p. 462).

### *Interactions between root and foliar attack*

In nature, plants are likely to be attacked by both root- and foliar-feeding insects. Few studies have considered both forms of attack and in those which have, one or both 'types of herbivory' has been applied manually. For example, Ridsdill-Smith (1977) reported that root feeding by scarab larvae only depressed foliage yield when the leaves were clipped. A similar effect was recorded by Graber, Fluke & Dexter (1931) and Davidson (1979). However, Gange & Brown (1989) have clearly demonstrated that root feeding by scarab larvae can significantly modify the performance of foliar-feeding aphids on the same plant. This was due to physiological changes in the host-plant resulting from the root feeding. On the other hand, foliar feeders can have a significant effect on root growth. For example, Pike & Schaffner (1985) and Burton (1986) have demonstrated the depressant effect on shoots and roots of foliar-aphid feeding. Rausher & Feeny (1980) recorded similar effects with a leaf chewer and Richards (1984) with manual defoliation. Detailed studies of the interactions between root and foliar feeders on plant growth are much needed.

## EXPERIMENTAL STUDIES

### *Root herbivory in natural plant communities*

Previous studies on agricultural pests provide a model system for estimating the effects of root herbivory in natural vegetation. In natural situations, however, the system is more complex with a wider range of herbivores and host-plants present. As a consequence, studies of root herbivory by insects in natural plant communities are rare, although this seems an exciting future area.

Karban (1980) recorded the effect of the periodical cicada (*Magicicada* spp.) on the growth of scrub oak, *Quercus ilicifolia*. Cicada nymphs feed on root xylem and can reach densities of up to 3 500 000 $\text{ha}^{-1}$ (Karban 1980). Adult emergence is cyclical: every seventeenth year in the north of their range and every thirteenth year in the south. Figure 3 shows the mean radial growth of oak trees with zero or three or more cicada egg nests at the roots. It can be seen that growth was significantly reduced by up to 30% in the years following the emergence year in trees with three or more egg nests. Non-significance before the emergence year was attributed to the different mechanism of feeding by older nymphs and the fact that by this stage in the life cycle populations have been greatly reduced. In this way, cicada feeding can impose periodical accumulation of scrub oak growth. To our knowledge, there are no studies which extend beyond an investigation of the effects of a single insect taxon on a plant community.

### *Experimental tests on the effects of root herbivory by insects on an early successional plant community*

Two manipulative field experiments assessed the effects of soil fauna on the establishment and development of a plant community. Experiments were carried out on vegetation establishing from bare ground (see Brown & Gange 1989) on an

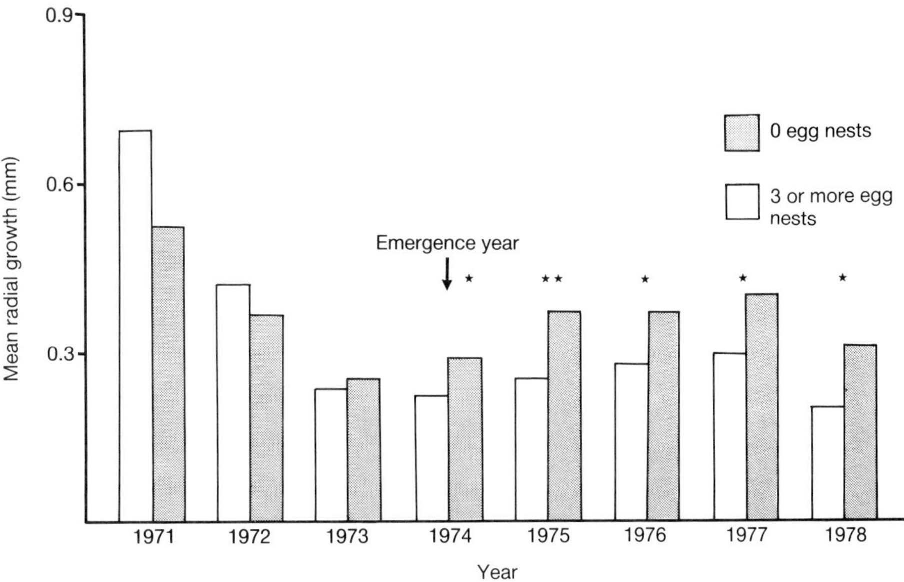


Fig. 3. Radial wood growth of *Quercus ilicifolia* with or without *Magicicada* egg nests at roots (from Karban 1980). $*P<0{\cdot}05$; $**P<0{\cdot}01$.

acid, sandy soil at Silwood Park, Berkshire, UK. The nature and density of the soil-dwelling insect macro-fauna were determined by taking three 15 cm × 10 cm soil cores from plots adjacent to where the vegetation was sampled. Samples were hand sorted and then placed in Tullgren funnels. The main root herbivores included Coleoptera and Lepidoptera larvae and particularly the dipteran, *Tipula oleracea* L.

An initial experiment addressed two questions.

1 What are the relative effects on vegetation of the soil and above-ground insect fauna?

2 What interactions occur between these two groups of herbivores?

An experiment with four treatments, each with five replicates — (i) control, (ii) treated with a soil insecticide (Dursban 5G) to reduce root herbivory, (iii) treated with foliar systemic insecticide (Dimethoate-40) to reduce foliar herbivory, and (iv) treated with both compounds to provide an overall reduction in herbivory — was established. The effect of these compounds on arthropod abundance was demonstrated by Hendrix, Brown & Gange (1988). Each replicate was a 3 × 3 m subplot within a randomized design and was separated from adjacent subplots by a 2 m 'walkway'. The experimental site was fenced to exclude rabbits and mollusc grazing was controlled by the application of slug pellets (see Brown & Gange 1989). Details of the chemicals and their application are also described in detail by Brown & Gange (1989). Dursban 5G (5% w/w chlorpyrifos) and Dimethoate-40 were applied at monthly and fortnightly intervals, respectively, throughout the growing season. In extensive tests, using the methodology described in Brown, Leijn &

Stinson (1987), neither compound was found to have any direct effects on a wide range of the plant species encountered in this study (unpublished results).

The vegetation was sampled from May to October during 1986–88. Samples were at fortnightly intervals during the first year, when rapid changes in the vegetation take place (Brown & Southwood 1987), and monthly thereafter. A total of ninety-eight species were recorded during the first 3 years of colonization. Species richness (the mean number of plant species per subplot) was marginally enhanced by the application of foliar insecticide in the first season ($F = 5{\cdot}14$, $P < 0{\cdot}05$). In the second and third years, the application of soil insecticide resulted in a significant increase in plant species richness (Year 2: $F = 20{\cdot}86$, $P < 0{\cdot}001$; Year 3: $F = 76{\cdot}23$, $P < 0{\cdot}001$) (Figs 4a and b). This trend was also reflected in the absolute number of plant species per treatment. Indeed, by the middle of the third season the mean number of species in the soil insecticide-treated plots was $17{\cdot}8 \pm 1{\cdot}0$ and the control plots $15{\cdot}2 \pm 0{\cdot}8$. Furthermore, in the foliar insecticide-treated plots, species richness was depressed from the middle of the second season with a mean of only $10{\cdot}0 \pm 0{\cdot}3$ by the middle of the third season. Ten of the ninety-eight species were only found in the soil insecticide-treated plots (e.g. *Glechoma*

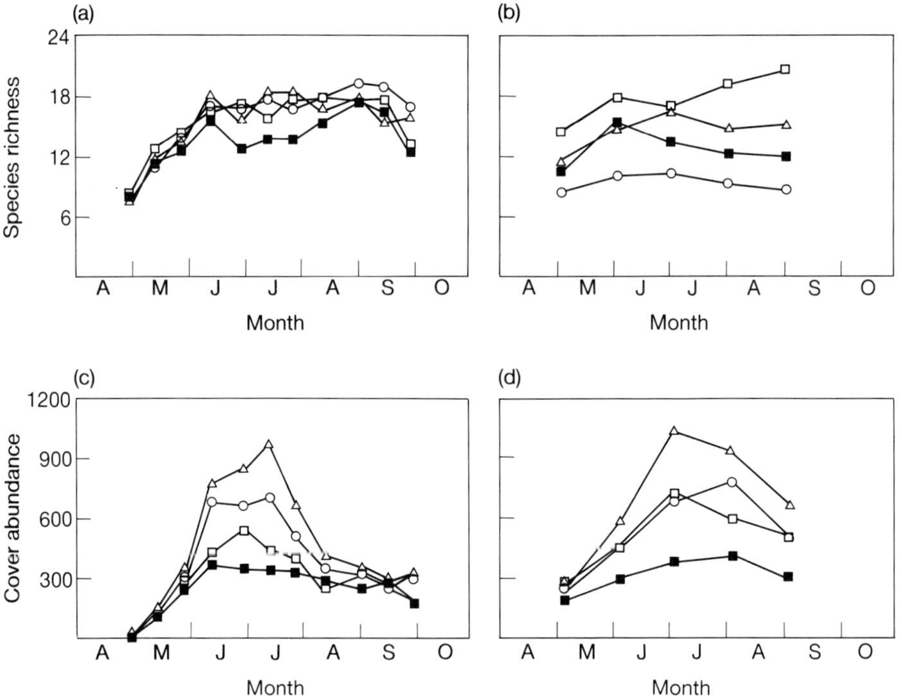


FIG. 4. Trends in plant community characteristics resulting from a reduction in foliar or root herbivory by insecticide application. Species richness: (a) year 1, (b) year 3. Total cover abundance: (c) year 1, (d) year 3. (■) Control; (○) foliar insecticide-treated; (□) soil insecticide-treated; (△) foliar and soil insecticide treated.

*hederacea* L., *Lamium album* L., *Lamium purpureum* L., *Stachys palustris* L.). This difference was visually apparent, particularly in mid-season, when many species were in flower.

The cover abundance also showed distinct trends (Figs 4c and d) with both foliar and soil compounds producing a significant increase over the controls. In the first year, the foliar insecticide had a greater effect. In the second and third seasons, the effects were similar for both compounds.

When plant life history groupings are considered (see Gibson, Brown & Jepsen 1987), there were clear successional trends with annual forbs dominant in the first year. (*Poa annua* L. was the only common annual grass.) In the second year, there was a dramatic increase in perennial grasses. Perennial forbs followed a similar pattern, but contributed <10% of the vegetation. These plant groupings responded differently to the different insecticide treatments. Annual forbs were enhanced by the application of both compounds in year 1 (see Fig. 4c), but only by the soil insecticide subsequently (Fig. 5a). Similarly, perennial grasses were increased by both compounds in the second and third seasons, although the effect of the foliar insecticide was greater (Fig. 5b). Although numbers were small, perennial forbs

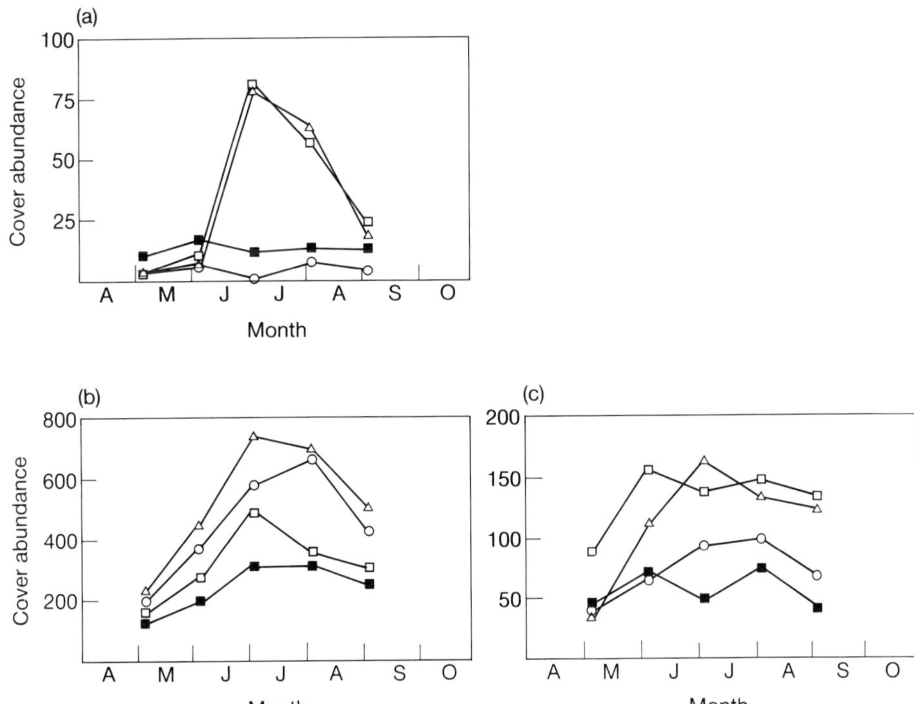


FIG. 5. Trends in cover abundance of three plant life history groupings in year 3, resulting from a reduction in foliar or root herbivory by insecticide application. (a) Annual forbs. (b) Perennial grasses. (c) Perennial forbs.

TABLE 2. Summary of ANOVA *F* values comparing cover abundance in common species resulting from a reduction in above- and below-ground herbivory by the application of a foliar or soil insecticide, respectively

| | Year 1 | | Year 2 | | Year 3 | |
|---|---|---|---|---|---|---|
| Species | Foliar insecticide | Soil insecticide | Foliar insecticide | Soil insecticide | Foliar insecticide | Soil insecticide |
| Herbs | | | | | | |
| *Capsella bursa-pastoris* | 5·83* | 5·12* | 0·09 | 6·05* | — | — |
| *Chenopodium album* | 5·32* | 2·10 | 1·06 | 2·26 | — | — |
| *Sonchus oleraceus* | 5·27* | 1·34 | 0·74 | 1·85 | — | — |
| *Stellaria media* | 4·76* | 10·46** | 0·10 | 2·02 | 2·66 | 49·75*** |
| *Tripleurospermum inodorum* | 16·13*** | 0·00 | 1·72 | 4·66* | 0·05 | 0·25 |
| *Polygonum persicaria* | 8·25* | 4·22* | 2·38 | 2·38 | — | — |
| *Veronica persica* | 2·06 | 0·27 | 1·54 | 5·18* | 0·15 | 49·65*** |
| Grasses | | | | | | |
| *Agrostis stolonifera* | 0·50 | 0·30 | 29·45*** | 1·69 | 5·88* | 0·30 |
| *Agropyron repens* | — | — | — | — | 4·79* | 2·50 |

* $P<0·05$; ** $P<0·01$; *** $P<0·001$; — species absent.

appear to have been suppressed by either insecticide in the first year. By the end of the second year, the application of soil insecticide resulted in an increase in the cover of perennial forbs which was significant in the third year ($F = 20·04$, $P<0·001$) (Fig. 5c).

As might be expected, single plant species varied in their response to the applications of the two compounds. These effects were not always consistent (Table 2). Plant species could be divided into four categories according to how their abundance was modified by herbivory. These were (i) species affected mainly by below-ground herbivory, (ii) species affected mainly by above-ground herbivory, (iii) species responding to both types of herbivory, although often at different stages in their establishment (this was the major group), and (iv) species which performed better in the control. The cover abundance of three of the commonest twelve species was increased by the application of soil insecticide. In *Stellaria media* L. (Vill), the response was particularly strong at the end of the first season and was also seen at the beginning of the second and third seasons before the species declined. *Capsella bursa-pastoris* L. (Medic) and *Veronica persica* L. responded in the same way. Foliar insecticide affected six species in the first year (including those enhanced by the soil compound), but only the perennial grass, *Agrostis stolonifera* L., in the second and third years. Another coarse grass, *Agropyron repens* L. (Beauv), although rare in early succession, increased dramatically in years 2 and 3 and responded similarly. Two, less common, forb species showed interesting trends in the second and third years: *Trifolium repens* L. generally grew better in control plots and there was also evidence of this in *Plantago media* L. *Ranunculus repens* L. was enhanced where the insecticides were applied separately, but depressed in both control and combined insecticide-treated plots. In *T. repens*, this response may occur because the species is not attacked by insect herbivores or because its growth benefits from a reduction in competition resulting from the impact of herbivory on other more vigorous species. As *T. repens* is subject to herbivory, mainly by weevils, the latter explanation is more likely. Herbivory thus appears to favour certain species. For example, *R. repens* is attacked by herbivores and, although it can withstand increased growth by other species, it fails to survive the competition resulting from the elimination of herbivory.

Results thus demonstrate substantial effects of community, life history grouping and single species levels. Throughout, there were no significant interactions

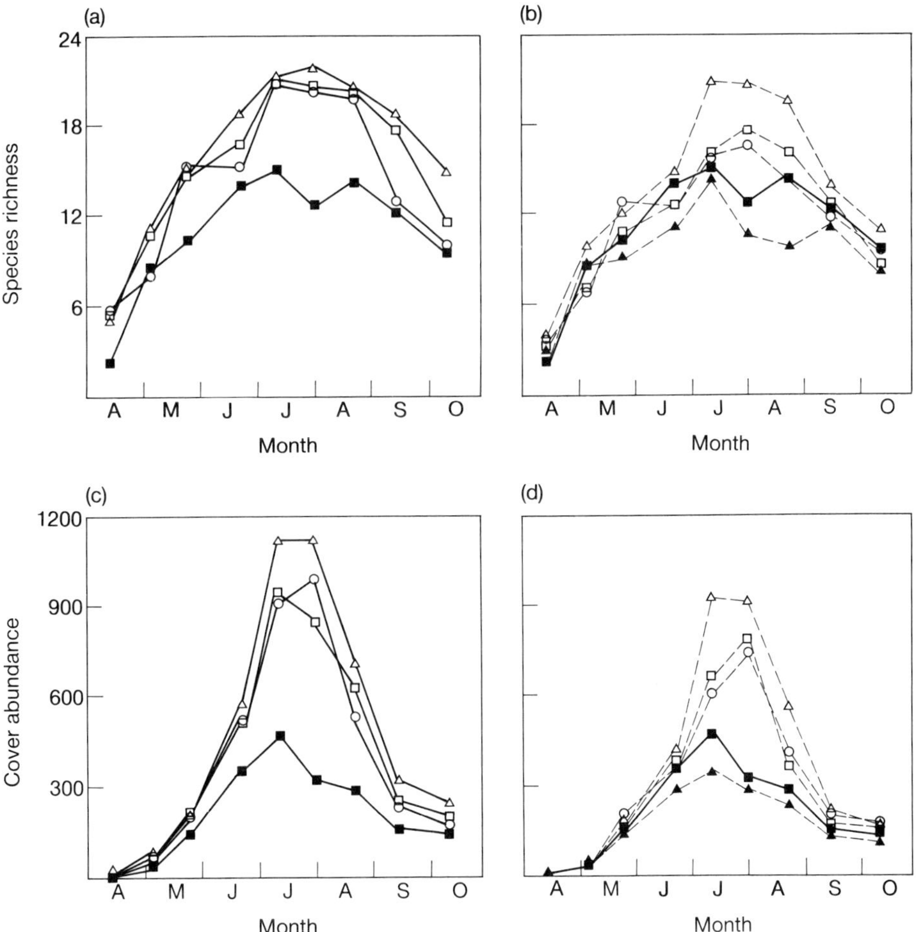


FIG. 6. Effects of reducing foliar herbivory, root herbivory or mycorrhizal infection by the application of insecticides or fungicide, respectively, during the first year of colonization of bare ground. (a), (b) Plant species richness; (c), (d) total cover abundance. (a), (c) Conventions as in Fig. 5; (b), (d) (■) control; (▲) soil fungicide-treated; (○) foliar insecticide and soil fungicide-treated; (□) soil insecticide and soil fungicide-treated; (△) foliar and soil insecticide- and fungicide-treated.

between the insecticides, suggesting effects of above- and below-ground herbivory to be additive. However, other factors, e.g. differences in nutrient cycling and the soil micro-flora induced by the treatments, may also have effects. Several of these effects have been studied, usually indirectly, by other workers (e.g. McGonigle & Fitter 1988; Newell 1984). The separation of effects of soil insecticides on insects, such as Collembola which feed on root mycorrhizas, and root herbivores is important. Under experimental conditions, Warnock, Fitter & Usher (1982), Moore, St. John & Coleman (1985) and McGonigle & Fitter (1988) have all demonstrated increased growth by *Holcus lanatus* L. when mycorrhizal feeding by Collembola is reduced. A second field experiment was initiated in 1988, with main treatments of soil insecticide, foliar insecticide and soil fungicide and all possible combinations to resolve the importance of these effects. Including an untreated control, the experiment had eight treatments with four replicates. The size of treatment plots and experimental design were similar to those in experiment 1. The fungicide used was Rovral (10% w/w iprodione), a contact fungicide with no systemic action (Ivens 1988). This was applied in granular form at monthly intervals. The vegetation was sampled every 3 weeks from mid-April 1988.

Fifty-eight plant species were recorded during the first year of treatment. Foliar and soil insecticides enhanced species richness, while fungicide significantly decreased the number (Figs 6a and b). By August, there were $12 \cdot 8 \pm 0 \cdot 8$ species in the control plots, $20 \cdot 3 \pm 1 \cdot 2$ and $20 \cdot 5 \pm 0 \cdot 8$ in foliar and soil insecticide-treated plots respectively but only $10 \cdot 5 \pm 1 \cdot 0$ in fungicide-treated plots. A similar effect was found for cover abundance (Figs 6c and d). The dominant life history grouping present was the annual forbs, and changes in the total vegetation were due to changes in this group. Assessed on a basis of cover abundance, reductions of insect herbivory and mycorrhizal infection seem to have opposite effects.

## CONCLUSIONS

Root herbivory can play a major role in the dynamics of plant communities as a result of differential effects on individual plant species. The results presented here suggest root herbivory is at least as important as foliar herbivory and in some species it is more important. These effects are significant in relation to changes in community organization. The effects of root herbivory are complex because species belonging to different life history groupings respond differently. Brown (1982, 1984), Brown, Jepsen & Gibson (1988), and Gibson, Brown & Jepsen (1987) have described how differential effects such as these can modify the rate and direction of plant succession, e.g. an enhancement of annual species in the second year of succession, as a result of a reduction in root herbivory, may slow the rate of succession. More marked, however, is the increase in the perennial grasses, resulting from a reduction in foliar herbivory, which leads to the rapid development of a grass-dominated sward. Concurrent with this is an increase in forb species richness and a reduction in seedling mortality. Recruitment, especially among annual and perennial forbs, is related to microsite availability. Here, the amount of bare ground in control and soil insecticide-treated plots was similar; in plots treated with foliar insecticide it was significantly reduced by the vigorous growth of grasses. Seedling mortality in control plots was high, and further mortality seems to occur after germination but before seedling emergence. Thus, the availability of microsites and a reduction in root herbivory, due to the soil insecticide treatment permits greater recruitment and this is manifest *inter alia* by an increase in plant species richness. The two experiments demonstrate the importance of root-feeding herbivores in seedling recruitment and the consequences for plant community development. The reduction in mycorrhizal associations by fungicide appears to depress plant colonization and growth. In a relatively poor soil, such as that at Silwood Park, mycorrhizas are likely to be important for plant growth (Harley & Harley 1987). The initial results presented here suggest that plants may overcome the loss of mycorrhizae if root and/or foliar herbivory are reduced.

Other considerations however, may be important in relation to the use of insecticides and fungicides in the manner employed here. The insects and fungi killed by the treatments may increase soil nitrogen. Dursban 5G is an organophosphate and may therefore increase soil phosphorus. However, application of a similar compound had no effect on shoot phosphorus content of *H. lanatus* in a field situation (McGonigle & Fitter 1988). Soil nutrient relations under these experimental treatments are obviously of potential importance and are currently under investigation.

## ACKNOWLEDGEMENTS

The experimental studies at Silwood Park were funded by NERC. We are grateful to D. Salt for assistance with field work.

## REFERENCES

**Aeschlimann, J.P. (1980).** The *Sitona* (Col.: Curculionidae) species occurring on *Medicago* and their natural enemies in the Mediterranean region. *Entomophaga*, **25**, 139–153.

**Anon (1984).** Lettuce aphids. Leaflet 392, Ministry of Agriculture, Fisheries and Food, Pinner.

**Bartlett, P.W. (1984).** Grape phylloxera. MAFF Plant Health Information Bulletin No. 3, 1–5.

**Bentley, S. & Whittaker, J.B. (1979).** Effects of grazing by a chrysomelid beetle, *Gastrophysa viridula*, on competition between *Rumex obtusi folius* and *R. crispus*. *Journal of Ecology*, **67**, 79–90.

**Brown, V.K. (1982).** The phytophagous insect community and its impact on early successional habitats. In *Proceedings of the 5th International Symposium on Insect Plant Relationships, Wageningen 1982* (Ed. by J.H. Visser & A.K. Minks), pp. 205–213. Pudoc, Wageningen.

**Brown, V.K. (1984).** Secondary succession: insect–plant relationships. *BioScience*, **34**, 710–716.

**Brown, V.K. & Southwood, T.R.E. (1987).** Secondary succession: patterns and strategies. In *Colonization, Succession and Stability* (Ed. by A.J. Gray, M.J. Crawley & P.J. Edwards), pp. 315–338. Symposia of the British Ecological Society, 26. Blackwell Scientific Publications, Oxford.

**Brown, V.K. & Gange, A.C. (1989).** Differential effects of above- and below-ground insect herbivory during early plant succession. *Oikos*, **54**, 67–76.

**Brown, V.K., Leijn, M. & Stinson, C.S.A. (1987).** The experimental manipulation of insect herbivore load by the use of an insecticide (malathion): the effect of application on plant growth. *Oecologia*, **72**, 377–386.

**Brown, V.K., Gange, A.C. & Gibson, C.W.D. (1988).** Effects of insect herbivory on vegetational structure. In *Proceedings of the International Symposium on Vegetational Structure* (Ed. by M.J.A. Werger, P.J.M. van der Aart, H.J. During & J.T.A. Verhoeven), pp. 263–279. S.P.B. Academic Publishing, The Hague.

**Brown, V.K., Jepsen, M. & Gibson, C.W.D. (1988).** Insect herbivory: effects on early old field succession demonstrated by chemical exclusion methods. *Oikos*, **52**, 293–302.

**Burton, R.L. (1986).** Effect of greenbug (*Schizaphis graminum*) (Homoptera: Aphididae) damage on root and shoot biomass of wheat seedlings. *Journal of Economic Entomology*, **79**, 633–636.

**Clements, R.O. (1984).** Control of insect pests in grassland. *Span*, **27**, 77–80.

**Clements, R.O. & Bentley, B.R. (1983).** The effect of three pesticide treatments on the establishment of white clover (*Trifolium repens*) sown with a slot-seeder. *Crop Protection*, **2**, 375–378.

**Clements, R.O., Bentley, B.R. & Nuttall, R.M. (1987).** The invertebrate population and response to pesticide treatment of two permanent and two temporary pastures. *Annals of Applied Biology*, **111**, 399–407.

**Cranshaw, W.S. (1985).** Clover root curculio (*Sitona hispidulus*) injury and abundance in Minnesota (USA) alfalfa of different stand age. *Great Lakes Entomology*, **18**, 93–95.

**Crawley, M.J. (1983).** *Herbivory: the Dynamics of Animal–Plant Interactions*. Blackwell Scientific Publications, Oxford.

**Davidson, R.L. (1979).** Effects of root feeding on foliage yield. In *Proceedings of the 2nd Australasian Conference on Grassland Invertebrate Ecology* (Ed. by T.K. Crosby & R.P. Pottinger), pp. 117–120. Government Printer, Wellington.

**Dixon, A.F.G. (1971).** Aphids as root feeders. In *Methods of Study in Quantitative Soil Ecology: Population, Production and Energy Flow* (Ed. by J. Phillipson), pp. 233–246. IBP Handbook No. 18.

**Dunn, J.A. (1959).** The biology of lettuce root aphid. *Annals of Applied Biology*, **47**, 475–491.

**Fleming, W.E. (1972).** Biology of the Japanese beetle. United States Department of Agriculture Technical Bulletin No. 1449, pp. 129. Washington, DC.

**Gange, A.C. & Brown, V.K. (1989).** Effects of root herbivory by an insect on a foliar-feeding species, mediated through changes in the host plant. *Oecologia*, **81**, 38–42.

**Gange, A.C., Brown, V.K., Evans, I.M. & Storr, A.L. (1989).** Variation in the impact of insect herbivory on *Trifolium pratense* through early plant succession. *Journal of Ecology*, **77**, 537–551.

**Gibson, C.W.D., Brown, V.K. & Jepsen, M. (1987).** Relationships between the effects of insect herbivory and sheep grazing on seasonal changes in an early successional plant community. *Oecologia*, **71**, 245–253.

**Gilbertson, R.L., Brown, W.M., Ruppel, E.G. & Capinera, J.L. (1986).** Association of corn stalk rot *Fusarium* spp. and western corn rootworm beetles in Colorado. *Phytopathology*, **76**, 1309–1314.

**Godfrey, L.D. & Yeargan, K.V. (1985).** Influence of soil moisture and weed density on clover root curculio *Sitona hispidulus*, larval stress to alfalfa (*Medicago sativa*). *Journal of Agricultural Entomology*, **2**, 370–377.

**Godfrey, L.D., Legg, D.E. & Yeargan, K.V. (1986).** Effects of soil-borne organisms on spring alfalfa establishment in an alfalfa rotation system. *Journal of Economic Entomology*, **79**, 1055–1063.

**Goldson, S.L. & French, R.A. (1983).** Age-related susceptibility of lucerne to *Sitona* weevil, *Sitona discoideus* Gyllenhal (Coleoptera: Curculionidae), larvae and the associated patterns of adult infestation. *New Zealand Journal of Agricultural Research*, **26**, 251–255.

**Goldson, S.L., Bourdot, G.W. & Proffitt, J.R. (1987).** A study of the effects of *Sitona discoideus* (Coleoptera: Curculionidae) larval feeding on the growth and development of lucerne (*Medicago sativa*). *Journal of Applied Ecology*, **24**, 153–161.

**Goldson, S.L., Dyson, C.B., Proffitt, J.R., Frampton, E.R. & Logan, J.A. (1985).** The effects of *Sitona discoideus* Gyllenhal (Coleoptera: Curculionidae) on lucerne yields in New Zealand. *Bulletin of Entomological Research*, **75**, 429–442.

**Goldson, S.L., Frampton, E.R. & Jamieson, P.D. (1986).** Relationship of *Sitona discoideus* (Coleoptera: Curculionidae) larval density to September–October potential soil moisture deficits. *New Zealand Journal of Agricultural Research*, **29**, 275–279.

**Graber, L.F., Fluke, C.L. & Dexter, S.T. (1931).** Insect injury to blue grass in relation to the environment. *Ecology*, **12**, 547–566.

**Harley, J.L. & Harley, E.L. (1987).** A check-list of mycorrhiza in the British flora. *New Phytologist* (Suppl.), **105**, 1–102.

**Harris, P. & Myers, J.H. (1981).** *Centaurea diffusa* Lam. and *C. maculosa* Lam. s. *lat.*, diffuse and spotted knapweed (Compositae). In *Biological Control Programmes Against Insects and Weeds in Canada 1969–80* (Ed. by J.S. Kelleher & M.J. Hulme), pp. 127–137. Commonwealth Agricultural Bureau, Slough.

**Harris, P., Wilkinson, A.T.S. & Myers, J.H. (1981).** *Senecio jacobaea* L., tansy ragwort (Compositae). In *Biological Control Programmes Against Insects and Weeds in Canada 1969–80* (Ed. by J.S. Kelleher & M.J. Hulme), pp. 195–201. Commonwealth Agricultural Bureau, Slough.

**Henderson, I.F. & Clements, R.O. (1974).** The effect of pesticides on the yield and botanical composition of a newly-sown rye grass ley and of an old mixed pasture. *Journal of the British Grassland Society*, **29**, 185–190.

**Henderson, I.F. & Clements, R.O. (1977).** Grass growth in different parts of England in relation to invertebrate numbers and pesticide treatment. *Journal of the British Grassland Society*, **32**, 89–98.

**Hendrix, S.D., Brown, V.K. & Gange, A.C. (1988).** Effects of insect herbivory on early plant succession: comparison of an English site and an American site. *Biological Journal of the Linnean Society*, **35**, 205–216.

**Hill, D.S. (1987).** *Agricultural Insect Pests of Temperate Regions and Their Control.* Cambridge University Press, 659 pp.

**Ivens, G.W. (Ed.) (1988).** *The U.K. Pesticide Guide.* C.A.B. International & British Crop Protection Council, Croydon, 434 pp.

**Karban, R. (1980).** Periodical cicada nymphs impose periodical oak tree wood accumulation. *Nature*, **287**, 326–327.

**Kard, B.M.R. & Hain, F.P. (1987).** White grub (Coleoptera: Scarabaeidae) densities, weed control practises, and root damage to Fraser Fir Christmas trees in the southern Appalachians. *Journal of Economic Entomology*, **80**, 1072–1075.

**Kinsman, S. & Platt, W.J. (1984).** The impact of a herbivore upon *Mirabilis hirsuta*, a fugitive prairie plant. *Oecologia*, **65**, 2–6.

**Ladd, Jr., T.L. & Buriff, C.R. (1979).** Japanese beetle: influence of larval feeding on bluegrass yields at two levels of soil moisture. *Journal of Economic Entomology*, **72**, 311–314.

**McDavid, G.E. (1981).** The Japanese beetle (*Popillia japonica*) in California. *Agrichemical Age*, **25**, 48–52.

**McGonigle, T.P. & Fitter, A.H. (1988).** Ecological consequences of arthropod grazing on VA mycorrhizal fungi. *Proceedings of the Royal Society of Edinburgh*, **94**B, 25–32.

**Moore, J.C., St. John, T.V. & Coleman, D.C. (1985).** Ingestion of vesicular–arbuscular mycorrhizal hyphae and spores by soil microarthropods. *Ecology*, **66**, 1979–1981.

**Newell, K. (1984).** Interaction between two decomposer Basidiomycetes and a Collembolan under Sitka spruce: grazing and its potential effects on fungal distribution and litter decomposition. *Soil Biology and Biochemistry*, **16**, 235–239.

**Pike, K.S. & Schaffner, R.L. (1985).** Development of autumn populations of cereal aphids, *Rhopalosiphum padi* (L.) and *Schizaphis graminum* (Rondani) (Homoptera: Aphididae), and their effects on winter wheat in Washington State. *Journal of Economic Entomology*, **78**, 676–680.

**Prestidge, R.A., Van der Zijpp, S. & Badan, D. (1985).** Effects of plant species and fertilizers on grass grub larvae, *Costelytra zealandica*. *New Zealand Journal of Agricultural Research*, **28**, 409–417.

**Quinn, M.A. & Hower, A.A. (1986).** Effects of root nodules and taproots on survival and abundance of *Sitona hispidulus* (Coleoptera: Curculionidae) on *Medicago sativa*. *Ecological Entomology*, **11**, 391–400.

**Radcliffe, J.E. (1970).** Some effects of grass grub (*Costelytra zealandica* (White)) larvae on pasture plants. *New Zealand Journal of Agricultural Research*, **13**, 87–104.

**Radcliffe, J.E. (1971a).** Effects of grass grub (*Costelytra zealandica* White) larvae on pasture plants. *New Zealand Journal of Agricultural Research*, **14**, 597–606.

**Radcliffe, J.E. (1971b).** Effects of grass grub (*Costelytra zealandica* White) larvae on pasture plants. II. Effect of grass grubs and soil moisture on perennial ryegrass and cocksfoot. *New Zealand Journal of Agricultural Research*, **14**, 607–617.

**Radcliffe, J.E. (1971c).** Effects of grass grub (*Costelytra zealandica* White) larvae on pasture plants. III.

Effect of grass grubs and plant density on perennial ryegrass and cocksfoot. *New Zealand Journal of Agricultural Research*, **14**, 618–624.

**Radcliffe, J.E. (1971d).** Effects of grass grub (*Costelytra zealandica* White) larvae on pasture plants. IV. Effect of grass grubs on perennial ryegrass and white clover. *New Zealand Journal of Agricultural Research*, **14**, 625–632.

**Rausher, M.D. & Feeny, P. (1980).** Herbivory, plant density and plant reproductive success: the effect of *Battus philenor* on *Aristolochia reticulata*. *Ecology*, **61**, 905–917.

**Richards, J.H. (1984).** Root growth response to defoliation in two *Agropyron* bunchgrasses: field observations with an improved root periscope. *Oecologia*, **64**, 21–25.

**Ridsdill-Smith, T.J. (1977).** Effects of root feeding by scarabaeid larvae on growth of perennial ryegrass plants. *Journal of Applied Ecology*, **14**, 73–80.

**Roberts, R.J. & Ridsdill-Smith, T.J. (1979).** Assessing pasture damage and losses in animal production caused by pasture insects. In *Proceedings of the 2nd Australasian Conference on Grassland Invertebrate Ecology* (Ed. by T.K. Crosby & R.P. Pottinger), pp. 124–125. Government Printer, Wellington.

**Warnock, A.J., Fitter, A.H. & Usher, M.B. (1982).** The influence of a springtail *Folsomia candida* (Insecta, Collembola) on the mycorrhizal association of the leek *Allium porrum* and the vesicular–arbuscular mycorrhizal endophyte *Glomus fasciculatus*. *New Phytologist*, **90**, 285–292.

**Whittaker, J.B. (1982).** The effect of grazing by a chrysomelid beetle, *Gastrophysa viridula*, on growth and survival of *Rumex crispus* on a shingle bank. *Journal of Ecology*, **70**, 291–296.

**Wiseman, B.R., Leach, D.B. & McMillan, W.W. (1973).** Effects of fertilizers on resistance of Antigua corn to fall-army worm and corn earworm. *Florida Entomologist*, **56**, 1–7.

**Witcosky, J.J. & Hansen, E.M. (1985).** Root-colonizing insects recovered from Douglas Fir (*Pseudotsuga menziesii*) various stages of decline due to black-stain root disease. *Phytopathology*, **75**, 399–402.

# Index